Galápagos Marine Invertebrates

Taxonomy, Biogeography, and Evolution in Darwin's Islands

TOPICS IN GEOBIOLOGY

Series Editors: F. G. Stehli, DOSECC, Inc., Gainesville, Florida

D. S. Jones, University of Florida, Gainesville, Florida

Volume 1 SKELETAL GROWTH OF AQUATIC ORGANISMS
Biological Records of Environmental Change
Edited by Donald C. Rhoads and Richard A. Lutz

Volume 2 ANIMAL-SEDIMENT RELATIONS
The Biogenic Alteration of Sediments
Edited by Peter L. McCall and Michael J. S. Tevesz

Volume 3 BIOTIC INTERACTIONS IN RECENT AND FOSSIL BENTHIC COMMUNITIES
Edited by Michael J. S. Tevesz and Peter L. McCall

Volume 4 THE GREAT AMERICAN BIOTIC INTERCHANGE
Edited by Francis G. Stehli and S. David Webb

Volume 5 MAGNETITE BIOMINERALIZATION AND MAGNETORECEPTION IN ORGANISMS
A New Biomagnetism
Edited by Joseph L. Kirschvink, Douglas S. Jones, and Bruce J. MacFadden

Volume 6 *NAUTILUS*
The Biology and Paleobiology of a Living Fossil
Edited by W. Bruce Saunders and Neil H. Landman

Volume 7 HETEROCHRONY IN EVOLUTION
A Multidisciplinary Approach
Edited by Michael L. McKinney

Volume 8 GALÁPAGOS MARINE INVERTEBRATES
Taxonomy, Biogeography, and Evolution in Darwin's Islands
Edited by Matthew J. James

Volume 9 TAPHONOMY
Releasing the Data Locked in the Fossil Record
Edited by Peter A. Allison and Derek E. G. Briggs

Galápagos Marine Invertebrates

Taxonomy, Biogeography, and Evolution in Darwin's Islands

Edited by
Matthew J. James
Sonoma State University
Rohnert Park, California

Plenum Press • New York and London

Library of Congress Cataloging-in-Publication Data

Galápagos marine invertebrates : taxonomy, biogeography, and evolution in Darwin's islands / edited by Matthew J. James.
p. cm. -- (Topics in geobiology ; v. 8)
Includes bibliographical references and index.
ISBN 0-306-43794-5
1. Marine invertebrates--Galapagos Islands. I. James, Matthew J. II. Series.
QL345.G2G35 1991
592.09866'5--dc20 91-19754
CIP

ISBN 0-306-43794-5

A Division of Plenum Publishing Corporation
233 Spring Street, New York, N.Y. 10013

Printed in the United States of America

To my parents

Janet Frances James, Ed.D.
(1932–1980)

Robert Francis James, M.D.

Contributors

William C. Banta Department of Biology, The American University, Washington, D.C. 20016

J. L. Barnard Department of Invertebrate Zoology, National Museum of Natural History, Smithsonian Institution, Washington, D.C. 20560

James A. Blake Science Applications International Corporation, Woods Hole, Massachusetts 02543

Richard C. Brusca Department of Marine Invertebrates, San Diego Museum of Natural History, San Diego, California 92112

Steven M. Chambers Division of Endangered Species, U.S. Fish and Wildlife Service, Albuquerque, New Mexico 87103. *Present address*: Southern California Field Station–Ventura Office, U.S. Fish and Wildlife Service, Ventura, California 93003

Francisco P. Chavez Monterey Bay Aquarium Research Institute, Pacific Grove, California 93950

Mitchell W. Colgan Earth Sciences Board of Studies, University of California–Santa Cruz, Santa Cruz, California 95064. *Present address*: Department of Geology, College of Charleston, Charleston, South Carolina 29424

Yves Finet Department of Invertebrates, Museum of Natural History, CH1211 Geneva 6, Switzerland

John S. Garth Allan Hancock Foundation, University of Southern California, Los Angeles, California 90089-0371

Terrence M. Gosliner Department of Invertebrate Zoology and Geology, California Academy of Sciences, Golden Gate Park, San Francisco, California 94118

Alan W. Harvey Shannon Point Marine Center, Anacortes, Washington 98221

Thomas M. Iliffe Department of Marine Biology, Texas A & M University at Galveston, Galveston, Texas 77553

Matthew J. James Department of Geology, Sonoma State University, Rohnert Park, California 94928

E. Alison Kay Department of Zoology, University of Hawaii, Honolulu, Hawaii 96822

David L. Malmquist Earth Sciences Board of Studies, University of California–Santa Cruz, Santa Cruz, California 95064

Linda Yvonne Maluf Department of Ecology and Evolutionary Biology, University of Arizona, Tucson, Arizona 85721. *Present address*: 4081 Minuteman, Boise, Idaho 83706

Sally E. Walker Department of Paleontology, University of California–Berkeley, Berkeley, California 94720. *Present address:* Department of Zoology, Oregon State University, Corvallis, Oregon 97331-2914

Wilfried Westheide Systematic Zoology, Department of Biology/Chemistry, University of Osnabrück, D-4500 Osnabrück, Federal Republic of Germany

Mary K. Wicksten Department of Biology, Texas A & M University, College Station, Texas 77843

Victor A. Zullo Department of Earth Sciences, University of North Carolina at Wilmington, Wilmington, North Carolina 28403-3297

Contents

Chapter 6 • Caridean and Stenopodid Shrimp of the Galápagos Islands

Mary K. Wicksten

Chapter 7 • Biogeographic Patterns of the Galápagos Porcelain Crab Fauna

Alan W. Harvey

Chapter 8 • Zoogeography of the Shallow-Water Cirriped Fauna of the Galápagos Islands and Adjacent Regions in the Tropical Eastern Pacific

Victor A. Zullo

Chapter 9 • Amphipoda of the Galápagos Islands

J. L. Barnard

V. Marine Cave Fauna

Chapter 10 • Anchialine Fauna of the Galápagos Islands

Thomas M. Iliffe

VI. Mollusca

Chapter 11 • The Marine Mollusks of the Galápagos: Determinants of Insular Marine Faunas

E. Alison Kay

Chapter 12 • The Marine Mollusks of the Galápagos Islands

Yves Finet

Chapter 13 • The Opisthobranch Gastropod Fauna of the Galápagos Islands

Terrence M. Gosliner

Chapter 14 • Biogeography of Galápagos Land Snails

Steven M. Chambers

Chapter 15 • Charles Darwin's Contribution to the Molluscan Fauna of the Galápagos Islands: Historical Perspective on Endemicity and Biogeography

Matthew J. James

VII. Echinodermata

Chapter 16 • Echinoderm Fauna of the Galápagos Islands

Linda Yvonne Maluf

VIII. Bryozoa

IX. Taphonomy and Paleoecology

Introduction

Marine Invertebrate Evolution in the Galápagos Islands

MATTHEW J. JAMES

1. Perspective of This Volume

Charles Darwin brought the Galápagos Islands to the attention of zoologists, botanists, and geologists following the six-week visit of H.M.S. *Beagle* to the islands in 1835. Since then published research on the biota of the islands, particularly in multiauthored volumes, has focused on terrestrial plants and animals. The present volume is designed specifically to provide a summary of work on the marine invertebrate fauna. One deviation from that objective was the inclusion of a chapter on land snails, which proved to be a good choice because the phylum Mollusca is now covered more thoroughly in this volume than in any single previous scholarly work on the Galápagos.

The academic bottom line with this book is to elucidate the evolutionary responses of shallow water, benthic marine invertebrates to the unique set of insular conditions that exist in the Galápagos Islands. The route taken to that objective has many paths including taxonomic revision, determining biogeographic affinities, and examining the ecological requirements of species. The information presented here is for some groups from the islands the first stage in a thorough process that can eventually lead to an understanding of the phylogenetic relationships of these species. Understanding the origin of species on oceanic islands is a fitting objective for researchers working in the archipelago known as 'Darwin's Islands.'

It is hoped that this volume will be useful to several diverse groups of persons with varying interests. Marine invertebrate zoologists with interests in the specific taxonomic groups discussed in the book will find current information about taxonomy and biogeography that can be applied to work on related groups in other geographic areas. Evolutionary biologists can obtain information about the extent of divergence between species of invertebrates for studies of evolutionary patterns on oceanic islands and for comparisons to the well known terrestrial biota of the

MATTHEW J. JAMES • Department of Geology, Sonoma State University, Rohnert Park, California 94928.

islands. Persons interested in resource planning and environmental conservation in the human-impacted areas of the islands can use these findings to help formulate decisions about how to best manage the fragile Galápagos ecosystem. The combined information of all the chapters in this book should go a long way towards improving our knowledge of Galápagos marine life.

2. Directions for Future Research

While organizing this book, I became aware of how little is known about some taxonomic groups of marine invertebrates in the Galápagos. In 1989, I compiled the second edition of a bibliography of Galápagos marine invertebrates which contained over 600 references (copies available upon request) and learned that virtually nothing is known about phyla such as Ctenophora, Rotifera, Priapulida, Sipuncula, Echiura, and Phoronida in the islands. Figure 1 summarizes published work on Galápagos marine invertebrates from pre-1900 to 1989. Two groups dominate the Galápagos invertebrate literature: mollusks and crustaceans. These groups are also well represented in the present volume. The increased publication activity from the 1960's onward (as seen in Figure 1) has not included all taxonomic groups. Significant work-in-progress on Galápagos marine invertebrates by numerous authors around the world would make a second volume easily accomplished, and such a volume is being contemplated.

A measure of success for this volume will be if it stimulates research on both the 'unknown' invertebrate phyla and additional research on the taxonomic groups discussed in the book. The improved balance in our knowledge between the terrestrial biota and the marine biota will strengthen our overall knowledge of life in the Galápagos Islands.

3. Plan of This Volume

The subtitle of this volume, "Taxonomy, Biogeography, and Evolution in Darwin's Islands," summarizes the main topics covered within and the perspective of the book. Taxonomy, the time-consuming and sometimes tortuous process of keeping track of the 'cast of characters,' is the solid foundation on which both biogeographic and evolutionary studies are based. Faunal lists (based largely on taxonomic revisions) are provided in Chapters 2, 3, 5, 6, 7, 8, 9, 10, 12, 13, 15, and 17, and will be invaluable in assessing the biota of the islands. A solid assessment of the fauna is necessary for both academic studies and for practical considerations such as planning and conservation. The complete taxonomic composition of some groups in the islands is summarized here for the first time in the scientific literature, including: meiofauna (covering several phyla, Chapter 2), polychaetes, (Chapter 3), land snails (Chapter 14) and echinoderms (Chapter 17). Extensive compilations such as these are the result of many years of work and will significantly expand our knowledge of the shallow water marine life (and in one instance of terrestrial life) in the islands.

The book is divided into nine sections with nineteen chapters, each section and/or chapter covering a separate taxonomic group or topic. Galápagos marine invertebrates display a spectrum of evolutionary responses to insular isolation,

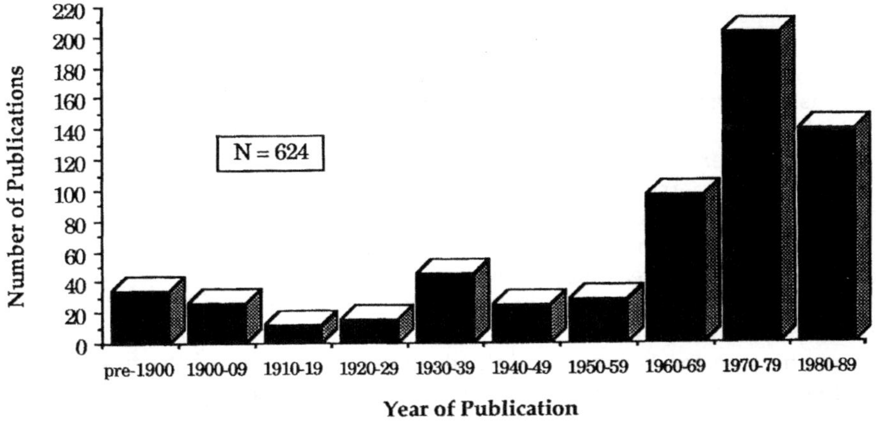

Figure 1. Summary of published contributions dealing with Galápagos marine invertebrates from pre-1900 to 1989. Data from "Studies of Galápagos Marine Invertebrates—A Bibliography" compiled by M. J. James (copies available upon request).

varying from relatively high species-level endemism to values typical for marine organisms with planktonic larvae experiencing gene flow between mainland and island populations.

The oceanographic setting of the islands is reviewed in Chapter 1, along with a section on the geologic setting of the islands to put them into their proper tectonic framework. Data on interannual environmental variability summarized here provides a crucial understanding of the workings of the Galápagos ecosystem.

Almost 40 papers have been published, mainly in German, on the interstitial fauna living on sandy beaches in the Galápagos. Chapter 2 summarizes this extensive body of research for the first time in English, thereby making this valuable work accessible to many more interested readers. The meiofauna show high species richness and low evolutionary divergence compared to other areas of the world.

Polychaete annelids represent an important component of many benthic ecosystems, and yet are very poorly known from the Galápagos. Chapter 3 is the first-ever compilation of this group from the islands. The patterns of taxonomic composition that emerge make it clear that additional sampling is necessary to provide a complete picture of all polychaete families in the islands. Relatively high endemism in the family Syllidae is an interesting evolutionary pattern.

The effect of El Niño as a mediating factor between carbonate accretion and erosion in the islands is examined in Chapter 4. The delicate balance between coral reef establishment and obliteration is seen to be the result of both biological and physical factors, many of which are unique to the geographic position of the Galápagos in the tropical eastern Pacific. While coral reefs do become established in the islands, they are held in check by the environmental consequences of periodic El Niño events.

Five chapters in the book deal exclusively with crustaceans (Chapters 5, 6, 7, 8, and 9), and Chapter 10 on the marine cave (anchialine) fauna includes ten crustacean taxa and one species of fish. The shallow-water brachyuran crab fauna, its most famous member being the brightly-colored "Sally Lightfoot" crab, is summarized in Chapter 5. Taxonomically, this is one of the best-studied groups of Galápagos marine invertebrates, with monographs on the group extending back to 1946.

Only 7% of the caridean and stenopodid shrimp of the Galápagos are endemic to the islands, as summarized in Chapter 6. This group of crustaceans is often overlooked due to small size and cryptic coloration and/or habitat.

Using a preexisting taxonomic scheme, Chapter 7 examines the biogeographic affinities of the porcelain crab fauna. The important point is made that when making biogeographic comparisons one should consider not merely the number of taxa found in common between two geographic areas, but also the species pool from which the two areas may draw.

Chapter 8 summarizes the barnacle fauna of the islands and notes that the absence of chthamaloids in the islands is particularly surprising given their commonness on the Panamic mainland. Barnacles are one of the few groups of Galápagos marine invertebrates with a significant fossil record.

Relatively high endemism is documented in Chapter 9 for Galápagos amphipods (38%) compared to other crustaceans in the islands. This pattern is largely due to the lack of planktonic dispersal stage in amphipods.

Marine caves in other areas of the world have been the sites of several significant taxonomic discoveries in recent years, including the discovery of a new class, two new orders, and two new families of crustaceans. Chapter 10 reviews the Galápagos anchialine fauna. Commensurate with the 'island within an island' viewpoint is the 100% endemism at the species level and at least three endemic genera. The unusual physical setting of marine caves has given rise to an evolutionary pattern more commonly encountered in the terrestrial realm.

The taxonomic composition of the phylum Mollusca is addressed in five chapters (Chapters 11, 12, 13, 14, and 15) with additional molluscan topics found in the two chapters on paleoecology and taphonomy (Chapters 18 and 19). The first thorough comparison of the molluscan faunas of the Hawaiian Islands and the Galápagos Islands is presented in Chapter 11. This comparison provides the basis for a unifying theory of the faunal composition of oceanic islands, namely, habitat, oceanographic conditions, and isolation acting in concert to produce the Hawaiian and Galápagos faunas. The fossil record of mollusks in the Galápagos provides the potential for studying changes in faunal composition over geological time.

Chapter 12 complements the information presented in Chapter 11 with additional insights into the taxonomic composition of the fauna and the factors that influence the affinities between the Galápagos and other geographic areas. With about 600 species comprising the fauna, mollusks are one of the most diverse groups of Galápagos marine invertebrates.

Opisthobranch gastropods, a group in which numerous additional discoveries are yet to be made in the islands, are summarized in Chapter 13. Anatomical details are presented here for several species, often for the first time in the literature. New distribution records for several species also add to our growing knowledge of this group in the islands.

Land snails (Chapter 14) are the one departure from the marine theme of this volume—but they are a significant group both for their interesting evolutionary diversity and their complicated biogeographic affinities. The inclusion of land snails rounds out coverage of the phylum Mollusca from the Galápagos as never before accomplished in the scientific literature. Endemism at the 100% level for species, as for the marine cave fauna, is seen once again in the land snails.

Chapter 15 is an historical study that concentrates on Charles Darwin. The effect of estimates of endemicity on Darwin's published work is examined to see how Darwin interpreted the extent of evolutionary divergence in the islands.

The Galápagos echinoderm fauna (Chapter 16) is recognized as a distinct subtropical biogeographic province. Exhibiting a 'typical' or 'expected' level of endemism for a group of marine organisms (17%), this group is summarized here for the first time in the scientific literature. A similar scenario exists for the Bryozoa (Chapter 17) with 18% endemism.

Taphonomy and paleoecology are two topics previously not discussed at length in the Galápagos literature. Chapters 18 and 19 address these topics with reference to subfossil and fossil mollusks, respectively. The Urvina Bay uplifted area, a remarkable site also discussed in Chapter 4, provided the ideal place to test hypotheses about fossil preservation and biases in the fossil record (Chapter 18). The implications of this study extend far beyond the shores of the Galápagos to the very core of how paleontologists view the fossil record. Tracing the post-mortem history of fossil mollusks from Vilamil on Isla Isabela is examined in Chapter 19 and it was found that a tremendous amount of additional information is contained on and in the shells of fossil mollusks that would not be observed by casual inspection. The inferred presence of hermit crabs (post-mortem occupation) and the absence of bioerosion (rapid post-depositional uplift) add significantly to our knowledge of the Galápagos fossil record.

In summary, with this volume we now have the first book devoted to summarizing the taxonomy, biogeography, and evolution of many Galápagos marine invertebrate groups. This work will join the growing literature on Galápagos marine invertebrates to improve our knowledge of life in the islands. The coverage of diverse topics on a variety of taxonomic groups adds greatly to our understanding of the islands made famous by Charles Darwin.

ACKNOWLEDGMENTS. In 1987, while preparing to serve as President of the Western Society of Malacologists (WSM) in 1988, I organized a symposium on the molluscan fauna of the Galápagos Islands. The symposium was held on the campus of Sonoma State University on July 18, 1988, and abstracts were published in Volume 21 of the Annual Report of WSM. I contacted Douglas S. Jones of Florida State University as a potential contributor to the symposium and although he declined because of other commitments, Doug extended an invitation to me to edit a volume of broader scope on the entire marine invertebrate fauna of the islands. This volume is the direct outgrowth of those early discussions.

I thank A. M. McNamara and M.-C. Cloutier of Plenum Publishing in New York for editorial and administrative support during the several years of manuscript preparation. M. Carigma-Sambilay at Plenum greatly aided final production.

My job as editor of reviewing and evaluating manuscripts has been made less burdensome by thoughtful and critical reviews by: R. C. Brusca, S. J. Carlson, H. W. Chaney, R. C. Dugdale, W. K. Emerson, K. Fauchauld, J. S. Garth, J. Haig, G. Hendler, E. A. Kay, J. W. Martin, S. Millen, W. A. Newman, J. D. Soule, D. W. Steadman, M. K. Wicksten, M. A. Wilson, and W. S. Wooster.

I thank C. S. Leung and E. P. James for devoted support during the several years it took to complete this volume.

This volume is Contribution 441 of the Charles Darwin Foundation.

I

Oceanographic Setting

Chapter 1

The Galápagos Islands and Their Relation to Oceanographic Processes in the Tropical Pacific

FRANCISCO P. CHAVEZ and RICHARD C. BRUSCA

1. Introduction

Equatorial regions are characterized by significant interannual variability in oceanographic processes that are linked to changes in climate throughout the globe (Rassmusson and Wallace, 1983). Due in part to their role in global climate variability, equatorial regions—and particularly the equatorial Pacific—have received considerable attention from meteorologists and physical oceanographers during past decades. Recent measurements made during National Oceanographic and Atmospheric Administration (NOAA)-sponsored programs such as the Equatorial Pacific Ocean Climate Studies (EPOCS) and the Tropical Ocean Global Atmosphere (TOGA), as well as National Science Foundation (NSF)-sponsored programs such as Tropic Heat, have led to significant advances in our understanding of the oceanographic processes at work in the equatorial Pacific.

FRANCISCO P. CHAVEZ • Monterey Bay Aquarium Research Institute, Pacific Grove, California 93950. RICHARD C. BRUSCA • Department of Marine Invertebrates, San Diego Museum of Natural History, San Diego, California 92112.

The Galápagos Islands are located on the equator in the eastern equatorial Pacific. The Galápagos are the only oceanic islands on the equator in the eastern equatorial Pacific, with the nearest other truly equatorial islands being Christmas and Jarvis Islands which are approximately 5000 kilometers to the west of the Galápagos. The Galápagos are subject to the influence of several major current systems and their associated water masses, and provide an ideal location for observing the large-scale interannual ocean variability which characterizes the equatorial Pacific. Their location is particularly interesting since the amplitude of the interannual variability is greatest in a swath from 5°N to 5°S. This contribution focuses on oceanographic conditions in the Galápagos region that are determined primarily by basinwide equatorial processes. A short description of the geographic–tectonic setting of the islands is also provided. Following this we describe mean oceanographic conditions and their seasonal variability. It is the mean conditions that make the islands and their ecosystem biologically vulnerable to interannual variability. Mean is used instead of 'normal,' since 'normal' implies that variations are abnormal and as discussed in the final section interannual variability is an integral part of the Galápagos ecosystem.

2. Tectonic Setting

Two principal geotectonic features have shaped the Galápagos region. The first is the proximity of the islands to two major rift systems, or spreading centers. The Galápagos Islands are situated almost midway between the South American coast and the roughly north–south line of the East Pacific Rise. They lie about 1000 kilometers west of South America and about the same distance east of that point where the East Pacific Rise makes its northwest turn to approach the mouth of the Gulf of California. The Galápagos Spreading Center takes off eastward from this bend in the East Pacific Rise (forming a 'triple junction') and passes several hundred kilometers north of the Galápagos Islands. South of the Galápagos Spreading Center, crustal movement is about 3 centimeters/year southward. The relatively fast eastward spread from the East Pacific Rise (about 7 cm/year) results in a net east–southeast crustal movement in the Galápagos region of approximately 6–7 cm/year. This directional spread was first proposed by Wilson (1963) over 25 years ago, and confirmed by various studies in the late 1970s and early 1980s (e.g., Morgan, 1972; Anderson *et al.*, 1975).

The second feature is a theorized 'hotspot' deep in the mantle that supplies lava to and through the crustal plates as they move slowly along. Although the Hawaiian Islands have become the classic model of the hotspot phenomenon, Wilson's (1963) original work also noted the Galápagos as an example of the process. The concentration of recent volcanism in the western islands corroborates and identifies the locality of the Galápagos hotspot. Also corroborating the theory is the fact that older islands are found in line with the crustal drift direction, to the southeast. It has been suggested that the Carnegie and Cocos Ridges may also be remnant products of this hotspot mechanism (Morgan, 1971, 1972; Holden and Dietz, 1972). Current interpretation views the Carnegie Ridge and the Galápagos archipelago forming as the Nazca Plate passed over the Galápagos hotspot on its slow but inexorable journey southeastward, where it is eventually subducted under the South American Plate by the Cordilleran Trench system. The Cocos Ridge is seen as a remnant of former hotspot activity, coupled with past and present northeast crustal spread of the Cocos Plate away from the Galápagos Spreading Center, and eventual subduction under the Caribbean Plate at the Central America Trench. The Malpelo Ridge may be a fragmented remnant of the Cocos Ridge. Dalrymple and Cox (1968) analyzed lava samples from Cocos Island, on the flank of the Cocos Ridge about 500 kilometers from Costa Rica, and found them to be clearly oceanic in character. Bellon *et al.* (1983) radiometrically dated the Cocos Islands at 1.9–2.4 million years.

Hey and Vogt (1977) and Hey *et al.* (1977) reasoned that the 45,000 square kilometer Galápagos Platform, which sits at the crux of the Cocos and Carnegie Ridges, matches the exact physiographic pattern predicted by 25 million years of fixed mantle hotspot activity in which one plate is moving to the northeast and the other to the southeast.

Refinements in dating methods, particularly paleomagnetic and potassium–argon (K–Ar) radiometric methods, have led to a general agreement among geologists, at least for the time being, that the Galápagos Platform is no more than 10–15 million years old, while the oldest exposed land of the archipelago is on the order of 2.5–4.0 million years old (Cox and Dalrymple, 1966; McBirney and Williams, 1969; Bailey, 1976; Bow, 1979; Baitis and Lindstrom, 1980; Hall *et al.*, 1980; Cox, 1983; Hall, 1983; Simkin, 1984; Geist *et al.*, 1985). Many exposed parts of the islands are, of course, much younger than this, and volcanic activity continues today with infrequent eruptions reported from the western islands of Fernandina and Isabela. These dates are in conflict with Durham's (1963, 1966) earlier paleontological study in which he reported late Miocene (5–10 million years) invertebrates from Santa Cruz Island. However, to our knowledge, no one has addressed the possibility that the strata in question may represent an old (Miocene) uplifted area deposited in shallow water prior to the emergence of that portion of Santa Cruz. Emergence of shallow subtidal areas in the Galápagos in recent times has been documented by Glynn and Wellington (1983). Further, we now know that sea level fluctuations with vertical excursions up to 130 meters probably took place during the Pliocene/Pleistocene ice ages (e.g., Fairbridge, 1973; Vail and Mitchum, 1979). Virtual agreement now exists among geologists that the islands have never been any closer to the mainland than they currently are, being the products of ocean floor volcanism. Most biogeographic studies have supported this view (see Brusca 1987 for a review). Southeast drift is apparently moving the islands ever closer to the South American continent.

3. Mean Conditions and Their Seasonal Variations

3.1. Introduction

On average the eastern equatorial Pacific is characterized by sea surface temperatures (SST) which are abnormally cool for a tropical region. The region is dominated by a high pressure system which is centered over Easter Island. The Easter Island High is the largest southern hemisphere high-pressure system (Wyrtki, 1982) and from it the southeast trade winds flow toward the Indonesian Low. The anticyclonic wind system associated with the Easter Island High drives the Peru Current which flows to the northwest and the South Equatorial Current which flows westward (Figures 1 and 2). The meridional winds along the west coast of South America, which flow from south to north, are responsible for coastal upwelling. The zonal winds which flow from east to west along the equator generate equatorial upwelling. These winds are persistent annually and interannually so that coastal and equatorial upwelling are continuous throughout the annual cycle with only short interruptions. According to Wyrtki (1981), "upwelling in the equatorial Pacific Ocean manifests itself by a tongue of cool water stretching from the coast of Peru to the [international] dateline." The cool tongue of the equatorial Pacific is maintained primarily by equatorial upwelling but is also supported by horizontal advection of waters from the east.

There are well-defined surface water masses in the Eastern Tropical Pacific: 1) recently upwelled waters with salinities between 34.95 and 35.10 parts per thousand; these waters can be further subdivided into waters of costal upwelling and waters of equatorial upwelling, with the former being significantly cooler; 2) tropical surface waters (TSW) with lower salinities and warmer SST, which are separated from waters of equatorial upwelling by the Equatorial Front which extends from the coast of Ecuador to the Galápagos Islands and westward (Wooster, 1969); 3) Subtropical Surface Waters (SSW), which are offshore of the coastally upwelled waters and to the south of the equatorially upwelled waters (Figure 1), and have higher salinities (greater than 35.2) and warmer SST than the upwelled waters.

3.2. Hydrography, Nutrients, and Phytoplankton

The mean annual cycle of sea surface temperature (SST) for a station at Baltra, Galápagos indicates that that region has a strong austral or southern hemisphere character even though it is on the equator (Figure 3). Maximum temperatures are found during February, March, and April, and the coolest temperatures are during August and September. In northern Peru, and probably at the Galápagos, the annual austral summer warming is generated by a combination of local heating and a decrease in the export of heat associated with weaker trade winds (Chavez, 1987).

A strong east–west gradient in sea surface temperature is found along the equator with cooler temperatures to the east and warmer temperatures to the west (Figure 4). An east–west gradient is also evident subsurface, with a sloping thermocline (Sverdrup *et al.*, 1942; Barber and Chavez, 1983, 1986) which comes closer to the surface at the eastern terminus (Figure 5). This east–west basinwide tilt in the thermocline is maintained by the trade winds (Cane, 1983; Mangum and Hayes, 1984) (Figure 6). The east–west gradient in sea surface temperatures is a result of a combination of the east–west tilt in the thermocline, equatorial upwelling, and horizontal advection of water from the east. Around the Galápagos Islands the east–west gradient is reversed. Coolest temperatures are found on the western side and warmest temperatures are found on the eastern side of the islands. The cool temperatures on the western side are probably a result of topographically driven upwelling associated with the surfacing of the Equatorial Undercurrent

a

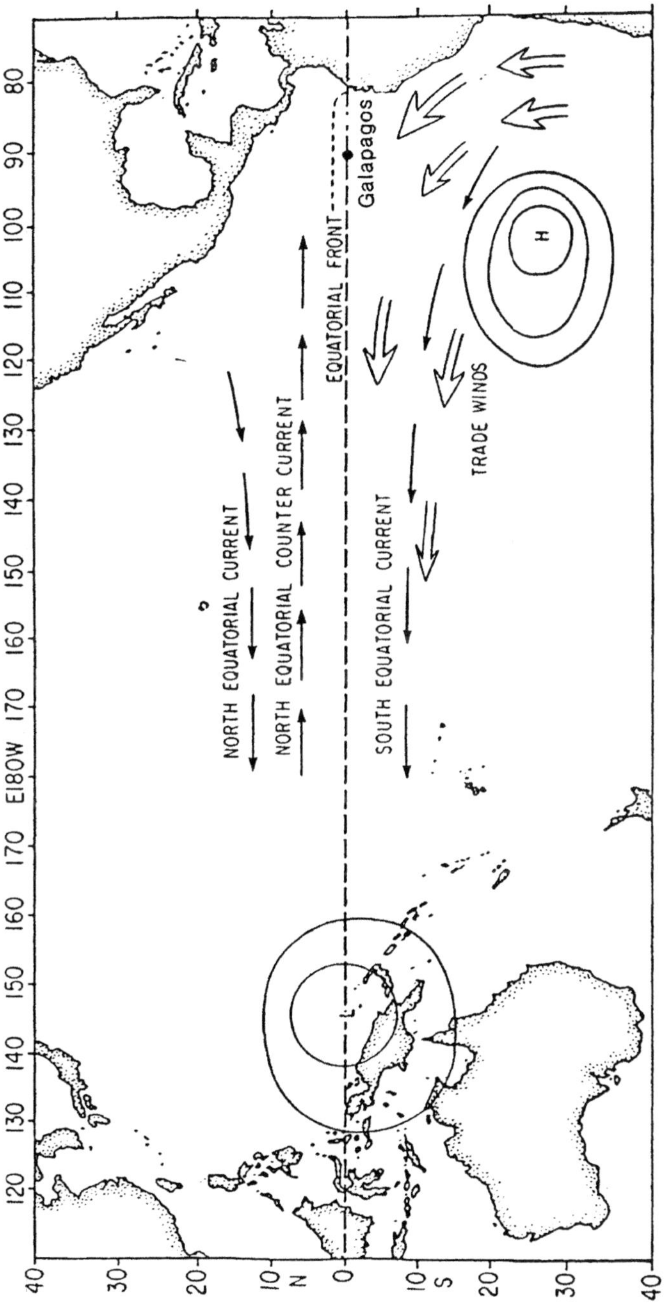

Figure 1. (a) Chart of the tropical Pacific showing the location of the major currents as well as the Easter Island High (H) and the Indonesian Low (L).

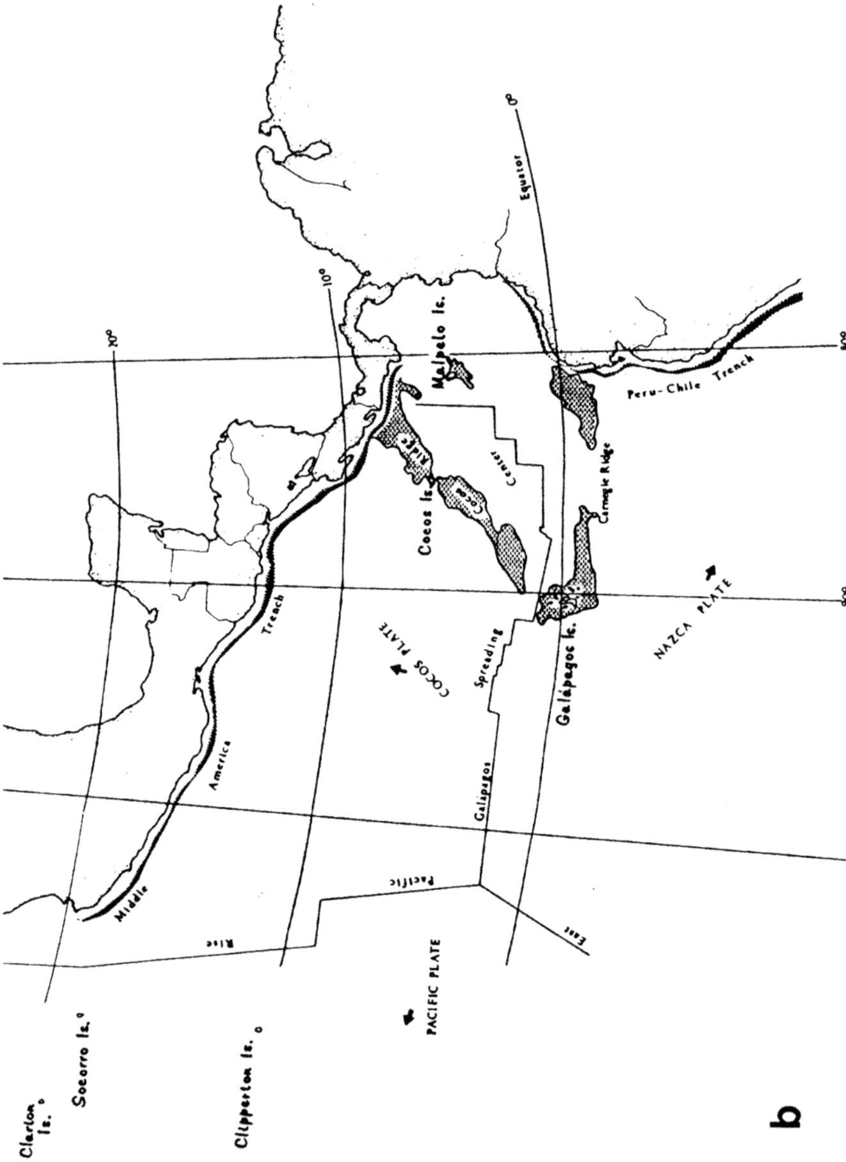

Figure 1. (b) Tectonic and geographic setting of the Galápagos Islands. Arrows indicate approximate direction of plate movements. Those portions of the Cocos and Carnegie Ridges above 1000 fathoms (1829 m) depth are stippled.

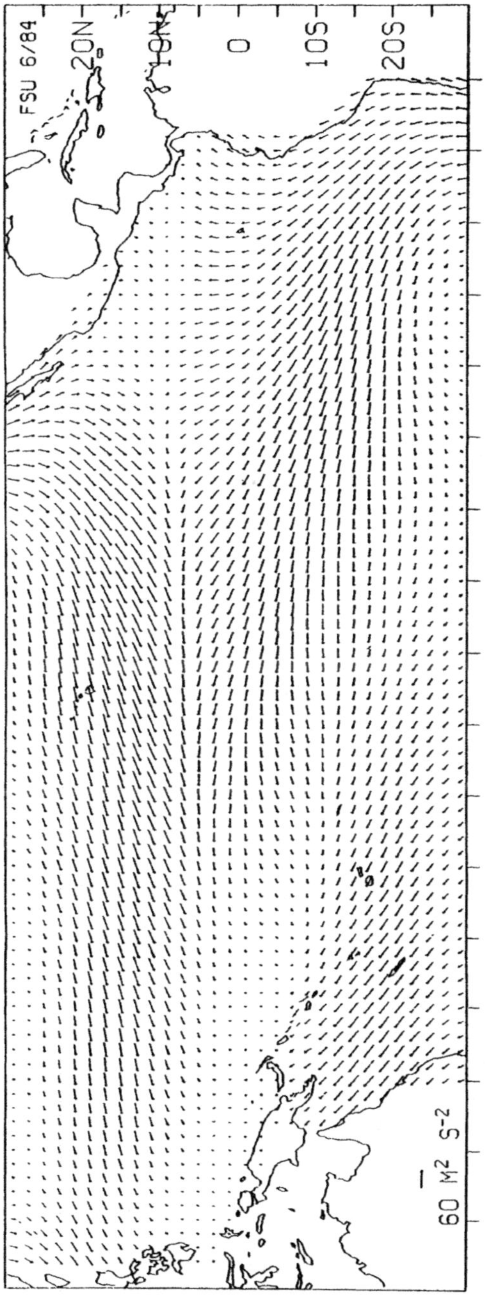

Figure 2. Twenty year mean (1961–1980) wind stress from the Florida State University ship wind analysis. The northeast and the southeast trades are clearly visible as is the wind stress minimum zone associated with the Intertropical Convergence Zone (ITCZ) around 10°N. The Galápagos are in a zone of weaker winds than to the south and west, and flow around the Islands is southerly. Data courtesy of Dr. James J. O'Brien.

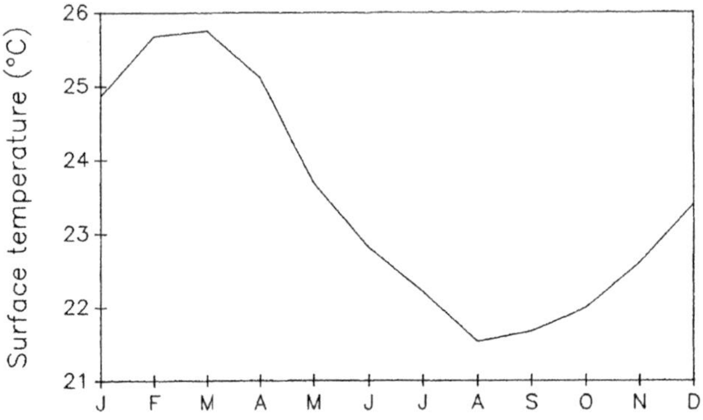

Figure 3. Twenty-two year mean (1960–1981) cycle of sea surface temperature taken from a pier at Isla Baltra, Galápagos. Although the Islands are on the equator, the cycle of sea surface temperature shows an austral or southern hemisphere character. Data courtesy of Dr. David Enfield.

(EUC). The east–west gradient found around the islands supports the notion that upwelling, rather than advection of waters from the east, is the primary mechanism for the cool sea surface temperatures along the equator.

Nitrate concentration at the surface, an index of ecosystem productivity, shows a similar pattern with highest values to the east (Figure 4). The sole exception to this pattern is found in and around the islands where concentrations are reduced. It is clear from the chlorophyll data that these low values are the result of uptake by phytoplankton. A sloping subsurface nutricline also shallows at the eastern boundary of the equatorial basin (Figure 5). Chlorophyll concentration, an index of phytoplankton biomass, shows a very different pattern with high levels in and around the islands and low and uniform levels to the east and west (Figure 4). A surprising observation is the lack of a gradient in chlorophyll to the west of the islands in the face of a substantial nutrient gradient. The high levels of phytoplankton biomass in and around the islands supports the hypothesis that phytoplankton require a certain proximity to land masses for bloom formation (Chavez, 1989).

There are conflicting reports on the seasonal variability in thermocline depth in the eastern equatorial Pacific (Meyers, 1979; Hayes and Halpern, 1984; Lukas, 1986). Lukas (1986) reported a deepening during January, February, and March, while Hayes and Halpern (1984) and Meyers (1979) reported that the deepening occurs from March to June. Observations along the coast of northern Peru seem to confirm that the deepening occurs in the March to June window (Chavez, 1987). The deepening of the thermocline has a duration of one-to-two months but varies each year in its timing. Seasonal subsurface variations in nutrients are similar to those in temperature (Chavez, 1987).

The meridional or north–south distribution of properties across the equator is very asymmetric, with warmer surface temperatures and lower nitrate concentrations to the north (Figure 7). The north–south asymmetry is evident from the coast of Peru to the dateline although it is more notable in the eastern Pacific. The Equatorial Front (EF) is the name given to the rather abrupt boundary between the cooler and more saline waters along the equator and the warmer, fresher tropical surface waters (TSW) to the north (Wooster, 1969). The EF also has a marked nutrient signature (Figure 7) with lower nutrients in the warmer water (Chavez,

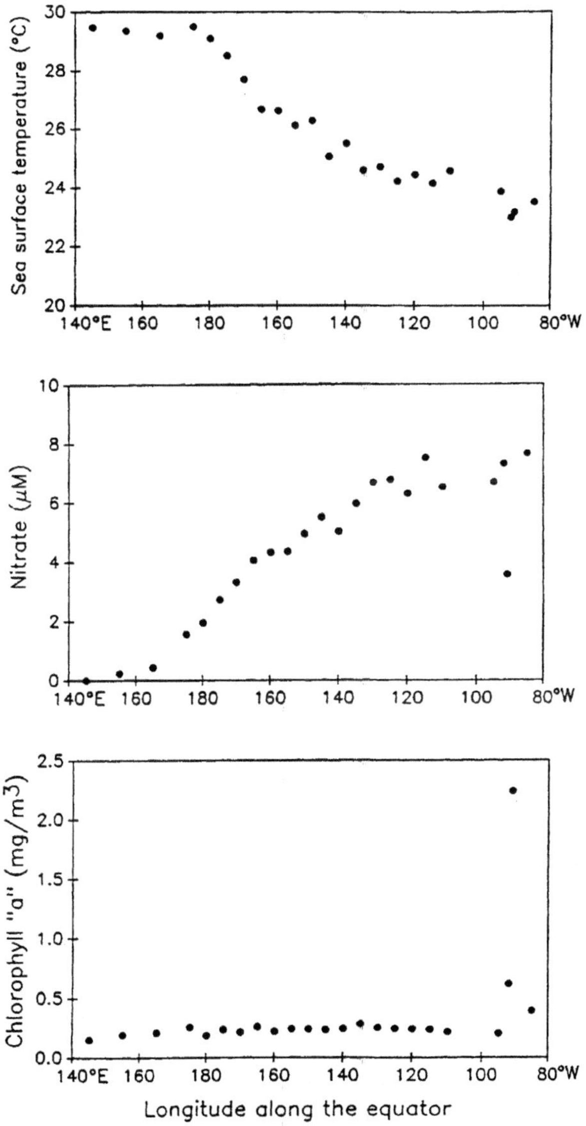

Figure 4. Means of surface temperature, surface nitrate, and surface chlorophyll from discrete observations along the equatorial Pacific on 40 cruises in 1980 and 1988. Measurements during the strong El Niño of 1982–83 were excluded from the calculations. The primary vehicles for the observations were NOAA ships working on the Equatorial Pacific Ocean Climate Studies (EPOCS) program. The lack of an east–west gradient chlorophyll despite gradients in temperature and nitrate is notable.

1987). The EF reaches its most southern extension at the western coast of South America where seasonally it may reach as far as 4°S. It then angles sharply northward, and at the Galápagos, where it is also known as the Galápagos Front, it is typically between the equator and 1°N. Further west it continues to angle slightly to the North and is found between 1°N and 3°N. The EF has a marked seasonal variation, with displacements to the north during the austral summer and to the south during winter and spring (Okuda *et al.*, 1983). During March, April, and May

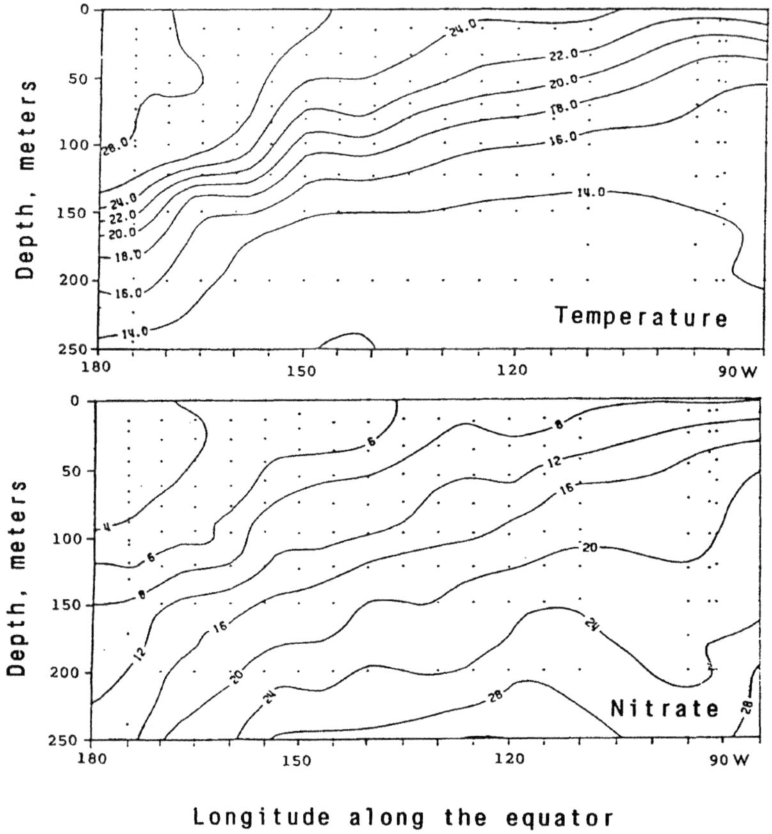

Figure 5. Mean profiles of temperature and nitrate along the equator. The east–west slope in the thermocline and nutricline is clear with cooler temperatures and higher nutrients closer to the surface at the eastern terminus.

the EF is to the north of the equator along 85°W but by June it has begun to migrate south and by November it is found around 2°S at which time it begins to return to the north (Chavez, 1987). The Islands extend from about 2°N to 2°S so that part of the Islands are immersed in the cool waters of the South Equatorial Current (SEC) and those which are north of the Equatorial Front (EF) are bathed by warmer waters and are much more tropical in nature.

3.3. Horizontal Currents

Reviews of ocean currents in the eastern and central tropical Pacific can be found in Sverdrup *et al.* (1942), Wyrtki (1966), Wooster (1970), and Wyrtki and Kilonsky (1984). The major surface current in the tropical Pacific is the South Equatorial Current (SEC) (Figure 1) which flows westward across the Pacific

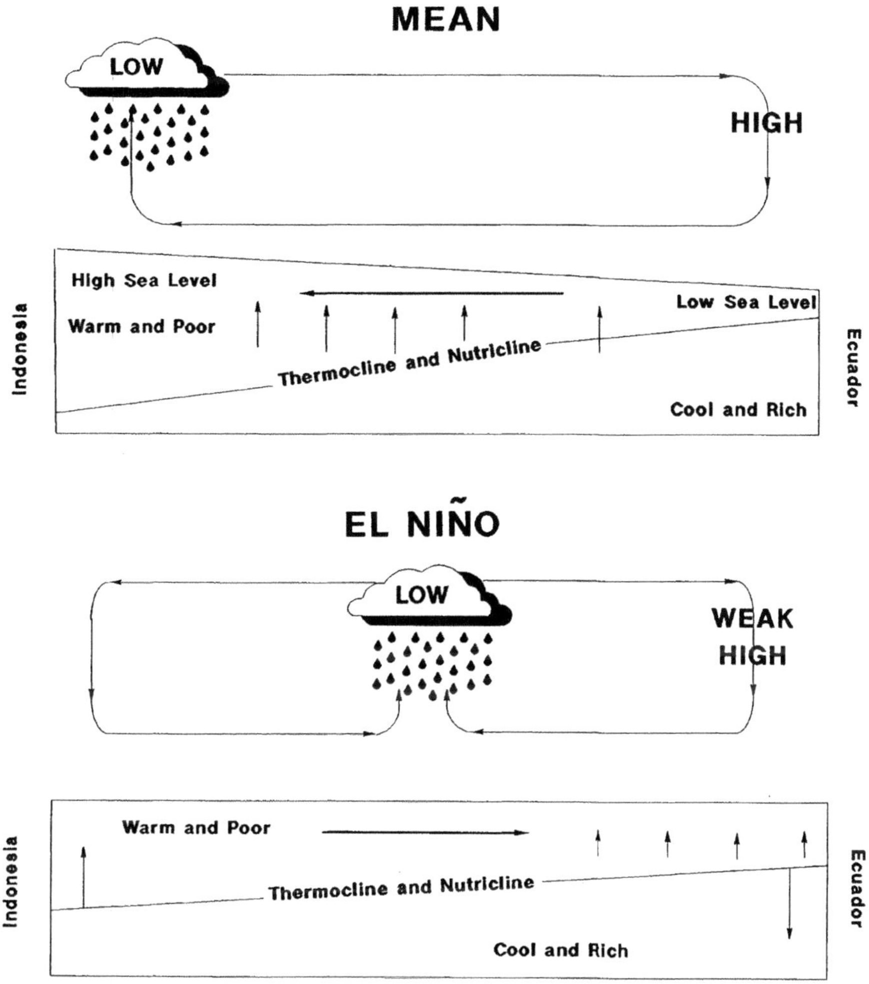

Figure 6. Oceanic and atmospheric conditions during the mean (above) and El Niño (below) states. During the mean, trade winds which blow from the Easter Island High to the Indonesian Low maintain a tilt in the thermocline and nutricline, drive equatorial upwelling, and pile up mass and heat in the western tropical Pacific. For reasons which are not well understood at present, westerly winds are generated in the western equatorial Pacific, as convective activity migrates eastward. The atmospheric forcing generates a series of oceanographic changes in the eastern equatorial Pacific; the most notable are higher sea level, a deeper thermocline and nutricline and abnormally warm sea surface temperatures. The oceanographic changes enhance the atmospheric effects so that these conditions remain in the eastern equatorial Pacific for 6 to 18 months. Less frequently than the El Niño condition there are extremely cool sea surface temperatures in the eastern Pacific associated with a higher than average Southern Oscillation Index (the difference between the Easter Island High and the Indonesian Low). This cool condition has been referred to as La Niña (Philander, 1985).

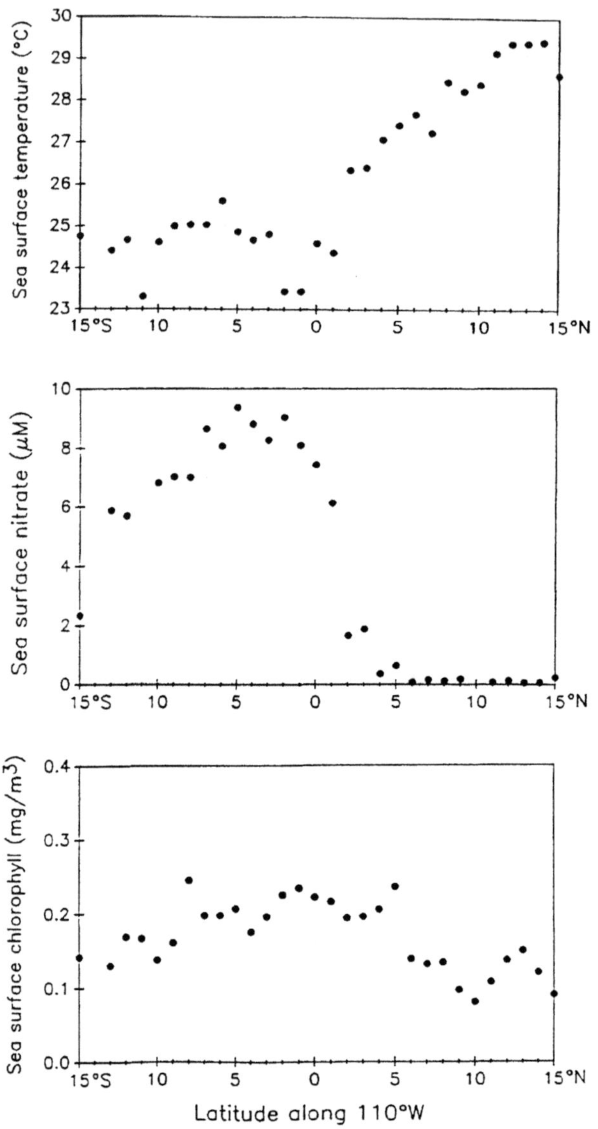

Figure 7. Means of surface temperature, surface nitrate, and surface chlorophyll from discrete observations along the 110 west meridian.

roughly between 5°N and 10°S with typical mean surface speeds of around 40 centimeters per second (Wyrtki, 1981), bathing the entire region of the Galápagos most of the year. The SEC is fed from the east by the Peru Current System (PCS). According to Wyrtki (1966), the PCS is composed of the Peru Coastal Current (PCC) and further offshore, the Peru Oceanic Current (POC), and these currents are usually separated by a weak and irregular flow to the southeast, the Peru Countercurrent (PC). The SEC is bounded to the north by the North Equatorial Countercurrent (NECC). The NECC has considerable seasonable variation, being most developed during August and September and weakest from February to April. During August and September the NECC reaches the South American continent while from February to April the current is difficult to detect past 120°W (Wyrtki, 1966). Mean flows in the core of the NECC are around 40 cm sec^{-1} (Wyrtki and Kilonsky, 1984). The South Equatorial Countercurrent (SECC) is much weaker than the NECC and has been observed occasionally around 10°S (Reid, 1959).

Temperature, nutrients, chlorophyll, and primary production along the equator are influenced by Long or Legeckis waves (Legeckis, 1977; Hansen and Paul, 1984; Chavez *et al.*, 1990). Theoretical studies (Philander, 1978) suggest that the Legeckis waves are generated by shear between the eastward flowing NECC and the westward flowing SEC. The waves are quite notable along the Equatorial Front in satellite sea surface temperature composites, have a scale of 1000 kilometers, and propagate westward at around 40 km per day (Legeckis, 1986).

Three major subsurface eastward flows have been identified in the equatorial region: the Equatorial Undercurrent (EUC) at the equator and the Northern and Southern Subsurface Countercurrents (NSCC and SSCC) at around 5°N and 5°S (Tsuchiya, 1985). Of these, the EUC is the most notable and of greatest consequence to the Galápagos. The EUC spans the entire equatorial Pacific and reaches maximum velocities in the core of over 100 cm sec^{-1} (2 knots) and mean speeds of around 80 cm sec^{-1} (Hayes *et al.*, 1987; Wyrtki and Kilonsky, 1984). In the eastern equatorial Pacific the EUC is strongest from March through May (Hayes and Halpern, 1984), and during this period it becomes evident in the hydrography and nutrient concentrations (Figure 8) to the east of the Galápagos (Lukas, 1986; Tsuchiya, 1985; Chavez, 1987; Pak and Zaneveld, 1973). Eastward surface flow is common along the equator during the March through May period and is probably associated with the surfacing of the EUC (Jones, 1969). The core of the EUC is commonly found in the thermocline so it is deepest in the west and shallows in the east (Wyrtki, 1981). At the Galápagos, the depth of the core of the EUC is approximately 60 m.

3.4. Upwelling

An excellent description of upwelling in the equatorial Pacific was given by Wyrtki (1981). An almost continuous westward surface wind flow, associated with the southeast and northeast trades, drives a poleward Ekman transport in both hemispheres, resulting in the upwelling of cool, nutrient-rich waters from below. Wyrtki (1981) suggested that water which is upwelled along the equator is recruited from 50 to 100 m depth, although geochemical observations (Broecker and Peng, 1982) and vertical velocities derived from current meter observations suggest that upwelling may be occurring down to 150 m.

Vertical velocities are very difficult to measure since they are on the order of 10^{-5}m sec^{-1}. Halpern *et al.* (1989) estimated that on time scales of months, vertical velocity in the equatorial Pacific between 165° and 90°W is around 2.5×10^{-5} m sec^{-1} (2 m day^{-1}), twice that reported by Wyrtki (1981). In addition, Halpern *et al.* (1989) suggested that upwelling is strongest around 140°W and decreases eastward. On time scales of days, upwelling rates of 3–5 m day^{-1} have been reported (Wyrtki and Eldin, 1982).

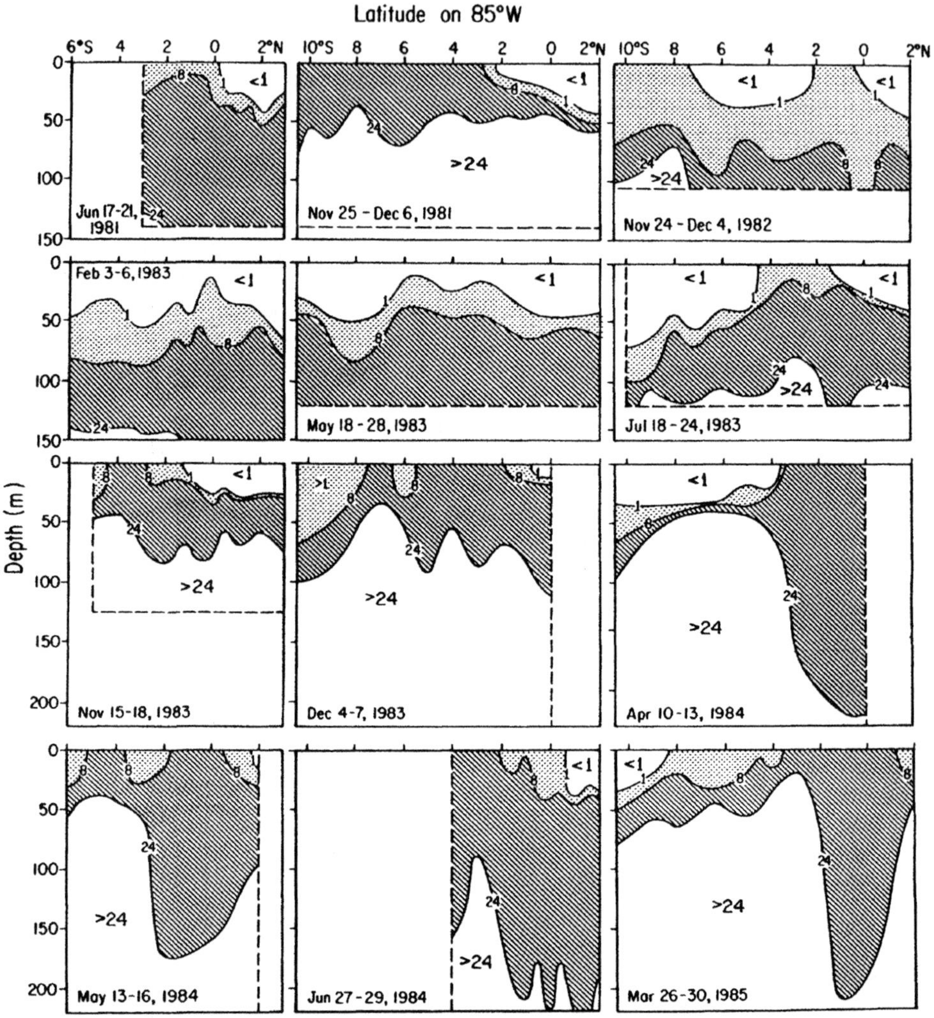

Figure 8. Twelve nitrate profiles along 85°W made between June 1981 and April 1985. The "nutricline" is defined here as the region with concentrations between 8 and 24 μm (hatched). The Equatorial Undercurrent (EUC) is recognized by a spreading of the nutricline, and during April 1984 and March 1985 it appears stronger and trapped along the equator. The contribution of the EUC is reduced during November and December. From November 1982 through July 1983, during El Niño, the nutricline is visibly deeper and the surficial nutrient rich equatorial zone disappears. The Equatorial Front (EF) is defined in these profiles by the northward transition between concentrations of nitrate that are less than 1 μM and greater than 1 μM. The Equatorial Front (EF) also varies seasonally and is further north during March, April, and May than in November and December. The EF disappeared during the 1982–83 El Niño.

The trade winds, which drive equatorial upwelling, are subject to sizeable seasonal variations. Wyrtki's (1981) estimate of upwelling, derived directly from the mean wind field, showed that upwelling is strongest between December and March and weakest from September to November. This result is counterintuitive because the southeast trades are strongest in August and September and weakest in February and March. The upwelling maximum, however, is driven primarily by northward Ekman transport which reaches a maximum in February as the northeast trades reach far south (Wyrtki, 1981).

Upwelling at the Galápagos has also been attributed to the surfacing of the Equatorial Undercurrent (EUC) as it collides with the westernmost islands, Islas Fernadina and Isabella. The EUC is a very fast subsurface eastward current which is "trapped" along the equator by the Coriolis effect. When the EUC reaches the Galápagos and collides with Isla Isabella Island, its flow becomes disorganized and further topographically driven upwelling may occur throughout the islands (Houvenaghel, 1984).

3.5. Sea Level

Low frequency variations in sea level reflect a variety of atmospheric and oceanic changes. In the equatorial Pacific, Wyrtki (1984) showed that nontidal sea level is associated with thermocline depth and dynamic height. As such there is a slope in sea level from the eastern equatorial ocean to the western equatorial ocean (Wyrtki, 1984). On average, sea level at the Galápagos is around 45 cm lower than sea level at the international dateline (Wyrtki, 1984).

Hayes and Halpern (1984) reported seasonal increases in sea level at the Galápagos on the order of 20 cm associated with a deepening of the thermocline and an increase in the strength of the Equatorial Undercurrent (EUC). The increases were observed between March and May. Smaller and more frequent oscillations in sea level, with periods of 40–60 days, are common in the tropics (Enfield *et al.*, 1987) and are linked to variations in the tropical wind fields (Madden and Julian, 1972).

Variations in sea level are also associated with changes in the nutricline (Chavez, 1987; Chavez, 1989). Data collected at the Galápagos during 1982 and 1983 (Kogelschatz *et al.*, 1985) show how sea level and nitrate concentration at 60 m, a depth from which upwelled water is recruited, are inversely related (Figure 9). In the March to June window, when the EUC is strongest and the thermocline and the nutricline are deeper, nutrient levels in the upwelling source water are lower and consequently biological production may be lower.

3.6. Climate

Local climate in the Galápagos reflects large-scale oceanographic and atmospheric conditions in the eastern tropical Pacific. The major low pressure zone affecting the Galápagos is the Intertropical Convergence Zone (ITCZ) which stretches from the coast of South America to Indonesia a few degrees north of the equator. The ITCZ is associated with a wind speed minimum and generally

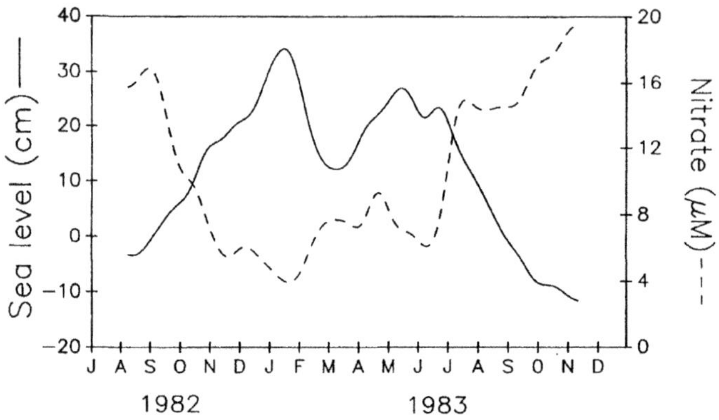

Figure 9. Relationship between sea level measured at a pier in Puerto Ayora, Galápagos, and nitrate at 60 m measured at a station offshore of Puerto Ayora (Kogelschatz *et al.*, 1985) during 1982 and 1983. The sea level was measured continuously and is courtesy of Dr. Klaus Wyrtki. Nitrate samples were collected three times a week with a small boat and is courtesy of Dr. R. T. Barber. The data has been smoothed with a normal filter in order to suppress periodicities of less than one week. The increase in sea level and the lower nitrate content of the 60 meter water are evident during the 1982–83 El Niño which lasted from September 1982 to July 1983 (Chavez *et al.*, 1984).

separates the northeast and the southeast trade winds (Rasmusson and Carpenter, 1982). Since the ITCZ is to the north of the Galápagos this puts the Islands within the regime of the southeast trades (Wooster and Hedgpeth, 1966). Trade wind speeds are in the order of 7 to 8 m sec^{-1} but at the Galápagos they are typically weaker than to the west. As a result of the locations of the Easter Island High and the ITCZ, the average wind direction around the Galápagos tends to be southerly (Figure 2). Further to the west along the equator the average wind direction has a more pronounced easterly component. The ITCZ exhibits a seasonal meridional migration reaching southern extremes in February and northern extremes in August. Precipitation patterns around the Galápagos Islands reflect the changes in the ITCZ, with highest rainfall from January to April (Robalino, 1985).

In general, the Galápagos climatic year may be divided into two seasons. A hot-wet period exists from about December to May and a cool-dry period (the "garua season") from about July to October. The Islands are probably best described as subtropical, although selected local areas would qualify as either tropical or warm-temperate based on their local thermal regime. Overall, the regime of the Galápagos rather closely resembles that of the northern Gulf of California or the northern Gulf of Mexico, two other regions generally regarded as subtropical.

4. Interannual Variability

4.1. Introduction

The waters of the South Equatorial Current are continually heated by the tropical sun and as a result both mass and heat tend to accumulate in the western tropical Pacific. The potential energy associated with increased mass and heat content is released periodically resulting in abnormally warm sea surface temperatures in the central and eastern tropical Pacific, and this phenomenon is known as El Niño. El Niño occurs aperiodically, once every three to eight years, and the magnitude of the anomalies varies significantly from event to event. The anomalies associated with El Niño persist in the eastern tropical Pacific for 6 to 18 months. The abnormally warm sea surface temperatures are associated with variations in the Eastern Island high and the Indonesian Low pressure systems which drive the southeast trades (Bjerknes, 1966). The seesaw in atmospheric pressure between the eastern and western Pacific is known as the Southern Oscillation (Walker, 1924). Although El Niño involves the entire tropical Pacific basin it is most notable along the coast of Peru where positive sea surface temperature (SST) anomalies of 10°C have been observed during these episodes.

Carrillo (1889) described a warm surface current which apparently came from the Gulf of Guayaquil and travelled southward up against the coast of northern Peru. The mariners of Paita called this current El Niño or "The Child" (Christ Child) since it appeared around Christmas. Schweigger (1945) showed there indeed is a weak and inoffensive current in northern Peru, which intensifies every year during the austral summer which he called the legitimate El Niño current (Chavez, 1987). As we shall see this current has little to do with the large scale oceanographic and meteorological events which are now known as El Niño. The "current" achieved notoriety as a result of the abnormally rainy and warm year of 1891. Carranza (1891), Eguiguren (1894), and Pezet (1896) suggested that the abnormal conditions in 1891 were the result of an unusually strong El Niño current. Later, Schott (1931) used the term El Niño for advection on a much larger scale. Schott (1931), who observed the 1925 El Niño, described a current that originated from around the Galápagos and reached the coast of Peru. Gunther (1936) and Mears (1943) referred to the El Niño current as responsible for the large scale oceanographic disturbances and their associated climatic repercussions. Thus, the name El Niño is today used to denote a complex set of processes that include more than just an intensification of a coastal current and advection of warm water from the north.

4.2. Hydrography, Nutrients, and Phytoplankton

Although El Niño appears as an amplification of the annual summer warming (Philander, 1985) it has a different origin. During El Niño, Kelvin waves forced by trade wind anomalies in the western equatorial Pacific raise sea level and deepen the thermocline and nutricline in the eastern tropical Pacific (Cane, 1983; Barber and Chavez, 1986; Enfield, 1988) (Figure 6). These anomalies are superimposed on the normal annual cycle so that there are significant warmings subsurface during the austral summer.

In addition to the rapid wave-propagated thermocline anomalies, there are slower advective disturbances during El Niño. The tropical surface waters (TSW) which normally are restricted to the north of 1°N are observed as far as 6°S as the Equatorial Front disappears. The Subtropical Surface Waters (SSW) impinge on the coast of South America associated with large scale changes in the South Pacific gyre (Wyrtki and Wenzel, 1984). Changes in the distribution of nitrate along 85°W prior to, during, and after the 1982–83 El Niño reflect some of the advective changes which occur during El Niño (Figure 8). Onset of the 1982–83 El Niño occurred in August 1982 around the Galápagos Islands and September 1982 at Paita, Peru (Chavez *et al.*, 1984). The nitrate section during November 1982 shows the depressed nutricline associated with the onset of El Niño. Soon afterward the fresher and nutrient poor tropical surface waters (TSW) from north of the Equatorial Front occupied the entire region. Anomalous conditions remained in the eastern tropical Pacific through May 1983, and partial recovery was observed in July 1983. By November 1983 conditions had returned to normal.

In the Galápagos anomalously high sea surface temperatures, enhanced fresh water input, reduced nutrient concentrations, deeper thermocline and nutricline, and lower levels of phytoplankton biomass and production were all observed during 1983 when compared to the cooler years of 1984 and 1985 (Figure 10).

4.3. Horizontal Currents

There are significant changes in circulation of the tropical Pacific during El Niño. The first advective effects are felt as Kelvin waves, generated in the western Pacific by westerly winds in this region, propagate along the equator from west to east (Harrison and Schopf, 1984; McPhaden *et al.*, 1988). Associated with the waves is an anomalous eastward current along the equator which may be on the order of 75 to 100 cm sec^{-1}. The duration of the anomalous currents is uncertain, however they probably continue as long as wind forcing in the western tropical Pacific is maintained.

During the 1982–83 El Niño there were significant changes in the position of the South Pacific gyre. The South Equatorial Current (SEC) was shifted southward so that its main flow was found between 10° and 20°S (Wyrtki, 1984). Observations by Hayes *et al.* (1987) showed that during the 1982–83 El Niño eastward flow was common along 95°N, just west of the Galápagos, as far as 6°S, with maximum current speeds of 60 cm sec^{-1}.

Transport by the NECC was enhanced significantly during 1982 and 1983 (Wyrtki and Kilonsky, 1984). Much of the redistribution of water from the western Pacific to the eastern Pacific may occur via the NECC. The flow described by Schott (1931) which appears to come from the Galápagos may be associated with the intensification of the NECC and result from the overflowing of water from the Panama basin. The Galápagos Islands are bathed by these warm, low salinity, and low nutrient waters during El Niño events (Figure 10).

Perhaps the most dramatic effect on currents during El Niño is the virtual disappearance of the EUC (Firing *et al.*, 1983). In the mean, the EUC is maintained by the sea level pressure gradient which is set up by the trade winds. When the trade winds relax and the excess water from the western Pacific is redistributed north, south and to the east the slope in sea level between eastern and western Pacific disappears (Wyrtki, 1984) (Figure 6) as does the EUC. Prior to direct observations of the disappearance of the EUC during El Niño it was believed that the redistribution of water from the western Pacific to the east occurred via this current.

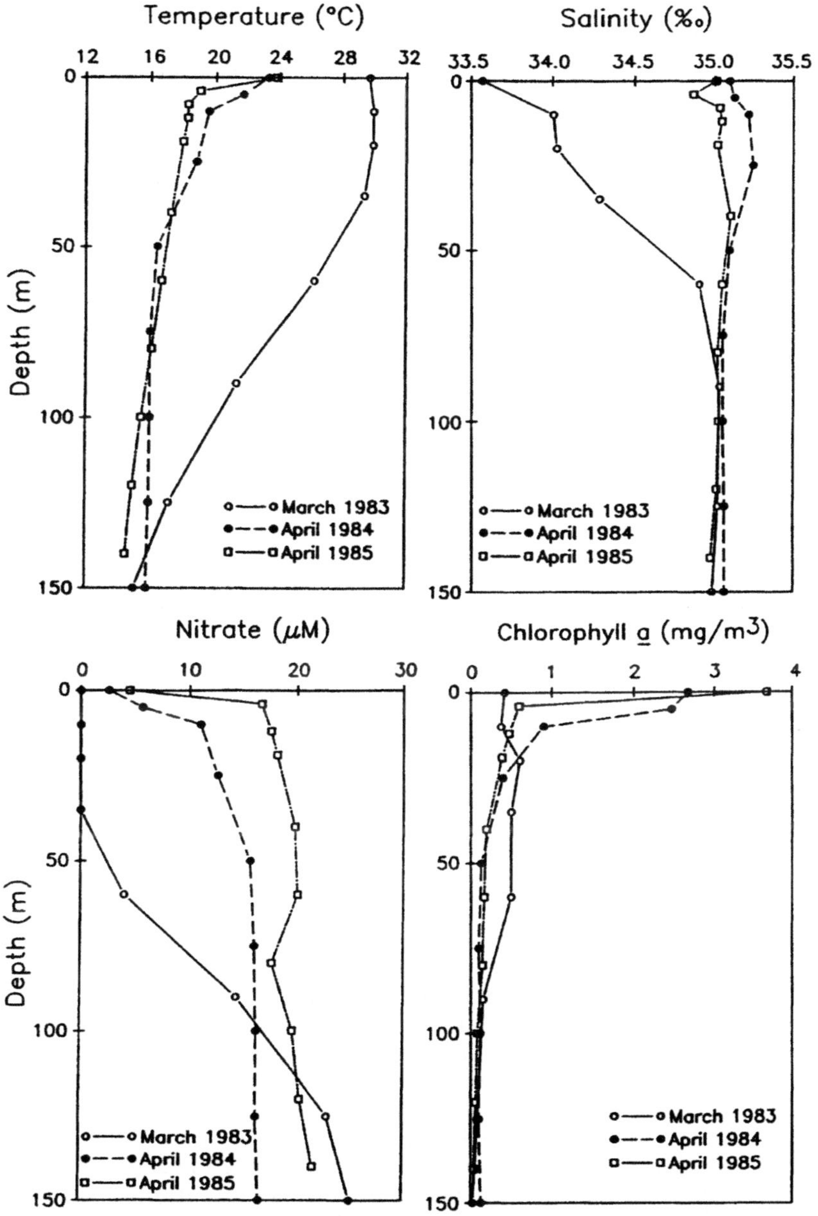

Figure 10. Vertical profiles of temperature, salinity, nitrate, and chlorophyll taken in the Bolivar Channel (0° 15′S, 91° 25′W) between Isla Isabella and Isla Fernadina Islands during March 1983, April 1984, and April 1985. The effects of El Niño are clearly visible during March 1983 with warmer sea surface temperature, lower surface nitrate, fresher surface water, lower surface chlorophyll, and deeper nutricline and thermocline. Data courtesy of Dr. R. T. Barber.

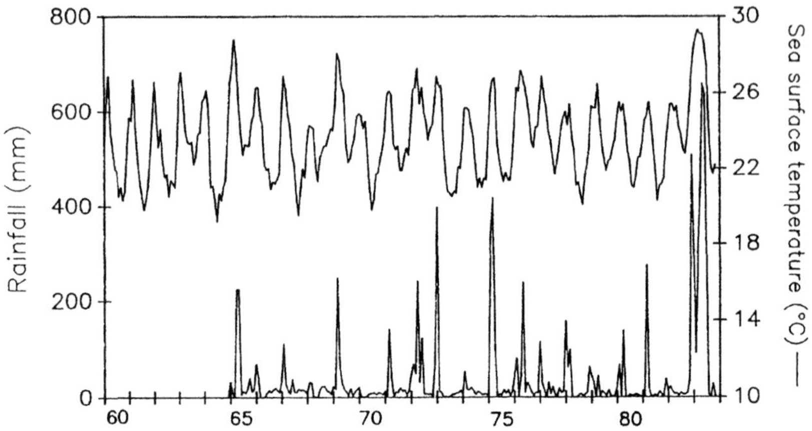

Figure 11. Time series of monthly sea surface temperature at Baltra, Galápagos, and precipitation at Puerto Ayora, Galápagos. The signal of the 1982–83 El Niño is prominent with large deviations in sea surface temperature and heavy precipitation.

4.4. Upwelling

Much the opposite of what was originally believed (Wooster and Guillen, 1974) the coastal winds of the southeastern tropical Pacific intensify during El Niño (Enfield, 1981; Enfield and Newberger, 1985; Bakun, 1987). Depending on the extent of the migration of the low pressure zone from the western tropical Pacific eastward (Gill and Rasmusson, 1983), upwelling is also probably maintained along the equator in the eastern Pacific. However in both coastal and equatorial systems the upwelling is of warm water as a result of a deeper thermocline. The southward migration of the ITCZ during El Niño contributes to reducing upwelling favorable winds along the equator.

4.5. Sea Level

Sea level is perhaps the best single feature that oceanographers have to follow the large scale changes which occur during El Niño. From propagating changes in sea level it has been possible to link events in the eastern equatorial Pacific with westerly winds in the western tropical Pacific (Wyrtki, 1975; Enfield *et al.*, 1983; Lukas *et al.*, 1984; Miller *et al.*, 1988). Theory shows that Kelvin waves excited by westerly winds in the western Pacific can propagate eastward at about 300 cm/sec along the equatorial waveguide deepening the thermocline and raising sea level (McCreary, 1976; Cane, 1983). Increases in sea level of 30 to 50 cm have been observed in the eastern tropical Pacific during El Niño with corresponding decreases in the western Pacific (Wyrtki, 1975; Wyrtki, 1984). In addition when the propagating disturbances reach the eastern boundary a portion of the energy propagates poleward; sea level increases associated with El Niño have been observed along the coast of the Americas from Alaska to Chile (Enfield and Allen, 1980).

The evolution of the 1982–83 El Niño at the Galápagos, in terms of sea level and nitrate at 60 meters, is shown in Figure 9. Wyrtki (1981) suggested that water upwelled along the equator comes from a depth of 50 to 100 meters, therefore nitrate concentration at 60 meters is a good proxy for the nitrate concentration of the upwelled water. Figure 9 shows that the concentration of nutrients in the upwelled source water along the equator is directly related to thermocline depth and sea level. The deepening of the thermocline and the nutricline in the eastern boundary during El Niño reduces the biological productivity of this region by lowering the level of nutrients in the upwelled water (Chavez and Barber, 1985).

4.6. Climate

Westerly winds in the western equatorial Pacific, generated in the lee of the Indonesian Low as it migrates eastward, are now believed to trigger the onset of El Niño (Cane, 1983; McPhaden *et al.*, 1988). The westerly winds are commonly associated with cyclone pairs on each side of the equator and intensive convective activity (Ramage, 1986). The oceanic response, in the form of warmer sea surface temperatures in the central and eastern Pacific, results in further atmospheric effects. The region of enhanced convective activity continues to migrate eastward and, as was the case during the 1982–83 event, may cross the entire Pacific and reach the Galápagos (Gill and Rasmusson, 1983). The ITCZ migrates south of its normal position as the Equatorial Front disappears (Rasmusson and Carpenter, 1982; Horel and Cornejo-Garrido, 1986). These changes bring enhanced precipitation to the Galápagos and the entire eastern tropical Pacific during El Niño (Figure 11).

5. Summary and Conclusions

1. The Galápagos Islands are located in the eastern equatorial Pacific, however the surrounding waters are considerably cooler and richer in nutrients than might be expected for a tropical region.
2. The cool and rich waters surrounding the Galápagos are a result of a combination of processes including:
   a. Trade winds, which blow primarily from east to west, set up a tilt in the thermocline and nutricline so that cool and rich waters are closer to the surface in the eastern terminus.
   b. The easterly component of the trade winds drives equatorial upwelling which "pumps" cool and rich water from around 50 to 100 meters to the surface.
   c. Horizontal advection of cool and rich waters by the Peru Current System.
   d. Local bathymetrically induced upwelling.
3. The seasonal cycle at the Galápagos has a distinct Southern Hemisphere character with warmer temperatures from February through April and cooler temperatures in August and September.
4. The Galápagos are located in a region where horizontal currents are particularly dynamic. The main current systems affecting the Galápagos are:

   a. The cool westward surface South Equatorial Current which bathes the Galápagos most of the year.
   b. The cool eastward subsurface Equatorial Undercurrent, which is "trapped" along the equator and is modified significantly by the Galápagos when it collides with Isla Isabella Island.
   c. The warm eastward surface North Equatorial Countercurrent which seasonally flows eastward north of the Galápagos.
5. The mean conditions described above are modified periodically, every three to eight years, by an ocean–atmosphere phenomenon known as the El Niño Southern Oscillation.
6. During El Niño years, the waters of the eastern equatorial Pacific are considerably warmer, sea level increases, and the major horizontal currents are modified as follows:
   a. The South Equatorial Current weakens considerably and is displaced southward. Eastward flow is found on occasion along the equator.
   b. The Equatorial Undercurrent weakens as the east–west slope in sea level is reduced.
   c. Transport of the North Equatorial Countercurrent increases considerably.

ACKNOWLEDGMENTS. The research of both authors was supported in part by grants from the U.S. National Science Foundation. The data in Figures 4, 5, 7, 8, 9, and 10 was collected under grants OCE-81-10702, OCE-86-13759, and OCE-89-01929 to R. T. Barber. This chapter benefitted from reviews by R. C. Dugdale, T. Simkin, and W. S. Wooster.

References

Anderson, R. N., Clague, D. A., Klitgord, K. D., Marshall, M., and Nishimori, R. K., 1975, Magnetic and petrologic variations along the Galápagos Spreading Center and their relations to the Galápagos melting anomaly., *Geol. Soc. Amer. Bull* **86**:683–694.

Bailey, K., 1976, Potassium–argon ages from the Galápagos Islands, *Science* **192**:465–467.

Baitis, H. W., and Lindstrom, M. M., 1980, Geology, petrography, and petrology of Pinzon Island, *Galápagos Archipelago, Contr. Mineral Petrol.* **72**:367–386.

Bakun, A., 1987, Monthly variability in the ocean habitat off Peru as deduced from maritime observations, 1953 to 1984, in: *The Peruvian Anchoveta and Its Upwelling Ecosystem: Three Decades of Change*, ICLARM Studies and Reviews 15 (D. Pauly and I. Tsukayama, eds.), International Center for Living Aquatic Resources Management, Manila, Philippines, pp. 46–74.

Barber, R. T., and Chavez, F. P., 1983, Biological consequences of El Niño, *Science* **222**:1203–1210.

Barber, R. T., and Chavez, F. P., 1986, Ocean variability in relation to living resources during the 1982/83 El Niño, *Nature* **319**:279–285.

Bellon, H., Saenz, R., and Tournon, J., 1983, K–Ar radiometric ages of lavas from Cocos Island (eastern pacific), *Mar. Geol.* **54**:M17–M23.

Bjerknes, J., 1966, A possible response of the atmospheric Hadley circulation to equatorial anomalies of ocean temperature, *Tellus* **18**:820–829.

Bow, C. S., 1979, The geology and petrogenesis of the lavas of Floreana and Santa Cruz islands, Galápagos Archipelago. Ph.D. Thesis, Univ. Oregon, Eugene, Oregon, 308 pp.

Broecker, W. S., and Peng, T. H., 1982, *Tracers in the Sea*, Eldiog Press, Palisades, New York, 690 pp.

Brusca, R. C., 1987, Biogeographic relationships of Galápagos marine isopod crustaceans, *Bull. Mar. Sci.* **41**:268–281.

Cane, M. A., 1983, Oceanographic events during El Niño, *Science* **222**:1189–94.

Carranza, L., 1891, Contra-corriente martima, observada en Paita y Pacasmayok, *Boln. Soc. Geogr. Lima* **2**:344–345.

Carrillo, C. N., 1889, Hidrografia oceanica, *Boln. Soc. Geogr. Lima* **1**:72–110.

Chavez, F. P., 1986, The legitimate El Niño current, *Tropical Ocean-Atmosphere Newsletter* **34**:1.

Chavez, F. P., 1987, The annual cycle of SST along the coast of Peru, *Tropical Ocean-Atmosphere Newsletter* **37**:4–6.

Chavez, F. P., 1989, Size distribution of phytoplankton in the central and eastern tropical Pacific, *Global Biochemic. Cycles* **3**:27–35.

Chavez, F. P., Barber, R. T., and Soldi, H. S., 1984, Propagated temperature changes during onset and recovery of the 1982/83 El Niño, *Nature* **309**:47–49.

Chavez, F. P., and Barber, R. T., 1985, Plankton production during El Niño, in: *International Conference on the TOGA Scientific Programme*. World Climate Research Publication, Geneva, pp. VI, 23–32.

Chavez, F. P., Buck, K. R., and Barber, R. T., 1990, Phytoplankton taxa in relation to primary production in the equatorial Pacific, *Deep-Sea Research* **37**:1733–1752.

Cox, A., 1983, Ages of the Galápagos islands: Patterns of evolution in Galápagos organisms, in: *Patterns of Evolution in Galápagos Organisms* (R. I. Bowman, M. Berson, and A. E. Leviton, eds.) American Association for the Advancement of Science, Pacific Division, San Francisco.

Cox, A., and Dalrymple, G. B., 1966, Palaeomagnetism and potassium–argon ages of some volcanic rocks from the Galápagos Islands, *Nature* **209**:776–777.

Dalrymple, G. G., and Cox, A., 1968, Palaeomagnetism, potassium–argon ages and petrology of some volcanic rocks from the Galápagos Islands, *Nature* **217**:1–8.

Durham, J. W., 1963, Paleogeographic conclusions in light of the biological data, in: *Pacific basin biogeography* pp. 355–365, Tenth Pacific Science Congress, Mishop Mus. Press, Honolulu, Hawaii.

Durham, J. W., 1966, Coelenterates, especially stony corals, from the Galápagos and cocos Islands, in: *The Galápagos* (R. I. Bowman, ed.), Proceedings of the Symposia of the Galápagos International Scientific Project, University of California Press, Berkeley, California.

Eguiguren, V., 1894, Las lluvias de Piura, *Boln. Soc. Geogr. Lima* **4**:241–258.

Enfield, D. B., 1981, Thermally driven wind variability in the planetary boundary layer above Lima, *J. Geophys. Res.* **86**:2005.

Enfield, D. B.,and Allen, J. S., 1980, On the structure and dynamics of monthly mean sea level anomalies along the Pacific coast of North and South America, *J. Phys. Oceanogr.* **10**:557–578.

Enfield, D. B., and Newberger, P. A., 1985, Peru coastal winds during 1982–83, *Proceedings of the 9th Annual Climate Diagnostics Workshop*, NOAA, Washington, D.C., pp. 147–156.

Enfield, D. B., Cornejo-Rodriguez, M. P., Smith, R. L., and Newberger, P. A., 1987, The equatorial source of propagating variability along the Peru Coast during the 1982–83 El Niño, *J. Geophys. Res.* **92**:14335–14346.

Fairbridge, R. W., 1973, Glaciation and plate migration, in: *Implications of Continental Drift to the Earth Sciences*, Vol. 1 (D. H. Tarling and S. K. Runcorn, eds.), Academic Press, New York, pp. 35–49.

Firing, E., Lukas, R., Sadler, J., and Wyrtki, K., 1983, Equatorial undercurrent disappears during 1982–1983 El Niño, *Science* **222**:1121–1123.

Geist, D. J., McBirney, A. R., and Duncan, R. A., 1985, Geology of Sante Fe Island: The oldest Galápagos Volcano, *J. Volcan. Geotherm. Res.* **26**:203–121.

Gill, A. E., Rasmusson, E., 1983, The 1982–83 climate anomaly in the equatorial Pacific, *Nature* **306**: 229–234.

Glynn, P. W., and Wellington, G. M., 1983, *Corals and Coral Reefs of the Galápagos Islands*, University of California Press, Berkeley, California, XVI, 330 pp.

Gunther, E. R., 1936, A report on oceanographical investigations in the Peru Coastal current, *Discover Rep.* **13**:107–276.

Hall, M. L., 1983, Origin of Española Island and the age of terrestrial life on the Galápagos Islands, *Science* **221**:545–547.

Hall, M. L., Ramon, P., and Yepes, H., 1980, The subaerial origin of Española (Hood) Island and the age of terrestrial life in the Galápagos, *Noticias de Galápagos* **31**:41–52.

Halpern, D., Knox, R. A., Luther, D. S., and Philander, S. G. H., 1989, Estimates of equatorial upwelling between 140° and 110°W during 1984, *J. Geophys. Res.* **94**:8018–8020.

Hansen, D. V., and Paul, C. A., 1984, Genesis of long waves in the equatorial Pacific, *J. Geophys. Res.* **89**:10431–10440.

Harrison, D. E., and Schopf, P. S., 1984, Kelvin-wave-induced anomalous advection and the onset of surface warming in El Niño Events, *Mon. Weather Rev.* **112**:923–933.

Hayes, S. P., 1985, Sea level and near surface temperature variability at the Galápagos Islands, 1979–1983, in: *El Niño in the Galápagos Islands: The 1982–1983 Event* (G. Robison and E. Del Piño, eds.), Charles Darwin Foundation for the Galápagos Islands, Quito, Ecuador, 49–81.

Hayes, S. P., and Halpern, D., 1984, Correlation of current and sea level in the eastern equatorial Pacific, *J. Phys. Oceanogr.* **14**:811–824.

Hayes, S. P., Mangum, L. J., Barber, R. T., Huyer, A., and Smith, R. L., 1987, Hydrographic variability west of the Galápagos Islands during the 1982/83 El Niño, *Prog. Oceanog.* **17**:137–162.

Hey, R. N., and Vogt, P. R., 1977, Rise axis jumps and sub-axial flow near the Galápagos Hotspot?, *Tectophysics* **37**:41–52.

Hey, R. N., Johnson, G. L., and Lowrie, A., 1977, Recent plate motions in the Galápagos area, *Geol. Soc. Amer. Bull.* **8**:1385–1403.

Holden, J. C., and Dietz, R. S., 1972, Galápagos gore, NazCoPac triple junction and Carnegie/Cocos Ridges, *Nature* **235**:266–269.

Horel, J. D.,and Cornejo-Garrido, A. G., 1986, Convection along the coast of northern Peru during 1983; Spatial and temporal variation of clouds and rainfall, *Mon. Weather Rev.* **114**:2091–2105.

Houvenaghel, G. T., 1984, Oceanographic setting of the Galápagos Islands, in: *Key environments, Galápagos* (R. Perry, ed.), pp. 43–54, Pergamon Press, New York.

Jones, J. H., 1969, Surfacing of Pacific Equatorial Undercurrent: direct observations, *Science* **163**: 1449–1450.

Kogelschatz, J., Solorzano, L., Barber, R. T., and Mendoza, P., 1985, Oceanographic conditions in the Galápagos Islands during the 1982/83 El Niño, in: *El Niño in the Galápagos Islands: The 1982/1983 Event* (G. Robison and E. M. del Piño, eds.), Charles Darwin Foundation for the Galápagos Islands, Quito, Ecuador, pp. 91–123.

Legeckis, R., 1977, Long waves in the eastern equatorial Pacific Ocean: A view from a geostationary satellite, *Science* **197**:1179–1181.

Legeckis, R., 1986, Long waves in the equatorial Pacific and Atlantic Oceans during 1983, *Ocean–Air Interactions* **1**:1–101.

Lukas, R., 1986, The termination of the equatorial undercurrent in the Eastern pacific, *Progress in Oceanography* **16**:63–90.

Lukas, R., Hayes, S. P., and Wyrtki, K., 1984, Equatorial sea level response during the 1982–1983 El Niño, *J. Geophys. Res.* #89:10,425–10,430.

Madden, R. A., and Julian, R., 1972, Description of global scale circulation scales in the tropics with a 40–50 day period, *J. Atmos. Sci.* **29**:1109–1123.

Mangum, L. J., and Hays, S. P., 1984, The vertical structure of the zonal pressure gradient in the eastern equatorial Pacific, *J. Geophys. Res.* **89**:10,441–10,449.

McBirney, A. R., and Williams, H., 1969, Geology and petrology of the Galápagos Islands, *Geol. Soc. Amer. Mem.* **118**:1–197.

McCreary, J. P., 1976, Eastern tropical ocean response to changing wind systems: with application to El Niño, *J. Phys. Oceanogr.* **6**:632–645.

McPhaden, M. J., Freitag, H. P., Hayes, S. P., Taft, B. A., Chien, Z., and Wyrtki, K., 1988, The response of the equatorial Pacific Ocean to a westerly wind burst in May 1986, *J. Geophys. Res.* **93**:10,589–10,603.

Mears, E. G., 1943, *The Ocean Current "The Child."* Smithsonian Inst., Washington, D.C., report for 1943.

Meyers, G., 1979, Annual variation in the slope of the 14#C isotherm along the equator in the Pacific Ocean, *J. Phys. Oceanog.* **9**:885–891.

Miller, L., Cheney, R. L., and Douglas, B., 1988, Geosat altimeter observations of Kelvin Waves and the 1986–87 El Niño, *Science* **239**:52–54.

Morgan, W. J., 1971, Convection plumes in the lower mantle, *Nature* (Lond.) **230**:42–43.

Morgan, W. J., 1972, Plate motions and deep mantle convection, *Geol. Soc. Amer. Mem.* **132**:7–22.

Okuda, T., Trejos de Suescum, R., Valencia, M., and Rodriguez, A., 1983, Variacion estacional de la posicion del frente ecuatorial y su efecto sobre la fertilidad de las aguas superficiales ecuatorianas, *Acta Oceanografica del Pacifico*, INOCAR, Ecuador, **2**:53–84.

Pak, H., and Zaneveld, J. R. V., 1973, The Cromwell Current on the east side of the Galápagos Islands, *J. Geophys. Res.* **78**:7845–7859.

Pezet, F. A., 1896, La contra-corriente "El Niño", en la costa norte del Peru, *Boln. Soc. Geogr. Lima* **5**: 457–461.

Philander, S. G. H., 1978, Instabilities of zonal equatorial currents: II, *J. Geophys. Res.* **83**:3679–3682.

Philander, S. G. H., 1985, El Niño and La Niña, *J. Atmos. Sci.* **42**:2652–2662.

Philander, S. G. H., and Seigel, A. D., 1985, Simulation of El Niño of 1982–83, in: *Hydrodynamics of the Equatorial Ocean* (J. C. J. Nihoul, ed.) pp. 517–542.

Rasmusson, E. M., and Carpenter, T. H., 1982, Variations in tropical sea surface temperature and surface wind fields associated with the Southern Oscillation/El Niño, *Mon. Weather Rev.* **110**:354–384.

Rasmusson, E. M., and Wallace, J. M., 1983, Meteorological aspects of the El Niño/Southern Oscillation, *Science* **222**:1195–1202.

Reid, J. L., 1959, Evidence of a South Equatorial Countercurrent in the Pacific Ocean, *Nature* **184**: 209–210.

Robalino, M., 1985, Registros Meteorologicos de la estacion cientifica Charles Darwin para 1982–1983, *El Niño en las Islas Galápagos: El Evento de 1982–1983*, Charles Darwin Foundation for the Galápagos Islands, Quito, Ecuador.

Schott, G., 1931, Der Peru-Strom und seine nordlichen Nachbargebiete in normaler und anormaler Ausbildung., *Ann. Hydrogr. Marit. Meteorol.* **59**:161–196; 200–213; 240–252.

Schweigger, E., 1945, La "legitima" corriente del Nino, *Boln. Comp. Administr. Guano* **21**:255–296.

Simkin, T., 1984, Geology of Galápagos Islands, *Key environments, Galápagos* (R. Perry, ed.), Pergamon Press, New York, pp. 15–42.

Sverdrup, H. U., Johnson, M. W., and Flemming, R. H., 1942, *The Oceans: Their Physics, Chemistry and General Biology*, Prentice-Hall, Englewood Cliffs, New Jersey, 1060 p.

Tsuchiya, M., 1985, Subsurface countercurrents in the Pacific Ocean, *J. Mar. Res.* **13**:145–175.

Vail, P. R., and Mitchum, R. M., Jr., 1979, Global cycles of relative changes of sea level from seismic stratigraphy, *Amer. Assoc. Petrol. Geol. Mem.* **29**:469–472.

Walker, G. T., 1924, Correlations in seasonal variations of weather. IX. A further study of world weather, *Mem. Indian Meteorol. Dep.* **24**:275–332.

Wilson, J. T., 1963, Hypothesis of Earth's behavior, *Nature* **198**:925–929.

Wooster, W. S., 1969, Equatorial front between Peru and the Galápagos, *Deep-Sea Research* (Suppl.) **16**:407–419.

Wooster, W. S., and Hedgepeth, J. W., 1966, The Oceanographic setting of the Galápagos, in: *The Galápagos. Proceedings of a Symposium of the Galápagos International Scientific Project* (R. I. Bowman, ed.) pp. 100–101, University of California Press, Berkeley, California.

Wooster, W. S., and Guillen, O., 1974, Characteristics of El Niño in 1972, *J. Mar. Res.* **32**:387–404.

Wyrtki, K., 1966, Oceanography of the eastern equatorial Pacific Ocean, *Oceanogr. Mar. Biol. Annu. Rev.* **4**:22–68.

Wyrtki, K., 1975, El Niño, the dynamic response of the equatorial Pacific Ocean to atmospheric forcing, *J. Phys. Oceanogr.* **5**:572–584.

Wyrtki, K., 1981, An estimate of equatorial upwelling in the Pacific, *J. Phys. Oceanogr.* **11**:1205–1214.

Wyrtki, K., 1982, The southern Oscillation, Ocean–Atmosphere interaction and El Niño, *Mar. Technol. Soc. J.* **16**(1):3–10.

Wyrtki, K., 1984, The slope of sea level along the equator during the 1982/1983 El Niño, *J. Geophys. Res.* **89**:10,419–10,424.

Wyrtki, K., 1985, Water displacements in the Pacific and the genesis of El Niño cycles, *J. Geophys. Res.* **90**:7129–7132.

Wyrtki, K., and Eldin, G., 1982, Equatorial upwelling events in the Central Pacific, *J. Phys. Oceanogr.* **12**: 984–988.

Wyrtki, K., and Kilonsky, B., 1984, Mean water mass and current structure during the Hawaii-to-Tahiti Shuttle Experiment, *J. Phys. Oceanogr.* **14**:242–254.

Wyrtki, K., and Wenzel, J., 1984, Possible gyre–gyre interactions in the Pacific Ocean, *Nature* **309**: 538–540.

II

Meiofauna and Annelida

Chapter 2

The Meiofauna of the Galápagos

A Review

WILFRIED WESTHEIDE

1. Introduction

About three hundred and ninety species of the marine meiofauna of the Galápagos Islands in fourteen higher invertebrate taxa are known. These animals are usually found living between sand grains on sandy beaches throughout the archipelago. The number of endemic genera and species is difficult to estimate due to inadequate comparative work in the eastern Pacific. It is supposed that only a smaller part of those approximately 300 taxa described from the Galápagos as new to

WILFRIED WESTHEIDE • Systematic Zoology, Department of Biology/Chemistry, University of Osnabrück, D-4500 Osnabrück, Federal Republic of Germany.

science are endemic to the islands. Almost 90% of the Galápagos meiofauna species were reported by the Göttingen Galapagos Meiofauna Expedition.

From February 1972 to March 1973 a group of zoologists from the University of Göttingen visited the Galápagos Islands and intensively investigated the meiofauna of their beaches. Inspiration for this project was provided by the comprehensive long-term investigations of Peter Ax and his students on the psammobiontic fauna of the beaches of the North Sea Island Sylt. Here, since 1962 the entire spectrum of species, their abundances, and the dynamics of total fauna and individual species populations were recorded, as well as abiotic and biotic environmental factors and their influences on species distribution pattern in time and space within intertidal habitats (e.g., Schmidt, 1968, 1969; Westheide, 1968; Hartwig, 1973; Ehlers, 1973; Sopott, 1973; Hoxhold, 1974). Studies in systematics and morphology were carried out as well as investigations in reproduction, development, and ecology of abundant and characteristic species from a large series of taxa (e.g., Westheide, 1967; Ax, 1969; Teuchert, 1968; Meineke and Westheide, 1979). Early in the 1970s the meiofauna community of Sylt's sandy beaches was considered to be the best known in the world with more than 400 species of ciliates and metazoans having been reported, many of them previously undescribed.

The work on Sylt allowed for a comprehensive comparison with meiofauna communities of warmer seas. In contrast to the temperate area of the North Sea with its considerable seasonal climatic changes, tropical seas exhibit higher water temperatures throughout the year. A comparison of this kind appeared to be a promising way to arrive at general statements about seasonal dynamics, reproductive periods, dominance of taxa, and zonation patterns of the meiofauna inhabiting tidal sand beaches.

The Galápagos Islands seemed to be an appropriate geographic locality for such comparative work—with their differently exposed sandy beaches and water temperatures around 25°C with seasonal variations of only 5 to 6°C. However, the islands were chosen for other reasons, too: like any zoologist who goes to the Galápagos, the members of the expedition especially wanted to learn more about evolution. All of us who participated hoped that meiofaunal organisms, which usually do not possess pelagic larval stages and are not suited for active dispersal within the water column, would have undergone a similarly independent evolution in the island beaches as had been found for the terrestrial vertebrates and plants on the islands. We expected to find examples of radiation and speciation between meiofauna populations of various islands, which might give some indications to the general pattern and rate of speciation in meiofaunal taxa. Because the islands are only 2–3 million years old, we were apprehensive that we might find Galápagos beaches low in diversity. It was thought that the young age of the islands would not have been sufficient for dispersal and diversification of meiofaunal taxa.

Concerns of this sort were not substantiated, but the far-reaching expectations of the project were not totally fulfilled, either. What we found was a species-rich and individual-rich fauna. Many of the species turned out to be familiar "old friend" taxa known from other meiofaunal communities. On first examination, these taxa did not reveal anything spectacular. There were no meiofaunal equivalents of giant tortoises or lava lizards. Nevertheless, careful sampling activities on the islands and subtle evaluation of the material in the following years allowed us to elucidate the entire taxonomic pattern of the meiofauna biocenosis in this group of isolated tropical oceanic islands. These organisms afforded a glimpse of the rate of speciation within individual meiofauna taxa, and they generally furthered our understanding of differences in island colonization between marine littoral and terrestrial faunas.

Peter Schmidt (now a Professor of Zoology at the Technical University of Aachen) and his wife Adelheid Schmidt stayed in the islands for one year (February 1972 to March 1973). They carried out by far the largest part of the investigations and collected most of the material. Their tremendous sampling program comprised the entire quantitative analysis (more than 1,000 samples with ca. 230,000 individuals), recording of abiotic factors in the beaches and careful sorting of specimens for most of the systematic investigations. Various groups were worked up by Schmidt himself or were coauthored with various colleagues (Macrostomida, Polycladida, Nemertini, Gastrotricha, Kinorhyncha, Aeolosomatidae, so-called archiannelid Polychaeta, Tardigrada, Halacarida, Isopoda). Unfortunately, he left unpublished or without discussion many of his valuable quantitative ecological investigations, leaving unresolved certain questions regarding the general objectives of the expedition.

Peter Ax and Renate Ax (Proseriata, Typhloplanoida), Ulrich Ehlers (Gnathostomulida, Typhloplanoida, Acoela), Siegmar Hoxhold (Kalyptorhyncha), and myself (Polychaeta), each stayed for two months in the islands in 1972, concentrating on sampling and processing of the respective taxa. A number of other specialists from the Göttingen Zoological Institute and from various other institutions, including taxonomists from France, Sweden, the United States, and Chile (see Section 3) subsequently participated in the investigation of the preserved material. In total about 390 species have been identified, the majority of which are new to science. A number of specimens remain to be examined and described in the literature.

The expedition was organized by the Akademie der Wissenschaften und der Literatur, Mainz, under supervision of Peter Ax. Funding was provided by the Volkswagenwerk Foundation, which also funded the salaries of some of the scientists who investigated preserved material in subsequent years.

From 1973 to the present, the Mainz Academy continuously published the results, first in the form of separate issues in the series *Mikrofauna des Meeresbodens* (Editor Peter Ax); and from 1984 on, the descriptions appeared as articles in the volumes of the journal *Microfauna Marina* (Akademie der Wissenschaften und der Literatur, Mainz, and Gustav Fischer Verlag, Stuttgart, New York). Altogether, about 40 articles with more than 1500 pages have been published on the results of the project. This makes the Göttingen Galápagos Meiofauna Expedition the best documented project dedicated to the study of marine meiofauna.

2. Sampling and Extraction

Fig. 1 shows the entire group of islands and the beaches investigated by the Schmidts (Ax and Schmidt, 1973; Schmidt, 1978). Most of these collecting stations were visited with difficulty by small Galapagonian fishing boats. For qualitative sampling, various volumes of sediment from different levels of intertidal and shallow subtidal areas were brought into the laboratory of the Charles Darwin Research Station. Qualitative sampling was done by full transects of the beaches' intertidal regions along a line perpendicular to the low water line at intervals of 1 or

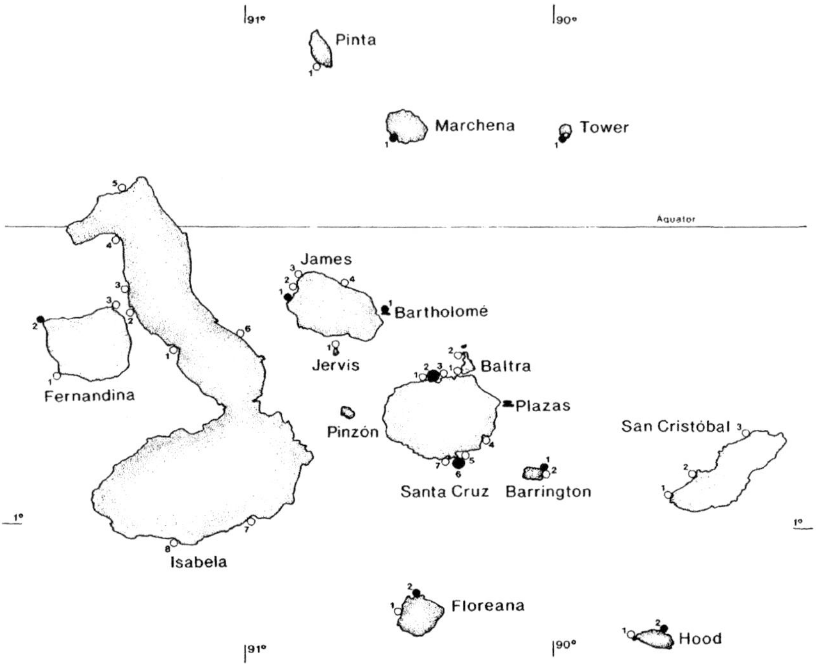

Figure 1. Schmidt's collection sites of sand beaches during the Göttingen Meiofauna Expedition. Large solid points on Santa Cruz indicate the two beaches (2 = Playa Borrero, 6 = Bahia Academy) where quantitative samples were taken regularly. Small solid points = one quantitative transect. Open circles = qualitative samples only. For more details see Ax and Schmidt (1973), Schmidt (1978).

2 m (tidal amplitudes between 1.2 to 1.9 m, see Houvenaghel and Houvenaghel, 1974). Here the Schmidts used small hand-held tubes for corers (inner diameter 3.5 cm, length about 11 cm) collecting a volume of 100 cm^3 of sediment and a surface area of 10 cm^2. In the upper region of the beach slope, pits of decreasing depths were dug till the groundwater level was reached, and core samples were taken from the vertical face. Time limitations did not allow investigation of the meiofauna below the ground-water level during low tide. This area is, however, generally known to be extremely poor in individuals and species.

Extraction of specimens occurred in the laboratory rooms of the Charles Darwin Research Station near Puerto Ayora on Santa Cruz, within one week after sampling. The seawater–ice technique was used, followed by one intensive washing of the remaining sediment. This resulted in an average extraction efficiency of 84% of the total fauna with minimal efficiency of 51% for Tardigrada and a maximum of 96.5% for Platyhelminthes (Schmidt, 1978). Soft-bodied meiofaunal elements were usually examined with a high quality compound microscope while still alive immediately after extraction from the sediment. This included taking photographs with a flash system, and taking measurements and drawings of the animals with a camera lucida. The "hard" fauna (e.g., nematodes, halacarids, copepods) were fixed in formalin after counting and transported to Göttingen and to other places for further investigation.

Sampling by the Schmidts was mainly concentrated on sand beaches. The other participants during their stay on the islands usually joined them when taking samples, but also tried to find as many species of their special groups in other habitats, such as mangrove sediments, rock pool sediments, stones and old corals of shallow sublitoral areas. Nevertheless by far the largest part of all animals investigated belong to the meiofauna, especially true interstitial species, with the exception of some polyclad Platyhelminthes and several polychaetes, which comprise a series of smaller non-interstitial and several macrofaunal species.

Special methods of extraction were used for non-sand inhabiting animals, e.g., endofaunal polychaetes living in stones and corals; they were forced to come to the surface within a day or so in water-filled buckets by depletion of oxygen.

Physicomorphological characters considered in Schmidt's investigations included grain size (Md-values between 743–1540 μm on Fernandina and 191–830 μm in Playa Borrero on Santa Cruz), nature of substratum (volcanic sand or fragments of corals, molluscs, echinoderms, etcetera, or a mixture of both), and amount of organic matter by the combustion method at high temperatures (between 0.2 and 0.4% in general), thus revealing all of the beaches to be relatively clean and exposed (Schmidt, 1978).

Water temperature fluctuated only slightly between 23°C and 29°C (1972 was an El Niño year) and probably did not influence the abundance dynamics observed (see below). During low tide, surface temperatures of the beaches reached values of 40°C and above on sunny days. Salinity values of about 40% were measured under such conditions.

3. Composition of Taxa

The following compilation comprises the meiofauna from the Galapagos Islands that was collected during the Göttingen expedition 1972/73, including a few scattered references to earlier collections.

3.1. Platyhelminthes (Fig. 2, 3, 9)

Intensive studies have been carried out on the free-living Platyhelminthes ("Turbellaria"), revealing a high number of different suprageneric taxa, the individuals of which locally may represent more than 50% of the total fauna in individual samples of sandy beaches, e.g., in one sample 339 out of a total of 399 animals were turbellarians. Almost all of the species found were undescribed, several of them belonging to new genera (Ax and Ehlers, 1973; Ax and Ax, 1974; Ehlers and Ax, 1974; Sopott-Ehlers and Schmidt, 1974a; Sopott-Ehlers and Schmidt, 1974b; Sopott-Ehlers and Schmidt, 1975; Schmidt and Sopott-Ehlers, 1976; Ax and Ax, 1977; Ehlers and Dörjes, 1979; Ehlers and Ehlers, 1981; Noldt and Hoxhold, 1984; Ehlers and Sopott-Ehlers, 1989).

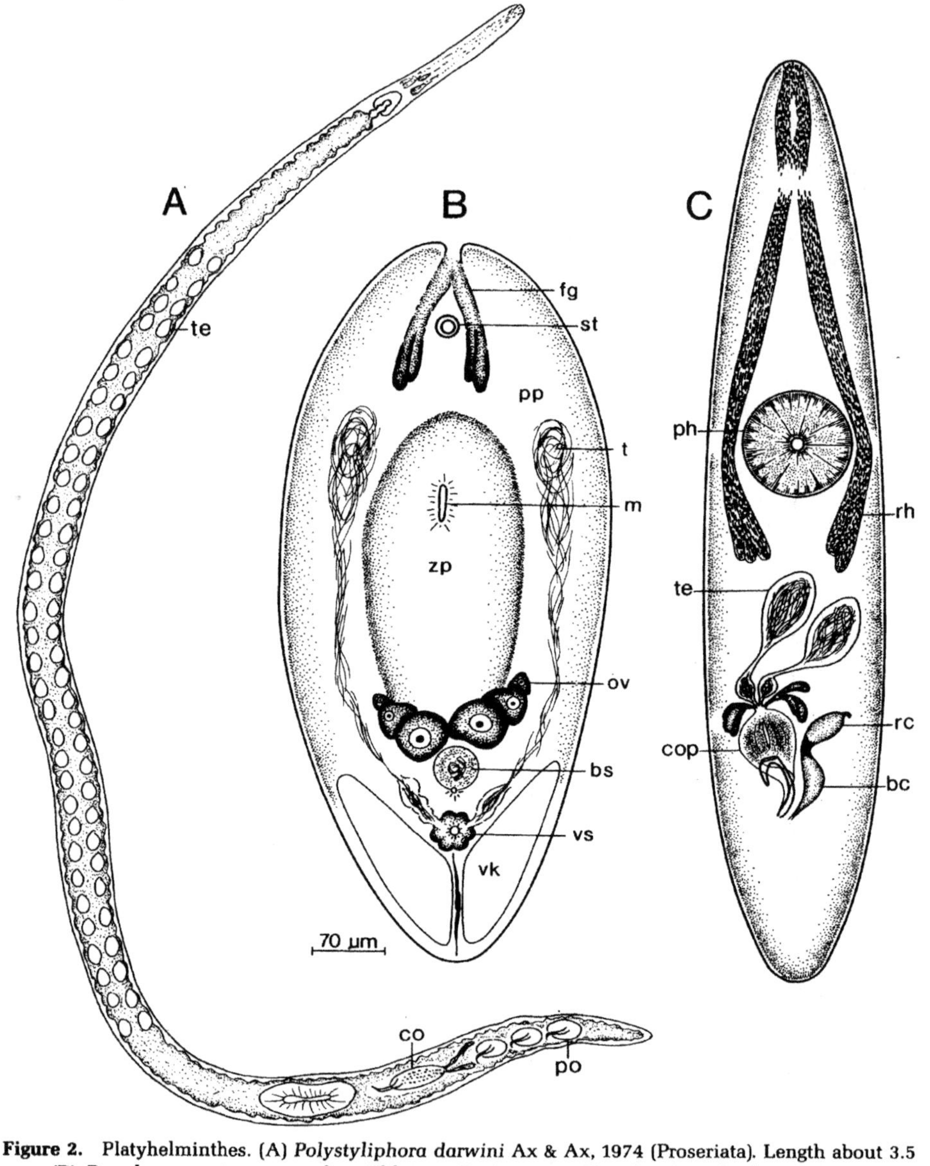

Figure 2. Platyhelminthes. (A) *Polystyliphora darwini* Ax & Ax, 1974 (Proseriata). Length about 3.5 mm. (B) *Pseudmecynostomum tardum* Ehlers & Dörjes, 1974 (Acoela). Length about 600 μm. (C) *Cryptostiopera cornuta* Ehlers & Ax, 1974 (Typhloplanoida). Length about 400 μm. (A) From Ax and Ax (1974), (B) from Ehlers and Dörjes (1974), (C) from Ehlers and Ax (1974).

3.1.1. Acoela

Sixteen species were found from five families:

Convolutidae: *Conaperta colorata* Ehlers & Dörjes, 1979; *Avagina polyvacuola* Ehlers & Dörjes, 1979; *Amphiscolops* sp.; *Praeaphanostoma thalassophilum* Ehlers & Dörjes, 1979; *Praeaphanostoma musculosum* Ehlers & Dörjes, 1979; *Praeaphanostoma vitreum* Ehlers & Dörjes, 1979.

Haploposthiidae: *Pseudohaplogonaria minima* Ehlers & Dörjes, 1979.

Otocelididae: *Otocelis phycophilus* Ehlers & Dörjes, 1979; *Philocelis karlingi pacifica* Ehlers & Dörjes, 1979; *Parotocelis luteopunctata* Ehlers & Dörjes, 1979; *Exocelis exopenis* Ehlers & Dörjes, 1979.

Childiidae: *Philactinoposthia tenebrosa* Ehlers & Dörjes, 1979; *Pseudactinoposthia parva* Ehlers & Dörjes, 1979*; *Paractinoposthia pseudovesicula* (Ehlers & Dörjes, 1979).

Mecynostomidae: *Pseudmecynostomum pellucidum* Ehlers & Dörjes, 1979; *Pseudmecynostomum tardum* Ehlers & Dörjes, 1979 (Fig. 2B).

New genera: *Parotocelis* Ehlers & Dörjes, 1979; *Exocelis* Ehlers & Dörjes, 1979; *Paractinoposthia* Ehlers & Dörjes, 1979 (see *Monoposthia* in Ehlers and Dörjes, 1979).

3.1.2. Macrostomida

Fourteen species were found from two families:

Macrostomidae: *Myozona psila* Sopott-Ehlers & Schmidt, 1974; *Myozona aerumnosa* Sopott-Ehlers & Schmidt, 1974; *Macrostomum astericis* Schmidt & Sopott-Ehlers, 1976; *Macrostomum obelicis* Schmidt & Sopott-Ehlers, 1976; *Macrostomum majesticis* Schmidt & Sopott-Ehlers, 1976; *Macrostomum troubadicis* Schmidt & Sopott-Ehlers, 1976; *Macrostomum prognosticis* Schmidt & Sopott-Ehlers, 1976; *Macrostomum miraculicis* Schmidt & Sopott-Ehlers, 1976; *Macrostomum galloprovinciale* Schmidt & Sopott-Ehlers, 1976; *Macrostomum ideficis* Schmidt & Sopott-Ehlers, 1976; *Siccomacrostomum triviale* Schmidt & Sopott-Ehlers, 1976.

Dolichomacrostomidae: *Myozonaria ascia* Sopott-Ehlers & Schmidt, 1974; *Paramyozonaria riegeri* Sopott-Ehlers & Schmidt, 1974; *Paromalostomum subflavum* Sopott-Ehlers & Schmidt, 1974 (Fig. 3C).

New genus: *Siccomacrostomum* Schmidt & Sopott-Ehlers, 1976.

3.1.3. Polycladida

The species listed below are interstitial forms of coarse intertidal sediments. Four other species could not be identified due to juvenile specimens or life observations without histological investigations. Altogether 19 polyclad species

*To distinguish between species authors and references, the symbol "&" is used for the former and the word "and" for the latter.

from five families were found during the investigations, several of them true macrofaunal organisms from non-interstitial habitats as those reported from earlier studies conducted on the Galápagos Islands (Woodworth, 1894; Bock, 1923; Plehn *in* Stummer-Traunfels, 1933; Hyman, 1939, 1953):

Planoceridae: *Amyris favis* Sopott-Ehlers & Schmidt, 1975.
Theamidae: *Theama occidua* Sopott-Ehlers & Schmidt, 1975.
Leptoplanidae: *Copidoplana virgae* Sopott-Ehlers & Schmidt, 1975; *Euplanina horrida* Sopott-Ehlers & Schmidt, 1975; *Mucroplana caelata* Sopott-Ehlers & Schmidt, 1975.
Cestoplanidae: *Cestoplana cuneata* Sopott-Ehlers & Schmidt, 1975; *Cestoplana nexa* Sopott-Ehlers & Schmidt, 1975.
Boniniidae: cf. *Traunfelsia* sp.

New genera: *Euplanina* Sopott-Ehlers & Schmidt, 1975; *Mucroplana* Sopott-Ehlers & Schmidt, 1975.

3.1.4. Proseriata

Seventeen species of this taxon were found belonging to five different families:

Otoplanidae: *Kata galapagoensis* Ax & Ax, 1974a (Fig. 3B); *Parotoplana turgida* Ax & Ax, 1974; *Philosyrtis santacruzensis* Ax & Ax, 1974; *Galapagoplana bituba* Ax & Ax, 1974.
Nematoplanidae: *Nematoplana* (?) *nigrocapitula* Ax, 1966.
Polystyliphoridae: *Polystyliphora darwini* Ax & Ax, 1974 (Fig. 2A).
Coelogynoporidae: *Vannuccia hastata* Ax & Ax, 1974; *Carenscoilia monostyla* Ax & Ax, 1974.
Monocelididae: *Monocelis galapagoensis* Ax & Ax, 1977; *Monocelis oculifera* Ax & Ax, 1977; *Archiloa papillosa* Ax & Ax, 1977; *Minona fernandinensis* Ax & Ax, 1977; *Minona stimula* Ax & Ax, 1977; *Duplominona galapagoensis* Ax & Ax, 1977; *Duplominona karlingi* Ax & Ax, 1977 (Fig. 9A); *Duplominona krameri* Ax & Ax, 1977; *Duplominona sieversi* Ax & Ax, 1977.

New genus: *Galapagoplana* Ax & Ax, 1974.

3.1.5. Typhloplanoida

Typhloplanoids comprise 19 species from five families:

Promesostomidae: *Promesostoma sartagine* Ax & Ehlers, 1973; *Promesostoma tenebrosum* Ax & Ehlers, 1973. *Wydula* (?) *simplex* Ehlers & Ehlers, 1981; *Pararhynchella fusca* Ehlers & Ehlers, 1981; *Kymocarens tibialis* Ehlers & Ehlers, 1981; *Kymocarens proxenetoides* Ehlers & Ehlers, 1981.
Trigonostomidae: *Ptychopera scutulifer* Ehlers & Ax, 1974; *Cryptostiopera cornuta* Ehlers & Ax, 1974 (Fig. 2C); *Ceratopera paragracilis* Ehlers & Ax, 1974; *Ceratopera bifida* Ehlers & Ax, 1974; *Messoplana falcata valida* Ehlers & Ax, 1974; *Trigonostomum setigerum* O. Schmidt, 1852.
Byrsophlebidae: *Maehrenthalia caligulachaena* Ehlers & Ehlers, 1981.
Kytorhynchidae: *Neokytorhynchus pacificus* Ehlers & Ehlers, 1981.

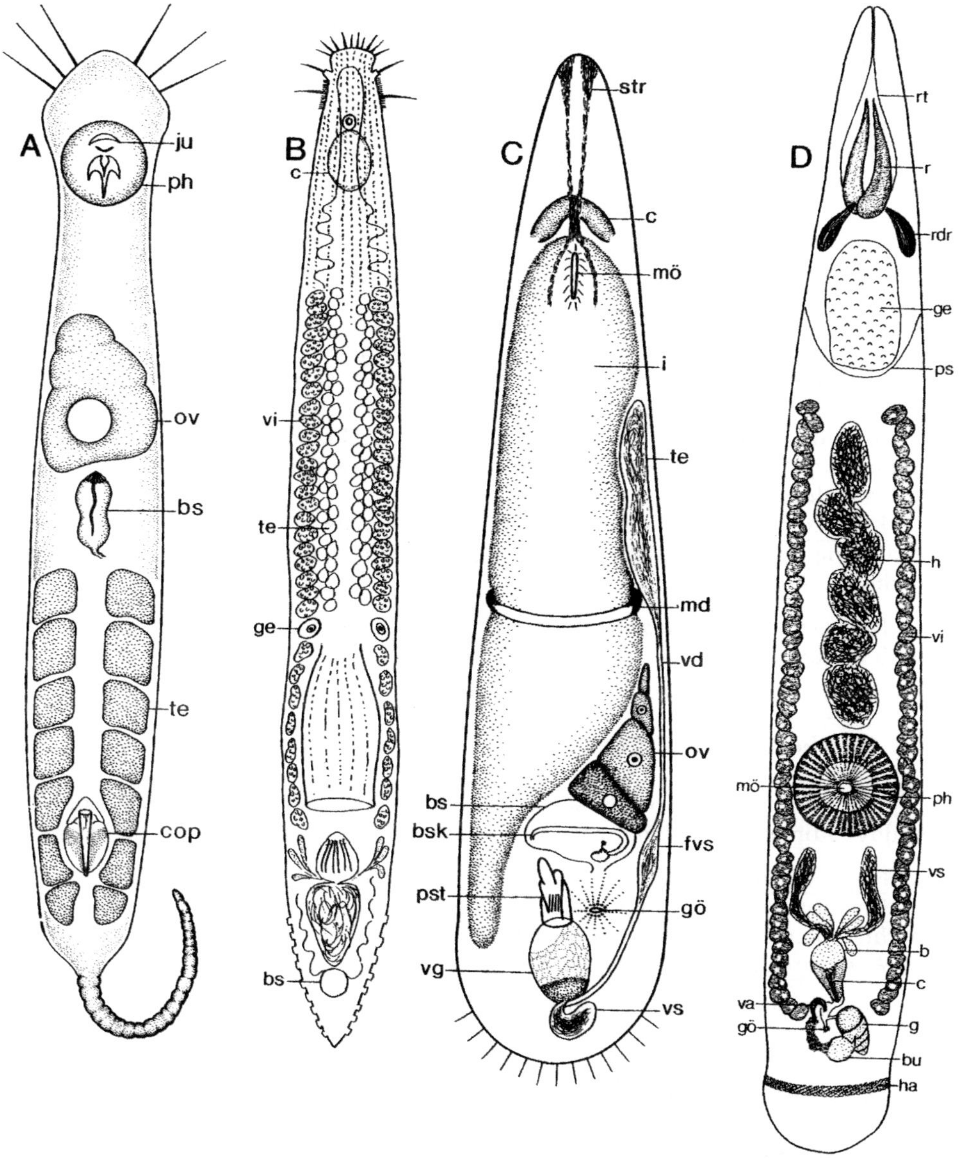

Figure 3. (A) *Corculognathia apennata* Ehlers & Ehlers, 1973 (Gnathostomulida). Length about 600 μm. (B) *Kata galapagoensis* Ax & Ax, 1974 (Platyhelminthes, Proseriata). Length 3 mm. (C) *Paramyozonaria riegeri* Sopott-Ehlers & Schmidt, 1974 (Platyhelminthes, Macrostomida). Length about 800 μm. (D) *Schizochilus santacruzensis* Noldt & Hoxhold, 1984 (Platyhelminthes, Kalyptorhynchia). Length 1.7 mm. (A) From Ehlers and Ehlers (1973), (B) from Ax and Ax (1974), (C) from Sopott-Ehlers and Schmidt (1974), (D) from Noldt and Hoxhold (1985).

Typhloplanidae: *Haloplanella semicircula semicircula* Ehlers & Sopott-Ehlers, 1989; *Haloplanella semicircula santacruzensis* Ehlers & Sopott-Ehlers, 1989; *Haloplanella semicircula sancristobalensis* Ehlers & Sopott-Ehlers, 1989; *Haloplanella conversa* Ehlers & Sopott-Ehlers, 1989; *Pratoplana ayorae* Ehlers & Sopott-Ehlers, 1989.

New genera: *Cryptostiopera* Ehlers & Ax, 1974; *Pararhynchella* Ehlers & Ehlers, 1981; *Kymocarens* Ehlers & Ehlers, 1981; *Neokytorhynchus* Ehlers & Ehlers, 1981.

3.1.6. Kalyptorhynchia

Ten species from two families were described:

Schizorhynchidae: *Carcharodorhynchus galapagoensis* Noldt & Hoxhold, 1984; *Carcharodorhynchus tabulaeferus* Noldt & Hoxhold, 1984; *Carcharodorhynchus longicirrus* Noldt & Hoxhold, 1984; *Carcharodorhynchus arista* Noldt & Hoxhold, 1984; *Schizochilus bruneti* Noldt & Hoxhold, 1984; *Schizochilus santacruzensis* Noldt & Hoxhold, 1984 (Fig. 3D); *Coagulescorhynchus virguladivina* Noldt & Hoxhold, 1984.
Karkinorhynchidae: *Cheliplana barringtonensis* Noldt & Hoxhold, 1984; *Cheliplana pacifica* Noldt & Hoxhold, 1984; *Archipelagoplana triplocirro* Noldt & Hoxhold, 1984.

New genera: *Coagulescorhynchus* Noldt & Hoxhold, 1984; *Archipelagoplana* Noldt & Hoxhold, 1984.

3.2. Gnathostomulida (Fig. 3A)

The four new Galápagos species from shallow sublittoral mostly sandy–muddy sediments were among the first records of this taxon from the Pacific area.

Onychognathiidae: *Onychognathia bractearotunda* Ehlers & Ehlers, 1973.
Gnathostomulidae: *Gnathostomula costata* Ehlers & Ehlers, 1973; *Corculognathia apennata* Ehlers & Ehlers, 1973 (Fig. 3A).
Austrognathiidae: *Austrognatharia atraclava* Ehlers & Ehlers, 1973.

New genera: *Corculognathia* Ehlers & Ehlers, 1973.

3.3. Nemertini (Fig. 4B)

The cosmopolitan interstitial hoplonemertean genus *Ototyphlonemertes* is represented by 7 species in the Galápagos Islands (Mock and Schmidt, 1975). They mainly occur in the lower intertidal and the shallow subtidal areas of most of the sand beaches, their individual numbers, however, always being very low (maximum number 11 *O. fila* in 100 cm^3, Isla Barrington, low water line): *Ototyphlonemertes fila* Corrêa, 1953; *O. erneba* Corrêa, 1950; *O. americana* Gerner, 1969; *O. cirrula* Mock & Schmidt, 1975 (Fig. 4B); *O. santacruzensis* Mock & Schmidt, 1975.

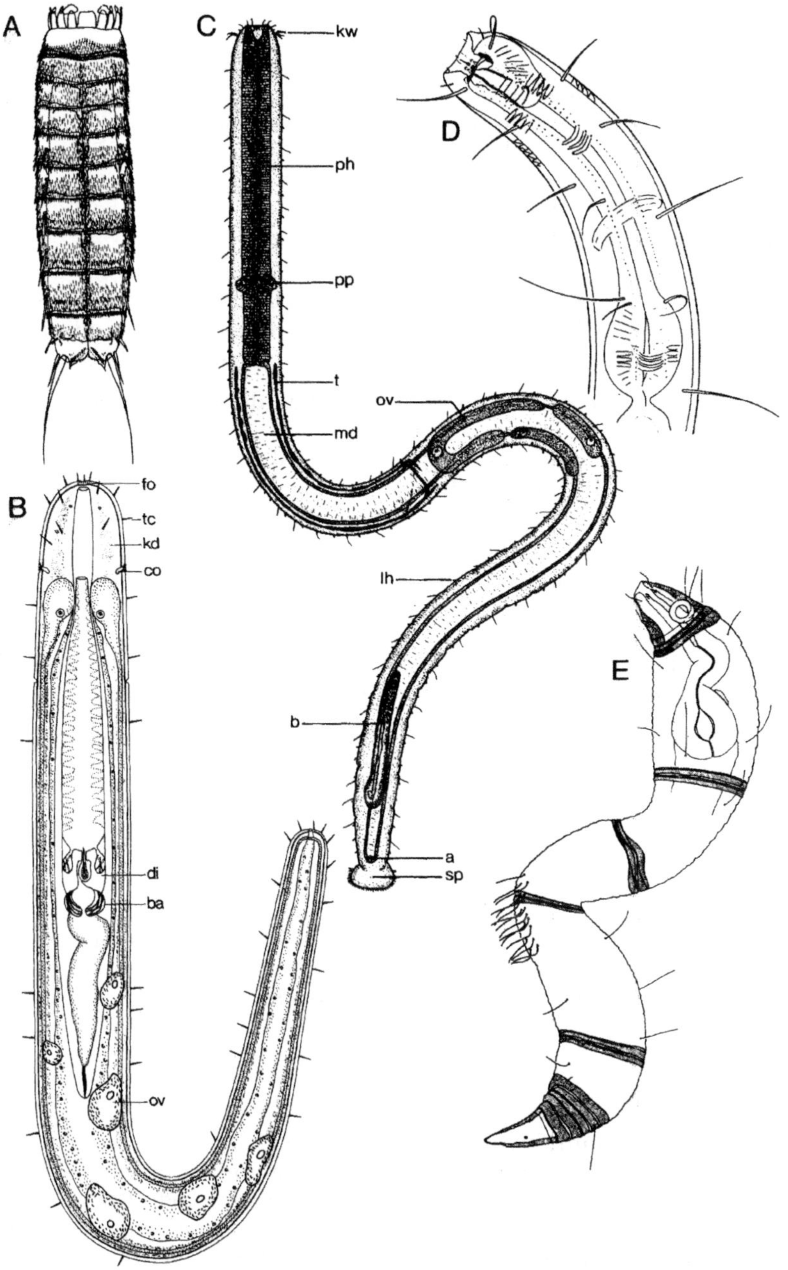

Figure 4. (A) *Echinoderes pacificus* Schmidt, 1974, Female; ventral view (Kinorhyncha, Cyclorhagida). Length 370 μm. (B) *Ototyphlonemertes cirrula* Mock & Schmidt, 1975 (Nemertini, Enopla). Length 3 mm. (C) *Megadasys pacificus* Schmidt, 1974 (Gastrotricha, Macrodasyoidea). Length 3.5 mm. (D) *Hypodontolaimus galapagensis galapagensis* Blome, 1985 (Nematoda, Chromadoridae). Anterior end. Total length about 470 μm. (E) *Bathyepsilonema lissum* Clasing, 1984 (Nematoda, Epsilonematidae). Length about 550 μm. (A) From Schmidt (1974), (B) from Mock and Schmidt (1975), (C) from Schmidt (1974), (D) from Blome (1985), (E) from Clasing (1984).

Two species of this genus possessing stylets composed of two single strands wounded round each other could not be identified because of lack of material.

Individuals of other genera, e.g., *Cephalothrix*, were rarely observed (Mock and Schmidt, 1975).

3.4. Gastrotricha (Fig. 4C)

Schmidt (1974), quite familiar with this typical interstitial group by his studies on the North Sea coast, described 18 species of Macrodasyoidea, including one new genus, *Megadasys* Schmidt, 1974, and 5 species of Chaetonotoidea:

Lepidodasyidae: *Cephalodasys pacificus* Schmidt, 1974; *Paradasys pacificus* Schmidt, 1974.

Macrodasyidae: *Macrodasys pacificus* Schmidt, 1974 (two other *Macrodasys* species were given no distinct specific status); *Megadasys pacificus* Schmidt, 1974 (Fig. 4C).

Dactylopodolidae: *Dendrodasys pacificus* Schmidt, 1974.

Turbanellidae: *Turbanella pacifica* Schmidt, 1974; *Paraturbanella pallida pacifica* Schmidt, 1974.

Planodasyidae: *Crasiella pacifica* Schmidt, 1974.

Thaumastodermatidae: *Tetranchyroderma pacifica* Schmidt, 1974; *Diplodasys ankeli pacificus* Schmidt, 1974; *Platydasys pacificus* Schmidt, 1974.

The chaetonotoids were not identified to species level. The 5 species belong to *Aspidiophus*, *Chaetonotus* (s.l.), and *Xenotrichula*.

3.5. Nematoda (Fig. 4D,E)

From the extremely high number of nematods sorted by the Schmidts on the Galápagos Islands, only the most species-rich taxon Chromadoridae and the morphologically aberrant Epsilonematidae have been published so far (Blome, 1985; Clasing, 1984). Only the cyatholaimid genus *Paracanthonchus* was found to be more abundant.

3.5.1. Chromadoridae

Chromadorinae: *Spiliphera dolichura* De Man, 1893; *Atrochromadora denticulata* Wieser & Hopper, 1967; *Chromadora macrolaimoides* Steiner, 1915; *Chromadora nudicapitata* Bastian, 1865; *Prochromadorella hexapapillata* Blome, 1985; *Prochromadorella paramucrodonta* (Allgen, 1929); *Prochromadorella salpingifera* Blome, 1985; *Prochromadorella zygophora* Blome, 1985.

Euchromadorinae: *Actinonema longicaudatum* (Steiner, 1918); *Endeolophos minutus* (Gerlach, 1967); *Endeolophos spinosus galapagensis* Blome, 1985; *Euchromadora atypica* Blome, 1985; *Rhips galapagensis* Blome, 1985; *Rhips gracilicauda* Blome, 1985; *Trochamus prosoporus* Blome, 1985.

Hypodontolaiminae: *Chromadorita nephramphida* Blome, 1985; *Chromadorita pallida* Blome, 1985; *Dichromadora* sp.; *Hypodontolaimus galapagensis*

galapagensis Blome, 1985 (Fig. 4D); *Hypodontolaimus galapagensis sanctaecrucis* Blome, 1985; *Innocuonema asymmetricum* Blome, 1985; *Spilophorella euxina* Filipjew, 1918 sensu Stekhoven, 1950; *Spilophorella paradoxa* (De Man, 1888).

3.5.2. Epsilonematidae

Bathyepsilonema compactum Clasing, 1984; *Bathyepsilonema lissum* Clasing, 1984 (Fig. 4E); *Epsilonema fernandinense* Clasing, 1984; *Epsilonema lasium* Lorenzen, 1973; *Epsilonema mangrovense* Clasing, 1984; *Leptepsilonema filiforme* Clasing, 1984; *Metepsilonema acanthum* Clasing, 1984; *Metepsilonema bermudae* Lorenzen, 1973; *Metepsilonema callosum* Lorenzen, 1973; *Perepsilonema papulosum* Lorenzen, 1973.

3.6. Kinorhyncha (Fig. 4A)

In upper sublittoral sands of beaches on Santa Cruz and Marchena, Schmidt (1974) found *Echinoderes pacificus* Schmidt, 1974 (Cyclorhagida, Echinoderidae). This is the kinorhynch genus with the highest number of species. A species of the family Semnoderidae could not be described due to inadequate material.

Typical interstitial species of **Rotifera** and opisthobranch **Gastropoda**, also collected by Schmidt (1978), were not investigated taxonomically.

3.7. Annelida—Polychaeta (Fig. 5)

Polychaete collections were repeatedly carried out on the Galápagos, but mainly comprised randomly sampled specimens of macrofaunal species (Kinberg, 1865; Chamberlin, 1919; Augener, 1933; Monro, 1933a,b; Hartman, 1939), with the exception of a more comprehensive collection of Spirorbidae by Bailey and Harris (1968).

More than 100 species (including the so-called archiannelids) were found during the Göttingen Galápagos Meiofauna Expedition, only 69 species of which have been reported in the literature (Westheide, 1974; Schmidt and Westheide, 1977; Westheide, 1977a; Westheide, 1982). Twenty-seven of these were described as new; no new genera were erected. Most species were found in shallow sublitoral areas, including species that inhabited old corals and crevices of stones. Only true macrofaunal representatives (body width 1 mm and wider) are not listed below.

Pisionidae: *Pisione oerstedi pulla* Westheide, 1974; *Pisione galapagoensis* Westheide, 1974 (Fig. 5C).

Phyllodocidae: *Hesionura* spp. (W. Westheide, unpublished results).

Nereidae: *Ceratonereis monronis* Westheide, 1977 (Fig. 5F).

Hesionidae: *Podarke* (?) sp.; *Microphthalmus indefatigatus* Westheide, 1974 (Fig. 5B); *Hesionides arenaria pacifica* Westheide, 1974; *Hesionides unilamellata* Westheide, 1974.

Pilargidae: *Synelmis albini* (Langerhans, 1881).

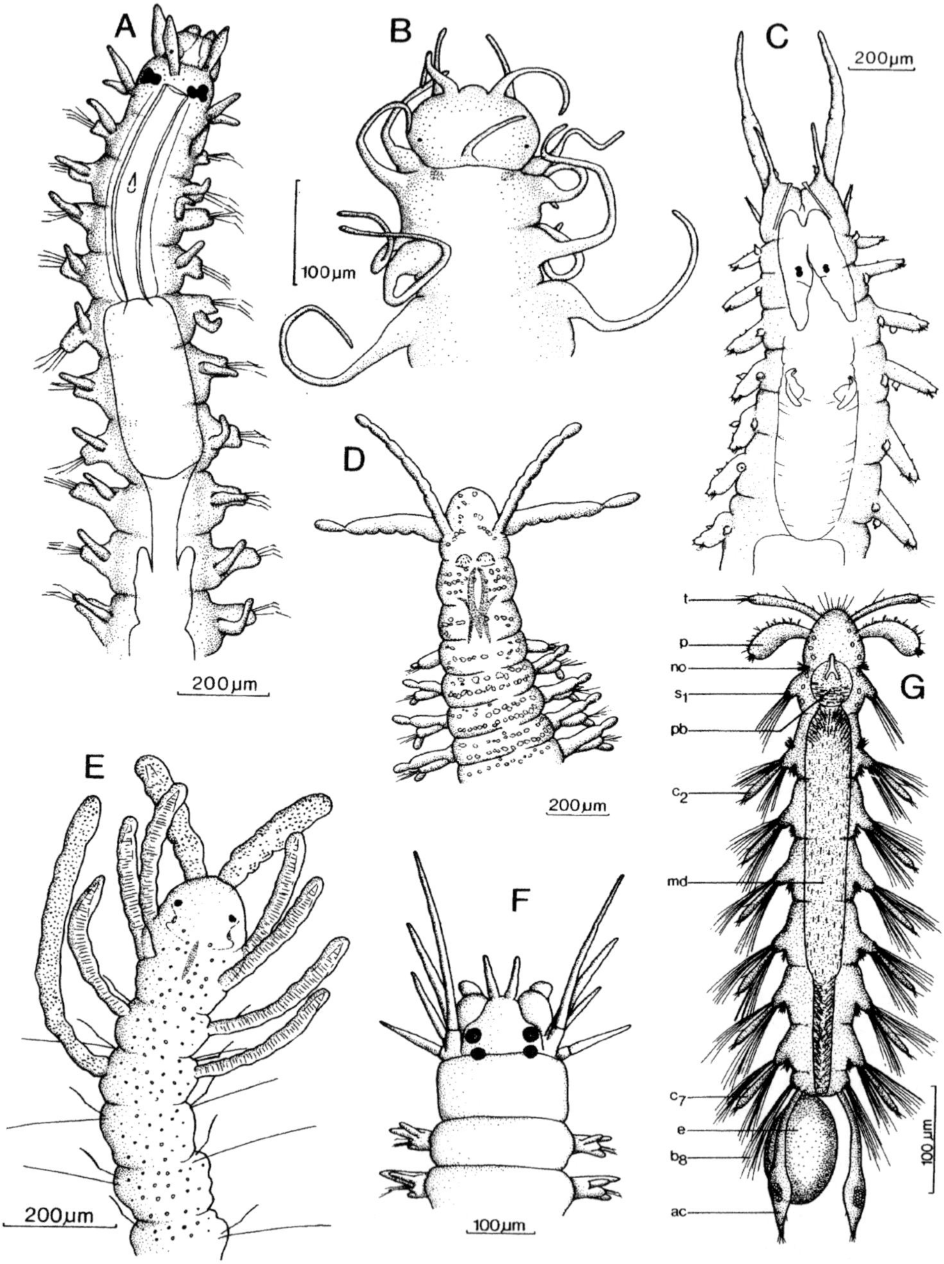

Figure 5. Polychaeta (Annelida). (A) *Brania mediodentata* Westheide, 1974 (Syllidae). (B) *Microphthalmus indefatigatus* Westheide, 1974 (Hesionidae). (C) *Pisione galapagoensis* Westheide, 1974. (D) *Dorvillea pacifica* (Westheide, 1977) (Dorvilleidae). (E) *Macrochaeta multipapillata* Westheide, 1981 (Acrocirridae). (F) *Ceratonereis monronis* Westheide, 1977 (Nereidae). (G) *Nerillidium lothari* Schmidt & Westheide, 1977 (Nerillidae). (A–C) From Westheide (1974), (D), (F), from Westheide (1977), (E) from Westheide (1981), (G) from Schmidt and Westheide (1977).

Eunicidae: *Nematonereis unicornis* (Grube, 1840).

Dorvilleidae: *Dorvillea pacifica* (Westheide, 1977) (Fig. 5D); *Protodorvillea* sp.; *Ophryotrocha* sp.

Dinophilidae: *Dinophilus* sp.

Syllidae: *Haplosyllis spongicola* (Grube, 1855); *Opisthosyllis corallicola* Hartmann-Schröder, 1965; *Trypanosyllis taeniaeformis* (Haswell, 1865); *Ehlersia rosea magna* Westheide, 1974; *Ehlersia* spec.; *Typosyllis hyalina* (Grube, 1863); *Typosyllis variegata* (Grube, 1860); *Typosyllis glarearia* Westheide, 1974; *Typosyllis* spec., *Branchiosyllis exilis* (Gravier, 1900); *Parasphaerosyllis indica* Monro, 1937; *Amblyosyllis granosa* Ehlers, 1897; *Eusyllis homocirrata* Hartmann-Schröder, 1958; *Odontosyllis fulgurans dolerens* Westheide, 1974; *Syllides edentula* (Claparède, 1868); *Syllides japonica edentata* Westheide, 1974; *Brania heterocirra* Rioja, 1941; *Brania subterranea* (Hartmann-Schröder, 1956); *Brania oculata* (Hartmann-Schröder, 1960); *Brania concinna* Westheide, 1974; *Brania mediodentata* Westheide, 1974 (Fig. 5A); *Sphaerosyllis hystrix* Claparède, 1863; *Sphaerosyllis centroamericana* Hartmann-Schröder, 1959; *Sphaerosyllis pumila* Westheide, 1974; *Exogone dispar* (Webster, 1979); *Exogone naidinoides* Westheide, 1974; *Exogone occidentalis* Westheide, 1974; *Exogone longicornis* Westheide, 1974; *Exogone microtentaculata* Westheide, 1974; *Autolytus multidenticulatus* Westheide, 1974; *Autolytus arboreus* Westheide, 1974; *Autolytus* sp.; *Procerea fasciata* Langerhans, 1879; *Procerea* sp.

Nerillidae: *Nerilla parva* Schmidt & Westheide, 1977; *Nerillidium lothari* Schmidt & Westheide, 1977 (Fig. 5G); *Mesonerilla ecuadoriensis* Schmidt & Westheide, 1977.

Protodrilidae: *Protodrilus pierantonii* Aiyar & Alikunhi, 1944; *Protodrilus infundibuliformis* Schmidt & Westheide, 1977; *Protodrilus spp.*

Saccocirridae: *Saccocirrus sonomacus* Martin, 1977; *Saccocirrus* sp.

Polygordiidae: *Polygordius pacificus floreanensis* Schmidt & Westheide, 1977.

Diurodrilidae: *Diurodrilus* sp.

Cirratulidae: *Caulleriella alata* (Southern, 1914); *Cirriformia violacea* Westheide, 1982; *Cirriformia* spec.; *Tharyx* spec.

Ctenodrilidae: *Ctenodrilus serratus* (O. Schmidt, 1857).

Questidae: *Questa media* Westheide, 1982.

Acrocirridae: *Macrochaeta multipapillata* Westheide, 1982 (Fig. 5E).

3.8. Annelida—Aeolosomatidae

From the aeolosomatids, not regarded anymore to be true clitellate annelids, a new subspecies of the only true marine species was described from the sandy intertidal of six islands: *Aeolosoma maritimum dubiosum* Westheide & Schmidt, 1974 (Westheide and Schmidt, 1974).

3.9. Annelida—Oligochaeta

Schmidt's rich collection of oligochaetes contained both Enchytraeidae and Tubificidae; unfortunately, most of these were lost. Of the remaining, mostly immature specimens, Erséus (1984) identified the tubificid species *Aktedrilus locyi* Erséus, 1984, *Aktedrilus cuneus* Erséus, 1980, and *Heterodrilus pentcheffi* Erséus, 1981; *Bacescuella parvithecata* Erséus, 1978, was found in another collection from the Galápagos Islands. Erséus (1984) also indicated that three species of the enchytraeid genus *Marionina* from Galápagos would be described by K. Coates in a future publication.

3.10. Tardigrada (Fig. 6)

Material for the description of marine Galápagos tardigrades from four families came from collections carried out by M. McGinty-Bayly in 1968, A. DeGauge in 1973, and P. Schmidt in 1972 and was published in a joint paper (McKirdy, Schmidt, and McGinty-Bayly, 1976). Besides a number of terrestrial tardigrades, Schuster and Grigarick (1966) reported two marine species, *Echiniscoides sigismundi* (Schultze, 1865) and *Archechiniscus marci* Schulz, 1953.

Halechiniscidae: *Tanarctus velatus* McKirdy, Schmidt, & McGinty-Bayly, 1976.

Batillipedidae: *Orzeliscus* cf. *belopus* Du Bois-Reymond Marcus, 1952.

Stygarctidae: *Stygarctus abornatus* McKirdy, Schmidt, & McGinty-Bayly, 1976; *Parastygarctus sterreri* Renaud-Mornant, 1970; *Pseudostygarctus triungulatus* McKirdy, Schmidt, & McGinty-Bayly, 1976 (Fig. 6B); *Megastygarctides orbiculatus* McKirdy, Schmidt, & McGinty-Bayly, 1976 (Fig. 6A).

Orcellidae: *Echiniscoides sigismundi* (M. Schultze, 1965); *Anisonyches diakidius* Pollock, 1975. Two specimens of halechiniscids were not identified due to inadequate material.

New genera: *Pseudostygarctus* McKirdy, Schmidt, & McGinty-Bayly, 1976; *Megastygarctides* McKirdy, Schmidt, & McGinty-Bayly, 1976.

3.11. Chelicerata—Acari (Fig. 7C)

Investigations of Schmidt's material by Bartsch (1977) revealed 42 halacarid species in 12 genera, the ecology of which was characterized by Bartsch and Schmidt (1978). Only four species were previously described, and no new genera were erected. Habitats of the species are subtidal and intertidal sandy areas. Several eurytopic species were found not to be restricted to the interstitial habitat.

Rhombognathinae: *Rhombognathus robustus* Bartsch, 1977; *Rhombognathus longiscutatus* Bartsch, 1977; *Rhombognathus semireticulatus* Bartsch, 1977; Rhombognathus ellipticus Bartsch, 1977; *Rhombognathus similis* Bartsch, 1977; *Rhombognathus heterosetosus* Bartsch, 1977; *Rhombognathus* spec.; *Isobactrus asper* Bartsch, 1977.

Actacarinae: *Actacarus mollis* Bartsch, 1977; *Actacarus uniscutatus* Bartsch, 1977; *Actacarus hastatus* Bartsch, 1977.

Halacarinae: *Copidognathus lineatus* Bartsch, 1977; *Copidognathus gibboides* Bartsch, 1977; *Copidognathus pauciporus* Bartsch, 1977; *Copidognathus lepidus*

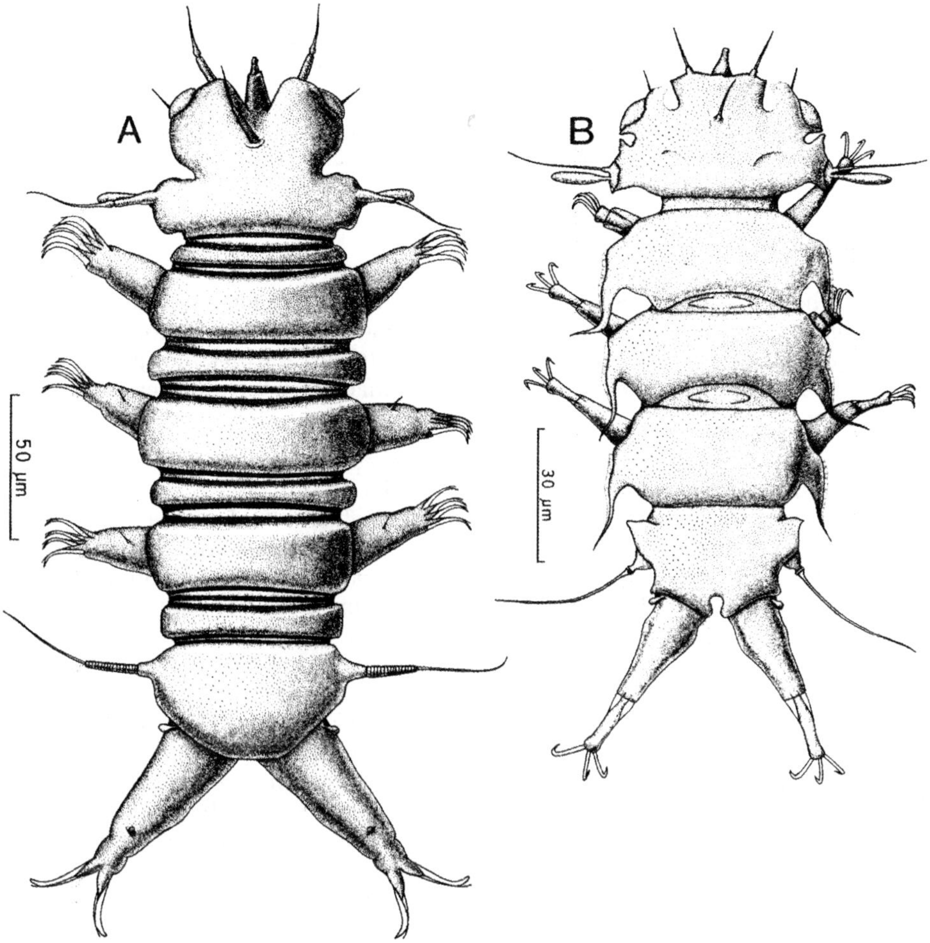

Figure 6. Tardigrada (Heterotardigrada). (A) *Megastygarctides orbiculatus* McKirdy, Schmidt & McGinty-Bayly, 1976. (B) *Pseudostygarctus triungulatus* McKirdy, Schmidt, & McGinty-Bayly, 1976. From McKirdy *et al*., 1976.

Bartsch, 1977; *Copidognathus scitus* Bartsch, 1977; *Copidognathus bairdi bairdi* Newell, 1947; *Copidognathus grandiculus* Bartsch, 1977; *Copidognathus tuberipes* Bartsch, 1977; *Copidognathus peregrinus* Bartsch, 1977; *Copidognathus gracilis* (Viets, 1936); *Copidognathus guttatus* Bartsch, 1977; *Copidognathus tenuirostris* Bartsch, 1977; *Copidognathus* spec.; *Arhodeoporus bonairensis* (Viets, 1936); *Arhodeoporus bucculentus* Bartsch, 1977; *Agauopsis ornata* (Lohmann, 1893); *Agauopsis denticulatus* Bartsch, 1977; *Agauopsis* spec., *Atelopsalis aliger* Bartsch, 1977.

Acarochelopodinae: *Acarochelopodia aduncispina* Bartsch, 1977; *Acarochelopodia cuneifera* Bartsch, 1977 (Fig. 7C).

Lohmannelinae: *Scaptognathus pauciporus* Bartsch, 1977; *Scaptognathus gibbosus* (Bartsch, 1977; *Scaptognathides bicornis* Bartsch, 1977.

Simognathinae: *Simognathus similis* Bartsch, 1977; *Simognathus disparilis* Bartsch, 1977; *Acaromantis subasper* Bartsch, 1977; *Acaromantis fastigatus* Bartsch, 1977; *Acaromantis armatus* Bartsch, 1977; *Acaromantis punctulus* Bartsch, 1977; *Acaromantis grandiculus* Bartsch, 1977.

3.12. Crustacea—Copepoda (Fig. 7A)

The harpacticoid copepods probably are the most species-rich taxon in the Galápagos meiofauna. Many were described in a series of voluminous contributions by Mielke (1979, 1981, 1982b, 1984, 1989a, 1989b). Besides the species listed below, the material contained several forms that are poorly understood taxonomically.

Longipediidae: *Longipedia helgolandica santacruzensis* Mielke, 1979.

Canuellidae: *Galapacanuella beckeri* Mielke, 1979.

Ectinosomatidae: *Ectinosoma pectinatum* Mielke, 1979; *Ectinosoma nonpectinatum* Mielke, 1979; *Arenosetella germanica galapagoensis* Mielke, 1979; *Hastigerella abbotti santacruzensis* Mielke, 1979; *Sigmatidium triarticulatum* Mielke, 1979; *Sigmatidium kunzi* Mielke, 1979; *Sigmatidium* spec.; *Noodtiella frequentior* Mielke, 1979; *Noodtiella hoodensis* Mielke, 1979.

Darcythompsoniidae: *Darcythompsonia fairliensis* (T. Scott, 1899).

Metidae: *Metis galapagoensis* Mielke, 1989; *Metis* spec.

Paramesochridae: *Apodopsyllus arcuatus* Mielke, 1984; *Diarthrodella parorbiculata pacifica* Mielke, 1984; *Diarthrodella neotropica* Mielke, 1984; *Diarthrodella galapagoensis* Mielke, 1984; *Leptopsyllus punctatus* Mielke, 1984; *Leptopsyllus platyspinosus* Mielke, 1984; *Paramesochra helgolandica galapagoensis* Mielke, 1984; *Paramesochra unaspina* Mielke, 1984; *Scottopsyllus langi* Mielke, 1984; *Kliopsyllus spiniger spiniger* Wells, Kunz, & Rao, 1975; *Kliopsyllus regulextans* Mielke, 1984; *Kliopsyllus similis* Mielke, 1984; *Kliopsyllus unguiseta* Mielke, 1984.

Tetragonicipitidae: *Tetragoniceps galapagoensis* Mielke, 1989; *Pteropsyllus trisetosus* Mielke, 1989; *Oniscopsis robinsoni* Chappuis & Delamare Deboutteville, 1956; *Phyllopodopsyllus angolensis* Kunz, 1984; *Phyllopodopsyllus thiebaudi santacruzensis* Mielke, 1989; *Phyllopodopsyllus furciger* Sars, 1907; *Phyllopodopsyllus galapagoensis* Mielke, 1989; *Phyllopodopsyllus kunzi* Mielke, 1989.

Cylindropsyllidae: *Leptastacus ctenatus* Mielke, 1982; *Leptastacus dispinosus* Mielke, 1982; *Leptastacus spatuliseta* Mielke, 1982; *Notopontia galapagoensis* Mielke, 1982; *Arenopontia peteraxi* Mielke, 1982; *Arenopontia trisetosa* Mielke, 1982.

Laophontidae: *Laophonte galapagoensis* Mielke, 1981; *Loureirophonte isabelensis* Mielke, 1981; *Heterolaophonte serratula* Mielke, 1981; *Paralaophonte pacifica galapagoensis* Mielke, 1981; *Paralaophonte problematica* Mielke, 1981; *Esola longicauda galapagoensis* Mielke, 1981; *Onychocamptus* spec.; *Echinolaophonte tetracheir* Mielke, 1981; *Afrolaophonte schmidti* Mielke, 1981; *Klieonychocamptoides itoi* Mielke, 1981; *Laophontina triarticulata* Coull & Zo, 1980.

Ancorabolidae: *Paralaophontodes exopoditus* Mielke, 1981.

New genus: *Galapacanŭella* Mielke, 1979.

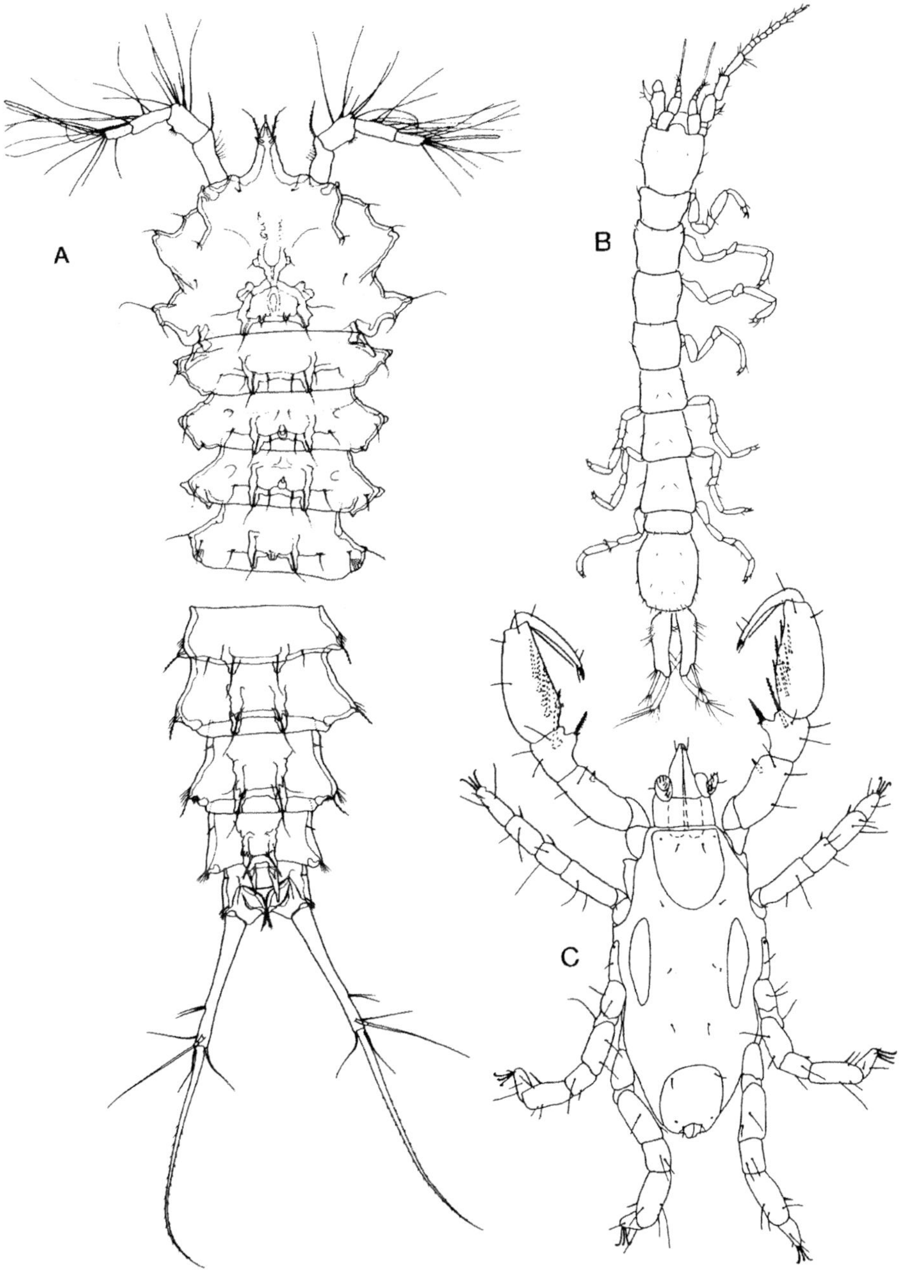

Figure 7. (A), (B): *Paralaophontodes exopoditus* Mielke, 1981 (Copepoda, Harpacticoida, Ancorabolidae). Female, dorsal view of cephalothorax and pereion (A) and abdomen (B). Length 600 μm. (B) *Microcharon galapagoensis* Coineau & Schmidt, 1979 (Isopoda, Asellota, Microparasellidae). Length about 1.2 mm. (C) *Acarochelopodia cuneifera* Bartsch, 1977 (Acari, Halacaridae). Length about 250 μm. (A) From Mielke (1981); (B) from Coineau and Schmidt (1979); (C) Courtesy Dr. I. Bartsch, Hamburg.

3.13. Crustacea—Ostracoda (Fig. 11)

Gottwald (1983) stated that he identified at least 60 different species of ostracods in Schmidt's material from sandy beaches of which he described 22 new species in the order Podocopida. Earlier collections from the Galápagos Islands considered species from brackish-water (Triebel, 1956), sublittoral (Pokorny, 1970, 1972), and rockpool habitats (Bate *et al.*, 1981), which, however, are not listed below.

Psammocytheridae: *Psammocythere santacruzensis* Gottwald, 1983.
Cytheromatidae: *Microloxoconcha santacruzensis* Gottwald, 1983; *Fernandinacythere arenicola* Gottwald, 1983.
Cobanocytheridae: *Cobanocythere subterranea* Hartmann, 1959; *Cobanocythere arenicola* Gottwald, 1983; *Cobanocythere lanceolata* Gottwald, 1983; *Cobanocythere psammophila* Gottwald, 1983; *Cobanocythere pacifica* Gottwald, 1983; *Cobanocythere fernandinensis* Gottwald, 1983; *Cobanocythere hoodensis* Gottwald, 1983; *Cobanocythere sublitoralis* Gottwald, 1983; *Cobanocythere santacruzensis* Gottwald, 1983; *Cobanocythere labiata* Hartmann, 1959; *Cobanocythere guttaeformis* Gottwald, 1983; *Cobanocythere elongata* Gottwald, 1983.
Xestoleberididae: *Xestoleberis arcturi darwini* Gottwald, 1983; *Xestoleberis galapagoensis* Gottwald, 1983.
Parvocytheridae (Fig. 11): *Parvocythere fernandinensis* Gottwald, 1983; *Parvocythere galapagoensis* Gottwald, 1983; *Parvocythere schmidti* Gottwald, 1983; *Parvocythere subterranea* Gottwald, 1983; *Parvocythere supralitoralis* Gottwald, 1983.

New genus: *Fernandinacythere* Gottwald, 1983.

3.14. Crustacea—Isopoda (Fig. 7B)

One typical interstitial isopod species, *Microcharon galapagoensis* Coineau & Schmidt, 1979 (Microparasellidae), was described from coarse sands in lower regions of intertidal areas on Santa Cruz, Isabela, Tower, and San Cristobal (Coineau and Schmidt, 1979).

4. Abundance and Dynamics (Fig. 8)

The immense quantitative data collected by Schmidt (1978) allowed him to give convincing abundance values for various beaches in the islands. The values range from an average of 118 ind/100 cm^3 (Fernandina) to 667 ind/100 cm^3 (Santa Cruz, Playa Borrero), with the highest number in an individual sample of more than 3000 individuals. The maximum number of metazoans under a surface of 10 cm^2 was 7640 (down to a depth of 1 m). These values are higher than in any other collection site with the exception of Sylt (North Sea), where Schmidt (1968) found a maximum of 10,972 metazoans under a surface of 10 cm^2. Population dynamics and reproductive periods are strongly correlated with water temperature in temperate sandy beaches, the maximum densities occurring after periods of increasing temperatures in spring and summer (Schmidt, 1968; 1972a; 1972b). Expectations of reduced seasonal fluctuations on the Galápagos archipelago with its relatively

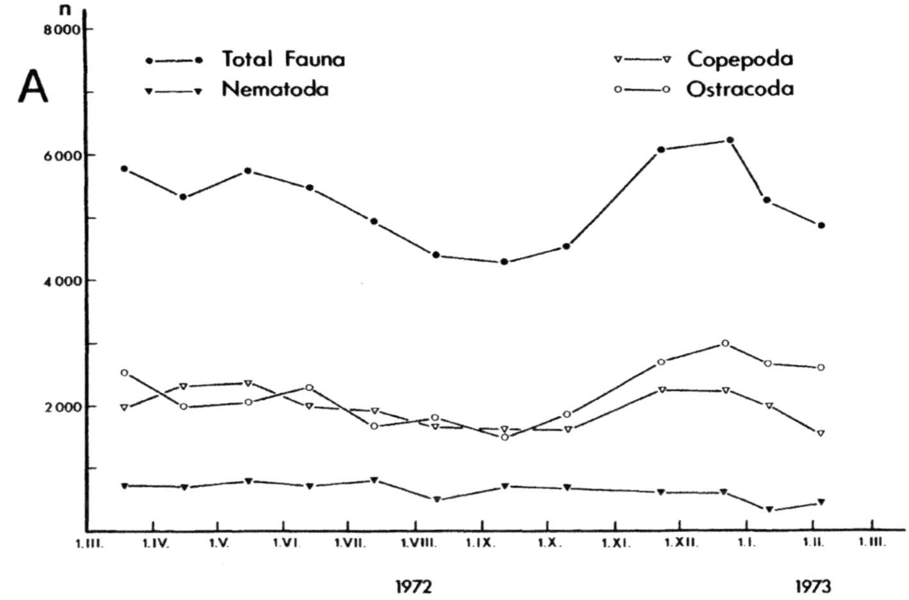

Figure 8. Abundance dynamics of total fauna and dominant taxa. (A) Santa Cruz, Bahia Academy.

constant water temperatures, however, were not supported. In the beach of Bahia Academy (Santa Cruz, southern coast) abundance values of the total fauna in a transect fluctuated between 4267 and 6198 in 50 cm^3 sand, with a minimum abundance occurring in August to October (Schmidt, 1978). The fauna from the Playa Borrero collection site on the north coast of Santa Cruz showed considerably higher seasonal variation, with 14,126 individuals in November and 5707 individuals per 50 cm^3 in February. Fluctuations especially were dependent upon differences in the dominant taxa Nematoda, Copepoda, and Ostracoda.

As Schmidt (1978) pointed out, these seasonal variations are as large or even larger than those found in the North Sea and the Baltic. The most remarkable result of these two density series is the fact that their minimum and maximum values do not coincide. Schmidt (1978) concluded that these fluctuations were not due to differences in water temperatures but could be explained by specific local events, such as heavy rainfall just before sampling or input of organic matter by dead macrofauna. Unfortunately, results on population dynamics and reproductive seasonality of smaller taxa or individual Galápagos species have been only sparsely published. The most common halacarids (*Actacarus mollis, Scaptognathides bicornis*) reproduce throughout the year (Bartsch and Schmidt, 1978). Clasing (1984) found conspicuous seasonal variation in the numbers of the nematode *Metepsilonema bermudae* but no marked differences in the ratio of juvenile and adult individuals throughout the year. She also assumed that reproduction is not seasonally restricted in these species. Mielke (1989a), however, presented evidence for a seasonality of reproduction in the copepod *Oniscopsis robinsoni*, based on a higher number of copepodites in May, June, and July.

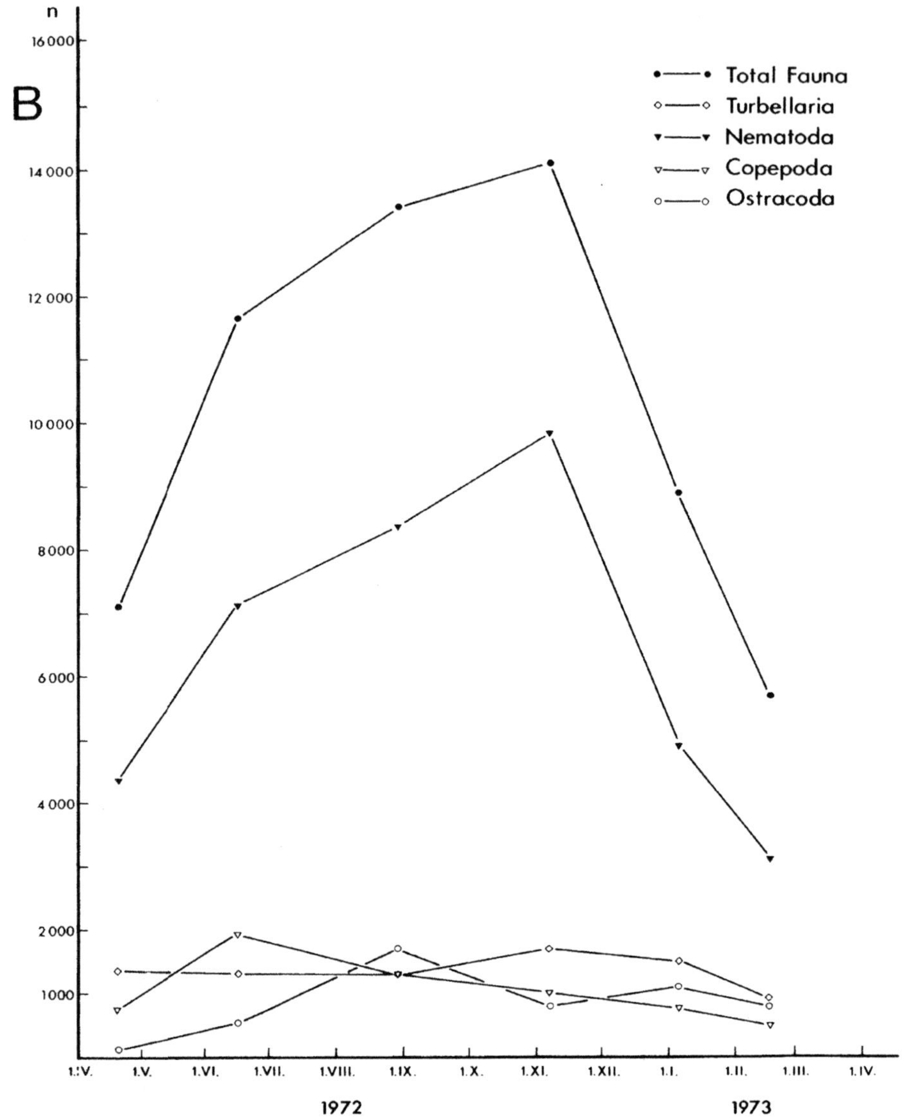

Figure 8. (B) Santa Cruz, Playa Borrero. After Schmidt, 1978.

5. Dominating Groups

Nematoda, Copepoda, and Ostracoda are the groups with the highest mean values of animals in individual island beaches. Platyhelminthes ("Turbellaria") and Polychaeta are present in all beaches. Each taxon comprises at least 2.6% of the total fauna of a given beach transect; in certain individual samples, more than 80% platyhelminths and more than 74% polychaetes were found. Gastrotricha (up to

4.8% on Tower; more than 31% in individual samples on Tower) and Acari (up to 2.6% on Marchena; up to 26% in individual samples from Barrington and Hood) comprise only about 1% in a few beaches. Mean values for the dominant taxa of all beaches investigated are: Nematoda 33.2%, Ostracoda 26.3%, Copepoda 21.2%, Turbellaria 8.9%, and Polychaeta 5.9% (Schmidt, 1978). Whereas nematodes and copepods are the dominant groups of marine sand beaches worldwide, the Galápagos pattern of dominance is surprising because of the high percentage of ostracods. These high numbers of individuals and species cannot be satisfactorily correlated with special environmental conditions within Galápagos beaches. Although sediments and exposition of sandy tidal areas from, for example, Fernandina, Hood, James, Marchena, Barrington, Tower, and Santa Cruz islands are rather different, the share of ostracods in all of the lotic beaches is never below 22.5%. On Hood, ostracods in the large and relatively long beach of Bahia Gardner present more than half of all metazoans (55.9%); in one sample of this beach Schmidt counted 2856 ostracod individuals in 100 cm^3, that is 95.6% of the total metazoan fauna. Only in the three moderately lotic beaches on Bartholomé, Floreana, and Santa Cruz (Playa Borrero) was the percent of ostracods comparatively low with values of 1.6, 5.5, and 12.5%, respectively.

But also these values are higher than those found in similarly investigated continental beaches. Neither in temperate—e.g., North Sea, Baltic, Norway (Schmidt, 1968; 1969; 1972a; 1972b)—nor in warm oceanic littoral areas have ostracods normally been found to be one of the dominating groups. One exception is a Delaware beach (Atlantic coast of North America), where Hummon *et al.* (1976) reported ostracods to be the second most abundant taxon.

It is therefore not unlikely that the dominance of ostracods has to do with the pattern of colonization of the sand beaches in the Archipelago. There is a general worldwide similarity in the interstitial fauna dominance pattern, with nematodes and copepods being almost always the dominant taxa. Let us assume this pattern of dominance represents the stable climax situation of sandy meiofaunal biocoenoses. One may expect that colonization of newly formed isolated islands by different taxa will not take place in the same abundance ratio as individuals found in continental beaches. Individual taxa just might reach the islands earlier than others by chance; some taxa might be better suited for drifting and rafting, or may survive different means of dispersal better than others. This certainly will lead to a gradually timed pattern of colonization by the individual taxon and will influence the dominance pattern of both the founder generations and the subsequent generations.

It appears, therefore, to be a completely open question, if the Galápagos situation should be considered a unique but stable community structure or a successional stage that will lead somehow to the normal nematode–copepod dominance. The considerable variation in the percentages of the three dominant groups within the different beaches (e.g., ostracods between 1.6%–55.9%) may support the latter possibility. In this case the Galápagos faunal structure would allow us to gain insight into the succession of meiofauna beach communities. We would then have to conclude that a period of 1 to 3 million years is not sufficient to establish the climax community structure of these biocoenoses on oceanic islands.

Schmidt (1978) found another peculiarity in the composition of the Galápagos meiofauna: Tardigrada never composed more than 1% of the total fauna in a beach transect; the highest value in an individual sample was 17% (Santa Cruz, Playa

Borrero). Especially in comparison with Schmidt's (1968, 1969) investigations on Sylt and Renaud-Debyser's (1963) analysis of the Arcachon beaches the low percent of tardigrades in Galápagos beaches is striking. Presence and abundance of tardigrade species, however, generally appears to be more irregular over time than that of other groups. Thus Schmidt's tardigrade values from Sylt have not been observed in recent years. My personal observations in continental warm water tropical beaches did not reveal high numbers of this taxon either. On the other hand one of the highest number of individuals of a single taxon ever observed in a littoral sand sample by the author were tardigrades in a Bermuda beach.

Different, however, from the overall dominance pattern is that the total number of individuals in animal groups may be surprisingly similar to other collection sites of the world. An example of this is the Polychaeta (Westheide 1977b), for which Schmidt (1978) found values between 2.2% (Santa Cruz, Playa Borrero) and 15.8% (Floreana), the mean value of all beaches being 6.1%. For comparison, in Sylt beaches (southern North Sea) the share of polychaetes nearly reaches 5%, in Mediterranean beaches near Tunis it is between 1.5 to 4.4%; a percentage of about 1% is known from Southeast India.

6. Zonation

Vertical and horizontal zonation of total meiofauna and individual taxa is typical for tidal beaches throughout the world. The Galápagos beaches are no exception (Ax and Schmidt, 1973; Bartsch and Schmidt, 1976; Schmidt, 1978; Gottwald, 1983; Clasing, 1984). As expected, the majority of individuals inhabited the lower and middle parts of the beach slopes, highest values being found in the upper 10–20 cm of sediment. High air temperatures during low tide, however, may force animals to migrate into deeper layers; heavy rainfalls may kill nearly the entire meiofauna down to a depth of 50 cm.

Three examples may demonstrate single species zonation pattern in the Playa Borrero beach on Santa Cruz. The epsilonematid *Perepsilonema papulosum* lives around the low water line (Clasing, 1984). The macrostomid *Siccomacrostomum triviale* was exclusively found in the middle part of the tidal area, and *Macrostomum ideficis* inhabits the uppermost part just below the high water line (Schmidt and Sopott-Ehlers, 1976). Distribution patterns of congeneric species occurring together in a beach are interesting: they partition the beach environment, with very little overlap (Fig. 10).

7. Number of Species

Extremely low species numbers is a well known characteristic of the terrestrial fauna of small islands. This phenomenon, so easily recognized in Galápagos terrestrial arthropods, is not, however, found in the marine intertidal meiofauna. Surprisingly, Galápagos beaches are not generally species-poor, nor do individual species occur in conspicuously high number [although in several groups, the number of species appear to be distinctly smaller than in the beaches of the North Sea islands of Sylt (e.g., Plathyhelminthes: Ax, 1977)]. It should be kept in mind that sampling in the Galápagos by the Göttingen group was restricted to a selection of small-scale habitats only, in a relatively short period of time. Thus additional species are to be expected with additional collecting. We also cannot exclude the possibility that the number of species in individual taxa is different in tropical areas, so that the Galápagos situation may simply reflect the normal subtropical and tropical situation (see also discussion in Ehlers and Dörjes, 1979, for Acoela).

It should be emphasized, however, that there are several taxa whose number of species obviously do not differ markedly from other tropical and temperate shorelines of the world. For instance, this seems to be true for the meiofaunal Polychaeta (Westheide, 1974; 1977a; 1982; Schmidt and Westheide, 1977). Bailey and Harris (1968) came to similar conclusions for the Galápagos spirorbid polychaetes. Blome (1985) found the nematode family Chromadoridae as rich in genera and species as in continental beaches. There is even one example of the number of species in a taxon being considerably higher than elsewhere. The nematode family Epsilonematidae is present on the Galápagos islands with 10 species in 5 genera, a number which according to Clasing (1984) has not yet been found in any other area.

Even more interesting is a comparison of the spectrum of supra-specific taxa. In several groups the identified species generally represent most of the taxa that normally can be found in comparable habitats. This holds true, for example in chromadorid Nematoda (Blome, 1985), Gastrotricha, and Polychaeta. The 13 macrodasyoid species from the Galápagos represent all 6 families within this gastrotrich order, and the 5 chaetonotoid gastrotrichs belong at least to 3 different genera (Schmidt, 1974). The spectrum of polychaete genera is almost the same as that revealed in the majority of sandy beaches throughout temperate and tropical seas. Even the distribution pattern of polychaete genera within beaches is widely consistent, although species may not be identical. For instance, the shallow subtidal and low tidal areas with coarse sediments are characterized by the genera *Pisione* (Pisionidae), *Hesionura* (Phyllodocidae), *Typosyllis*, *Exogone*, *Brania* (Syllidae), *Microphthalmus* (Hesionidae), *Dorvillea* (Dorvilleidae), *Macrochaeta* (Acrocirridae), and *Nerillidium* (Nerillidae). Lower wave-swept parts of the beach slope are inhabited by *Saccocirrus* (Saccocirridae) and *Polygordius* (Polygordiidae), the upper parts by *Hesionides* (Hesionidae), *Protodrilus* (Protodrilidae) and *Diurodrilus* (Diurodrilidae), together with one syllid species (it is usually *Eusyllis homocirrata* in the Galápagos beaches). Each of these genera found in the Galápagos Islands is represented by at least one species in an identical transect position within Mediterranean or Indian beaches (Rao, 1972; Westheide, 1972; 1977b; and unpublished data). The majority of these genera also occur in temperate areas, e.g., in beaches of the North Sea.

Nevertheless there are several widely distributed supraspecific taxa which obviously were not able to disperse to the Galápagos Islands. The well known interstitial cnidarian genus *Halammohydra* Remane is lacking as well as the turbellarian taxon Dalyellioidea and the Mystacocarida (Crustacea) (Ax and Schmidt, 1973). Species-rich genera of polyclad platyhelminths like *Stylochus* and *Notoplana* have not yet been recorded from the islands (Sopott-Ehlers and Schmidt, 1975). Within the Polychaeta it was not possible to find the two typical interstitial genera *Protodriloides* (Protodriloidae) and *Trilobodrilus* (Dinophilidae), both occurring on the North American Pacific coast. Blome (1985) mentioned that the common nematode genus *Neochromadora* was not detected within the tremendous number of nematode specimens collected from the Galápagos Islands. McKirdy *et al.* (1976) found no *Batillipes* species within the tardigrades. It is not possible to speculate whether this is due to taxon-specific problems of dispersal or simply a matter of chance.

8. Origin of Species

Almost 95% of the species identified belong to known genera; no suprageneric category had to be created. Evolutionary distance to animals outside the Galápagos is generally small. All investigators point to the fact that most of the species or their direct ancestors must have reached the islands independently. This leads to the important question how they managed to do this.

Interstitial meiofauna organisms in general are not suited for active migration within the free water column. Migration on the sea bottom is hardly probable because unfavorable non-sandy sediments will act as unbrigdeable barriers. Only very few of the meiofauna taxa concerned possess pelagic larvae, e.g., the polychaete genera *Pisione* and *Polygordius*. The nauplius larvae of harpacticoid copepods normally do not leave the sediment. It was concluded, therefore, that widespread or even cosmopolitan distribution patterns of species and genera, as well as their widely consistent pattern of distribution within beaches, generally go back to an old supercontinent that already possessed these faunal elements before continental drift occurred (Rao, 1972; Sterrer, 1973; Westheide, 1977b).

The Galápagos Islands, however, are completely oceanic; a land bridge to the American continent, though often discussed, never existed. Thus their entire littoral meiofauna must have been transported or drifted across the ocean within less than 3 million years of the present.

Meiofauna organisms are regularly found drifting in the water column (e.g., Hagerman and Rieger, 1981; Gerlach, 1977), recruited by passive erosion from sediments with subsequent colonization of bottom areas, or by active entry into the water with subsequent settlement or even active site selection (Palmer, 1988; Armonies, 1988). Calculations of Hagerman and Rieger (1981) of the sinking velocity, however, showed long-distance dispersal across oceans of suspended meiofauna to be rather unlikely. This may also be true for the shortest distance between the American continent and the Galápagos archipelago of 1000 km. Other potential means of meiofauna dispersal listed by Gerlach (1977), e.g., transport by birds, by floating materials, by ballast sand in sailing vessels and in the water attached to suspended sediment during periods of heavy storms have to be considered. Besides many accidental observations of rafting, considerable advances have been made in documenting the great potential for dispersal of benthic meiofauna on drifting material (e.g., Hicks, 1988). This suggests that raftborne individuals may well have colonized beaches of oceanic islands like the Galápagos. Under the light of these new findings, it has to be asked if arrival of meiofauna species is as infrequent and genetic isolation as complete on oceanic islands as is usually believed.

The rafting hypothesis implies that there should be some degree of consistency between the direction of the sea currents passing the islands and the geographic origin of the faunal elements. According to Abbott (1966) the South Equatorial current and the Humboldt current come from the South American west coast, the latter sweeping up the coast to the North before it swings out into the Pacific to pass through the islands. From December to May the warm El Niño from the Central American area displaces the Humboldt current. The influence of the California current may be of less importance. From the open Pacific the islands are reached by the North Equatorial current, the South Equatorial current, and the Equatorial counter current. Currents from the American continent may be considered to have served as transportation vehicles for floating and rafting animals. However, in general one can state that faunistic consistency between the Galápagos and the American Pacific coast line is relatively low. This holds true even if relationships to the Caribbean region are included, which are interpreted as a result of the former connection between the Atlantic and Pacific when the land bridge of Panama was broken up. Mielke (1985) presented evidence that the zoogeographic affinities of copepods between the Panama coast and Galápagos are closer than between the South American coastal areas passed by the Humboldt current. This may be true also for other taxa.

The following discussion contains some of the examples of American west coast affinities to Galápagos meiofauna species.

Of the 15 acoelous turbellarians, *Conaperta colorata* has a distinct sister species relationship to a Pacific North American species, *C. actuosa* Kozloff, whereas most of the closest relations of the other species are known from Europe (Ehlers and Dörjes, 1979). Sopott-Ehlers and Schmidt (1975) stated that some of the Galápagos polyclads may have close affinities to species from the Californian coast. The otoplanid *Philosyrtis sanjuanensis* from the North American Pacific coast is considered to be the sister species of *P. santacruzensis* from Galápagos (Ax, 1977). Another turbellarian, *Nematoplana nigrocapitula* Ax, 1966, from the North American Pacific coast, was also reported from the Galápagos (Ax and Ax, 1974).

Two nemerteans, *Ototyphlonemertes fila* Corrêa and *O. erneba* Corrêa are conspecific to populations from the North and South American east coast (Florida, Brazil); only *O. americana* Gerner was described from the North American Pacific coast (Washington) (Mock and Schmidt, 1975).

The nematodes *Metepsilonema callosum* Lorenzen and *Perepsilonema papulosum* Lorenzen are known from southern Chile (Clasing, 1984). The copepod *Afrolaophonte schmidti* described by Mielke (1981) from the Galápagos, and also found on the Pacific coast of Panama (Mielke, 1982a), is closely related to *Afrolaophonte chilensis* Mielke from the Chilean coast (Mielke, 1985). There are two other Galápagos copepods, *Klienychocamptoides itoi* and *Laophontina triarticulata*, which Mielke (1982a) recorded from the Panamanian Pacific and Atlantic coast, respectively.

Within Polychaeta, the nominate subspecies of *Pisione oerstedi pulla* Westheide was reported from Chile (Castro, 1958). The cosmopolitan nominate subspecies of *Hesionides arenaria pacifica* Westheide is a common representative in various American west coast beaches (see figure 1A in Westheide, 1977b). Rioja (1941) described *Brania heterocirra* from the Mexican Pacific coast, where he also reported *Typosyllis hyalina* Grube. A close relative of *Questa media* Westheide from the Galápagos is *Q. caudicirra* Hartman from the southern Californian coast. One of the *Saccocirrus* species was considered to be identical with *S. sonomacus* Martin from northern California. Unpublished material from South American beaches may well extend this list of Galápagos–American west coast affinities within the Polychaeta. The oligochaete *Aktedrilus locyi* Erséus, 1980, conforms well with the type specimens from California and Oregon (Erséus, 1984).

Populations of the podocopid ostracods *Cobanocythere subterranea*, *C. labiata*, and *C. elongata* were found to be conspecific to populations on the Pacific coast of Central America (El Salvador, Panama). The sister species of *C. labiata* is the Caribbean *C. mielkei* (Gottwald, 1983). *Tourconcha lapidiscola* (Hartmann, 1959) and *Coxoconcha lenticuloides* (Swain & Gilby, 1974), two ostracod species living on green algae, were detected on the Galápagos as well as on the Central American coast and in the Caribbean (Bate *et al.*, 1981; Gottwald, 1983).

Generally, zoogeographic relations to nearly all faunistic areas can be stated [e.g., see discussion of zoogeographic distribution of polychaetes (Westheide, 1974) and podocopid ostracods (Gottwald, 1983)]. Discussions of this kind, however, suffer from incomplete or total lack of meiofauna taxonomic investigations in the Pacific area and the generally limited knowledge of meiofauna species distribution. Thus, for the time being, one can neither be relatively sure of the endemic character of the species described from the Galápagos, nor is one able to discuss sister species relationships with a high degree of accuracy. This is demonstrated by the fact that within polyclads and polychaeates (taxa whose systematic investigation occurred prior to that of taxa completely restricted to meiofauna dimensions) the number of endemic species is considerably lower than within the other groups. Even this low percentage of endemicity may be reduced in the near future, as can be shown by the following examples: *Hesionides unilamellata* Westheide, 1974, from the Galápagos was also identified in beach sands from Hawaii and must be removed from the list of endemic species. Russell (1991) found *Exogone naidinoides* Westheide, 1974, and *Exogone occidentalis* Westheide, 1974, in Belize (Central American Atlantic coast). The degree of endemism certainly reflects sampling activities in a certain taxon.

Two other problems have to be mentioned in this context. (1) Species descriptions, especially older ones, are often less detailed than is necessary for elucidation of small-scale differences in morphological characters. (2) In small-sized, so-called soft-bodied animals, availability of exact biometric data is problematic because of small structural diversity and inconstant dimensions of many characters; thus identification of species differences is more difficult than in hard-bodied taxa, e.g., tardigrades, crustaceans, halacarids.

This latter fact, in particular, takes on an increased importance under the light that the percentage of morphologically highly similar but truly distinct species (= "sibling species") in the meiofauna obviously is much higher than previously expected [e.g., see the polychaete *Microphthalmus–listensis*-species-group (Westheide and Rieger, 1985; Specht and Westheide, 1988)]. This affects also the subspecies problem. In taxonomic procedure, small-scale differences in one or very few characters between populations will often lead to the erection of a subspecies, or, as Ax (1977) defined more precisely, Galápagos platyhelminth populations were considered to be subspecies if differences between them and populations outside the archipelago were smaller than between sympatric continental species. Theoretically, however, subspecies populations should be recognized by their ability to interbreed, a criterion, however, that is practically impossible to prove in the majority of meiofauna organisms. The subspecies label on a Galápagos taxon may thus be nothing more than an indication of relatively slight morphological differences compared to a non-Galápagos population or between geographically separated populations within the archipelago.

9. Meiofauna Speciation in the Galápagos Islands

Meiofaunal species found in the Galápagos can be grouped into the following evolutionary categories (see also Ax, 1977): (1) Species showing virtually no differences compared to populations from other parts of the world. (2) Endemic species and subspecies, the closest relatives of which live outside the archipelago. (3) Groups of two or more congeneric endemic species, which, however, are less related to each other than to species outside the archipelago. This means that their ancestors must have reached the islands independently in separate events not going back to a common stem species from the Galápagos. Obviously the majority of congeneric species belong to this category (see Composition of Taxa). (4) In a few cases which are listed below there is evidence for species multiplication within the Galápagos Islands.

Four new highly similar species of the proseriate platyhelminth family Monocelididae, *Duplominona galapagoensis*, *D. karlingi* (Fig. 9), *D. krameri*, and *D. sieversi*, were shown by Ax (1977) to form a monophyletic group of species that evolved within the Galápagos Islands. *Duplominona* contains 8 species, of which the Galápagos forms present the most evolved members (Fig. 9). The sister species of the Galápagos four-species group is *D. kaneohei* from Hawaii. A synapomorphic feature of the Galápagos group is an enclosed cuticular stylet of the male copulatory organ; the common derived character of *D. kaneohei* with the four Galápagos species is the union of the vagina with the male pore (Fig. 9B). Slight but consistent morphological differences in populations of the typhloplanoid *Haloplanella semicircula* caused Ehlers and Sopott-Ehlers (1989) to erect the three subspecies *H. s. semicircula* (from James), *H. s. santacruzensis* (from Santa Cruz) and *H. s. sancristobalensis* (from San Cristobal). Blome (1985) found the same situation within a nematode species, which he separated into *Hypodontolaimus galapagensis galapagensis* and *H. g. sanctaecrucis*. Although their true subspecies character is not yet proven, these morphologically different allopatric taxa undoubtedly evolved within the archipelago.

Mielke (1979) found evidence for intensive processes of speciation within the islands in the harpacticoid genus *Ectinosoma*. Apparently two *Ectinosoma* species reached the islands independently and gave rise to two different evolutionary lines, each consisting of a series of species or subspecies distributed over the entire archipelago, the systematic arrangement of which, however, could not yet be exactly elucidated. Slight morphological differences between populations of *Phyllopodopsyllus galapagoensis* on Fernandina, Isabela, Santa Cruz and San Cristobal suggest incipient speciation within another harpacticoid taxon (Mielke, 1989).

Ostracods appear to indicate the most obvious examples for actual speciation within the archipelago (Gottwald, 1983). A comprehensive biometric analysis revealed the existence of two forms of *Cobanocythere labiata*, which differ in substrate preference and body length on the different islands.

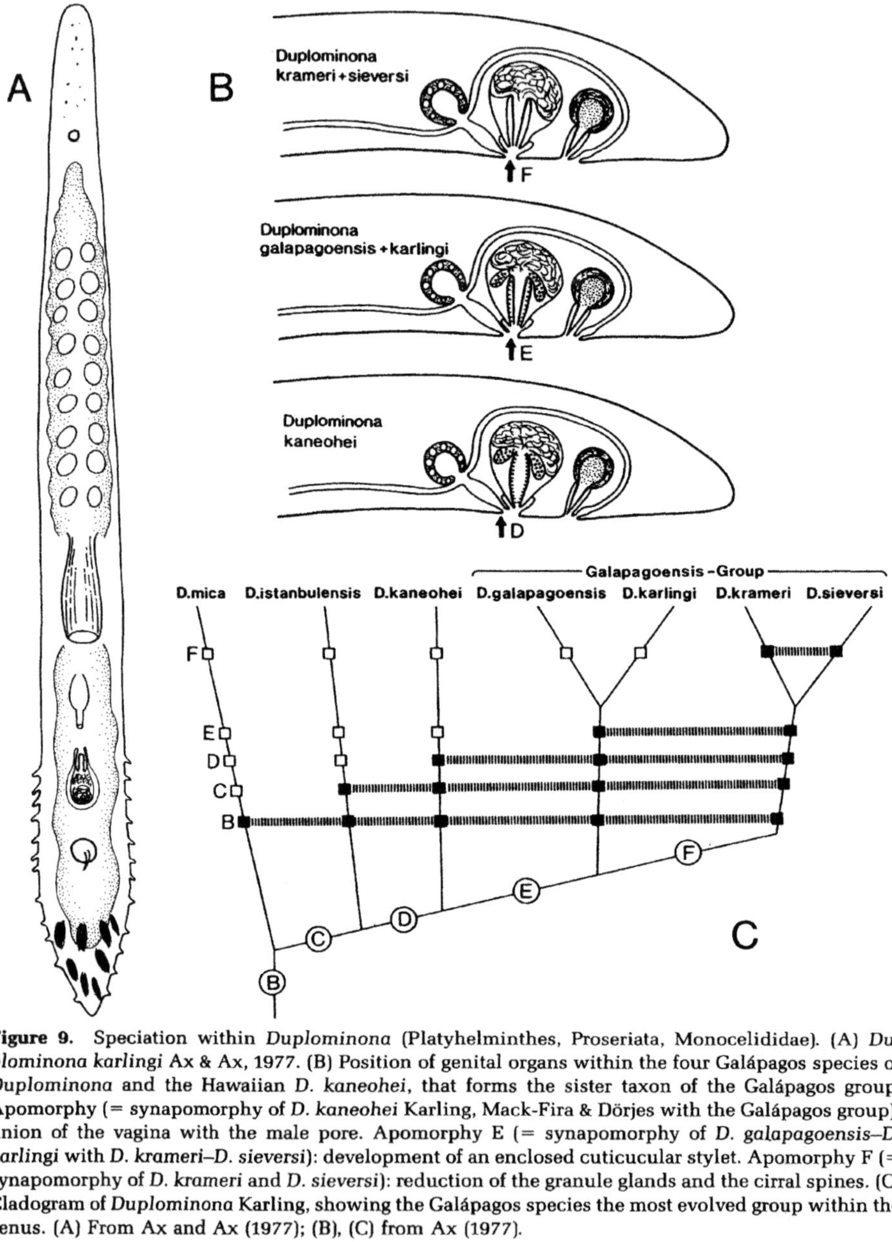

Figure 9. Speciation within *Duplominona* (Platyhelminthes, Proseriata, Monocelididae). (A) *Duplominona karlingi* Ax & Ax, 1977. (B) Position of genital organs within the four Galápagos species of *Duplominona* and the Hawaiian *D. kaneohei*, that forms the sister taxon of the Galápagos group. Apomorphy (= synapomorphy of *D. kaneohei* Karling, Mack-Fira & Dörjes with the Galápagos group): union of the vagina with the male pore. Apomorphy E (= synapomorphy of *D. galapagoensis–D. karlingi* with *D. krameri–D. sieversi*): development of an enclosed cuticucular stylet. Apomorphy F (= synapomorphy of *D. krameri* and *D. sieversi*): reduction of the granule glands and the cirral spines. (C) Cladogram of *Duplominona* Karling, showing the Galápagos species the most evolved group within the genus. (A) From Ax and Ax (1977); (B), (C) from Ax (1977).

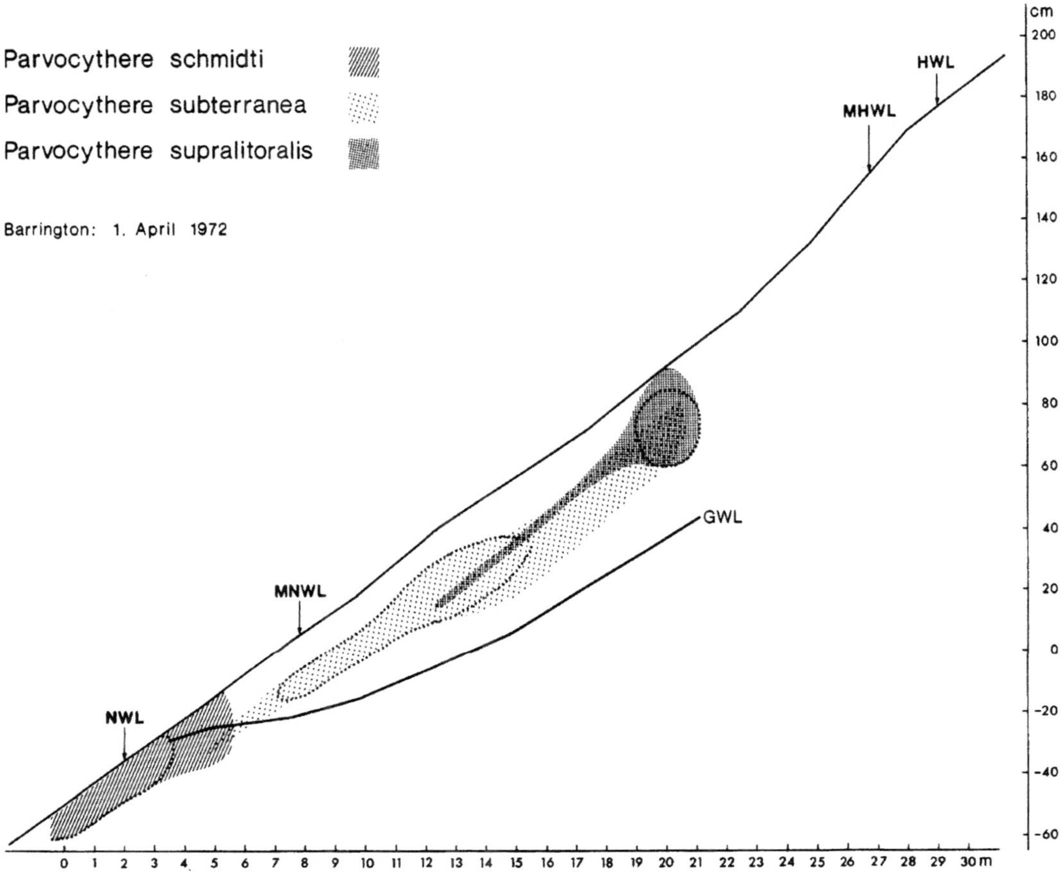

Figure 10. Distribution pattern of 3 species of the ostracod genus *Parvocythere* within the beach of Barrington (April 1, 1972). The dotted lines surround the area where 90% of the individual species were found. From Gottwald (1983). HWL: high water level. NWL: low water level. GWL: ground water level at low tide.

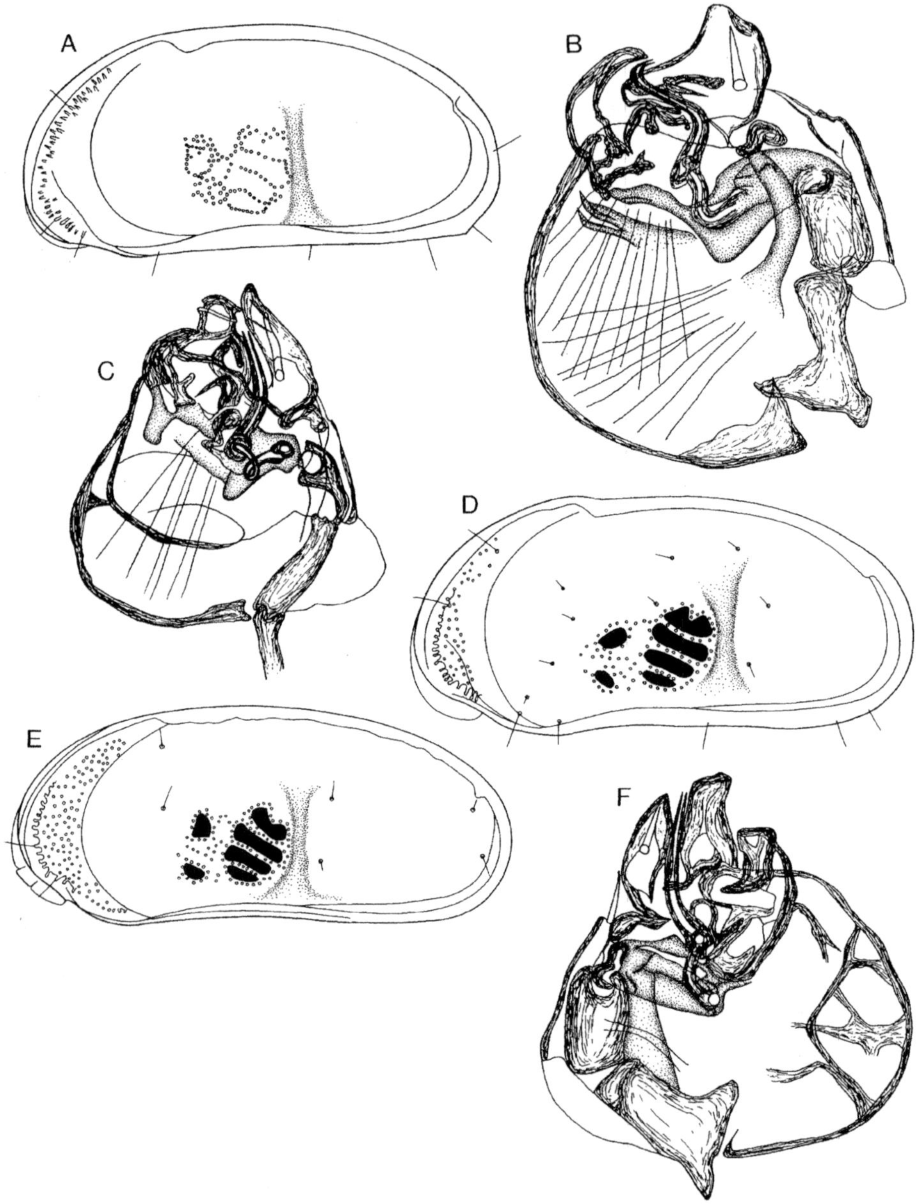

Figure 11. The three most closely related species of the ostracod genus *Parvocythere* in the Galápagos Islands. (A–B) *P. galapagoensis* (A) Left shell, female. (B) Copulatory organ. (C–D) *P. schmidti*. (C). Copulatory organ. (D) Left shell, female, (E–F) *P. subterranea* (E) Left shell, female. (F) Copulatory organ. From Gottwald (1983).

Not less than six species of the *Cobanocythere lanceolata* group are considered to have arisen within the archipelago, their species character being proven by the fact that five of them occur together in one locality without showing any overlap of morphological characters.

There is also much evidence that speciation of *Parvocythere galapagoensis* Gottwald, *P. schmidti* Gottwald, and *P. subterranea* Gottwald occurred within the archipelago, although a synapomorphic character for the three species could not be found. They were sampled on different islands (Fig. 11), but also occur together in two beaches occupying different areas of the beach slope (Fig. 10).

The examples of speciation presented above suggest that a period of about 3 million years (the geological age of the islands), is sufficient for recognizable evolutionary processes in meiofauna organisms. This suggests a relatively rapid rate of speciation on the islands compared to the high degree of meiofaunal species similarity between continents. For instance, the rate of speciation in the many amphiatlantic species and transallopatric meiofauna species pairs on both sides of the Atlantic (Sterrer, 1973; Ruppert, 1977; Westheide and Rieger, 1987; Westheide, 1977b; 1987; Ax and Armonies, 1987) would be comparable only in the case that a supposed "Thule land bridge" broke up not earlier than 2 million years ago—which indeed is believed by a few geologists. Species similarity between other continents, however, generally suggests low rates of speciation within continental beaches considering the long periods of separation from an old supercontinent.

More rapid speciation in Galápagos meiofauna may be due to the fact that the first colonists found empty beaches with nearly the entire spectrum of sand beach niches and resources favoring small-scale evolutionary processes. The crucial difference between potential sand beach colonists and colonists of the terrestrial realm might have been that the latter found only a very small number of suitable habitats allowing survival of only a few forms. Most of the psammobiontic meiofauna species however reaching the islands accidentally, certainly might have found their appropriate habitats. Thus, the beaches were soon filled with a species-rich community resulting in a considerably shorter initial phase of colonization that generally did not allow conspicuous adaptive radiation and speciation. Better "survival during rafting" and better "survival after landing" compared to the terrestrial fauna may be the reason for high species-richness of the Galápagos meiofauna on the one hand and their relatively little evolutionary divergence on the other.

References

Abbott, D. P., 1966, Factors influencing the zoogeographic affinities of the Galápagos inshore marine fauna, in: *The Galápagos* (R. Bowman, ed.) p. 108–122.

Armonies, W., 1988, Active emergence of meiofauna from intertidal sediment, *Mar. Ecol. Prog. Ser.* **43**:151–159.

Augener, H., 1933, Die Polychaeten von den Galapagos-Inseln. The Norwegian Zoological Expedition to the Galápagos Islands 1925, conducted by Alf Wollebaeck, VI, *Nyt. Mag. Naturvidensk* (Oslo) **73**:55–66.

Ax, P., 1969, Populationsdynamik, Lebenszyklen und Fortpflanzungsbiologie der Mikrofauna des Meeressandes, *Verh. Dt. Zool. Ges.* (1968) **32**:66–113.

Ax, P., 1977, Problems of speciation in the interstitial fauna of Galápagos, in: The Meiofauna Species in Time and Space (W. Sterrer and P. Ax, eds.), *Mikrofauna Meeresboden* **61**:29–43.

Ax, P., and Armonies, W., 1987, Amphiatlantic identities in the composition of the boreal brackish water community of Plathelminthes. A comparison between the Canadian and European Atlantic coast, *Microfauna Mar.* **3**:7–80.

Ax, P. and Ax, R., 1974a, Interstitielle Fauna von Galapagos. V. Otoplanidae (Turbellaria, Proseriata), *Mikrofauna Meeresboden* **27**:1–28.

Ax, P., and Ax, R., 1974b, Interstitielle Fauna von Galapagos. VII. Nematoplanidae, Polystyliphoridae, Coelogynoporidae (Turbellaria, Proseriata), *Mikrofauna Meeresboden* **29**:1–28.

Ax, P., and Ax, R., 1977, Interstitielle Fauna von Galapagos. XIX. Monocelididae (Turbellaria, Proseriata), *Mikrofauna Meeresboden* **64**:1–44.

Ax, P., and Ehlers, U., 1973, Interstitielle Fauna von Galapagos. III. Promesostominae (Turbellaria, Typhloplanoida), *Mikrofauna Meeresboden* **34**:1–16.

Ax, P., and Schmidt, P., 1973, Interstitielle Fauna von Galapagos. I. Einführung, *Mikrofauna Meeresboden* **20**:1–38.

Bailey, J. H., and Harris, M. P., 1968, Spirorbinae (Polychaeta: Serpulidae) of the Galápagos Islands, *J. Zool.* (Lond.) **155**:161–184.

Bartsch, I., 1977, Interstitielle Fauna von Galapagos. XX. Halacaridae (Acari), *Mikrofauna Meeresboden* **65**:1–108.

Bartsch, J., and Schmidt, P., 1978, Interstitielle Fauna von Galapagos. XXII. Zur Ökologie der Halacaridae (Acari), *Mikrofauna Meeresboden* **69**:1–38.

Bate, R. H., Whittaker, J. E. and Mayes, C. A., 1981, Marine Ostracoda of the Galápagos Islands and Ecuador, *Zool. J. Linn. Soc.* **73**:1–79.

Blome, D., 1985, Interstitielle Fauna von Galapagos. XXXV. Chromadoridae (Nematoda), *Microfauna Mar.* **2**:271–329.

Bock, S., 1923, *Boninia*, a new polyclad genus from the Pacific, *Nov. Act. Reg. Soc.* (Uppsala), ser. **4**, **6**:1–32.

Castro, C., 1958, Estudios sobre poliquetos chilenos *Pisione oerstedi* Gr., *Rev. Biol. Mar.* (Valparaiso) **7**:183–187.

Clasing, E., 1984, Interstitielle Fauna von Galapagos XXXII. Epsilonematidae (Nematodes), *Microfauna Mar.* **1**:149–190.

Coineau, N., and Schmidt, P., 1979, Interstitielle Fauna von Galapagos. XXIV. Microparasellidae (Isopoda, Asellota), *Mikrofauna Meeresboden* **73**:1–19.

Ehlers, B., and Ehlers, U., 1973, Interstitielle Fauna von Galapagos. II. Gnathostomulida, *Mikrofauna Meeresboden* **22**:1–27.

Ehlers, U., 1973, Zur Populationsstruktur interstitieller Typhloplanoida und Dalyellioida (Turbellaria, Neorhabdocoela), *Mikrofauna Meeresboden* **19**:1–105.

Ehlers, U., and Ax, P., 1974, Interstitielle Fauna von Galapagos. VIII. Trigonostominae (Turbellaria, Typhloplanoida), *Mikrofauna Meeresboden* **30**:1–33.

Ehlers, U., and Dörjes, J., 1979, Interstitielle Fauna von Galapagos. XXIII. Acoela (Turbellaria), *Mikrofauna Meeresboden* **72**:1–75.

Ehlers, U., and Ehlers, B., 1981, Interstitielle Fauna von Galapagos. XXVII. Byrsophlebidae, Promesostomidae, Brinkmanniellinae, Kytorhynchidae (Turbellaria, Typhloplanoida), *Mikrofauna Meeresboden* **83**:1–35.

Ehlers, U., and Sopott-Ehlers, B., 1989, Interstitielle Fauna von Galapagos, XXXVIII. *Haloplanella* Luther und *Pratoplana* Ax (Typhloplanoida, Plathelminthes), *Microfauna Mar.* **5**:191–198.

Erséus, C., 1984, Interstitielle Fauna von Galapagos XXXIII. Tubificidae (Annelida, Oligochaeta), *Microfauna Mar.* **1**: 191–198.

Gerlach, S. A., 1977, Means of meiofauna dispersal, in: The Meiofauna Species in Time and Space (W. Sterrer and P. Ax, eds.), *Mikrofauna Meeresboden* **61**:89–103.

Gottwald, J., 1983, Interstitielle Fauna von Galapagos. XXX. Podocopida 1 (Ostracoda), *Mikrofauna Meeresboden* **90**:1–187.

Hagerman, G. M., and Rieger, R. M., 1981, Dispersal of benthic meiofauna by wave and current action in Bogue Sound, N.C., USA, *P.S.Z.N. Mar. Ecol.* **2**:245–270.

Hartman, O., 1939, The polychaetous annelids collected by the Presidential cruise of 1938, *Smithson. Misc. Coll.* **98**:1–22.

Hartwig, E., 1973, Die Ciliaten des Gezeiten-Sandstrandes der Nordseeinsel Sylt. II. Ökologie, *Mikrofauna Meeresboden* **21**:1–171.

Hicks, G. R. G., 1988, Sediment rafting: a novel mechanism for the small-scale dispersal of intertidal estuarine meiofauna, *Mar. Ecol. Prog. Ser.* **48**:69–80.
Houvenaghel, G. T., and Houvenaghel, N., 1974, Aspects écologiques de la zonation intertidale sur les côtes rocheuses des îles Galapagos, *Mar. Biol.* **26**:135–152.
Hoxhold, S., 1974, Zur Populationsstruktur und Abundanzdynamik interstitieller Kalyptorhynchia (Turbellaria, Neorhabdocoela), *Mikrofauna Meeresboden* **41**:1–134.
Hummon, W. D., Fleeger, J. W., and Hummon, M. R., 1976, Meiofauna–macrofauna interactions. I. Sand beach meiofauna affected by maturing *Limulus* eggs, *Chesapeake Sci.* **17**:297–299.
Hyman, L. H., 1939, Polyclad worms collected on the Presidential cruise of 1938, *Smithson. Misc. Coll.* **98**:1–13.
Hyman, L. H., 1953, Some polyclad flatworms from the Galápagos Islands, *Allan Hancock Pacific Exped.* **15**:183–210.
Kinberg, J. G. H., 1865, Annulata nova, *Öfv. Vet. Akad. Stockholm Förh.* **22**:167–179.
McKirdy, D., Schmidt, P., and McGinty-Bayly, M., 1976, Interstitielle Fauna von Galapagos. XVI. Tardigrada, *Mikrofauna Meeresboden* **68**:1–43.
Meineke, T., and Westheide, W., 1979, Gezeitenabhängige Wanderungen der Interstitialfauna in einem Sandstrand der Insel Sylt (Nordsee), *Mikrofauna Meeresboden* **75**:1–36.
Mielke, W., 1979, Interstitielle Fauna von Galapagos. XXV. Longipediidae, Canuellidae, Ectinosomatidae (Harpacticoida), *Mikrofauna Meeresboden* **77**;1–107.
Mielke, W., 1981, Interstitielle Fauna von Galapagos XXVIII. Laophontinae (Laophontidae), Ancorabolidae (Harpacticoida), *Mikrofauna Meeresboden* **84**:1–106.
Mielke, W., 1982a, Einige Laophontidae (Copepoda, Harpacticoida) von Panamá, *Crustaceana* **42**:1–11.
Mielke, W., 1982b, Interstitielle Fauna von Galapagos. XXIX. Darcythompsoniidea, Cylindropsyllidae (Harpacticoida). *Mikrofauna Meeresboden* **87**:1–52.
Mielke, W., 1984, Interstitielle Fauna von Galapagos XXXI. Paramesochridae (Harpacticoida), *Microfauna Mar.* **1**:63–148.
Mielke, W., 1985, Interstitielle Copepoda aus dem zentralen Landesteil von Chile: Cylindropsyllidae, Laophontidae, Ancorabolidae (Harpacticoida), *Microfauna Mar.* **2**:181–270.
Mielke, W., 1989a, Interstitielle Fauna von Galapagos, XXXVI. Tetragonicipitidae (Harpacticoida), *Microfauna Mar.* **5**:95–172.
Mielke, W., 1989b, Interstitielle Fauna von Galapagos, XXXVII. Metidae (Harpacticoida), *Microfauna Mar.* **5**:173–188.
Mock, H., and Schmidt, P., 1975, Interstitiella Fauna von Galapagos. XIII. *Ototyphlonemertes* Diesing (Nemertini, Hoplonemertini), *Mikrofauna Meeresboden* **51**:1–40.
Monro, C. C. A., 1933, The Polychaeta Errantia collected by Dr. C. Crossland at Colon, in the Panama region and the Galápagos Islands during the expedition of the S.Y. "St. George", *Proc. Zool. Soc. Lond.* **pt. 1**:1–96.
Noldt, U., and Hoxhold, S., 1984, Interstitielle Fauna von Galapagos. XXXIV. Schizorhynchia (Plathelminthes, Kalyptorhynchia), *Microfauna Mar.* **1**:199–256.
Palmer, M. A., 1988, Dispersal of marine meiofauna: a review and conceptual model explaining passive transport and active emergence with implications for recruitment, *Mar. Ecol. Prog. Ser.* *48*: 81–91.
Pokorny, V., 1970, The genus *Radimella* Pokorny, 1969 (Ostracoda, Crustacea) in the Galapagos Islands, *Acta Univ. Carolinae, Geologica* (1969) **4**:293–334.
Pokorny, V., 1972, The genus *Caudites* Corgell & Fields, 1937 (Ostracoda, Crustacea) in the Galapagos Islands, *Acta Univ. Carolinae, Geologica* (1970) **4**:267–302.
Rao, G. Ch, 1972, On the geographical distribution of interstitial fauna of marine beach sand, *Proc. Ind. Nat. Sci. Acad.* **38B**:164–178.
Renaud-Debyser, J., 1963, Recherches écologiques sur la fauna interstitielle des sables (Bassin d'Archachon, île de Bimini, Bahamas), *Vie Milieu* **15**(suppl.):1–157.
Rioja, E., 1941, Estudios anelidologicos III. Datos para el conocimiento de la fauna de poliquetos de las costas del Pacifico de Mexico, *An. Inst. Biol. Mex.* **12**:669–746.
Ruppert, E. E., 1977, Zoogeography and speciation in marine Gastrotricha, in: The Meiofauna Species in Time and Space (W. Sterrer and P. Ax, eds.), *Mikrofauna Meeresboden* **61**:231–251.
Russell, D. E., 1991, Exogoninae (Polychaeta: Syllidae) from the Belizean Barrier Reef with a key to species of *Sphaerosyllis*, *J. Nat. Hist.* **25**:49–74.
Schmidt, P., 1968, Die quantitative Verteilung und Populationsdynamik des Mesopsammons am

Gezeiten-Sandstrand der Nordseeinsel Sylt. I. Faktorengefüge und biologische Gliederung des Lebensraumes, *Int. Revue ges. Hydrobiol.* **53**:723–779.

Schmidt, P., 1969, Die quantitative Verteilung und Populationsdynamik des Mesopsammons am Gezeiten-Sandstrand der Nordseeinsel Sylt. II. Quantitative Verteilung und Populationsdynamik einzelner Arten, *Int. Revue ges. Hydrobiol.* **54**:95–174.

Schmidt, P., 1972a, Zonierung und jahreszeitliche Fluktuationen des Mesopsammons im Sandstrand von Schilksee (Kieler Bucht), *Mikrofauna Meeresboden* **10**:1–60.

Schmidt, P., 1972b, Zonierung und jahreszeitliche Fluktuationen der interstitiellen Fauna in Sandstränden des Gebiets von Tromso (Norwegen), *Mikrofauna Meeresboden* **12**:1–86.

Schmidt, P., 1974a, Interstitielle Fauna von Galapagos. IV. Gastrotricha, *Mikrofauna Meeresboden* **26**: 1–76.

Schmidt, P., 1974b, Interstitielle Fauna von Galapagos. X. Kinorhyncha, *Mikrofauna Meeresboden* **43**: 1–15.

Schmidt, P., 1978, Interstitielle Fauna von Galapagos. XXI. Lebensraum, Umweltfaktoren, Gesamtfauna, *Mikrofauna Meeresboden* **68**:1–52.

Schmidt, P., and Sopott-Ehlers, B., 1976, Interstitielle Fauna von Galapagos. XV. *Macrostomum* O. Schmidt, 1848 und *Siccomacrostomum triviale* nov. gen. nov. spec. (Turbellaria, Macrostomida), *Mikrofauna Meeresboden* **57**:1–45.

Schmidt, P., and Westheide, W., 1977, Interstitielle Fauna von Galapagos. XVII. Polygordiidae, Saccocrridae, Protodrilidae, Nerillidae, Dinophilidae (Polychaeta), *Mikrofauna Meeresboden* **62**:1–38.

Schuster, R. O., and Grigarick, A. A., 1966, Tardigrada from the Galápagos and Cocos Islands, *Proc. Calif. Acad. Sci.* (4th ser.) **34**:315–328.

Sopott, B., 1973, Jahreszeitliche Verteilung und Lebenszyklen der Proseriata (Turbellaria) eines Sandstrandes der Nordseeinsel Sylt, *Mikrofauna Meeresboden* **15**:1–106.

Sopott-Ehlers, B., and Schmidt, P., 1974a, Interstitielle Fauna von Galapagos. IX. Dolichomacrostomidae (Turbellaria, Macrostomida), *Mikrofauna Meeresboden* **34**:1–20.

Sopott-Ehlers, B., and Schmidt, P., 1974b, Interstitielle Fauna von Galapagos. XII. *Myozona* Marcus (Trubellaria, Macrostomida), *Mikrofauna Meeresboden* **46**:1–19.

Sopott-Ehlers, B., and Schmidt, P., 1975, Interstitielle Fauna von Galapagos. IXV. Polycladida (Turbellaria), *Mikrofauna Marina* **54**:1–32.

Specht, A., and Westheide, W., 1988, Intra- and interspecific ultrastructural character variation: The chaetation of the *Microphthalmus-listensis*-species-group (Polychaeta: Hesionidae) *Zoomorphology* **107**:371–376.

Sterrer, W., 1973, Plate tectonics as a mechanism for dispersal and speciation in interstitial sand fauna, *Neth. J. Sea Res.* **7**:200–220.

Stummer-Traunfels, R. von, 1933, Polycladida, in: Klassen und Ordnungen des Tierreichs (H. G. Bronn, ed.) 4, Abt. 1c, Liefg. **179**:3485–3596.

Teuchert, G., 1968, Zur Fortpflanzung und Entwicklung der Macrodasyoidea (Gastrotricha), *Z. Morph. Tiere* **63**:343–418.

Triebel, E., 1956, Brackwasser-Ostracoden von den Galapagos-Inseln, *Senck. Biol.* **37**:447–476.

Westheide, W., 1967, Monographie der Gattungen *Hesionides* Friedrich und *Microphthalmus* Mecznikow (Polychaeta, Hesionidae). Ein Beitrag zur Organisation und Biologie psammobionter Polychaeten, *Z. Morph. Tiere* **61**:1–159.

Westheide, W., 1968, Zur quantitativen Verteilung von Bakterien und Hefen in einem Gezeitenstrand der Nordseeküste, *Mar. Biol.* **1**:336–347.

Westheide, W., 1972, La faune des Polychètes et des Archiannélides dans les plages sableuses à versac de la côte méditerranéenne de la Tunisie, *Bull. Inst. Oceanogr. Pêche, Salammbô* **2**:449–468.

Westheide, W., 1974, Interstitielle Fauna von Galapagos. XI. Pisionidae, Hesionidae, Pilargidae, Syllidae (Polychaeta), *Mikrofauna Meeresboden* **44**:1–146.

Westheide, W., 1977a, Interstitielle Fauna von Galapagos. XVIII. Nereidae, Eunicidae, Dorvilleidae (Polychaeta), *Mikrofauna Meeresboden* **63**:1–40.

Westheide, W., 1977b, The geographical distribution of interstitial polychaetes, in: The Meiofauna Species in Time and Space (W. Sterrer and P. Ax, eds.), *Mikrofauna Meeresboden* **61**:287–302.

Westheide, W., 1981, Interstitielle Fauna von Galapagos. XXVI. Questidae, Cirratulidae, Acrocirridae, Ctenodrilidae (Polychaeta), *Mikrofauna Meeresboden* **82**:1–24.

Westheide, W., 1987, The interstitial polychaete *Hesionides pettiboneae* n. sp. (Hesionidae) from the U.S. east coast and its transatlantic relationships, *Bull. Biol. Soc. Wash.* **7**:131–139.

Westheide, W., and Rieger, R. M., 1987, Systematics of the amphiatlantic *Microphthalmus listensis*-species-group (Polychaeta: Hesionidae): Facts and concepts for reconstruction of phylogeny and speciation, *Z. f. zool. Syst. Evolutionsforsch.* **25**:12–39.

Westheide, W., and Schmidt, P., 1974, Interstitielle Fauna von Galapagos. VI. *Aeolosoma maritimum dubiosum* nov. ssp. (Annelida, Oligochaeta), *Mikrofauna Meeresboden* **28**:1–10.

Woodworth, W., McM, 1894, Reports on the dredging operations off the west coast of central America to the Galápagos, to the west coast of Mexico, and in the Golf of California, in charge of Alexander Agassiz, carried on by the U.S. Fish Commission steamer "Albatros" during 1891. IX. Report on the Turbellaria, *Bull. Mus. Comp. Zool.* (Harvard College) **25**:49–52.

Chapter 3

The Polychaete Fauna of the Galápagos Islands

JAMES A. BLAKE

1. Introduction

In general, the polychaete fauna of the Galápagos Islands is poorly known. In summarizing the species that have been recorded from the islands since the middle of the nineteenth century several facts are obvious. First, collections have mostly been random or concentrated in special habitats or substrata. Second, most papers that have described species of polychaetes from the Galápagos Islands have done so without any effort to synthesize or interpret the records with available biogeographic data. Third, based upon a review of the available records, a misleading picture of the fauna emerges because families that contain numerous species on other coastlines, have rarely been recorded from the Galápagos. The lack of records of these families suggests that some habitats have not been adequately sampled.

Most of the early records have not been verified with subsequent examination of specimens and when new collections have been examined a fauna with its share of endemic species is present. In reviewing the published records of Galápagos polychaetes, a composite list has been compiled. However, because I have not examined the specimens that earlier investigators described, there is no way that the identifications can be confirmed. In those cases where subsequent workers have reexamined earlier collections, their revisions have been incorporated. I have, therefore, attempted to update the nomenclature as best as I could given the lack of direct observation of material. In those cases where former names are obviously obsolete, they have been replaced. There are cases, however, where I have left the older designation because we really have no idea what species the original investi-

JAMES A. BLAKE • Science Applications International Corporation, Woods Hole, Massachusetts 02543.

gator actually examined, given recent revisionary work. Pelagic polychaetes are not included, whereas the archiannelid families are included.

2. History of Polychaete Collections in the Galápagos Islands

The earliest records of polychaetes from the Galápagos Islands were published by Kinberg (1865, 1866). These specimens were collected by the *Eugenie* expedition, and published in a series of papers under the title of "Annulata Nova" by Kinberg. Kinberg's collection of polychaete types was reexamined by Hartman (1948). Of 34 species considered by Hartman to be valid, only three were from the Galápagos.

Chamberlin (1919) reported on the results of the *Albatross* expeditions to the Pacific. Sixteen different species were reported from the vicinity of the Galápagos, most of which were either deep-water benthic species or pelagic polychaetes. A few polychaetes from the Galápagos were reported by Treadwell (1928) and by Augener (1933) as part of collections taken by the *Arcturus* and Norwegian Zoological Expeditions, respectively.

The most important works on the polychaete fauna of Panama and the Galápagos were by Monro. He published three papers that described 58 species of polychaetes from the Galápagos Islands (Monro, 1928, 1933a, 1933b). Most of these species were collected by C. Crossland.

The Allan Hancock Foundation undertook four expeditions to the Galápagos between 1933 and 1938. The polychaetes were reported by Olga Hartman as part of her series of monographs on polychaetes from the tropical eastern Pacific. The stations that were sampled and the polychaetes that were identified as part of the Allan Hancock Expeditions are listed in the Appendix. Additional species were identified by Dr. Hartman, but have remained unpublished. The published and unpublished materials are lodged in the Hancock collections that have recently been transferred to the Los Angeles County Museum of Natural History.

The R/V *Anton Bruun* stopped in the Galápagos Islands as part of the Southeastern Pacific Biological and Oceanographic Program, but except for isolated records published by Blake (1983, 1986), very little of this material has been published. Some of these collections are accessioned with the National Museum of Natural History, Smithsonian Institution, in Washington, D.C. M. P. Harris collected spirorbids throughout the Galápagos between November 1965 and February 1967. These records were published by Bailey and Harris (1968). An extensive survey of the interstitial fauna of the Galápagos Islands was conducted by the University of Göttingen (1972–1973). The polychaetes have been partially described by Westheide (1974, 1977, 1981; see also Chapter 2, this volume). The archiannelid families were described by Schmidt and Westheide (1977).

3. Results

A total of 192 species have been recorded from the Galápagos Islands. These species are distributed in 35 families. The Syllidae are the best represented with 40 species. The Polynoidae has 16 species, the Eunicidae has 15 species, the Spirorbidae has 13 species, the Nereididae has 10 species, and the Cirratulidae has seven species. The remaining families are each represented by fewer than seven species. It is remarkable that families that are dominant on other coastlines are so poorly represented. For example, the Spionidae are typically a dominant polychaete family of infaunal benthic communities and are only represented by 4 species. The Paraonidae are not yet recorded from the fauna and the Capitellidae, Sabellidae, and Ampharetidae are each only represented by a single species.

In the list of species that follows, the collections appear to have mostly come from intertidal and shallow subtidal collections of sediment, rocks, algae, and coral. Some species recorded by Nybakken (1978) were from the stomach contents of *Conus* spp. and indicate that additional species are available to these predators that have not been taken in casual collecting.

A summary of the polychaete families recorded from the Galápagos Islands along with a count of total species and endemic species are presented in Table 1.

4. List of Species with Reference Citations

Family Phyllodocidae

Eulalia personata Gravier, 1907.—Augener, 1933:60–61. Floreana Island, Post Office Bay, intertidal on beach.

Eumida sanguinea (Oersted, 1843).—Monro, 1933a:21. James Island, James Bay, 30 m, clean sand and weed.

Genetyllis castanea (Marenzeller, 1879).—Monro, 1933a:21. Isabella Island, Tagus Cove, intertidal.

Phyllodoce madeirensis Langerhans, 1880.—Monro, 1933a:21–22. James Island, James Bay, 27 m, clean sand with weed; Isabella Island, Tagus Cove, intertidal; Gorgona Island, from coral and dredged in 27 m; Taboga Island, from floats.—As *Phyllodoce varia* Treadwell, 1928:467, fig. 179, figs. 69–71 (*fide* Hartman, 1956).

Additional records: "Phyllodocidae" recorded as prey of *Conus diadema* by Nybakken (1978).

Family Glyceridae

?*Glycera oxycephala* Ehlers, 1887.—Hartman, 1940:248, Pl. 37, figs. 74–75, Pl. 43, figs. 122–124; Pl. 44, fig. 125. AHPE* Sta. 786-38.

Family Goniadidae

Glycinde armigera Moore, 1911.—Hartman, 1950:49–51, Pl. 6, figs. 1–12.—As *Glycinde multidens*.—Hartman, 1940:249–250, Pl. 126-131. AHPE Sta. 173-34. (not Müller, 1858, *fide* Hartman, 1950).

Family Nephtyidae

Aglaophamus dicirrus Hartman, 1950:122–124, Pl. 18, figs, 1–8.—As *Nephtys dibranchis*.—Hartman, 1940:237. AHPE Sta. 185-34, 187-34 (not Grube, 1877, *fide* Hartman, 1950).

*AHPE, Allan Hancock Pacific Expeditions.

Table 1. Summary of Number of Species in Each Family of Polychaetes Recorded from the Galápagos Islands

Family	No. of species	No. of endemic species	Comments
Phyllodocidae	4	0	Undersampled
Glyceridae	1	0	–
Goniadidae	1	0	–
Nephtyidae	1	0	Undersampled
Hesionidae	6	3	–
Pilargidae	2	0	–
Syllidae	40	13	Highly diverse
Nereididae	10	3	–
Aphroditidae	1	indeterminate	–
Polynoidae	16	3	–
Acoetidae	1	0	–
Sigalionidae	5	0	–
Chrysopetalidae	3	0	–
Pisionidae	2	2	–
Amphinomidae	4	0	–
Onuphidae	3	1	Undersampled
Eunicidae	15	1	–
Lumbrineridae	6	0	–
Arabellidae	2	1	–
Dorvilleidae	4	1	Undersampled
Orbiniidae	4	1	Undersampled
Spionidae	4	2	Undersampled
Chaetopteridae	2	0	–
Questidae	1	1	–
Cirratulidae	7	1	Undersampled
Ctenodrilidae	1	0	–
Acrocirridae	1	1	–
Flabelligeridae	3	1	–
Capitellidae	1	0	Undersampled
Maldanidae	1	0	Undersampled
Sabellariidae	2	0	Undersampled
Ampharetidae	1	indeterminate	Undersampled
Terebellidae	7	0	Undersampled
Sabellidae	1	0	Undersampled
Serpulidae	5	1	–
Spirorbidae	13	8	–
Polygordiidae	1	1	–
Saccocirridae	2	0	–
Protodrilidae	3	1	–
Nerillidae	3	3	All endemic
Dinophilidae	2	1	–
Totals	192	50	–

Family Hesionidae

Hesione intertexta Grube, 1878.—Hartman, 1940:212, Pl. 33, figs. 30–31. AHPE Sta. 59–33, 148–34, 167–34, 169–34, 310–34, 336–34, 811–38.

Hesione splendida Savigny, 1818.—Monro, 1933a:25–25. Isabella Island, Tagus Cove, intertidal.

Hesionides arenaria pacifica Westheide, 1974:25–27, figs. 8–9. James Island;

Bartholomew Island; Santa Cruz Island, Santa Cruz North, Playa Borrero, Bahia Borrero, Puerto Nunez, Academy Bay; Barrington Island; Floreana Island; San Cristobal, Wreck Bay, Punta Bassa; Hood Island, Gardner Bay. Intertidal in sand.

Hesionides unilamellata Westheide, 1974:27–31, figs. 10–11. Fernandina Island, Isabela Island, Marchena Island, Tower Island, James Island, Bartholomew Island, Santa Cruz Island, Baltra Island, Barrington Island, Floreana Island, San Cristobal Island, and Hood Island. Intertidal beaches.

Microphthalmus indefatigatus Westheide, 1974:21–25, fig. 7. Santa Cruz Island, Academy Bay. Intertidal to shallow subtidal in sediment with shells.

Podarke sp. Westheide, 1974:20–21, fig. 6. Santa Cruz Island, Academy Bay, north side, shallow subtidal in coarse shell sediment.

Family Pilargidae

Ancistrosyllis sp. Hartman, 1947: 504–505. AHPE, Sta. 156-34, Fragments only.

Synelmis albini (Langerhans, 1881).—Westheide, 1974:31–34, figs. 12–13. Santa Cruz Academy Bay, Santa Cruz Island, Academy Bay, subtidal in sediments rich in detritus.

Family Syllidae
Subfamily Syllinae

Branchiosyllis exilis (Gravier, 1900).—Westheide, 1974:60–64, fig. 26. Santa Cruz Island, Academy Bay. Intertidal, under rocks.

Ehlersia rosea magna Westheide, 1974:41–45, figs. 17–18. Santa Cruz Island, Academy Bay. Intertidal in sediments with detritus.

Ehlersia sp. Westheide, 1974:45–47, fig. 19, Santa Cruz Island, Academy Bay. Shallow subtidal.

Haplosyllis spongicola (Grube, 1855).—Westheide, 1974:35–37, fig. 14. Santa Cruz Island, Academy Bay. Intertidal, under rocks and in old coral.

Opisthosylis corallicola Hartmann-Schröder, 1965.—Westheide, 1974:37–39, fig. 15, Santa Cruz Island, Academy Bay. Intertidal, under rocks.

Parasphaerosyllis indica Monro, 1937.—Westheide, 1974:64–67, figs. 27–29, Santa Island, Academy Bay. In old coral.

Syllis longissima Gravier, 1900.—As *Syllis pacifica* Ehlers: Treadwell, 1928: 465 (*fide* Hartman, 1959).

Syllis fuscosuturata Augener, 1922.—Monro, 1933a:32–34, fig. 14. Isabella Island, Tagus Cove, intertidal (*fide* Fauchald, 1977).

Trypanosyllis gigantea (McIntosh, 1885).—As *Trypanosyllis taeniaeformis* (Haswell).—Monro, 1933a:35–36. Charles Island, Onslow Island, 7.3 m, from a stone.—Westheide, 1974:39–41, fig. 16. Santa Cruz Island, Academy Bay. In old coral.

Trypanosyllis vittigera Ehlers, 1887.—Treadwell, 1928:465.

Typosyllis caeca (Monro, 1933).—As *Syllis caeca* Monro, 1933a:30–32. Isabella Island, Tagus Cove, from rock shelf on south side (*fide* Fauchald, 1977).

Typosyllis corallicola Verrill, 1900.—Treadwell, 1928:466

Typosyllis glarearia Westheide, 1974:55–58, figs. 23–34. Santa Cruz Island, Academy Bay. Intertidal to shallow subtidal in sediment with detritus and shell.

Typosyllis hyalina (Grube, 1863).—Treadwell, 1928:466.—Westheide, 1974:47–51, fig. 20. Santa Cruz Island, Academy Bay. Subtidal to 3 m, in sediments with detritus and shell and in old coral.

Typosyllis variegata (Grube, 1860).—Westheide, 1974:51–54, figs. 21–22. Santa Cruz Island, Academy Bay. In old coral in the surf zone.—As *Syllis variegata* Grube. Monro, 1933a: 28–29. Isabella Island; James Island, Taboga island.

Typosyllis zonata (Haswell, 1886).—As *Syllis zonata* (Haswell): Monro, 1933a:29–30. James Island, James Bay, from weeds.

Typosyllis sp. Westheide, 1974:58–59, fig. 25. Santa Cruz Island, Academy Bay. Intertidal, associated with chaetopterid tubes and old coral.

Subfamily Eusyllinae

Amblyosyllis granosa Ehlers, 1897.—Westheide, 1974:68–71, figs. 30–31. Santa Cruz Island, Academy Bay. Surf zone in lava rock.

Eusyllis homocirrata Hartmann-Schröder, 1958.—Westheide, 1974:71–73, figs. 32, 38A. Santa Cruz Island, Bahia Borrero; Hood Island, Gardner Bay. Intertidal in beaches.

Odontosyllis fulgurans dolerans Westheide, 1974:73–77, figs. 33–34. Santa Cruz Island, Academy Bay. Intertidal to 2 m, under rocks.

Syllides edentula (Claparède, 1868).—Westheide, 1974:77–80, fig. 35, 36A–D. Tower Island, Darwin Bay; Santa Cruz Island, Academy Bay; San Cristobal Island, Punta Bassa. Intertidal in sediment with shells.

Syllides japonica edentata Westheide, 1974:81–83, fig. 36E, 37. Santa Cruz Island, Academy Bay.

Subfamily Exogoninae

Brania concinna Westheide, 1974: 91–93, fig. 41. Santa Cruz Island, Academy Bay. Intertidal on rocks, subtidal to 3 m with algae on rocks.

Brania heterocirra Rioja, 1941.—Westheide, 1974:83–87, fig. 38C–D,39. Fernandina Island, Punta Espinosa; Santa Cruz Island, Academy Bay; San Cristobal Island, Punta Bassa. Intertidal to shallow subtidal in sediment with large shells and sand.

Brania mediodentata Westheide, 1974:93–97, figs. 42A–C, 43. Fernandina Island, Punta Espinosa; Santa Cruz Island, Academy Bay. Shallow subtidal, in sand.

Brania oculata (Hartmann-Schröder, 1960).—Westheide, 1974:89–90. Tower Island, lagoon; Santa Cruz Island, Academy Bay.

Brania subterranea (Hartmann-Schröder, 1956).—Westheide, 1974:87–89, fig. 40, 42D–F. Tower Island, Darwin Bay; Santa Cruz Island, Academy Bay; San Cristobal Island, Punta Bassa. Intertidal in beach sand.

Sphaerosyllis centroamericana Hartmann-Schröder, 1959.—Westheide, 1974: 101–103, 45, 46D-E. Tower Island, Darwin Bay; Santa Cruz Island, Academy Bay; San Cristobal Island, Punta Bassa. In Mangroves, in sediment with detritus.

Sphaerosyllis hystrix Claparède, 1863.—Westheide, 1974:97–101, figs. 44, 46A–G. Santa Cruz Island, Academy Bay; San Cristobal Island, Punta Bassa. Intertidal in sand, with detritus.

Sphaerosyllis pumila Westheide, 1974:103–105, figs. 46F, 47. Santa Cruz Island, Academy Bay; San Cristobal Island, Punta Bassa. Shallow subtidal in organically enriched sediment.

Exogone dispar (Webster, 1879).—Westheide, 1974:106–109, figs. 48–49. Santa

Cruz Island, Academy Bay. Shallow subtidal in sediments sand and shell sediments with detritus.

Exogone longicornis Westheide, 1974:117–121, figs. 54–55. Santa Cruz, Academy Bay. Subtidal in sand with shells and detritus.

Exogone microtentaculata Westheide, 1974:121–123. Santa Cruz Island, Academy Bay. Shallow subtidal in sediment with detritus.

Exogone naidinoides Westheide, 1974:109–113, fig. 50, 51E–F. Santa Cruz Island, Academy Bay. Intertidal on rocks and on old coral.

Exogone occidentalis Westheide, 1974:113–117, figs. 52–53. Isabela Island; Marchena Island; Santa Cruz Island, Academy Bay. Intertidal to shallow subtidal in sand with detritus.

Subfamily Autolytinae

Autolytus multidenticulatus Westheide, 1974:123–126, figs. 57–58. Santa Cruz Island, Academy Bay. Intertidal in old coral in surf zone.

Autolytus arboreus Westheide, 1974:127–128, fig. 59. Santa Cruz Island, Academy Bay. Intertidal under rocks.

Autolytus sp. Westheide, 1974:129–130, fig. 60. Santa Cruz Island, Academy Bay. Subtidal, 3 m with algae on rocks.

Proceraea fasciata Langerhans, 1879.—Westheide, 1974:131–133, figs. 61–62. Santa Cruz Island, Academy Bay. Intertidal, under rocks.

Proceraea sp. Westheide, 1974:133–135, fig. 63. Santa Cruz Island, Academy Bay. Subtidal to 3 m with algae on rocks.

Additional Records: Unknown "Syllidae" recorded as prey of *Conus diadema* by Nybakken (1978).

Family Nereididae

Ceratonereis monronis Westheide, 1977:8–14, figs. 1–3. Santa Cruz Island, Academy Bay. Shallow subtidal in sediments with large shells and enriched with detritus.—As "Heteronereis" sp. Monro, 1933a:49–51, fig. 21. James Island, James Bay, from dip net (*fide* Westheide, 1977).

Ceratonereis tentaculata Kinberg, 1866.—Monro, 1933a:45. James Island, James Bay, 27 m, clean sand and weed.—Hartman, 1940:218, Pl. 35, fig. 47, AHPE Sta. 197-34.

Cirronereis gracilis Kinberg, 1866.—Hartman, 1948:53–55, Pl. VIII, figs. 1–4. Charles Islands, 2 of Kinberg's specimens.

Nereis jacksoni Kinberg, 1866.—Nybakken, 1978:275. Recorded as principal prey item of *Conus diadema*.

Nereis pseudonereis Hartman, 1940:223–224, Pl. 36, figs. 63–64, Pl. 37, figs. 65–66. AHPE Sta. 326-34 (holotype).

Nereis riisei Grube, 1857.—Monro, 1933a:43–44. Isabella Island, Tagus Cove, dredged in 22 m, coarse clean sand with weed.—Hartman, 1940:221–222, Pl. 33, fig. 37. AHPE Sta. 99-33, 148-34, 788-38.

Platynereis bicanaliculata (Baird, 1863).—As *Uncinereis agassizi* (Ehlers): Hartman, 1940:231–232. AHPE 66-33, 148-34, 169-34, 177-34, 182-34, 193-34.—As *Platynereis dumerilii* (Audouin and Milne-Edwards) var. *agassizi* Ehlers: Monro, 1933a:44–45. Isabella Island, dredged in 22 m, coarse clean sand with weed; James Island, James Bay, 27 m, clean sand and abundant weed (*fide* Hartman, 1959).

Platynereis dumerilii (Audouin and Milne-Edwards, 1833).—As *Leptonereis maculata* Kinberg: Treadwell, 1928:469. Seymour Bay (*fide* Hartman, 1956).

Pseudonereis gallapagensis Kinberg, 1866.—Hartman, 1940:231. AHPE Sta. 154-34.—Hartman, 1948:68–69. Holotype from Santa Cruz Island.

Pseudonereis variegata (Grube, 1857).—Westheide, 1977:7–8, Santa Cruz Island, Academy Bay. In old coral in surf zone.

Additional records: *Neanthes* sp. C, *N.* sp. D, *Ceratonereis* sp., and an "unknown nereid" recorded as prey of some *Conus* species by Nybakken (1978).

Family Aphroditidae

Laetmonice sp. Chamberlin, 1919:83. Off the Galápagos Islands, *Albatross* Sta. 3400, 2418 m, light grey ooze.

Family Polynoidae

Lepidasthenia gigas (Johnson, 1897).—As *Lepidametria gigas* (Johnson): Hartman, 1939:47–48, Pl. 8, figs. 99–104. AHPE, Sta. 148-34 (*fide* Hartman, 1959).

Lepidasthenia minikoiensis Potts, 1910.—Monroe, 1928:561–562. Santa Cruz Island, Conway Bay, shore pools.—Augener, 1933:58–60, fig. 1. Floreana Island, Post Office Bay.

Lepidasthenia retrodentata Monro, 1933a:13–14, fig. 6. Santa Cruz Island, Conway Bay, shore pools from peat.

Lepidasthenia varius Treadwell, 1917.—As *Lepidasthenia picta* Treadwell, 1928:456–457, fig. 177 (10–13). 5 m (*fide* Hartman, 1956).

Lepidasthenia virens (Blanchard, 1849).—As *Lepidametria virens* (Blanchard): Hartman, 1939:46–47, Pl. 8, figs. 105–110. AHPE, Sta. 167-34, 169-34, 182-34, 315-34 (*fide* Hartman, 1959).

Lepidonotus nesophilus Chamberlin, 1919:75–78, Pl. 4, figs. 1–7, Pl. 5, fig. 13. Chatham Island, intertidal.—Hartman, 1939:38–39, Pl. 7, figs. 83–95. AHPE Sta. 148-34, 814-38.

Lepidonotus furcillatus Ehlers, 1901.—Hartman, 1939:42, Pl. 5, figs. 57–58. AHPE Sta. 171-34.

Harmothoe hirsuta Johnson, 1897.—Monro, 1928:558–559, fig. 8. James Island, James Bay, shore pools in lava.

Harmothoe lunulata (Delle Chiaje, 1828).—Monro, 1928:559–560.

Halosydna parva Kinberg, 1855.—Hartman, 1939:33–34, Pl. 21, figs. 265–267. AHPE Sta. 152-34.

Harmothoe sylliformia Treadwell, 1928:452–454, fig. 177 (1–4). 549–1281 m.—Monroe, 1933a:14. Charles Island, Onslow Island.—Hartman, 1956:273.

?*Halosydna johnsoni* (Darboux, 1899).—As *Halosydna reticulata* Johnson.—Monro, 1928:563–565. Taboga Island from floats off the stage at end of Hotel Pier; Tagus Cove, Isabella Island.

Halosydnella elegans (Kinberg, 1855).—Monro, 1928:567–568, figs. 15–18. Isabella Island, Tagus Cove, fine gravel in pools. [According to Hartman, 1948, these specimens may refer to a different species than Kinberg's].—Hartman, 1948:19–20, Pl. I, figs. 1–4. Type specimen from Chatham Island (*Eugenie* Expedition).—As ?*Halosydna galapagensis* Monro, 1928:565–566, figs. 12–14. Isabella Island, Tagus Cove, fine gravel in pools (*fide* Hartman, 1948:19–20).

Hololepida sp. Hartman, 1939:50. AHPE Sta. 148-43.
Iphione ovata Kinberg, 1855.—Hartman, 1939:27, Pl. 3, figs. 31–32. AHPE Sta. 343-34; 789-38.
Thormora johnstoni (Kinberg, 1855).—Monro, 1928:556–557, figs. 5–7. Galápagos, James Bay, 9–11 m.—Hartman, 1939:50–51, Pl. 7, figs. 96–98. AHPE Sta. 194-34, 789-38.
Additional records: "Polynoidae" recorded as prey of *Conus diadema* by Nybakken (1978).

Family Acoetidae (=Polyodontidae)
Polyodontes panamensis (Chamberlin, 1919).—Hartman, 1939:84, Pl. 24, figs. 289–292. AHPE Sta. 198-34.

Family Sigalionidae
Sthenelais caerulea (Schmarda, 1861).—Monro, 1933a:17. James Island, James Bay, 9–11 m, clean sand with weed.
Sthenelais fusca Johnson, 1897.—Hartman, 1939:61–62, Pl. 13, figs. 153–162. AHPE Sta. 28-33, 343-34.
?*Sthenelais malayana* Horst, 1917.—Monro, 1933a:16. Santa Cruz Island, Conway Bay, shore pools.
Thalenessa lewissi (Berkeley and Berkeley, 1939).—*Eusigalion hancocki* Hartman, 1939:59–60, Pl. 12, figs. 141–145, 148–152. AHPE Sta. 66-33, 74-33 (*fide* Fauchald, 1977).
Thalenessa spinosa (Hartman, 1939).—As *Eusigalion spinosum* Hartman, 1939:57–59, Pl. 11, figs. 134–140, Pl. 12, figs. 146–147. AHPE Sta. 198-34, 814-38 (*fide* Hartman, 1959).

Family Chrysopetalidae
Bhawania goodei Webster, 1884.—Monro, 1933a:18–19. Isabella Island, Tagus Cove, along shore; Taboga Island, low water (Fauchald, 1977 confirmed Monro's identifications of this species from Panama).
Bhawania riveti Gravier, 1908.—As *Chrysopetalum riveti*: Treadwell, 1928:466. Tagus Cave, Galápagos.
Chrysopetalum occidentale Johnson, 1897.—Monro, 1933a:19. Charles Island, Onslow Island; Gorgona Island

Family Pisionidae
Pisione oerstedi pulla Westheide, 1974:11–14, figs. 1–2. Santa Cruz Island, Academy Bay, north side, shallow subtidal.
Pisione galapagoensis Westheide, 1974:14–19, figs. 3–5. Fernandina Island, Punta Espinosa; Santa Cruz Island, Academy Bay; San Cristobal Island, Punta Bassa.

Family Amphinomidae
Eurythoe complanata Pallas, 1766.—Chamberlin, 1919:28–30, Pl. 14, figs. 3–8. Chatham Island, intertidal.—Treadwell, 1928:450. Tower Island, Hood Island.—Augener, 1933:57. Floreana Island. Intertidal, beach.—Monro, 1933a:4–5. Charles Island; Isabella Island; Santa Cruz Island.—Hartman,

1940:202–203, Pl. 31, figs. 1–4. AHPE, 27-33, 33-33, 41-33, 42-33, 48-33, 52-33, 69-33, 73-33, 74-33, 94-33, 96-33, 101-33, 146-34, 161-34, 163-34, 194-34, 196-34, 315-34, 357-34, 782-38, 784-38, 789-38, 796-38, 811-38.—As *Eurythoe* sp. Nybakken, 1978:275. Recorded as prey of *Conus diadema* (Probably *E. complanata* (JAB)).

Chloeia viridis Schmarda, 1861.—Hartman, 1940:205.AHPE Sta. 161-34, 177-34, 182-34, 196-34, 198-34, 202-34, 204a-34, 814-38.

Notopygos ornata Grube, 1856.—Hartman, 1940:207. AHPE Sta. 202-34, 783-38.

Pherecardia striata (Kinberg, 1857).—Hartman, 1940:207. AHPE Sta. 146-34.

Family Onuphidae

Mooreonuphis litoralis (Monro, 1933).—As *Onuphis litoralis* Monro 1933a:78–80, fig. 33. Santa Cruz Island, Conway Bay, intertidal.

Hyalinoecia tubicola (Müller, 1776).—Chamberlin, 1919:315–317. Fragments found off the Galápagos Islands at *Albatross* Stations 3400 at 2419 m, and 3407 at 1620 m.

Rhamphobrachium longisetosum Berkeley and Berkeley, 1938.—Hartman, 1944a:48–49, Pl. 1, figs. 1–8. AHPE, Sta. 812a-38.

Additional records: "Onuphidae" recorded as prey of *Conus tiaratus* by Nybakken (1978).

Family Eunicidae

Eunice afra Peters, 1854.—Nybakken, 1978:275. Recorded as prey of *Conus lucidus*, *C. tiaratus*, and *C. nux*.

Eunice antennata (Savigny, 1820).—Treadwell, 1928:475.—Monro, 1933a:59–60. Isabella Island, Tagus Cove, intertidal.—Hartman, 1944a:115–117, Pl. 7, figs. 154–156. AHPE Sta. 99-33, 143-34, 148-34.

Eunice biannulata Moore, 1904.—Nybakken, 1978:275. Recorded as prey of *Conus tiaratus*.—As *Eunice longicirrata*: Hartman, 1944a:104–107, Pl. 6, figs. 75–80. AHPE, Sta. 147-34. (Fauchald, 1977 referred Panamanian specimens of this species to *E. biannulata* Moore, but did not mention Hartman's Galápagos specimens).

Eunice contingens (Chamberlin, 1919).—As *Leodice contingens* Chamberlin, 1919:260–263, Pl. 57, figs. 1–5. Off the Galápagos Islands nr. Hood Island, Ripple Point, *Albatross* Sta. 4642, 549 m, bottom of broken shells and ooze (*fide* Hartman, 1959).

Eunice filamentosa Grube, 1856.—Monro, 1933a:65–66. Isabella Island, Tagus Cove, intertidal; Santa Cruz Island, Conway Bay, intertidal.

Eunice mutilata Webster, 1884.—Hartman, 1944a:113–114, Pl. 6, figs. 140–141. AHPE Sta. 41-33.

Eunice pauroneurata (Chamberlin, 1919).—As *Leodice pauroneurata* Chamberlin, 1919:249–253, Pl. 57, figs. 8–9, Pl. 58, figs. 1–9, Pl. 59, figs. 1–3. Off the Galápagos Islands, *Albatross* Sta. 3401, 723 m, ooze (*fide* Hartman, 1959).

Eunice vittata (Delle Chiaje, 1828).—Hartman, 1944:118, AHPE Sta. 157-34.

Eunice sp. Westheide, 1977:14–16, fig. 4. Santa Cruz Island, Academy Bay. Under rocks in the surf zone.

Lysidice collaris Grube, 1870.—Monro, 1933a:69–70. Santa Cruz Island, Conway Bay, shore pools; Isabella Island, Tagus Cove, dredged in 22 m in coarse, clean sand with weed (*fide* Westheide, 1977).

Lysidice ninetta Audouin and Milne-Edwards, 1833.—Westheide,1977:17–18. Santa Cruz Island, Academy Bay. Under rocks and in old coral in the surf zone.

Lysidice robusta Kinberg, 1856.—Augener, 1933:61: Floreana Island.

Marphysa aenea Blanchard, 1849.—Monro, 1933a:67–68. Isabella Island, Tagus Cove, intertidal; Charles Island, Post Office Bay, intertidal; Santa Cruz Island, Conway Bay, from peat bed.

Marphysa furcellata Crossland, 1903.—Westheide, 1977:18–22, figs. 5–7. Santa Cruz Island, Academy Bay. Intertidal.

Nematonereis unicornis (Grube, 1840).—Westheide, 1977:22–24, fig. 8. Santa Cruz Island, Academy Bay. Shallow subtidal in sediments with shell and detritus.

Palola siciliensis (Grube, 1840).—Hartman, 1944a:131. AHPE, Sta. 62-33.—Westheide, 1977:16–17. Santa Cruz Island, Academy Bay. Surf zone in old coral.—As *Palola edentulum* (Ehlers, 1901): Hartman, 1948:81, examination Kinberg's specimens from Chatham island (*fide* Fauchald, 1977),—As *Eunice siciliensis* Grube, var. *edentulum* Ehlers: Monro, 1933a:62–63. Isabella Island, Tagus Cove, intertidal.—As *Nicidion gallapagensis* Kinberg, 1865:43, Chatham Island (*fide* Hartman, 1944a:130).

Family Lumbrineridae

Lumbrinereis brevicirra (Schmarda, 1861).—Monro, 1933a:83–84. Santa Cruz Island, Conway Bay, in peat; Charles Island, Post Office Bay, intertidal (not Schmarda, 1861 according to Fauchald, 1977).

Lumbrinereis tetraura (Schmarda, 1861).—As *Lumbriconereis tetraura*: Augener, 1933:61–62. Floreana Island, Post Office Bay.

Lumbrinereis heteropoda (Marenzeller, 1879).—Monro, 1933a:82–84. Charles Island, Post Office Bay, intertidal; Isabella Island, Tagus Cove, intertidal.

Lumbrinereis latreilli Audouin and Milne-Edwards, 1833.—Monro, 1933a:84–85. James Island, James Bay, 9–11 m, in clean sand with weed; Isabella Island, Tagus Cove, dredged in 22 m, in coarse sand with weed.—Hartman, 1944a:158–159, Pl. 9, figs. 213–216. AHPE, Sta. 169-34, 814-38.

Lumbrineris inflata Moore, 1911.—As *Lumbrinereis albifrons* Crossland,—Monro, 1933a:85. Isabella Island, Tagus Cove, intertidal (*fide* Hartman, 1944a:160).

Lumbrinereis monroi Fauchald, 1970:99–103, Pl. 16: figs. e–i.—As *Lumbrinereis africana* (Augener, 1918): Monro, 1933a:86. Charles Island, Post Office Bay, intertidal.

Additional records: *Lumbrinereis* sp. recorded as prey of *Conus tiaratus* by Nybakken (1978).

Family Arabellidae

Arabella iricolor Montague, 1804.—Augener, 1933:62–63. Floreana Island.—As *Arabella dubia* Treadwell, 1922: Treadwell, 1928:477 (*fide* Hartman, 1956).

Arabella monroi Colbath, 1989:292, figs. 3e, 4a,b.—As *Arabella mutans* (Chamberlin, 1919): Monro, 1933a:88–89. Santa Cruz Island, Conway Bay, intertidal; Charles Island, Post Office Bay, intertidal; Isabella Island, Tagus Cove, intertidal.—Nybakken (1978:275). Recorded as prey of *Conus tiaratus*.

Family Dorvilleidae

Dorvillea rudolphi (Delle Chiaje, 1833).—Monro, 1933a:92–93. Santa Cruz Island, intertidal.

Ophryotrocha sp. Westheide, 1977:30–33, figs. 12A–B, 13. Santa Cruz Island, Academy Bay. Shallow subtidal, in interstices of old coral and rocks.

Protodorvillea sp. Westheide, 1977:28–29, fig. 11. Santa Cruz Island, Academy Bay. Shallow subtidal, in sediment with detritus and shell.

Schistomeringos pacifica Westheide, 1977:24–28, figs. 9, 10, 12C. Santa Cruz Island, Academy Bay. Intertidal to shallow subtidal in organically enriched sediments.

Family Orbiniidae

New genus and species. Blake, in preparation, Santa Cruz Island, Puerto Nunez, intertidal.

Naineris grubei (Gravier, 1908).—As *Naidonereis grubei*: Augener, 1933:63. Floreana Island. Intertidal, beach.

Naineris laevigata (Grube, 1855).—Hartman, 1957:297–298, Pl. 35, figs. 1–8. Does not detail which AHPE stations, but Galápagos Islands, are mentioned.—Monro, 1933b:1045. Charles Island, Post Office Bay, intertidal (*fide* Hartman, 1957).

Family Spionidae

Prionospio (Minuspio) sp. Blake, 1983:228. Santa Cruz Island, Academy Bay, *Anton Bruun* Sta. 66122, 17–18 m.

Microspio paradoxa Blake, 1983:237–240, fig. 17. Santa Cruz Island, Academy Bay. *Anton Bruun* Sta. 66122, 17–18 m.

Microspio minuta (Hartmann-Schröder, 1962).—Blake, 1983:240–241, fig. 18. Santa Cruz Island, Academy Bay, *Anton Bruun* Sta. 66122, 17–18 m.

Boccardia galapagense Blake, 1986:16–18, fig. 1. Santa Cruz Island, Academy Bay, *Anton Bruun* Sta. 66128, in coral head.

Family Chaetopteridae

Chaetopterus variopedatus (Renier, 1804).—Monro, 1933b:1050. James Island, James Bay, dredged 9–11 m, on clean sand with weed.

Mesochaetopterus minutus Potts, 1914.—Monro, 1933b:1052. Santa Cruz Island, Conway Bay, intertidal.

Family Questidae

Questa media Westheide, 1981:6–8, fig.1. Santa Cruz Island, Academy Bay. Shallow subtidal, in sand with shell.

Family Cirratulidae

Caulleriella alata (Southern, 1914).-Westheide, 1981:8–10, fig. 2C. Santa Cruz Island, Academy Bay. Intertidal in Mangroves in fine sediments.

Cirriformia violacea Westheide, 1981:10–12, figs. 2A–B, 3. Santa Cruz Island, Academy Bay. In organically enriched sediment.

Cirriformia sp. Westheide, 1981:12–13. Santa Cruz Island, Academy Bay.

Cirrhineris nesiotes Chamberlin, 1919:373–375, Pl. 70, figs. 5–6. Chatham Island, intertidal. (Possibly a species of *Chaetozone* (JAB)).

Cirriformia tentaculata (Montagu, 1808).—As *Cirratulus (Audouinia) polytricha* Schmarda.—Augener, 1933:63–65. Floreana Island, Post Office Bay.—Monro, 1933b:1054–1055. James Island, intertidal; Charles Island, Post Office Bay, intertidal; Isabella Island, Tagus Cove, intertidal (*fide* Fauchald, 1977).

Timerete perbranchiata Chamberlin, 1918.—As *Ambo americana* Monro, 1933b:1055–1056. James Island, intertidal (*fide* Hartman, 1959).

Tharyx sp. Westheide, 1981:13. Santa Cruz Island, Academy Bay. Intertidal under rocks.

Additional records: *Cirriformia* sp., *Chaetozone* sp., and "Cirratulidae" recorded as prey of *Conus diadema* by Nybakken (1978).

Family Ctenodrilidae

Ctenodrilus serratus (O. Schmidt, 1857).—Westheide, 1981:18–19, fig. 7. Santa Cruz Island, Academy Bay. In detritus with shells.

Family Acrocirridae

Macrochaeta multipapillata Westheide, 1981:14–18, figs. 4–6. Santa Cruz Island, Academy Bay; Fernandina Island, Punta Espinosa; San Cristobal Island, Punta Bassa.

Family Flabelligeridae

Ilyphagus ascendens Chamberlin, 1919:403–404, No. figs. Off Hood Island, 12 mi SE of Ripple Point, *Albatross* Sta. 4649, 1158 m, grey ooze.

Semiodera cariboa (Grube, 1859).—As *Semiodera glabra* Treadwell, 1928:479–480, fig. 179 (62–65) (*fide* Hartman, 1956).

Pherusa arenosa americana (Monro).—As *Stylaroides capensis* (McIntosh), var. *americana* Monro.—Monro, 1933b:1057–1058, fig. 6. Charles Island, Post Office Bay, intertidal; Isabella Island, Tagus Cove, intertidal pools (*fide* Hartman, 1959).

Family Capitellidae

Dasybranchus lumbricoides Grube, 1878.—*Dasybranchus caducus* Grube, var. *lumbricoides* Grube.—Monro, 1933b:1059–1060, fig. 8. Santa Cruz Island, Conway Bay, intertidal. *fide* Hartman, 1959.

Additional records: "Capitellidae" recorded as prey of *Conus lucidus* and *C. tiaratus* by Nybakken (1978).

Family Maldanidae

Euclymene annandalei Southern, 1921.—As *Clymene (Euclymene) annan-*

dalei: Monro, 1933b:1062. Santa Cruz Island, Conway Bay, intertidal pools (*fide* Hartman, 1959).

Family Sabellariidae

Idanthyrsus pennatus (Peters, 1855).—Hartman, 1944b:336, Pl. 31, fig. 35. AHPE, Sta. 30-33, 56-33, 69-33, 80-33, 94-33, 101a-33.—As *Idanthyrsus cretus* Chamberlin, 1919: Treadwell, 1928:481. Probable synonymy. See Hartman, 1944b. —*Idanthyrsus regalis* Chamberlin, 1919:487–490, Pl. 74, figs. 1–8. Chatham Island, intertidal (*fide* Hartman, 1944b).

Lygdamis nesiotes (Chamberlin, 1919).—Hartman, 1944b:332–333. AHPE, Sta. 28-33. *Lygdamis* sp.—Nybakken, 1978:275. Recorded as prey of *Conus diadema*.

Family Ampharetidae

Ampharete sp.—Nybakken, 1978:275. Recorded as prey of *Conus lucidus*.

Family Terebellidae

Amphitrides bruneocomata (Ehlers, 1887).—As *Terebella bruneocomata*: Monro, 1933b:1071–1072. Charles Island, Post Office Bay, intertidal (*fide* Hartman, 1959).

Eupolymnia insulana Chamberlin, 1919:434–436, Pl. 79, fig. 4–6. Off the Galápagos Islands, 4 mi SE of Hood Island, *Albatross* Sta. 4642, 549 m on bottom of broken shell and ooze.

Eupolymnia nebulosa (Montague, 1818).—As *Polymnia nebulosa*: Monro, 1933b:1072. James Island, James Bay, dredged, 27 m, in sand with weed; Santa Cruz Island, Conway Bay, intertidal pools.

Lanicides taboguillae (Chamberlin, 1919).—Monro, 1933b:1069-1070, fig. 17. Santa Cruz Island, Conway Bay, intertidal pools; Charles Island, Post Office Bay, intertidal,—As *Nicolea cetrata* Ehlers variety *galapagensis*: Augener, 1933:65–66. Floreana Island, Post Office Bay (*fide* Hartman, 1959).—*Nicolea galapagensis* Chamberlin, 1919:426–429, Pl. 80, figs. 1–3. Chatham Island, intertidal (*fide* Hartman, 1959).

Pista alata Moore, 1909.—Monro, 1933b:1066–1068, fig. 15. Santa Cruz Island, Conway Bay, intertidal pools.

Streblosoma verrilli Treadwell, 1911.—Monro, 1933b:1073–1074, fig. 20. Santa Cruz Island, Conway Bay; Charles Island, Post Office Bay, intertidal. (This species was referred to *Thelepus setosus* by Hartman, 1956. The identity of Monro's specimens remains to be determined).

Terebella gorgonae Monro, 1933b:1070–1071, fig. 18. Santa Cruz Island, Conway Bay; Isabella Island, Tagus Cove, intertidal.

Additional records: Thelepinae sp. A, *Eupolymnia* sp., *Terebella* sp., and *Polycirrus* sp. recorded as prey of *Conus* sp. by Nybakken (1978).

Family Sabellidae

Branchiomma mushaensis Gravier, 1906.—Monro, 1933b:1078. James Island, James Bay, intertidal pools; Isabella Island, Tagus Cove, intertidal and dredged in 22 m. (This species was referred to *Magelomma quadrioculata*

(Willey) by Hartman (1959), but the Galápagos specimens have not been reexamined).

Additional records: "Sabellinae" recorded as prey of *Conus* by Nybakken (1978).

Family Serpulidae

Spirobranchus giganteus Pallas, 1766.—Monro, 1933b:1080. James Island, James Bay, intertidal pools; Charles Island, Onslow Island, 7 m; Isabella Island, Tagus Cove, intertidal in pools.

Eupomatus inermis (Monro, 1933).—As *Hydroides inermis* Monro, 1933b: 1083–1085, fig. 27. James Island, James Bay, dredged in 9–11 m, bottom of clean sand and weeds (*fide* Hartman, 1959).

Vermiliopsis acanthophora Augener, 1914.—Monro, 1933b:1085. James Island, James Bay, dredged in 9–11 m, on bottom of clean sand and weeds.

Josphella sp. Monro, 1933b:1090. Santa Cruz Island, Conway Bay, dredged in 9 m on a bottom of sand thinly spread over flat rock; Charles Island, Post Office Bay, intertidal.

Salmacina tribranchiata (Moore, 1923).—As *Salmacina dysteri* (Huxley), var. *tribranchiata* (Moore).—Monro, 1933b:1090–1091, fig. 31. Isabella Island, Tagus Cove, intertidal in pool (*fide* Hartman, 1959).

Family Spirorbidae*

Spirorbis sp. Monro, 1933b:1091. James Island, James Bay, dredged in 9–11 m, on bottom of clean sand with weed.

Spirorbis translucens Bailey and Harris, 1968:165–167, figs. 3a, 4, 6.

Spirorbis claparedei (Caullery and Mesnil, 1897).—Bailey and Harris, 1968: 167–169, figs. 3b, 7.

Spirorbis bidentatus Bailey and Harris, 1968:169–170, figs. 3c, 8.

Spirorbis placophora Bailey and Harris, 1968:170, figs. 3d, 9.

Spirorbis berkeleyana Rioja, 1942.—Bailey and Harris, 1968:170–172, figs. 3e, 10.

Spirorbis regalis Bailey and Harris, 1968:172–174, figs. 3f, 11.

Spirorbis tuberculatus Bailey and Harris, 1968:174–175, fig. 12.

Spirorbis bushi Rioja, 1942.—Bailey and Harris, 1968:175, fig. 5h, 13.

Spirorbis tricornigerus Rioja, 1942.—Bailey and Harris, 1968:175, figs. 5i, 14.

Spirorbis marioni (Caullery and Mesnil, 1897).—Bailey and Harris, 1968:175–177, figs. 5j, 15.

Spirorbis pagenstecheri Quatrefages, 1865.—Bailey and Harris, 1968:178–180, fig. 5k.

Spirorbis unicornis Bailey and Harris, 1968:180, figs. 5l, 16.

Family Polygordiidae

Polygoridius pacificus floreanensis Schmidt and Westheide, 1977:5–8, figs. 1–2, 3A–B. Floreana Island, Black Beach, intertidal in coarse sand.

*No attempt is made here to correct the generic classification of these spirorbids.

Family Saccociridae

Saccocirrus sonomacus Martin, 1977.—Schmidt and Westheide, 1977:8–11, figs. 3E–F, 4, 5. Santa Cruz Island, Academy Bay, surf zone in coarse sediment with shells.

Saccocirrus sp. Schmidt and Westheide, 1977:11–13, figs. 3C–D, 6. Santa Cruz Island, Academy Bay, shallow subtidal, in sediment with shells.

Family Protodrilidae

Protodrilus pierantonii Aiyar and Alikunhi, 1944.—Schmidt and Westheide, 1977:13–16, figs. 7–8. Hood Island, Gardner Bay, subtidal in organically enriched sediment.

Protodrilus infundibuliformis Schmidt and Westheide, 1977;16–18, figs. 9–11. Santa Cruz Island, Academy Bay, subtidal in sediment with shell; Hood Island, Gardner Bay, subtidal in sand.

Protodrilus sp. Schmidt and Westheide, 1977:18–20, figs. 12–13. Santa Cruz Island, Playa Borreo, surf zone, with organically enriched sediment.

Family Nerillidae

Nerilla parva Schmidt and Westheide, 1977:21–27, figs. 14, 15, 16B, 17. Santa Cruz Island, Academy Bay, intertidal in coarse sand with shells.

Nerillidium lothari Schmidt and Westheide, 1977:27–31, figs. 18–20. Santa Cruz Island, Puerto Nuñez, shallow subtidal.

Mesonerilla ecuadoriensis Schmidt and Westheide, 1977:31–33, figs. 16A, 21. Santa Cruz Island, Academy Bay, intertidal in detritus rich sediments.

Family Dinophilidae

Dinophilus sp. Schmidt and Westheide, 1977:33. Recorded from Tower Island, Isabella Island, and Fernandina Island in intertidal sediments.

Diurodrilus spp. Schmidt and Westheide, 1977:33–34, fig. 22. Recorded from Santa Cruz Island and Floreana Island.

5. Discussion

The most obvious observation to be made as a result of this compilation of the recorded species of polychaetes from the Galápagos Islands is that many families are either not recorded or very poorly represented. Most of these families are ones that are representative of the benthic infauna, such as the Paraonidae, Spionidae, Capitellidae, Ampharetidae, and the subfamily Fabricinae of the Sabellidae. These polychaetes tend to be ones that burrow or build tubes in the sediment. Most are deposit feeders. Free-living or crawling polychaetes such as the syllids, nereidids, eunicids in contrast are well represented.

The absence of many infaunal species is undoubtedly the result of not sampling for them. Some insight into these missing groups is evident in evaluating the list of prey items of *Conus* spp. published by Nybakken (1978). Additional genera of terebellids are evident in his list along with the only ampharetid thus far recorded from the Galápagos. It is evident to this writer that a considerably larger fauna of polychaetes would be evident if grabs and fine mesh screens were used to sample the subtidal benthos. Additionally, only a single species of the *Polydora*-complex has been recorded (*Boccardia galapagense* Blake, 1986). These spionids are a dominant group along most coastlines of the world, including the western coast of South America (Blake, 1983). It is likely that collections taken in the cryptic habitats where they occur, such as under coralline algae, in shells, in coral, and sediments, will yield numerous species.

Out of 192 species recorded from the Galápagos Islands since the first papers by Kinberg over 125 years ago, 50 species or 31% are endemic. Most of the endemic species are ones that are small and inhabit soft bottom benthic habitats. As already noted, the number of species and probably endemic species is expected to increase once additional sampling in the subtidal benthos is undertaken.

6. Appendix: List of Stations Sampled by the Allan Hancock Pacific Expeditions*

1933

27-33. 22 Jan. 1933. Galápagos Islands, Gardner Bay, Hood Island, Rocky pit.
Eurythoe complanata

28-33. 22 Jan. 1933. Galápagos Islands, Hood Island, Gardiner Bay, diving, 3.6 m.
Sthenelais fusca, Lygdamis nesiotes

30-33. Jan. 1933. Galápagos Islands, Osborn Island, Osborn Bay, intertidal.
Idanthyrsus pennatus

33-33. 27 Jan. 1933. Galápagos Islands, Black Beach Anchorage, Charles Island. Shore.
Eurythoe complanata

41-33. 30 Jan. 1933. Galápagos Islands, Chatham Island. Dredged east of Wreck Bay, 7.3 m.
Eurythoe complanata, Eunice mutilata

42-33. 31 Jan. 1933. Galápagos Islands, Chatham Island, opposite Kicker Rock. Shore.
Eurythoe complanata

48.33. 2 Feb. 1933. Galápagos Islands, Barrington Island. In bay, along shore.
Eurythoe complanata

52-33. 4 Feb. 1933. Galápagos Islands, Santa Cruz Island, Academy Bay. Shore.
Eurythoe complanata

56-33. Feb. 1933. Galápagos Islands, Charles Island, Flamingo Bay. Shore.
Idanthyrsus pennata

59-33. 6 Feb. 1933. Galápagos Islands, Charles Islands, Cormorant Bay. Dredged, in 24 m.
Hesione intertexta

62-33. Feb. 1933. Galápagos Islands, Isabella Island, rocky shore.
Palola siciliensis

*Species listed for each station include only those recorded in the publications of O. Hartman.

66-33. 9 Feb. 1933. Galápagos Islands, Isabella Island, Tagus Cove. Dredged, in 18–36 m.
Uncinereis agassizi, Thalenessa lewissi
69-33. 11 Feb. 1933. Galápagos Islands, Isabella Island, Isabella Point. Shore.
Eurythoe complanata, Idanthyrsus pennata
73-33. 13 Feb. 1933. Galápagos Islands, Isabella Island, Cartago Bay. North Beach.
Eurythoe complanata
74-33. 14 Feb. 1933, Galápagos Islands, Isabella Island, Cartago Bay. Dredged.
Eurythoe complanata, Thalenessa lewissi
76-33. 14 Feb. 1933. Galápagos Islands, Isabella Island, Cartago Bay, intertidal.
Thormora johnstoni
80-33. Feb. 1933. Galápagos Islands, Duncan Island, shore.
Idanthyrsus pennata
85-33. Feb. 1933. Galápagos Islands, North Seymour Island, shore.
Eunice filamentosa
94-33. 22 Feb. 1933. Galápagos Islands, Tower Island, Darwin Bay. Coral, from seal beach.
Eurythoe complanata, Eunice filamentosa, Idanthyrsus pennata
96-33. 24 Feb. 1933. Galápagos Islands, Tower Island, Darwin Bay. Coral, from seal beach.
Eurythoe complanata
99-33. 25 Feb. 1933. Galápagos Islands, Tower Island, Darwin Bay. In tangles.
Nereis riiseri, Eunice antennata
101-33. 26 Feb. 1933. Galápagos Islands, Tower Island, Seal Beach, on shore.
Eurythoe complanata
101a-33 Feb. 1933. Galápagos Islands, Tower Island, Darwin Bay, intertidal.
Idanthyrsus pennata

1934

143-34. Jan. 1934. Galápagos Islands, Wenman Island, 183–275 m coral.
Eunice antennata
146-34. 12 Jan. 1934. Galápagos Islands, Isabella Island, north end. Shore collecting south of point.
Eurythoe complanata, Pherecardia striata
147-34. 13 Jan. 1934. Galápagos Islands, Isabella Island, Tagus Cove, dredged in 55 m, coral and rocks.
Eunice biannulata
148-34. 13 Jan. 1934. Galápagos Islands, Isabella Island, Tagus Cove. Dredged, 22–27 m.
Lepidonotus nesophilus, Lepidasthenia gigas, Hololepida sp., *Hesione intertexta, Nereis riisei, Uncinereis agassizi, Eunice antennata*
152-34. 13 Jan. 1934. Galápagos Islands, Isabella Island, Tagus Cove, shore, in coral.
Eunice filamentosa, Halosydna parva
154-34. 15 Jan. 1934. Galápagos Islands, Isabella Island, Tagus Cove. Reef, north of hill. Shore.
Pseudonereis gallapagensis

156-34. 15 Jan. 1934. Galápagos Islands, Isabella Island, Tagus cove, dredged in 146–183 m.
Ancistrosyllis sp.

157-34. 15 Jan. 1934. Galápagos Islands, Isabella Island, Tagus Cove, dredged in 18–346 m, sand and shells.
Eunice vittata

161-34. 17 Jan. 1934. Galápagos Islands, Charles Island. Taylor Rock. Dredged in 0–5.5 m.
Chloeia viridis, Eurythoe complanata, Notopygos ornata

163-34. 18 Jan. 1934. Galápagos Islands, Charles Island. Black Beach. Shore.
Eurythoe complanata

167-34. 19 Jan. 1934. Galápagos Islands, Charles Island, Post Office Bay. Dredged, in 27 m. Rock.
Hesione intertexta, Lepidasthenia virens

169-34. 20 Jan. 1934. Galápagos Islands, Santa Cruz Island, Academy Bay. Dredged. Rock, mostly covered with algae.
Lepidasthenia virens, Hesione intertexta, Unicinereis agassizi, Lumbrineris latreilli

171-34. 21 Jan. 1934. Galápagos Islands, Chatham Island, E of Wreck Bay, dredged, 64–73 m, coarse sand and corallines.
Lepidonotus furcillatus

173-34. 22 Jan. 1934. Galápagos Islands, South Seymour Island, Velero Bay. Dredged, in 9 m. Sand, with rock patches.
Glycinde armigera

177-34. 23 Jan. 1934. Galápagos Islands, James Island, Sullivan Bay. Dredged, in 9–37 m. Rock, with sand patches.
Chloeia viridis, Uncinereis agassizi

182-34. 24 Jan. 1934. Galápagos Islands, James Island, James Bay. Dredged, in 55 m. Coarse sand.
Lepidasthenia virens, Chloeia viridis, Uncinereis agassizi

185-34. 25 Jan. 1934. Galápagos Islands, Isabella Island, Cartago Bay. 2 miles from white rock. Dredged,in 59 m. Mud.
Aglaophamus dicirrus

187.34. 25 Jan. 1934. Galápagos Islands, Isabella Island, Cartago Bay, southwest part. Dredged in 15–18 m, in sand, with occasional rock patches.
Aglaophamus dicirrus

193-34. 27 Jan. 1934. Galápagos Islands, Charles Island, Post Office Bay. Dredged in 15–18 m. Sand and Rock with algae.
Uncinereis agassizi

194-34. 27 Jan. 1934. Galápagos Islands, Charles Island, Post Office Bay, Onslow Island, coral inside crater.
Thormora johnstoni, Eurythoe complanata

196-34. 29 Jan. 1934. Galápagos Islands, Charles Island, north of island. Dredged, 15–18 m. Rough Rock.
Chloeia viridis, Eurythoe complanata

197-34. 29 Jan. 1934. Galápagos Islands, Charles Island, dredged off point, in 64–73 m. Rocky.
Ceratonereis tentaculata

198-34. 29 Jan. 1934. Galápagos Islands, Charles Island, dredged in 101–119 m. Sand.
Polyodontes panamensis, Thalenessa spinosa, Chloeia viridis

202-34. 31 Jan. 1934. Galápagos Islands, Hood Island, Gardner Bay, Osborn Island. Shore.
Chloeia viridis, Notopygos ornata

204a-34. 8 Feb. 1934. Galápagos Islands, Isabella Island, Tagus Cove. From fish trap.
Chloeia viridis

310-34. 3 Dec. 1934. Galápagos Islands, Marchena Island, North Bay. Tangles, in 17 m.
Hesione intertexta

315-34. 8 Dec. 1934. Galápagos Islands, Santa Cruz Island. In coral, opposite Gordon Rocks.
Lepidasthenia virens, Eurythoe complanata

326-34. 10 Dec. 1934. Galápagos Islands, Isabella Island, Tagus Cove, south side. Dredged, in 27 m.
Nereis pseudonereis

336-34. 11 Dec. 1934. Galápagos Islands, James Island, Sullivan Bay. Dredged in 37 m. Red algae.
Hesione intertexta

343-34. 12 Dec. 1934. Galápagos Islands, James Island, Sullivan Bay, intertidal.
Iphione ovata, Sthenelais fusca

357-34. 17 Dec. 1934. Galápagos Islands, Hood Island. In coral.
Eurythoe complanata

1938

782-38. 16 Jan. 1938. Galápagos Islands, Tower Island, Darwin Bay. Shore. Rock, at Seal Beach.
Eurythoe complanata, Marphysa aenea

783-38. 16 Jan. 1938. Galápagos Islands, Tower Island, Darwin Bay. Dredged, in 73–128 m. White sand, rock.
Notopygos ornata

784-38. 17 Jan. 1938. Galápagos Islands, Tower Island, Darwin Bay. Shore, rock, at Middle Beach.
Eurythoe complanata

786-38. 18 Jan. 1938. Galápagos Islands, NE of Santa Cruz Island, dredged, in 717 m, in sand.
?Glycera oxycephala

788-38. 19 Jan. 1938. Galápagos Islands, south and east of Daphne Major Island, dredged, in 101 m, coral, shell.
Nereis riisei

789-38. 19 Jan. 1938. Galápagos Islands, South Seymour Island, shore, rocky.
Iphione ovata, Thormora johnstoni, Eurythoe complanata

796-38. 21 Jan. 1938. Galápagos Islands, James Island, Sullivan Bay, shore, rocky.
Iphione ovata, Eurythoe complanata

810-38. Jan. 1938. Galápagos Islands, Barrington Island, dredged in 88–134 m, sand and rock.
Eunice vittata

811-38. 26 Jan. 1938. Galápagos Islands, Barrington Island, in coral.
Eurythoe complanata, Hesione intertexta

812a-38. Jan. 1938. Galápagos Islands, Chatham Island, off Freshwater Bay, dredged in 732 m, coarse sand.
Rhamphobrachium longosetosum

814-38. 28 Jan. 1938. Galápagos Islands, N of Hood Island, dredged in 37–74 m, sand and shell.
Lepidonotus nesophilus, Thalenessa spinosa, Chloeia viridis, Lumbrineris latreilli

References

Augener, H., 1933, Polychaeten von den Galapagos-Inseln. The Norwegian Zoological Expedition to the Galápagos Islands 1925, conducted by Alf Wollebaek. VI, *Medd. f. Zool. Mus., Oslo*, No. **25**:55–66.

Bailey, J. H., and Harris, M. P., 1968, Spirorbinae (Polychaeta: Serpulidae) of the Galápagos Islands, *J. Zool.* **155**:161–184.

Blake, J. A., 1983, Polychaetes of the Family Spionidae from South America, Antarctica, and adjacent seas and islands, *Antarctic Res. Ser.* **39**:205–287.

Blake, J. A., 1986, A new species of *Boccardia* (Polychaeta: Spionidae) from the Galápagos Islands and a redescription of *Boccardia basilaria* Hartman from Southern California, *Bull. South. Calif. Acad. Sci.* **85**:16–21.

Chamberlin, R., 1919, The Annelida Polychaeta, *Mem. Mus. Comp. Zool., Harvard* **48**:1–514, 80 pls.

Colbath, G. K., 1989, A revision of *Arabella mutans* (Chamberlin, 1919) and related species (Polychaeta: Arabellidae), *Proc. Biol. Soc. Wash.* **102**:283–299.

Fauchald, K., 1977, Polychaetes from intertidal areas of Panama, with a review of previous shallow-water records, *Smiths. Contribs. Zool.* No. 221:1–81.

Hartman, O., 1940, Polychaetous Annelids Part II. Chrysopetalidae to Goniadidae, *Allan Hancock Pac. Exped.* **7**(3):173–288, plates 31–44.

Hartman, O., 1944a, Polychaetous Annelids Part V. Eunicea, *Allan Hancock Pacif. Exped.* **10**(1):1–238, plates 1–18.

Hartman, O., 1944b, Polychaetous Annelids Part VI. Paraonidae, Magelonidae, Longosomidae, Ctenodrilidae, and Sabellariidae, *Allan Hancock Pacif. Exped.* **10**(3):311–390, plates 27–42.

Hartman, O., 1947, Polychaetous Annelids Part VIII. Pilargiidae, *Allan Hancock Pac. Exped.* **10**(5):483–524, plates 59–63.

Hartman, O., 1948, The marine annelids erected by Kinberg with notes on other types in the Swedish State Museum, *Arkiv f. Zool.* **42**A(1):138, plates I–XVIII.

Hartman, O., 1950, Goniadidae, Glyceridae, and Nephtyidae, *Allan Hancock Pacif. Exped.* **15**(1):1–182, plates 1–19, text figs. 1–3.

Hartman, O., 1956, Polychaetous annelids erected by Treadwell, 1891 to 1948, together with a brief chronology, *Bull. Amer. Mus. Nat. Hist.* **109**(2):239–310, plate 21.

Hartman, O., 1957, Orbiniidae, Apistobranchidae, Paraonidae and Longosomidae, *Allan Hancock Pacif. Exped.* **15**(3):211–394, plates 20–44, 1 chart.

Hartman, O., 1959, Catalogue of the Polychaetous Annelids of the World, *Allan Hancock Found., Occ. Pap.* No. 23: 1–628.

Kinberg, J. G. H., 1865, Annulata nova, *Oefv. Vet. Akad. Stockholm Förh.* **21**:559–574.

Kinberg, J. G. H., 1866, Annulata nova, *Oefv. Vet. Akad. Stockholm Förh.* **22**:167–179, 239–258.

Monro, C. C. A., 1928, Polychaeta of the families Polynoidae and Acoetidae from the Vicinity of the Panama Canal, *J. Linn. Soc. London* **36**:555–576.

Monro, C. C. A., 1933a, The Polychaeta Errantia collected by Dr. C. Crossland at Colon, in the Panama Region, and the Galápagos Islands during the expedition of the S. Y. "St. George," *Proc. Zool. Soc. Lond.* Part **1**:1–96.

Monro, C. C. A., 1933b, The Polychaeta Sedentaria collected by Dr. C. Crossland at Colon, in the Panama Region, and the Galápagos Islands during the expedition of the S. Y. "St. George," *Proc. Zool. Soc. Lond.* Part **2**:1039–1092.

Nybakken, J., 1978, Population characteristics and food resource utilization of *Conus* in the Galápagos Islands, *Pacif. Sci.* **32**:271–280.

Schmidt, O., and Westheide, W., 1977, Interstielle Fauna von Galapagos XVII. Polygordiidae, Saccocirridae, Protodrilidae, Nerillidae, Dinophilidae (Polychaeta), *Mikrofauna Meeresbodens* **62**:1–38.

Treadwell, A. L., 1928, Polychaetous annelids from the Arcturus Oceanographic Expedition, *Zoologica* **8**(8):449–485.

Westheide, W., 1974, Interstielle Fauna von Galapagos XI. Pisionidae, Hesionidae, Pilargidae, Syllidae (Polychaeta), *Mikrofauna Meeresbodens* **44**:195–338.

Westheide, W., 1977, Interstielle Fauna von Galapagos XVIII. Nereidae, Eunicidae, Dorvilleidae (Polychaeta), *Mikrofauna Meeresbodens* **63**:1–40.

Westheide, W., 1981, Interstielle Fauna von Galapagos XXVI. Questidae, Cirratulidae, Acrocirridae, Ctenodrilidae (Polychaeta), *Mikrofauna Meeresbodens* **82**:1–24.

III

Coelenterata

Chapter 4

El Niño and Coral Reef Development in the Galápagos Islands

A Study of the Urvina Bay Uplift

MITCHELL W. COLGAN

1. Introduction

Among the unusual sights in the Galápagos Islands is Urvina Bay where trees spring from coral heads and iguanas roam among corals. In 1954, the upward movement of magma suddenly raised a portion of Urvina Bay more than 6 m above sea level and drove the shoreline 1.2 km seaward. The uplifted coral communities at Urvina Bay provide an opportunity to determine how Galápagos reefs develop.

The Galápagos Islands are a part of the eastern Pacific reef-building province, which occurs discontinuously from the tip of Baja California to the coast of Ecuador (Glynn, 1974). Charles Darwin, in 1835 during his voyage on the HMS *Beagle*, probably collected the first coral specimen from the Galápagos Islands, yet he failed to detect the small reefs that fringe the volcanic islands (Darwin, 1842; Glynn and Wellington, 1983). The large oceanic reefs surrounding other Pacific Islands (e.g., Marshall Islands and Mariana Islands) have volcanic foundations similar to those of the Galápagos Islands. The age of the Galápagos Islands [3.2 million years (Hickman and Lipps, 1982; Hall, 1983)] provides ample time for reefs

MITCHELL W. COLGAN • Earth Sciences Board of Studies, University of California—Santa Cruz, Santa Cruz, California 95064. *Present address:* Department of Geology, College of Charleston, Charleston, South Carolina 29424.

to flourish, yet corals in the Galápagos Islands construct thin, small reefs unlike those in the central and western Pacific.

Reef construction is a long-term, historically contingent process, and is therefore susceptible to temporal environmental changes such as sea level fluctuations and/or climatic shifts. Today's reefs result from a post-Pleistocene "urban renewal" (Goreau, 1969; Stoddart, 1969), after melting ice and rising sea levels enabled corals to colonize and form reefs at their present locations. Some 18,000 years ago, lowered sea levels restricted coral growth throughout the world, and older high sea level build-ups were eroded under subaerial conditions (Hoffmeister and Ladd, 1944). These karst structures became the foundation for many modern reefs (Purdy, 1974). In the western and central Pacific, corals colonized these newly flooded karstic surfaces and grew to form today's large reefs (Davies and Marshall, 1980; Davies, 1983). In the Galápagos Islands, corals build reefs on basalt and are unable to form large reef structures. They appear mired in the earliest phases of colonization and development.

The Galápagos Islands experience seasonal, and occasionally intense upwelling of cool nutrient-rich waters that can negatively influences reef growth (Hallock, 1988; Hallock and Schlager, 1986). Consequently reefs are best developed in the nonupwelling locations of the eastern Pacific (Glynn and Stewart, 1973; Dana, 1975; Glynn, 1977; Wellington, 1984). Furthermore, volcanism, cooler waters, an abundance of bioeroding organisms, and distances from high diversity areas in the Indo-Pacific contribute to restrict reef growth (Stehli and Wells, 1971; Glynn *et al.*, 1979; Glynn and Wellington, 1983; Wellington, 1984). And until 1982, these physical and biological perturbations seemed adequate to explain the small size of Galápagos reefs.

An intense El Niño event in 1982–1983 increased sea levels, raised sea surface temperatures, and reduced equatorial and coastal upwelling of cool, nutrient-rich waters (Cane, 1983; 1986; Philander, 1983; Feldman *et al.*, 1984; Enfield, 1989). This intense occurrence raised mean sea surface temperatures 3°–4°C (Glynn *et al.*, 1988; Glynn and D'Croz, 1990) and stressed reef-building corals, turning vast stretches of reef from golden-brown to white (Glynn, 1983, 1984; Lessios *et al.*, 1983; Glynn *et al.*, 1988). In the Galápagos Islands, the catastrophic disturbance caused up to 97% coral mortality (Glynn, 1984; Robinson, 1985; Glynn *et al.*, 1988), and added new insight concerning the cause of the small size and poorly developed reefs of the Galápagos Islands (Glynn, 1988a; Colgan, 1990a,b).

An uplifted coral community at Urvina Bay, Isabela Island offers an opportunity to study the history of reef development in the Galápagos Islands. Because of the almost unprecedented access that this uplift provides to a normally submerged coral community, a detailed examination can be made of the growth and destruction of Galápagos Island reefs.

1.1. Urvina Bay

In the Galápagos Islands, cool, nutrient-rich upwelling waters commonly restrict coral reefs to the east sides of the islands (Glynn and Wellington, 1983). Urvina Bay is located on the west-central coast of Isabela Island at the foot of the Alcedo Volcano (Fig. 1), and is in thermal zone 5 (Harris, 1969), typically a warm-temperate region with little reef growth (Glynn and Wellington, 1983). Fernandina

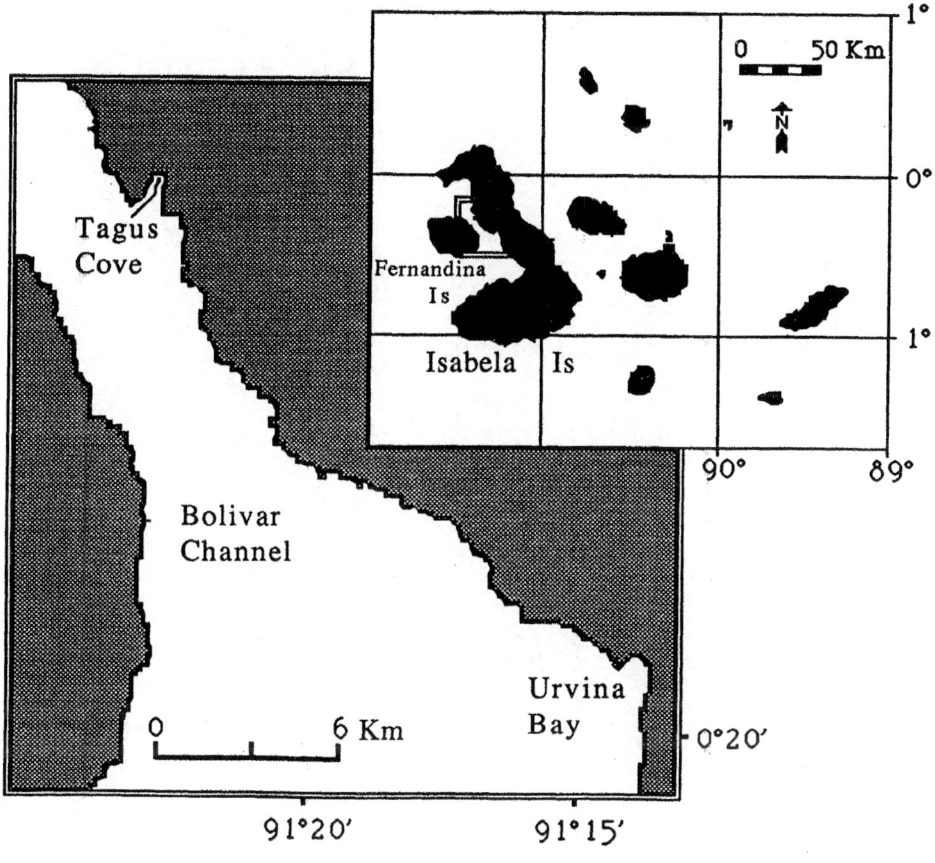

Figure 1. Location map of Urvina Bay, Isabela Island, Galápagos Islands.

Island protects Urvina Bay from intense upwelling, thus enabling corals to live along this section of Isabela Island (Glynn and Wellington, 1983). Just south of Urvina Bay, fleshy algae, more reminiscent of the temperate coast of California than of an equatorial region, dominate the hard bottom community. Small coral reefs were present in the waters of Urvina Bay until 1954 when this community was suddenly thrust above water.

1.2. The Uplift of Urvina Bay

In 1954, along the west-central coast of Isabela Island, the injection of magma suddenly raised a portion of Urvina Bay (Fig. 2) more than 6 m above sea level and drove the shoreline 1.2 km seaward (Couffer, 1956; Colgan and Malmquist, 1987; Malmquist, this volume). Such local uplifts, often caused by the intrusion of magma, commonly accompany island growth in the Galápagos (McBirney and Williams, 1969). The receding waters of Urvina Bay exposed several km^2 of a marine shelf with scattered coral communities. Left behind, when the waters disappeared,

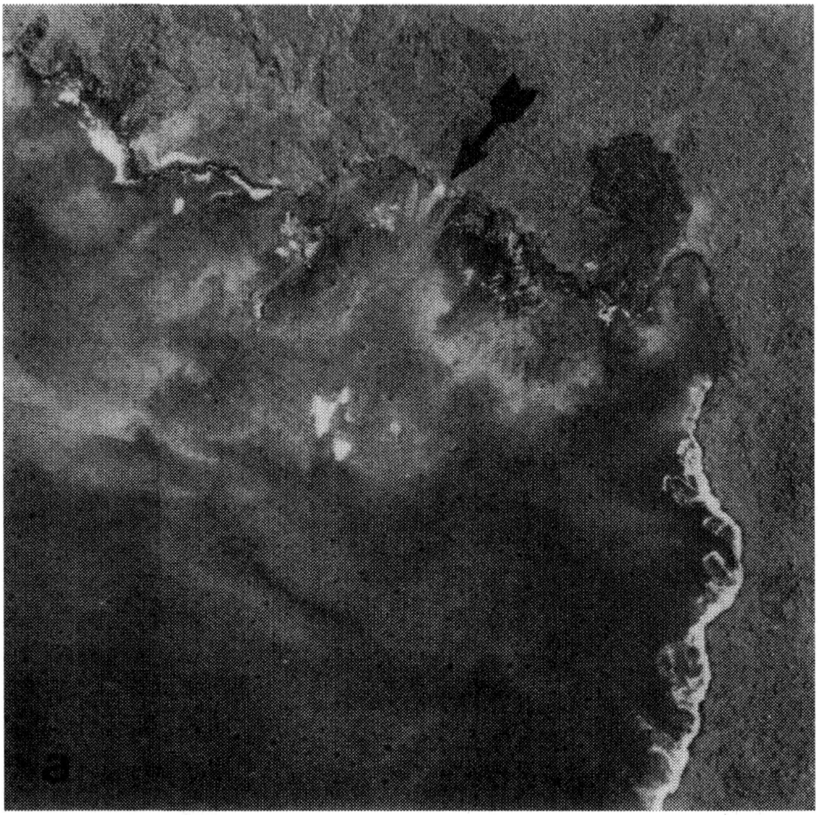

Figure 2. (a): Aerial photograph of Urvina Bay taken in 1946 before the uplift. The arrow points to a white, carbonate sand beach.

was a rich collection of invertebrate and vertebrate remains preserved nearly in place as they were in 1954 (Couffer, 1956). Evidence for the rapid uplift comes from the sedimentologic and paleontologic records (Colgan, 1990a; Malmquist, this volume). For example, the steep-angle, high energy beaches (now more than a kilometer from shore) appear unaltered by the receding waters. No small terraces or step-wise lateral displacement of the shoreline were found. Unable to escape during the uplift, sea turtles died in the subtidal pools and their skeletons now litter the uplifted pools (Chapter 18, this volume). One can walk through a 1954 coral community that bears witness to the effects of previous El Niño events (Couffer, 1956; Colgan and Malmquist, 1987; Chapter 18, this volume).

1.3. Urvina Bay Facies

A young "aa" lava flow from Volcán Darwin (at least 1000 years old, according to relative lava flow stratigraphy; K. Howard, pers. comm.) is the foundation of Urvina Bay. The irregular lava topography, water depth, and wave exposure produced several microhabitats, including beaches (sandy and rocky), rocky intertidal

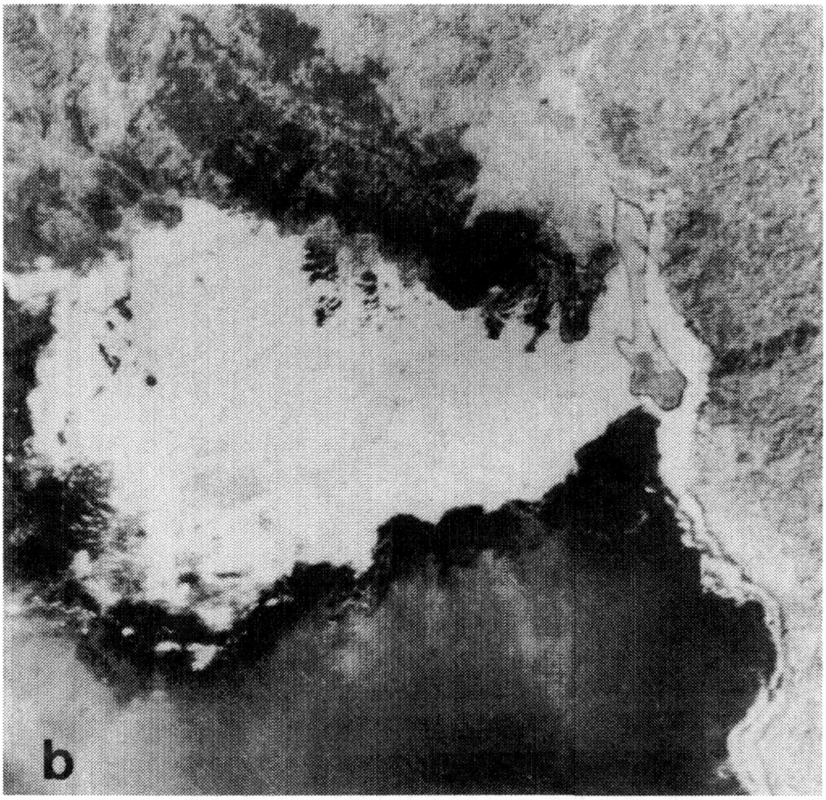

Figure 2. (b): Aerial photograph of Urvina Bay taken after the 1954 uplift, the beach is now nearly a kilometer from shore.

areas, mangrove thickets, a sandy bay, high-relief basalt ridges, low-relief basalt fields, sand channels, coral-filled channels, and small isolated coral communities and reefs (Fig. 3). A rich invertebrate community abounds, containing abundant sponges, hermatypic and ahermatypic corals, annelids, mollusks, arthropods, echinoderms, and tunicates. Some of the most abundant invertebrates are bioeroding sea urchins (e.g., *Eucidaris thouarsii* and *Echinometra vanbrunti*), and boring clams (e.g., *Lithophaga* spp.).

2. Methods

During three field seasons, the uplift was mapped using stadia rod, alidade, plane table, and measuring tape. All mapping was conducted in feet rather than meters. Since no maps of the uplift were available, the entire uplifted area of Urvina Bay was mapped (1:4800) and 11 detailed maps (1:600 and 1:1200) were produced. This survey method allowed us to locate accurately and measure many of the coral colonies. With the ease of walking, rather than diving, on a coral reef, we were able

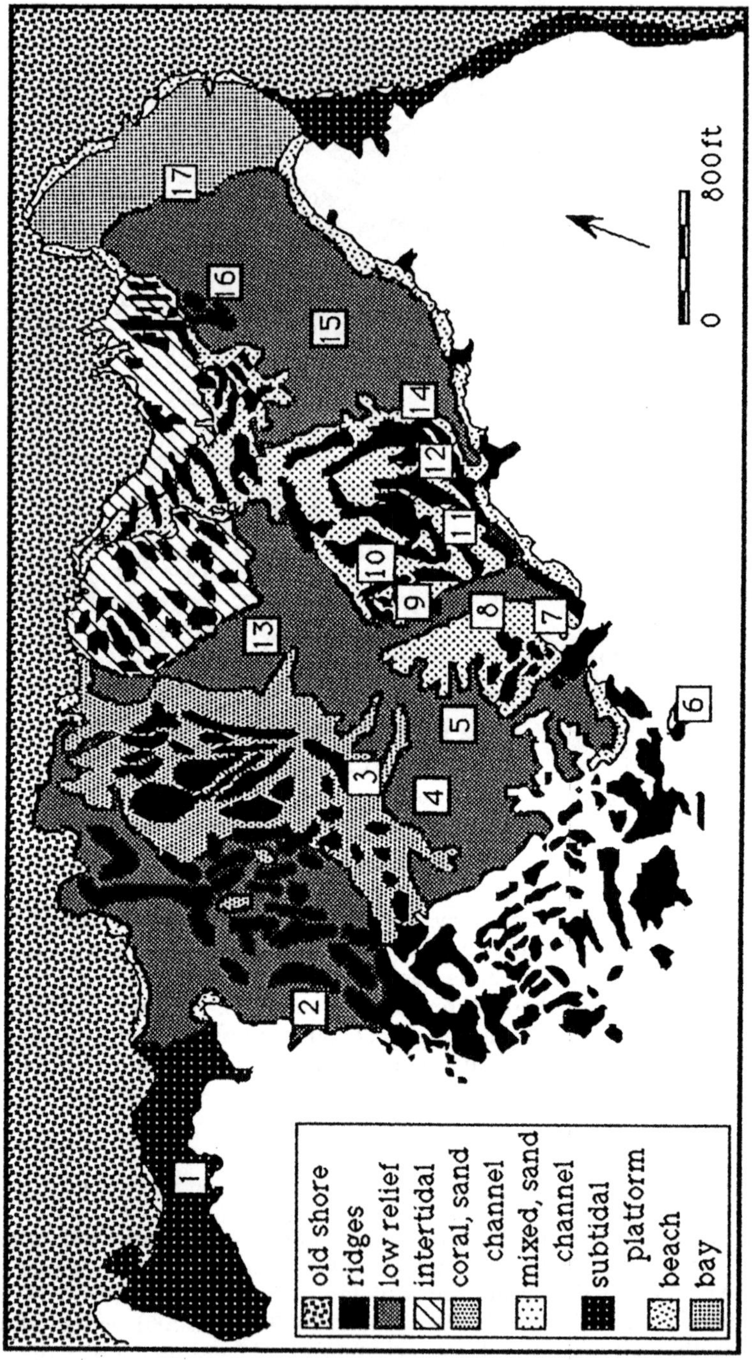

Figure 3. Map of the facies found on the Urvina Bay uplift. The numbers mark the location of 17 coral communities on the uplift. Detailed maps of site 3 (Fig. 5), site 7 (Fig. 9), site 8 (Fig. 6), and site 10 (Fig. 8) are presented later in the text.

to measure and map most of the individual coral colonies on the uplifted area. Fourteen transect surveys across the uplifted area augmented the mapping data (see Malmquist, this volume). A white line of calcareous algae marked the former mean low tide line, and we determined the paleowater depths by measuring distances below that marker.

We determined in the field if the coral was alive at the time of the uplift by examining the condition of the corallites (Fig. 4). If the corallites were not abraded, or covered by algae or other organisms and still in growth position, we assumed that the coral was alive.

Surface sediment samples (n = 132) consisting of approximately 500 g were collected in depressions and channels along the 14 transect lines. Channel stratigraphy and thickness were determined from cores, 4 cm wide plastic pipes, trenches, and steel probes.

Coral and red algal ages were determined using carbon 14 (± 70 years) with an atomic mass spectrometer at the Arizona AMS Facility.

3. Results and Discussion

3.1. Species Richness

Thirteen hermatypic ("reef building": Schuhmacher and Zibrowius, 1985) coral species belonging to six genera live in the waters around the Galápagos Islands (Wells, 1983). All genera have an Indo-Pacific distribution. Five of the 13 species are found only in the extreme eastern Pacific, although no hermatypic corals are endemic to the Galápagos Islands. Seven hermatypic coral species inhabited the shallow waters of Urvina Bay before the uplift, and with the exception of *Pavona varians*, all species [*Pocillopora damicornis, Pocillopora elegans, Psammocora* (*Stephanaria*) *stellata, Pavona clavus, Pavona gigantea*, and *Porites lobata*] were common on the uplifted area. The shallow paleowater depth of the uplifted community (approximately 7 m depth) may explain the absence of some species (e.g., *Cycloseris* spp.). However, recent (1987) examination of the submerged section of Urvina Bay uncovered no additional coral species. The corals found on the uplifted area are the major reef building species in the archipelago (i.e., *Pavona clavus, Pocillopora damicornis*, and *Porites lobata*) and also in the eastern Pacific (Dana, 1975; Glynn, 1976; Glynn and Wellington, 1983).

Unlike other reef building locations, the Galápagos Islands have a higher diversity of ahermatypic corals, 31 species, than hermatypic corals. Ten of the ahermatypic species are apparently endemic, and four others reflect a remnant Tethyan distribution with conspecifics found in the Mediterranean and Caribbean (Wells, 1983). The remaining 17 species have either a Panamic, Indo-Pacific, or cosmopolitan distribution (Wells, 1983). Many Galápagos ahermatypic corals live at depths greater than the 7 m exposed by the uplift. Therefore, few ahermatypic corals were found on the uplift. At least three species lived cryptically in crevices of the basalt ridges and rubble (*Astrangia equatorialis, Phyllangia consagensis*, and *Tubastraea coccinea*).

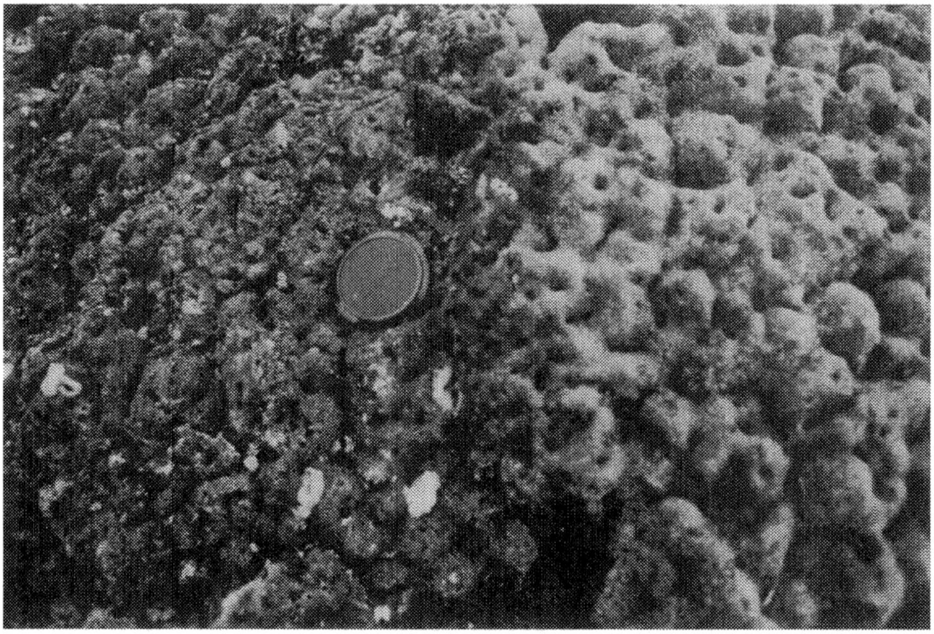

Figure 4. Close-up view of a colony of *Pavona clavus*. On the left, the portion of the colony that was dead at the time of uplift. The corallites are abraded and covered with vermetid mollusks and red algae. The right-hand portion of the colony was alive at the time of uplift. The corallites are clean and well defined. *Lithophaga*, a boring mussel, riddles the coral skeleton.

3.2. Coral Communities

The small, thin reefs of the Galápagos Islands are usually composed of a single predominant species. In 17 reef building localities surveyed by Glynn and Wellington (1983), 10 were dominated by *Pocillopora damicornis*, six by *Pavona clavus*, and one by *Porites lobata*. All these reefs, often less than 100 m^2 in size, were, like other eastern Pacific reefs, poorly cemented and weakly zoned (Glynn et *al.*, 1972; Glynn and Macintyre, 1977). Corallivores such as the sea urchin *Eucidaris thouarsii* were common at all locations. At Urvina Bay, a similar pattern was found. Coincidentally, our surveys also found 17 coral communities (Fig. 3). Each community was dominated by a single species, and in five locations only one species was found: *Pocillopora elegans* at sites 1, 2, and 6 (Fig. 3) and *Psammocora* at sites 7 and 18 (Fig. 3).

3.3. Branching Corals

One of three species of branching coral (*Pocillopora damicornis*, *Pocillopora elegans*, and *Psammocora stellata*) were found in each of the 17 communities. Small colonies of *Psammocora* were scattered throughout the uplifted area (Fig. 3).

3.3.1. *Pocillopora damicornis* Communities

As in many locations in the Galápagos Islands and the eastern Pacific, *Pocillopora damicornis* formed the most extensive, interlocking coral framework on the uplifted area (Glynn, 1976; Glynn and Wellington, 1983; Wellington, 1984). *P. damicornis* colonies were found in seven locations (Fig. 3; sites 3, 4, 7, 8, 9, 10, and 11) usually at paleowater depths greater than 2.5 m. *Pocillopora damicornis* often colonized depressions and the edges of basalt ridges.

The most extensive growth of *Pocillopora damicornis* occurred in the central portion of the uplift (Fig. 3; site 3) where it grew in and filled up the central channel that led to a coralline sand beach (Fig. 2). Greatest coral growth occurred in 3–4 m below mean low paleotide. In a survey covering 14,900 m^2 of the channel, *P. damicornis* colonies covered approximately 70% of the area and nearly filled the central channel (Fig. 5). Larger *Pocillopora* colonies measured more than 25 m across and stood over 1.0 m above the surrounding sediment. Scattered colonies of *Pavona clavus*, *Pocillopora elegans*, and *Psammocora* grew on the basalt surrounding the channel.

Probes and excavations of the channel revealed that an average of 93.4 cm of carbonate debris and *in situ* coral growth had accumulated ($s = 27.92$; max = 185 cm; $n = 29$). The thickness of the deposit varied because of the irregularities in the underlying basalt. Corals and algae started to colonize the area approximately 1000 years ago (985 ± 70 and 1080 ± 70 years B.P., respectively) when the lava foundation submerged. Trenches and cores disclosed a simple three layer stratigraphy of the channel fill: 1) The foundation of "aa" basalt was encrusted with calcareous algae and buried by algal-coated grains and molluscan and urchin debris. At the top of this layer, broken branches of *Psammocora* were found. 2) The next higher layer contained broken branches of *Pocillopora damicornis*, infiltrated with carbonate mud. 3) The *in situ* growth of *P. damicornis* forms the uppermost layer.

At the time of uplift, none of the *Pocillopora damicornis* colonies was alive. This is evidenced by upper and outer branches that were broken, urchin-eroded, over-turned, and algal encrusted. The widespread death of the *P. damicornis* colonies points to the 1941 El Niño warming event as the possible cause of mortality. The 1982–1983 El Niño event demonstrated the susceptibility of these fast growing colonies to warming events and *Pocillopora* colonies died in large numbers (Glynn, 1983; 1984). In the Galápagos Islands, extensive mortality resulted in the local extinction of *P. damicornis* (Glynn, 1988b).

If new corals were recruited to and grew on the dead *Pocillopora*, then framework development would have continued. However, at Urvina Bay, coral recruits did not colonize the dead framework, and without new growth, bioeroders (echinoderms, sponges, and mollusks) and currents tore down the previous growth. When a coral colony dies, algae colonize the dead skeleton attracting grazing herbivores that scrape the colonies' surface and remove carbonate (Glynn, 1990). Also, damselfishes (e.g., *Stegastes* spp.) cultivated algal lawns on the dead colonies and protected them by excluding sea urchins (Williams, 1979; Sammarco and Williams, 1982; Glynn and Wellington, 1983; Eakin, 1987). This protection resulted in a local decrease in bioerosion creating an irregular corallum topography with towers and pits (Eakin, 1988; Glynn, 1990) commonly seen on the uplift

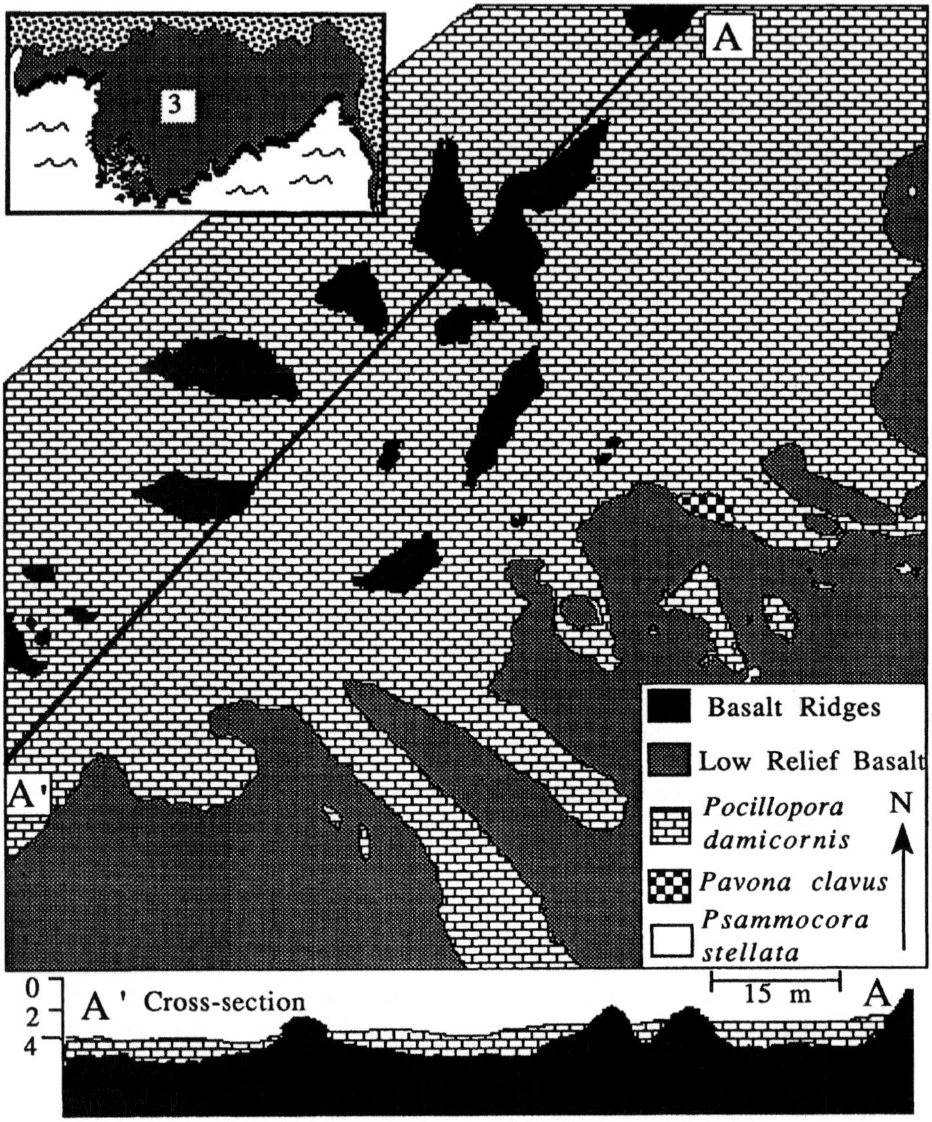

Figure 5. Map and cross-section of a part of the central channel that is filled with *Pocillopora damicornis* colonies (site 3). Inset shows the location of site 3 on the Urvina Bay uplift (Fig. 3).

(Colgan, 1990). However, sea urchins can potentially attack these damselfish towers from below, eventually toppling the protected sections of the colony. Three grazing urchins, *Eucidaris thouarsii*, *Echinometra vanbrunti*, and *Lytechinus semituberculatus* feed on dead *Pocillopora* colonies, and undercut and topple them (Malmquist, this volume). In one survey area covering 200 m^2, all 22 colonies examined were either urchin-pitted, undercut, or overturned. Currents and waves further tumbled the collapsed colonies to form an unconsolidated rubble of gravel sized (mean size > mm diameter), poorly sorted ($\sigma_\phi = 1.03$) *Pocillopora* branches and carbonate sands.

3.3.2. *Pocillopora elegans* Communities

At the time of the uplift, in contrast to the *Pocillopora damicornis* communities, most of the *Pocillopora elegans* colonies were alive and small in size. *P. elegans* colonies are found in seven locations (Fig. 3; sites 1, 2, 3, 6, 7, 11, 12, and 16), and in five areas they form discrete clusters of over 60 colonies. In these five locations, the colonies are uniformly small with mean widths ranging from 18.9 cm (s = 9.9, n = 60) to 22.5 cm (s = 12.1, n = 131), and there were no significant differences between sites (Colgan, 1990). Given an average growth rate of 2.4 mm/month (Glynn and Wellington, 1983), all of these colonies likely grew after the 1941 El Niño event.

One *Pocillopora elegans* community (site 16) is typical of these clusters, and is composed of an isolated group of 131 colonies that grew within a low-relief basalt depression in approximately 3 m paleowater depth. Along with *P. elegans*, two colonies of *Porites lobata* and a few scattered, small *Psammocora stellata* colonies share this 47.4 m^2 basin. The small *P. elegans* colonies, average width 18.8 cm, did not coalesce to form a framework, and cover only 1% of the depression. None of the *P. elegans* colonies grew on top of dead coral colonies, possibly because intense grazing may have excluded coral recruits from previous build-ups (Glynn, 1990). Instead, corals were recruited to the top of basalt outcrops forcing reef building to shift to new locations after coral death.

3.3.3. *Psammocora stellata* Communities

Psammocora stellata colonies were the most ubiquitous species on the uplifted area, and had the widest paleowater depth range, from 0.5 m to 7 m. Colonies occur in all but three locations (Fig. 3; sites 1, 2, and 6) and were the only species found in the shallow portion of the low relief basalt (Fig. 3; site 13) and on the rocky edge of the bay (Fig. 3; site 17). *Psammocora* colonies were smallest on the uplift with a mean width of 10.7 cm (s = 7.63, n = 379). On the basis of the location of the colony attachment sites on prominences and vertical surfaces I surmised that most of these small colonies were sexually reproduced recruits.

Within the low relief basalt (Fig. 3; site 13) a cluster of 38 *Psammocora* colonies, attached to the basalt, is typical of many shallow water clusters. In one survey of 11 m^2 area, 38 corals with an average width of 12.2 cm (s = 6.62, n = 38) covered less than 1% of the area. Most of the corals had settled on the upper portion of the basalt, and show a clustered distribution.

3.4. Massive Corals

Three species of massive corals (*Pavona clavus*, *Pavona gigantea*, *Porites lobata*) were found in 12 of the communities, and they tended to be more common in locations deeper than 3 m below mean low paleotide. Nearly monospecific stands of massive corals were found in two locations: *Pavona clavus* at site 8 (Fig. 3) and *Porites lobata* at site 14 (Fig. 3). Massive corals were commonly part of mixed communities; although, a single species usually dominated even in these locations. In no community was *Pavona gigantea* the predominate coral, and only one colony of *Pavona varians* is found on the uplifted area (Fig. 3; site 11).

Among the massive corals several themes are common: bioerosion, either relatively low sexual reproductive success or low recruitment success, and a skewed size class with a few large colonies and many smaller fragments. Several causes are central to these observations: including upwelling of nutrients, an abundance of bioeroders, and periodic El Niño events.

Bioerosion is a problem for the establishment of coral reefs in upwelling locations like Urvina Bay, and periodic El Niño events that lead to partial or complete colony death will accelerate bioerosion's effects (Glynn, 1990). An abundance of grazing sea urchins (e.g., *Eucidaris thouarsii* and *Echinometra vanbrunti*) and boring bivalves (e.g., *Lithophaga*) results in extensive bioerosion among all the massive corals. Some colonies, often smaller colonies, have been barely affected, but most colonies have undergone extensive bioerosion (Table 1). Without new recruits to dead colonies, bioerosion reduced 39 large (74 cm to 1249 cm wide), massive colonies to thin traces (Table 1; type 7) rising between 9 to 17 cm above the sediment (Colgan, 1990b). Bioerosion also leads to fragmentation, especially in *Pavona clavus*. Colonies of *Pavona clavus* are columnar, although they appear to be a massive, hemispherical colony like *Porites*. Bioerosion of these columnar branches can cause a colony to crumble. For example, a single large colony of *Pavona clavus* (3.25 m^2) collapsed and fragmented into 121 smaller pieces (mean width = 28.8 cm, s = 19.7) scattered over 11.05 m^2.

3.4.1. *Pavona clavus* Communities

Two coral locations (Fig. 3; sites 8 and 10) were used to examine the structure and development of *Pavona clavus* communities.

Site 8 is located between the low-relief basalt rubble and a coarse, bioclastic-rich sand channel (Fig. 3), and consists of a long, linear (30 m by 235 m) community of *P. clavus* colonies. In the survey area (7050 m^2), 424 separate colonies of *Pavona clavus* and 10 colonies of *Psammocora stellata* covered approximately 11% of the area. The average length of the *Pavona clavus* colonies showed a very skewed distribution with average widths ranging from 0.17 m to 13.1 m with mean size of 1.02 m (s = 1.18, n = 424). Of the *Pavona clavus* colonies, only 9% of the colonies were attached, apparently successful recruits, and the remainder were apparently clones. No unexpectedly, the attached colonies are larger (mean width = 3.4 m) than the fragments (mean width = 0.78 m).

The high amount of bioerosion and fragmentation, which may lead to asexual reproduction, partially compensate for the low success of sexual reproduction. At the time of the uplift, 35% of the colonies propagated through fragmentation were partially alive. One section (440 m^2) of the survey area illustrates the relationship between sexual reproduction, bioerosion, and fragmentation (Fig. 6). There, seven *Pavona* colonies settled and grew on a small rocky outcrop in approximately 4.5 m deep water. Partial or complete colony death, possibly related to an El Niño warming event, and bioerosion, primarily from *Eucidaris thouarsii*, fragmented

Table 1. Types of Bioerosion Seen in Massive Corals.[a]

TYPE		X-SECTION	DESCRIPTION
1			Skeleton intact, with little or no bioerosion
2			Skeleton intact, some bioerosion
3			Outer surface alive, but urchins have hollowed out the interior
4			The top and interior riddled, the outer edge is alive.
5			Most of the internal structure is eroded. Only small portions on the edge of colony are alive.
6			Entire colony riddled. No portion of the colony alive.
7			Planed down.
			Skeleton Intact; Skeleton Riddled

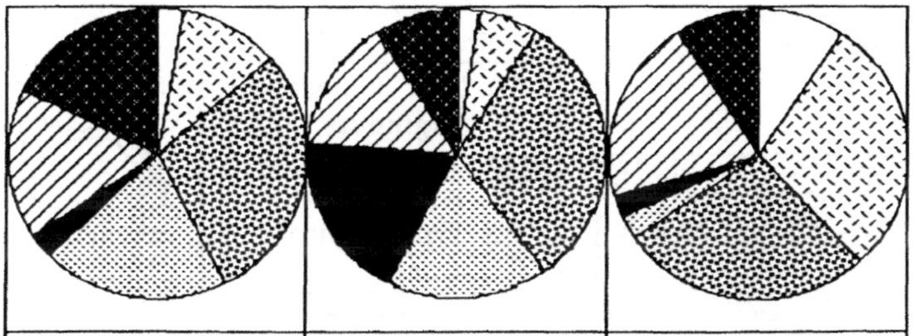

Porites lobata n = 138	*Pavona clavus* n = 239	*Pavona gigantea* n = 81

[a]The proportions of each bioerosion type within massive corals species and the numbers of colonies sampled are given below.

these colonies and produced at least 40 colonies. Currents transported some of these fragments to a nearby depression (5.5 m water depth) where some clones grew and fused forming a small, incipient reef framework (Fig. 6). The hemispherical colonies stand about 1 m high and appear to be solid, but sea urchins, attacking from under the colonies have hollowed out between 50 and 80% on the interiors leaving thin outer shells (Table 1; type 3; Colgan, 1990a). Between 6 and 12 urchins are found within each coral colony and, as a consequence, these hollowed corals would have eventually collapsed and fragmented.

Along with sea urchins, boring mussels (e.g., *Lithophaga* spp.) can weaken the *Pavona clavus* skeletons. At this site, the density of *Lithophaga* holes reach as high as approximately 100 holes/100 cm^2 (Fig. 4).

At another site (Fig. 3; site 10), an area of approximately 9680 m^2, a single large colony of *Pavona clavus* dominated the location. This coral community was framed by basalt ridges, submerged at low tide, and low relief "aa" basalt (Fig. 7). The large *Pavona clavus* colony grew approximately 4.3 m below mean low tide, and stood

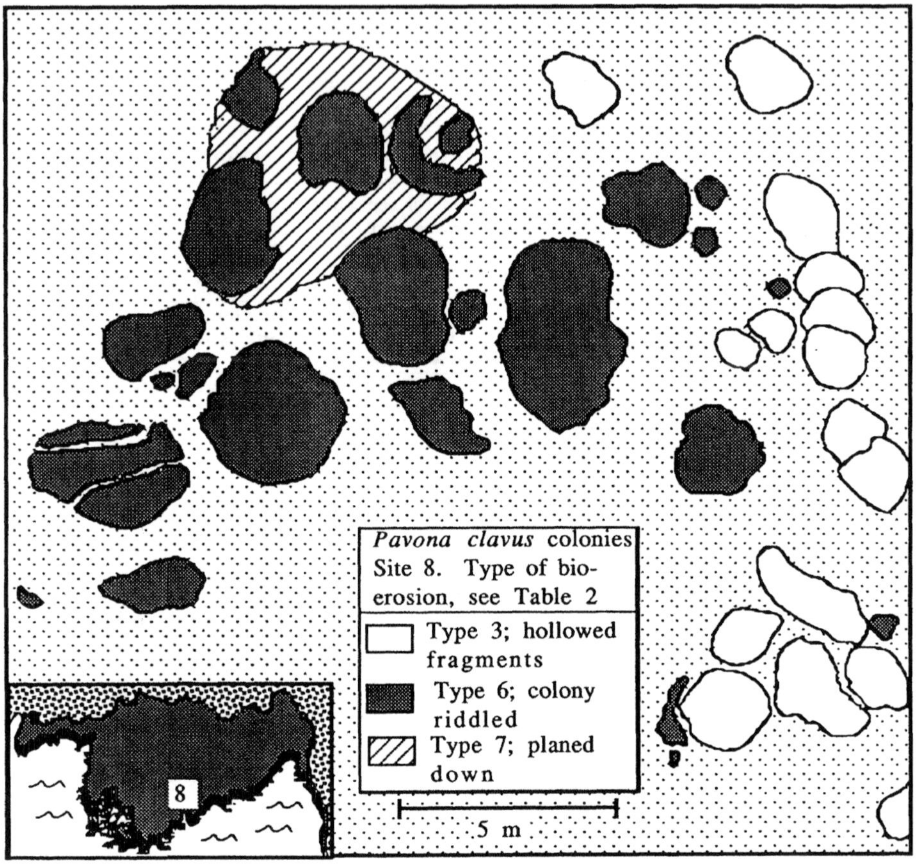

Figure 6. Planar view of a small section of the coral community mapped at site 8 (Fig. 3). All the colonies are *Pavona clavus*. The unshaded colonies are fragments, probably from the nearby larger colonies, transported to a small depression where some colonies have fused to form a small reef-like cluster. Inset shows the location of site 8 on the Urvina Bay uplift (Fig. 3).

3.1 m high and 11 m wide. The coral initially settled and grew on a basalt outcrop, now visible in the center of the colony, in the latter part of the 17th century (Dunbar *et al.*, 1988). The coral is a microatoll with the perimeter alive and the center dead and eroded (Fig. 8). The death of the upper portion of the colony may have resulted from either unusually low spring tides or an El Niño warming event that killed the upper portion of the colony. Once the upper surface of the coral was killed, bioeroders easily gained access to the colony's interior weakening the skeleton (*Eucidaris thouarsii* density = 8 tests/m^2; n = 20). This deterioration spawned 56 fragments around the base of the colony (mean length = 47.7 cm, s = 46.64), and toppled a portion of the colony (Fig. 8). Besides the coral fragments around the periphery of the colony, no other *Pavona clavus* colonies are found nearby. The large colony lived for nearly 350 years yet this apparent lack of success at recruiting corals to neighboring locations once again underscores the low rate of sexual reproductive success of this species. The failure of recruitment may be partially caused by the high concentration of urchins in the ridges and rocky outcrops surrounding the coral (*Eucidaris thouarsii* density = 24 tests/m^2; n = 20 quadrats).

3.4.2. *Porites lobata* Communities

On the edge of the uplift in approximately 5 m paleowater depth, a coral community consisting of a cluster of *Porites lobata* stands isolated from any other reef growing locations (Fig. 3; site 14). The 52 *Porites lobata* colonies grew in a 1546 m^2 sandy-rocky depression between the high ridge and low-relief basalt. *Porites lobata* colonies, concentrated in four clusters, cover 3.19% of the depression, and are the only coral species in the area. The average colony extends 90.53 cm wide (s = 64.9; n = 52) and 55.8 cm high (s = 35.4) with the largest colony standing 307 cm wide and 130 cm high. Most of the colonies (54%) apparently reproduced sexually and the remainder are loose fragments of nearby colonies. The fragments are significantly smaller than those that were apparently sexually reproduced (mean width 120 cm vs 54 cm; $t_{49} = 4.097$, $p < 0.01$). At the time of the uplift, 8% of the colonies were dead.

3.4.3. Mixed Massive Communities

Two coral communities had rich and evenly distributed collections of corals (Fig. 4; sites 7 and 11). One community (site 7), located near the "lagoon," containing colonies of *Pavona clavus*, *Pavona gigantea*, *Porites lobata*, and *Pocillopora damicornis* and a few scattered colonies of *Psammocora* were used to analyze these communities. Although diverse, the colonies are separated and have not coalesced to form a true framework (Fig. 9). Two *Pavona clavus* colonies gained an enormous size (12.5 m and 11.0 m across, respectively), however these, like all of the larger *Pavona* colonies, were fragmented and a halo of smaller colonies was found near them. *Pavona gigantea* colonies were the smallest massive colonies at this site and nearly half are attached and many had an encrusting growth form (n = 148; Table 2).

3.4.4. Corallith Community

At site 5 (Fig. 3), 42 attached colonies and 89 loose fragments of *Pavona clavus* were counted in three sampling areas. These colonies cover approximately 4.0% of 162 m^2. Among the fragments are rounded (average sphericity = 0.79; sphere = 1.00), apparently mobile colonies, with coral growth occurring on all sides. These rounded massive colonies of *Pavona clavus* or coralliths (Glynn, 1974), account for 36% of the fragments. Branching colonies of *Psammocora* also formed coralliths.

As in other locations, successful sexual recruitment appeared sporadic with a few coral recruits surviving at higher elevations on the basalt boulders. Site 5 had a high concentration of the sea urchins *Eucidaris thouarsii* and *Echinometra vanbrunti* (*Eucidaris thouarsii* density = 13.6 tests/m^2; n = 20), and these bioeroders had hollowed out all of the larger colonies (> 0.5 m length) causing portions of the colony to collapse (Table 1; types 3 and 5). There are no large colonies as a consequence of the high density of bioeroders and fragmentation. These colonies

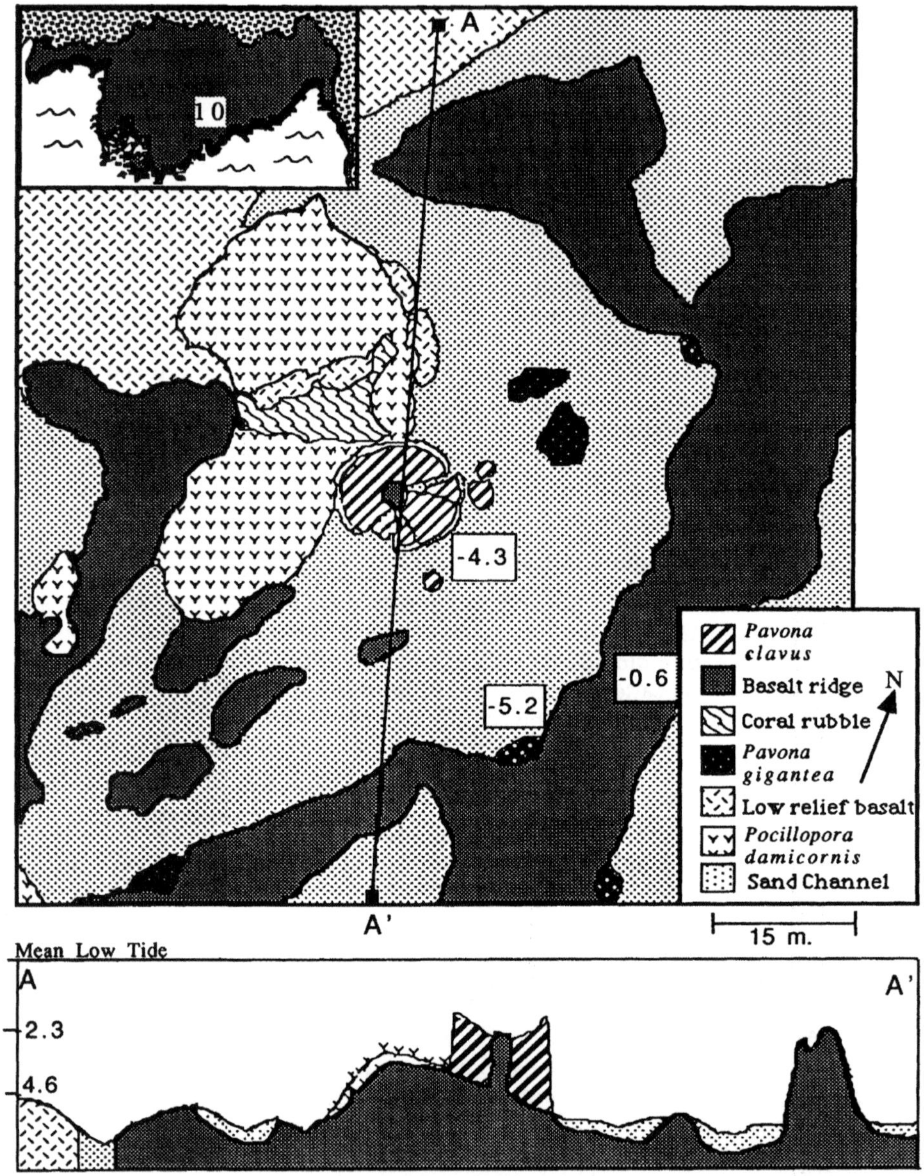

Figure 7. Planar (above) and vertical cross-section (below) views of site 10 (Fig. 3). In the center is a large *Pavona clavus* colony (Fig. 8). Inset shows the location of site 10 on the Urvina Bay uplift (Fig. 3).

Figure 8. A large colony of *Pavona clavus* (max. diameter, 12 m) located at site 10. Bioerosion weakened the skeleton causing a portion of the colony to fragment before being uplifted.

are the apparent source of the fragments that were transported to the lower lying areas of this coral community. Currents and perhaps fishes (Glynn, 1974) turned the coral fragments allowing growth to occur on all sides thus becoming coralliths. Some of these nomad colonies wedged into cracks and continued to grow, which may account for the relatively high concentration of attached *Pavona clavus* colonies.

4. Conclusion

The marine community at Urvina Bay may have existed for at least 1000 years, and the oldest living colony was recruited to Urvina Bay at least 350 years before the uplift. Yet, the coral reefs are poorly developed and scattered. In a few locations, small incipient reef frameworks were formed, but they were not zoned reef structures typical of other Pacific locations.

Coral reefs develop as a compromise between carbonate accretion and erosion. At Urvina Bay, as in other Galápagos reef settings, where there is intense erosional pressure from internal and external bioeroders, the balance between construction and destruction can be easily upset. When disturbances slow or halt coral growth and recruits fail to colonize successfully the dead coral colonies, those portions of the reef erode (Glynn, 1990). During the 1982–1983 El Niño event, when coral mortality was widespread, entire reef communities began to erode away.

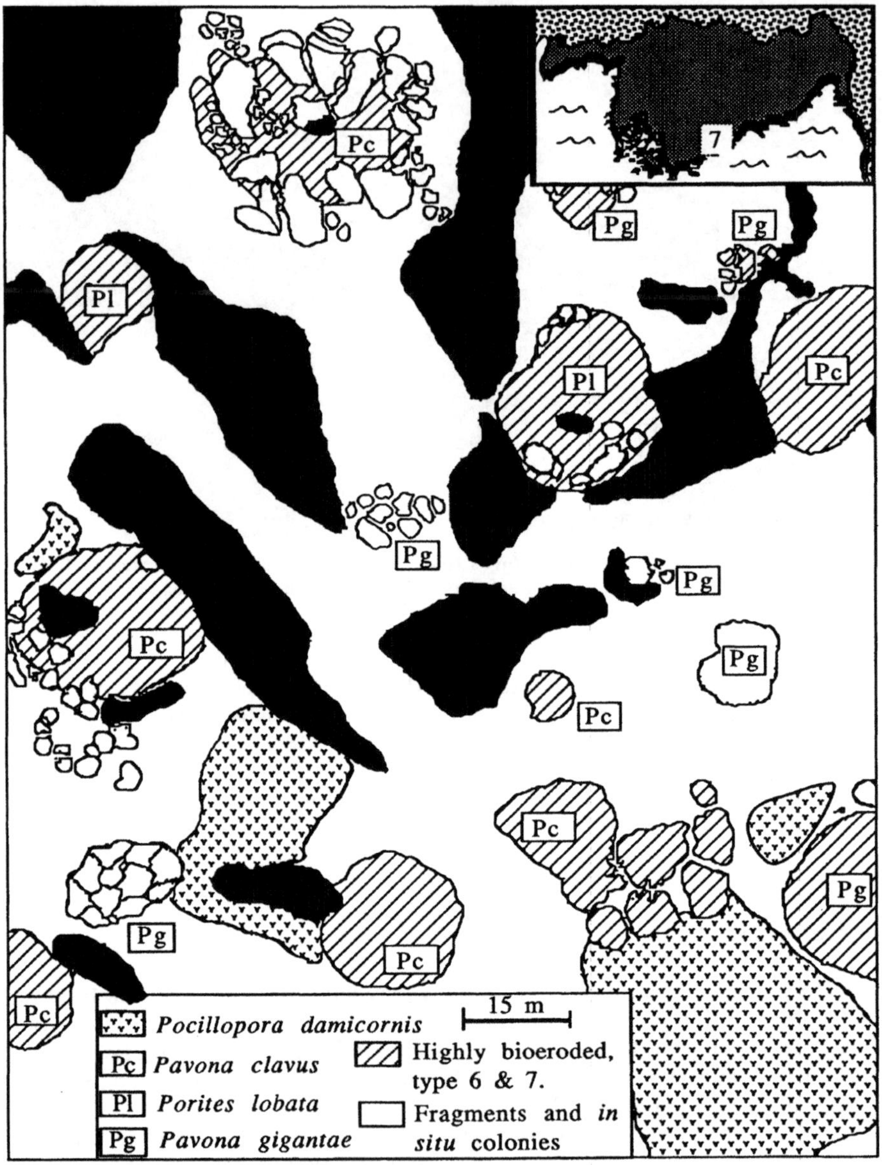

Figure 9. Planar view of a portion of the mixed coral community at site 7 (Fig. 3). Most massive colonies were bioeroded and fragmented. The *Pocillopora damicornis* colonies were dead at the time of the uplift. Inset shows the location of site 3 on the Urvina Bay uplift (Fig. 3).

Table 2. Coral Size Distribution in the Lagoon Area[a]

Species	Mean colony length (cm)	Range (cm)	Number (n)
Pavona clavus	85.4	4.9–1250	156
Attached	495.6		19
Fragment	34.1		137
Pavona gigantea	68.0	8.7–578	148
Attached	85.2		72
Fragment	49.8		76
Porites lobata	125.5	8.9–1173	72
Attached	224.8		29
Fragment	58.5		43

[a]See Fig. 3; Site 7; Fig. 9.

The small reefs of Urvina Bay in particular and the Galápagos Islands in general formed as a response to seasonal upwelling of cool nutrient-rich waters, low reproductive success, intense bioerosion, and periodic El Niño warming events. Cool, nutrient-rich upwelling waters retard coral growth (Stoddart, 1969; Jokiel and Coles, 1977; Kinsey and Davies, 1979; Glynn and D'Croz, 1990), stimulate the growth of benthic algae and other organisms that compete for space with surviving coral colonies and new recruits (Birkeland, 1988), and favor suspension and filter feeding bioeroders (Birkeland, 1977; Highsmith, 1980; 1981; Hallock and Schlager, 1986). These filter and suspension feeders (e.g., boring bivalves and sponges) along with grazing organisms, such as sea urchins, erode coral colonies.

Intense El Niño warming events stress and kill many coral colonies (Glynn, 1988a; 1990). On other Pacific locations, reefs recover rapidly following extensive, catastrophic coral mortality, similar in severity to the El Niño perturbation, through the growth of surviving corals, recruitment of new corals, and survival of the reef framework (Pearson, 1981; Colgan, 1987). In the Galápagos Islands, however, the return of upwelling nutrient-rich waters (Hansen, 1990) and the survival of *E. thousarsii* caused secondary mortality further slowing reef recovery (Glynn, 1988b).

Unlike some areas in the western Pacific, Galápagos Islands' sea urchin grazing excludes coral recruits (Glynn, 1990). The near elimination of coral recruits from dead colonies impedes carbonate accumulation and abandons the existing build-ups to biological and mechanical erosion. While some build-ups are being eroded, coral recruits may settle at new locations, where grazing pressures are less, thereby moving the site where corals grow and reefs develop.

Over the last 6500 years, from 18 to 65 El Niño events of the 1982–1983 magnitude could have disturbed the eastern Pacific reef building region (Colgan, 1990b). The recurrence of intense El Niño events brings a new understanding of the causes for the small reefs. Since the onset of glacial cycles, the eastern Pacific has faced two different climatic states that restricted reef growth and development: one during glacial periods with cooler waters, more intense upwelling (possibly doubling productivity), and lowered sea levels (De Vries and Schrader, 1981; Reimers and Suess, 1983; Schramm, 1985; Lyle *et al.*, 1988), and the other during interglacial periods with higher sea levels, warmer waters, and El Niño events (Colgan, 1990b).

During interglacial times, corals recolonize the continental shelf. In the western and central Pacific, where previous high stand reefs survived subaerial exposure, new reefs were built on antecedent structures enabling carbonate accumulation to continue (Goreau, 1969; Stoddart, 1973; Purdy, 1974; Davies, 1983). Whereas in the Galápagos, recruits settled on basaltic outcrops and reef building had to start from scratch. Recruits to the barren basalt faced not only harsh reef building conditions (e.g., upwelling and intense grazing), but recurrent intense El Niño events, which caused repeated coral death. Coralline carbonate accumulates between El Niño events, and reefs start to form. But after coral mortality, bioerosion removes much of the coral build-up. This repeated process prevents coral communities from developing large reef structures that can withstand El Niño related erosion. Thus, one generation's growth is not transferred to the next. Reefs in the Galápagos Islands are short-lived structures growing between major environmental perturbations without a continuity of space and time necessary for the formation of large reefs.

ACKNOWLEDGMENTS.. I thank Linda Anderson, William Anderson, Saïn Chai Colgan, Rene Espinosa, David Hollander, E. Allison Kay, Melissa Keppel, Léo F. Laporte, Margaret Laporte, David Malmquist, Olga Odenitez-Collart, and Tom Smalley for field assistance. For their help in the Galápagos Islands, I thank Sylvia Harcourt, Hank Kasteleijin, and Günther Reck at the Charles Darwin Research Station, Juan Black at the Charles Darwin Foundation, and Miguel Cifuentes and Humberto Ochoa at Parque Nacional Galápagos. Gene Gonzales, Sarah Griscom, Keith Howard, Eduardo Lopez, and R. Larry Phillips gave research support. TAME, an Ecuadorian national airline, subsidized air travel to the Galápagos Islands.

Discussions with Margaret Delaney, Peter Glynn, Matthew James, Léo Laporte, David Malmquist, and Gerard Wellington greatly improved the clarity of this manuscript.

This research was funded by NSF award EAR 8508966, a Chancellor's Special Research Fund from the University of California–Santa Cruz to Léo F. Laporte, and supported by NSF award OCE 8415615 to Peter W. Glynn. NSF Arizona AMS Facility NSF EAR 8512761 subsidized the dating of coral and algal samples.

References

Birkeland, C., 1977, The importance of biomass accumulation in the early successional stages of benthic communities to the survival of coral recruits, *Proc. 3rd Int. Coral Reef Symp.* **1**:15–21.

Birkeland, C., 1988, Geographic comparisons of coral-reef community processes, *Proc. 6th Int. Coral Reef Symp.* **1**:211–220.

Cane, M. A., 1983, Oceanographic events during El Niño, *Science* **222**:1189–1194.

Cane, M. A., 1986, El Niño, *Annu. Rev. Earth Planet. Sci.* **14**:43–70.

Colgan, M. W., 1987, Coral reef recovery on Guam (Micronesia) after catastrophic predation by *Acanthaster*, *Ecology* **68**:1592–1605.

Colgan, M. W., 1990a, Geology, paleontology, and the effects of El Niño on the development of an uplifted coral community at Urvina Bay, Isabela Island, Galápagos Islands. Ph.D. Dissertation, Univ. Calif., Santa Cruz, 118 pp.

Colgan, M. W., 1990b, El Niño events and the history of reef building in the eastern Pacific, in: *Global Ecological Consequences of the 1982–83 El Niño-Southern Oscillation* (P. W. Glynn, ed.), pp. 183–232, Elsevier Oceanography Series, Elsevier, New York.

Colgan, M. W., and Malmquist, D. L., 1987, The Urvina Bay Uplift: a dry trek through a coral community. *Oceanus* **30**:61–66.

Couffer J. C., 1956, The disappearance of Urvina Bay, *Nat. Hist.* **65**.:78–83.

Dana, T. F., 1975, Development of contemporary eastern Pacific coral reefs, *Mar. Biol.* **33**:355–374.

Darwin, C., 1842, *The Structure and Distribution of Coral Reefs*, Smith, Elder & Co., London. 214 p.

Davies, P. J., 1983, Reef growth, in: *Perspectives on Coral Reefs* (D. J. Barnes, ed.), pp. 69–106, Brain Clouston Pub., Manuka, Australia.

Davies, P. J., and Marshall, J. F., 1980. A model of epicontinental reef growth, *Nature* **287**:37–38.

De Vries, T. J., and Schrader, H., 1981, Variation of upwelling/oceanic conditions during the latest Pleistocene through Holocene off the Central Peruvian coast: diatom record, *Mar. Micropaleo.* **6**:157–167.

Dunbar, R. B., Wellington, G., Colgan, M., and Glynn, P. W., 1988, Galápagos massive corals: an annual record of eastern tropical Pacific climate during the past 400 years, *EOS* **68**:1743.

Eakin, C. M., 1987, Damselfish and their algal lawns: a case of plural mutualism, *Symbiosis* **4**:275–288.

Eakin, C. M., 1988, Avoidance of damselfish lawns by the sea urchin *Diadema mexicanum* at Uva Island, Panama, *Proc. 6th Int. Coral Reef Symp.* **2**:21–26.

Enfield, D. B., 1989, El Niño, past and present, *Rev. Geophys.* **27**:159–187.

Feldman, G., Clark, D., and Halpren, D., 1984, Satellite color observations of the phytoplankton distribution in the Eastern Equatorial Pacific during the 1982–1983 El Niño, *Science* **226**: 1069–1071.

Glynn, P. W., 1976, The impact of *Acanthaster* on corals and coral reefs in the eastern Pacific, *Environ. Conserv.* **1**:295–304.

Glynn, P. W., 1976, Some physical and biological determinants of coral community structure in the eastern Pacific, *Ecol. Monogr.* **46**:431–456.

Glynn, P. W., 1977, Coral growth in upwelling and nonupwelling areas off the Pacific coast of Panama, *J. Mar. Res.* **35**:567–587.

Glynn, P. W., 1983, Extensive 'bleaching' and death of coral reefs on the Pacific coast of Panama, *Environ. Conserv.* **10**:149–154.

Glynn, P. W., 1984, Widespread coral mortality and the 1982–83 El Niño warming event, *Environ. Conserv.* **11**:133–146.

Glynn, P. W., 1988a, El Niño-Southern Oscillation 1982–1983: nearshore population, community and ecosystem responses, *Annu. Rev. Ecol. Syst.*, **19**:309–345.

Glynn, P. W., 1988b, El Niño warming and reef framework destruction by echinoid bioerosion in the eastern Pacific, *Galaxea* **7**:129–160.

Glynn, P. W., 1990, Coral mortality and disturbances to the coral reefs of the tropical eastern Pacific, in: *Global Ecological Consequences of the 1982–83 El Niño-Southern Oscillation*, (P. W. Glynn, ed.), pp. 55–126, Elsevier Oceanography Series, Elsevier, New York.

Glynn, P. W., and D'Croz, L., 1990, Experimental evidence for high temperature stress as the cause of El Niño-coincident coral mortality, *Coral Reefs* **8**:181–191.

Glynn, P. W., and Macintyre, I. G., 1977, Growth rate and age of coral reefs on the Pacific coast of Panama, *Proc. 3rd Int. Coral Reef Symp.* **2**:251–259.

Glynn, P. W., and Stewart, R. H., 1973, Distribution of coral reefs in the Pearl Islands (Gulf of Panama) in relationship to thermal conditions, *Limnol. Oceangr.* **18**:367–379.

Glynn, P. W., and Wellington, G. M., 1983, *Corals and Coral Reefs of the Galápagos Islands*, University of California Press, Berkeley, California.

Glynn, P. W., Stewart, R. H., and McCosker, J. E., 1972, Pacific coral reefs of Panama: Structure, distribution and predators, *Geol. Rundschau* **61**:483–519.

Glynn, P. W., Wellington, G. M., and Birkeland, C., 1979, Coral reef growth in the Galápagos: Limitation by sea urchins. *Science* **203**:47–49.

Glynn, P. W., Cortes, J., Guzman, H. M., and Richmond, R. H., 1988, El Niño (1982–1983) associated coral mortality and relationship to sea surface temperature deviations in the tropical eastern Pacific, *Proc. 6th Int. Coral Reef Symp.* **3**:237–243.

Goreau, T. F., 1969, Post Pleistocene urban renewal in coral reefs, *Micronesica* **5**:323–326.

Hall, M. L., 1983, The origin of Española Island and the age of terrestrial life on the Galápagos Islands, *Science* **221**:545–547.

Hallock, P., 1988, The role of nutrient availability in bioerosion: Consequences to carbonate buildups, *Palaeogeogr., Palaeoclimat., Palaeoecol.* **63**:275–291.

Hallock, P., and Schlager, W., 1986, Nutrient excess and the demise of coral reefs and carbonate platform, *Palaios* **1**:389–398.

Hansen, D. V., 1990, Physical aspect of the El Niño event of 1982–1983, in: *Global Ecological Consequences of the 1982–83 El Niño-Southern Oscillation* (P. W. Glynn, ed.), pp. 1–20, Elsevier Oceanography Series, Elsevier, New York.

Harris, M. P., 1969, Breeding seasons of sea-birds in the Galápagos Islands, *J. Zool.* **27**:1–24.

Hickman, C. S., and Lipps, J. H., 1984, Geological youth of Galápagos Islands confirmed by marine stratigraphy and paleontology, *Science* **227**:1578–1580.

Highsmith, R. C., 1980, Geographic patterns of coral bioerosion: A productivity hypothesis, *J. Exp. Mar. Biol. Ecol.* **117**:193–198.

Highsmith, R. C., 1981, Coral bioerosion: Damage relative to skeletal density, *Am. Nat.* **117**:193–198.

Hoffmeister, J. E., and Ladd, H. S., 1944, The antecedent platform theory, *J. Geol.* **53**:388–402.

Jokiel, P. L., and Coles, S. L., 1977, Effects of temperature on the mortality and growth of Hawaiian reef corals, *Mar. Biol.* **43**:201–208.

Kinsey, D. W., and Davies, P. J., 1979, Effects of elevated nitrogen and phosphorus on coral reef growth, *Limnol. Oceanogr.* **24**:935–940.

Lessios, H. A., Glynn, P. W., and Robertson, D. R., 1983, Mass mortalities of coral reef organisms, *Science* **222**:715.

Lyle, M., Murray, D. W., Finney, B. P., Dymond, J., Robbins, J. M., and Brooksforce, K., 1988, The record of Late Pleistocene biogenic sedimentation in the eastern tropical Pacific Ocean. *Paleoceanography* **3**:39–59.

McBirney, A. R., and Williams, H., 1969, Geology and petrology of the Galápagos Islands, *Geol. Soc. Am. Mem.* **118**:1–197.

Pearson, R. G., 1981, Recovery and recolonization of coral reefs, *Mar. Ecol. Prog. Ser.* **4**:105–122.

Philander, S. G. H., 1983, El Niño Southern Oscillation phenomena, *Nature* **302**:295–301.

Purdy, E. G., 1974, Reef configurations: cause and effect, in: *Reefs in Time and Space* (L. F. Laporte ed.), *Soc. Econ. Paleo. Mineral. Spec. Pub.* **18**:9–76.

Reimers, C. E., and Suess, E., 1983, Late Quaternary fluctuations in the cycling of organic matter off central Peru: A proto-kerogen record, in: *Coastal Upwelling: Its Sedimentary Record, Part B: Sedimentary Record of Ancient Coastal Upwelling* (J. Thiede and E. Suess, ed.), pp. 497–526, Plenum Press, New York.

Robinson, G., 1985, The influence of the 1982–1983 El Niño on Galápagos marine life, in: *El Niño in the Galapágos Islands: the 1982–1983 Event* (G. Robinson and E. M. Píno, eds.), pp. 153–190, Charles Darwin Foundation for the Galápagos Islands, Quito, Ecuador.

Sammarco, P. W., and Williams, A. H., 1982, Damselfish territoriality: Influence on *Diadema* distribution and implication for coral community structure, *Mar. Ecol. Prog. Ser.* **8**:53–59.

Schramm, C. T., 1985, Implication of radiolarian assemblages for the late Quaternary paleoceanography of the eastern equatorial Pacific, *Quatern. Res.* **24**:204–218.

Schuhmacher, H., and H. Zibrowius, 1985, What is hermatypic? A redefinition of ecological groups in corals and other organisms, *Coral Reefs* **4**:1–9.

Stehli, F. G., and Wells, J. W., 1971, Diversity and age patterns in hermatypic corals, *Syst. Zool.* **20**: 115–126.

Stoddart, D. R., 1969, Ecology and morphology of recent coral reefs, *Biol. Rev.* **44**:433–498.

Stoddart, D. R., 1973, Coral reefs: the last two million years, *Geography* **58**:313–323.

Wellington, G. M., 1984, Marine environment and protection, in: *Key Environments, Galápagos*, (R. Perry, ed.), pp. 247–265, Pergamon Press, Oxford, England.

Wells, J. W., 1983, Annotated list of the scleractinian corals of the Galápagos Islands, in: *Corals and Coral reefs of the Galápagos Islands* (P. W. Glynn and G. W. Wellington), pp. 212–295, Univ. Calif. Press, Berkeley, California.

Williams, A. H., 1979, Interference behavior and ecology of threespot damselfish (*Eupomacentrus planiforns*). *Oecologia* **38**:223–230.

IV

Arthropoda (Crustacea)

Chapter 5

Taxonomy, Distribution, and Ecology of Galápagos Brachyura

JOHN S. GARTH

1. Taxonomy of Galápagos Brachyura

1.1. Introduction

This chapter constitutes an update of my "Littoral brachyuran fauna of the Galápagos Archipelago" (Garth, 1946a). Written 45 years ago, it dealt with a total of 120 species, 87 genera, and 15 families of "short-tailed" crabs, of which 44 species, 32 genera, and five families were reported for the first time from the Galápagos, largely as the result of five Hancock Expeditions (1932, 1933, 1934, 1935, and 1938) of which I was a member. Although I have not returned to the islands meanwhile, I have reviewed two important Galápagos collections: one made by Victor A. Zullo, resident systematist, Marine Biological Laboratory, Woods Hole, Massachusetts, in January–February 1964, as a member of GISP (Galápagos International Scientific Program); the other by Gerard M. Wellington [University of Houston, Houston, Texas (formerly Charles Darwin Research Station, Isla Santa Cruz, Galápagos, from November 1973 to August 1975)], while laying the groundwork for the establishment of a marine national park (Wellington, 1976). As a result of these collections, generously donated to the Allan Hancock Foundation (AHF), eight species and four genera were added to the insular fauna.

Of importance equal to the discovery of species and genera of brachyurans new to the Galápagos Islands was the rediscovery, in the University Museum, Oxford,

JOHN S. GARTH • Allan Hancock Foundation, University of Southern California, Los Angeles, California 90089-0371.

England, of Bell's lost types (DiMauro, 1982). Among them were syntypes of *Herbstia edwardsii*, *Thoe erosa*, *Pisoides edwardsii*, *Microphrys aculeatus*, *Mithrax ursus* (= *Mithrax bellii* Gerstaecker, 1857), *Mithrax nodosus*, *Mithrax denticulatus*, *Pitho sexdentata*, *Pericera* (= *Stenocionops*) *ovata* (all Bell, 1835); *Persephona edwardsii* and *Leucosilia jurinei* (both Bell, 1855). All were collected by Hugh Cuming, and were attributed to the Galápagos Islands. Because Zullo and Wellington failed to find eight of the ten species described by Bell (1835, 1855), as had Hancock and other expeditions before them, it was agreed (by Wellington and the writer, in correspondence) that they should be dropped from the Galápagos list, leaving it, as before, at 120 species. Corrected type localities for several species of spider crabs were given in my monograph on Pacific American Oxyrhyncha (Garth, 1958).

The Spedizione Lodovico Mares–G.R.S.T.S. (Gruppo Ricerche Scientifiche e Tecniche Subacquee, Florence, Italy) to the Galápagos Islands (December 1971 to January 1972) assembled a small but interesting collection of decapod Crustacea. Included were a dozen species of Brachyura, all of them previously reported. This material is preserved in the collections of the Museo Zoologico del-l'Universita, Florence (MF), with duplicate specimens in the Rijksmuseum van Natuurlijke Historie, Leiden, The Netherlands (RMNH) (Holthuis, 1979).

Although the number of species of brachyuran crabs found in the Galápagos remains the same, the number of genera has increased phenomenally. This is due to the tendency among taxonomists to break up, or "split," the larger genera into smaller ones. Such has been the fate of the older, more inclusive genera of Xanthidae: in particular, *Actaea*, *Medaeus*, *Micropanope*, and *Pilumnus*, well represented in the Galápagos. At least seven genera have been added to the Galápagos fauna as a result of such proliferation. These, when added to the genera of newly discovered species, which total four, make eleven genera to be added to the earlier list; however, the deletion of the eight species described by Bell, but not since found in the Galápagos, requires the deletion of several genera as well. The list (Table 1) now numbers 120 species, 91 genera, and 15 families of brachyuran crabs. Table 2 lists name changes, additions, and deletions from the original list.

Two additional changes have been proposed that affect the Galápagos brachyuran fauna. The oxystomatous genus *Osachila*, represented in the archipelago by *O. levis* and *O. galapagensis*, has been transferred from the family Calappidae to the family Parthenopidae (Guinot, 1966, p. 841). The genus *Daira*, represented in the islands by *D. americana*, also has been transferred to the Parthenopidae (Guinot, 1967b, p. 550), but on more tenuous grounds. I, therefore, follow Sakai (1976, p. 550) in considering *Daira* a xanthid of uncertain status.

Of the species marked "*genus incertae sedis*," *Actaea angusta* is insufficiently known to have been included in any of the smaller genera into which the old genus *Actaea* was partitioned (Guinot, 1976, p. 204). Like *Micropanope fraseri* and *Paraxanthias insculptus*, it remains without an acceptable generic assignment. *Lophopanopeus maculatus* was eliminated from that genus by Menzies (1948, p. 24), with the suggestion that it might be accommodated in *Micropanope*. As presently restricted (Guinot, 1967a, p. 349), *Micropanope* contains only its type species and one other. In such cases, the old generic name is used in quotation marks, as evidence of its unsuitability.

My unpublished studies on Gulf of California spider crabs will show that I was in error (Garth, 1946a) in synonymizing *Mithrax mexicanus* Glassell 1936 with *M.*

Table 1. Distribution of Galápagos Brachyura[a]

	Endemic/insular	Gulf of California	Panamic	Atlantic	Chile–Peru	Indo–Pacific	Origin and Affin.
Family Raninidae							
Notosceles ecuadorensis (Rathbun 1935)		x	x				EP
Ranilia fornicata (Faxon 1893)		x	x	an			AT
Family Dromiidae							
Dromidia larraburei Rathbun 1910		x	x	an			AT
Hypoconcha panamensis Smith 1869		x	x	an			AT
Family Dynomenidae							
Dynomene ursula Stimpson 1860		x	x				EP
Family Dorippidae							
Ethusa lata Rathbun 1893		x	x	an			AT
Clythrocerus laminatus Rathbun 1935	(Clarion)						EP
Family Leucosiidae							
Ebalia hancocki Rathbun 1933	(Socorro)						EP
Lithadia cumingii Bell 1855		x	x	an			AT
Uhlias ellipticus Stimpson 1871		x	x	an			AT
Leucosia jurinei (Saussure 1853)			x				EP
Randallia agaricias Rathbun 1898		x	x				EP
Family Calappidae							
Calappa convexa Saussure 1853		x	x	an			AT
Mursia gaudichaudii (Milne Edwards 1837)		x	x		x		WP
Cycloes bairdii Stimpson 1860		x	x	x			AT
Family Majidae							
Stenorhynchus debilis Smith 1871		x	x	an	x		AT
Anomalothir hoodensis Garth 1939	x			an			AT
Podochela margaritaria Rathbun 1902	x						EP
Podochela schmitti Garth 1939	x						EP
Euprognatha granulata Faxon 1893	(Cocos)						EP
Acanthonyx petiverii Milne Edwards 1834		x	x	x	x		AT
Eupleurodon rathbunae Garth 1939	x						EP
Eupleurodon trifurcatus Stimpson 1871		x	x		?		EP
Taliepus margnatus (Bell 1835)					x		EP
Sphenocarcinus agassizi Rathbun 1893			x	an			AT
Herbstia edwardsii Bell 1835	x	x					EP
Herbstia pyriformis (Bell 1835)	x						EP
Lissa aurivilliusi Rathbun 1898		x	x	an			AT
Pitho quinquedentata Bell 1835		x	x	an			AT
Pitho sexdentata Bell 1835			x	an			AT
Mithrax spinipes (Bell 1835)		x	x	an			AT
Mithrax mexicanus Glassell 1936			x				EP
Mithrax belli Gerstaecker 1856	x			an			AT
Mithrax pygmaeus Bell 1835		x					EP
Mithrax nodosus Bell	x	x					EP
Teleophrys cristulipes Stimpson 1860		x	x	an			AT
Stenocionops triangulata (Rathbun 1892)		x	x	an			AT

continued

Table 1. *(continued)*

	Endemic/insular	Gulf of California	Panamic	Atlantic	Chile–Peru	Indo–Pacific	Origin and Affin.
Microphrys aculeatus (Bell 1835)					x		EP
Microphrys trangulatus (Lockington 1876)		x	x				EP
Microphrys platysoma (Stimpson 1860)		x	x	an			AT
Tyche galapagensis Garth 1958	x						EP
Family Parthenopidae							
Parthenope exilipes (Rathbun 1893)		x	x	an			AT
Parthenope triangula (Stimpson 1860)	(La Plata)	x					WP
Daldorfia garthi Glassell 1940		x	x				WP
Solenolambrus arcuatus Stimpson 1871		x	x	an			AT
Mesorhoea bellii (A. Milne Edwards 1878)		x	x	an			AT
Aethra scruposa scutata Smith 1869		x					WP
Osachila galapagensis Rathbun 1935	x						EP
Osachila levis Rathbun 1898		x	x				EP
Family Portunidae							
Portunus stanfordi Rathbun 1902	x						EP
Portunus angustus Rathbun 1898	x			an			AT
Portunus tuberculatus (Stimpson 1860)		x	x				EP
Callinectes arcuatus Ordway 1863		x	x	an			AT
Callinectes toxotes Ordway 1863		x	x	an			AT
Cronius ruber (Lamarck 1818)		x	x	x	x		AT
Euphylax dovii Stimpson 1860			x		x		WP
Family Atelecyclidae							
Kraussia americana Garth 1939		x	x				WP
Family Xanthidae							
Liomera cinctimana (White 1847)		x	x			x	WP
Platypodiella gemmata (Rathbun 1902)			x				EP
Platypodiella rotundata (Stimpson 1860)		x	x	an			AT
Platyactaea dovii (Stimpson 1871)		x	x	an			AT
Paractaea sulcata (Stimpson 1860)		x	x	an			AT
"Actaea" angusta Rathbun 1898	x			an			AT
Glyptoxanthus hancocki Garth 1939	x						EP
Daira americana Stinpson 1860		x	x				WP
Lipaesthesius leeanas Rathbun 1898		x	x				EP
Edwardsium crosslandi (Finnegan 1931)	x						EP
Edwardsium lobipes (Rathbun 1898)		x	x	an			AT
"Medaeus" spinulifer (Rathbun 1898)		x	x				EP
Cycloxanthops vittatus (Stimpson 1860)		x	x				EP
Cataleptodius occidentalis (Stimpson 1871)		x		an			AT
Cataleptodius snodgrassi (Rathbun 1902)	x						EP
Xanthodius cooksoni (Miers 1877)	x	(Clarion) (Socorro)		an			AT
Lophoxanthus lamellipes (Stimpson 1860)		x	x				EP
"Lophopanopeus" maculatus Rathbun 1898		x	x				EP

continued

Table 1. *(continued)*

	Endemic/insular	Gulf of California	Panamic	Atlantic	Chile–Peru	Indo–Pacific	Origin and Affin.
Eurypanopeus transversus (Stimpson 1860)			x		x		EP
Hexapanopeus cartagoensis Garth 1939	x						EP
Eurytium affine (Streets & Kingsley 1877)		x					EP
Microcassiope xantusii (Stimpson 1871)		x	x	an			AT
Nanocassiope polita (Rathbun 1893)		x	(Cocos)	an			AT
"*Micropanope*" *fraseri* Garth 1946	x						EP
Tetraxanthus rathbunae Chase 1940				x			AT
Ectaesthesius bifrons Rathbun 1898	x						EP
"*Paraxanthias*" *insculptus* (Stimpson 1871)		x					EP
Menippe obtusa Stimpson 1860			x				EP
Globopilumnus xantusii (Stimpson 1860)		x	x				EP
Pilumnus pygmaeus Boone 1927		x	x				EP
Acidops fimbriatus Stimpson 1871		x					WP
Ozius verreauxii Saussure 1853			x				EP
Ozius perlatus Stimpson 1860		x	x	an			AT
Epixanthus tenuidactylus (Lockington 1877)		x	x				EP
Eriphia squamata Stimpson 1859		x	x	an			AT
Eriphia granulosa A. Milne Edwards 1880	x						EP
Eriphides hispida (Stimpson 1860)		x					EP
Domecia hispida Eydoux and Souleyet 1842		x	x			x	WP
Trapezia ferruginea Latreille 1825		x	x			x	WP
Trapezia digitalis Latreille 1825		x	x			x	WP
Quadrella nitida Smith 1869		x	x				WP
Jonesius triunguiculatus (Borradaile 1903)			x			x	WP
Family Pinnotheridae							
Parapinnixa glasselli Garth 1939	x						EP
Pinnixa transversalis (Milne Edwards and Lucas 1843)		x	x	an			AT
Pinnixa darwini Garth 1960	x			an			AT
Pinnaxodes chilensis (Milne Edwards 1837)					x		EP
Pinnotheres aff. *pubescens* (Holmes 1900)		x					EP
Family Palicidae							
Palicus cortezi (Crane 1937)		x					EP
Palicus fragilis (Rathbun 1893)		x	x				EP
Palicus velerae (Garth 1939)	x						EP
Palicus lucasii Rathbun 1893		x	x	an			AT
Family Grapsidae							
Grapsus grapsus (Linnaeus 1758)		x	x	x			AT
Geograpsus lividus (Milne Edwards 1837)		x	x	x			AT
Pachygrapsus transversus (Gibbes 1850)		x	x	x	x		AT
Planes cyaneus Dana 1852		x	x			x	WP
Euchirograpsus pacificus Türkay 1975	x			an			AT

continued

Table 1. (*continued*)

	Endemic/insular	Gulf of California	Panamic	Atlantic	Chile–Peru	Indo–Pacific	Origin and Affin.
Plagusia immaculata Lamarck 1818			x			x	WP
Percnon gibbesi (Milne Edwards 1853)		x	x	x	x		AT
Family Ocypodidae							
Ocypode gaudichaudii Milne Edwards and Lucas 1843			x		x		EP
Uca galapagensis Rathbun 1902	x						EP
Uca helleri Rathbun 1902	x						EP
Family Cryptochiridae							
Hapalocarcinus marsupialis Stimpson 1858			x			x	WP
Pseudocryptochirus crescentus (Edmondson 1933)		x				x	WP

	No.	%
[a]EP—Eastern Pacific	55	45.8
WP—Western Pacific	18	15.0
AT—Atlantic	8	6.6
an—Analogue	39	32.4
	120	99.8

spinipes (Bell, 1835), and that the former is a valid species. Reexamination of Galápagos material of *M. spinipes* in Hancock collections shows that all are *M. mexicanus* instead. As Bell's material was from two localities, Galápagos, 16 fathoms, and Santa Elena, 6 fathoms, the question arises: is not *M. spinipes* another mainland species wrongly attributed to the Galápagos? And, like others of Cuming's collecting, should it be attributed to the mainland alone? A subsequent paper will address these questions.

Working with material collected form corals at Gorgona Island, Colombia, Castro (1982) raised from synonymy the names *Trapezia corallina* Gerstaecker, 1857 and *T. formosa* Smith, 1869. Although he recognized these among Hancock material of *Trapezia*, there was a residue of unidentifiable specimens, owing to loss of color. Until fresh material becomes available, and pending studies underway by Bella Galil (pers. comm.), I continue to recognize only two species, *T. ferruginea* and *T. digitalis*, from Galápagos waters.

1.2. New Records since 1946

1.2.1. New to the Galápagos Islands

Eupleurodon trifurcatus Stimpson, Caleta Iguana, Isla Isabela, from *Corallina pinnatifolius* just below low water line, December 6, 1974, 1♀, 4♂ (3 ov), G.M. Wellington, collector (B-130).

Table 2. List of Name Changes, Additions, and Deletions

Garth, 1946	Changed to, or deleted	Authority
Family Raninidae		
Raninoides ecuadorensis Rathbun	*Notosceles ecuadorensis* (Rathbun)	Goeke, 1981
Family Leucosiidae		
Persephona edwardsii Bell	(delete)	Garth, 1991
Family Majidae		
Dasygyius depressus (Bell)	(delete) *Paradasygyius depressus* (Bell)	Garth, 1958
Dasygyius gibbosus (Bell)	(delete) *Collodes gibbosus* (Bell)	Garth, 1958
Pelia pulchella Bell	(delete)	Garth, 1991
Pisoides edwardsii (Bell)	(delete)	Garth, 1958
Thoe erosa Bell	(delete)	Garth, 1991
Mithrax denticulatus Bell	(delete)	Garth, 1958
Mithrax spinipes (Garth, not Bell)	*Mithrax mexicanus* Glassell	Garth, MS
Stenocionops ovata (Bell)	(delete)	Garth, 1991
Tyche lamellifrons Bell	*Tyche galapagensis* Garth	Garth, 1958
Family Xanthidae		
Carpilodes cinctimanus (White)	*Liomera cinctimana* (White)	Forest and Guinot, 1961
Platypodia gemmata Rathbun	*Platypodiella gemmata* (Rathbun)	Guinot, 1967b
Platypodia rotundata (Stimpson)	*Platypodiella rotundata* (Stimpson)	Guinot, 1967b
Actaea dovii Stimpson	*Platyactaea dovii* (Stimpson)	Guinot, 1967b
Actaea angusta Rathbun	*Genus incertae sedis*	Guinot, 1976
Actaea sulcata Stimpson	*Paractaea sulcata* (Stimpson)	Guinot, 1969b
Actaea crosslandi (Finnegan)	*Edwardsium crosslandi* (Finnegan)	Guinot, 1967a,b
Medaeus lobipes Rathbun	*Edwardsium lobipes* (Rathbun)	Guinot, 1967a,b
Medaeus spinulifer (Rathbun)	*Genus incertae sedis*	Guinot, 1971
Leptodius occidentalis (Stimpson)	*Cataleptodius occidentalis* (Stimpson)	Guinot, 1968
Leptodius snodgrassi Rathbun	*Cateleptodius snodgrassi* (Rathbun)	Guinot, 1968
Leptodius cooksoni Miers	*Xanthodius cooksoni* (Miers)	Guinot, 1968
Lophopanopeus maculatus Rathbun	*Genus incertae sedis*	Menzies, 1948
Micropanope xantusii (Stimpson)	*Microcassiope xantusii* (Stimpson)	Guinot, 1967a
Micropanope polita Rathbun	*Nanocassiope polita* (Rathbun)	Guinot, 1967a
Micropanope fraseri Garth	*Genus incertae sedis*	Guinot, 1967a
Paraxanthias insculptus (Stimpson)	*Genus incertae sedis*	Guinot, 1968
Pilumnus xantusii Stimpson	*Globopilumnus xantusii* (Stimpson)	Garth, 1968
Ozius tenuidactylus (Lockington)	*Epixanthus tenuidactylus* (Lockington)	Monod, 1956
Maldivia galapagensis Garth	*Jonesius triunguiculatus* (Borradaile)	Galil and Takeda, 1986
Family Grapsidae		
Euchirograpsus americanus A. Milne Edwards	*Euchirograpsus pacificus* Türkay	Türkay, 1975
Family Cymopoliidae	Family Palicidae	Holthuis and Gottlieb, 1958
Cymopolia fragilis Rathbun	*Palicus fragilis* (Rathbun)	Garth, 1991
Cymopolia cortezi Crane	*Palicus cortezi* (Crane)	Garth, 1991
Cymoplia velerae Garth	*Palicus velerae* (Garth)	Garth, 1991
Cymopolia lucasii (Rathbun)	*Palicus lucasii* (Rathbun)	Garth, 1991

The genus *Eupleurodon* was established by Stimpson (1971, p. 98) for the then unique species, *E. trifurcatus*, of which the female holotype, no longer extant, was collected by John Xantus at Cabo San Lucas, Baja California. Two species have been added since that date: *E. rathbunae* Garth (1939, p. 13), collected by *Velero III* at Gardner Bay, Hood Island (Isla Española), Galápagos, of which the unique holotype was also a female; and *E. peruvianus* (Rathbun, 1910, p. 535), transferred from *Epialtus* by Garth (1958, p. 243), of which the male holotype was collected by R. C. Murphy at Chinchas Islands, Peru. Because two of the three species were known from the female sex only, it was impossible to say, in the absence of males, whether *E. rathbunae* was distinct from *E. trifurcatus*. The discovery by Gerard Wellington of males of two types that can be matched with females of *E. rathbunae* and *E. trifurcatus*, respectively, supports the view that they are indeed separate species, and that both occur in the Galápagos Islands.

Callinectes arcuatus Ordway, S. Channel, Isla Santa Cruz, campsite, February 12, 1964, 1♀, V. A. Zullo, collector (Z-336); Turtle Cove, Isla Santa Cruz, February 14, 1964, 1 yg♂ , C. Angemeyer, collector (Z-341).

Callinectes toxotes Ordway, Villamil, Isla Isabela, from tidal lagoon in mangrove area, April 15, 1975, 1♂, G. M. Wellington, collector (B-147).

Swimming crabs of the genus *Callinectes* were not encountered in the Galápagos by Hancock or earlier expeditions. The male *C. arcuatus* found by Victor A. Zullo while a member of the Galápagos International Scientific Project (GISP) was reported by Garth and Stephenson (1966). The female *C. toxotes* found by Gerard M. Wellington is here reported for the first time.

Pinnotheres cf. *pubescens* Holmes, Caleta Iguana, Isla Isabela, internal parasite of *Lytechinus semituberculatus*, April 15, 1975, 4♀, G. M. Wellington, collector (B-149).

Although the pinnotherid specimens above compare well enough with the original description and illustration of *P. pubescens*, described by S. J. Holmes from the Gulf of California (Holmes, 1900), no specimen comparison is possible, as the type no longer exists. Also, according to F. C. Ziesenhenne (pers. comm.), a different species of *Lytechinus* occurs in the Gulf of California, *L. semituberculatus* being a Galápagos endemic. It therefore seems best to keep a little taxonomic distance between the two until specimens from the Gulf of California are found.

Hapalocarcinus marsupialis Stimpson, Academy Bay, Isla Santa Cruz, April 24, 1975, from *Pocillopora* coral, G. M. Wellington, collector.

The coral gall crab was considered an Indo-Pacific species until discovered at Port Utria, Colombia, by *Velero III* collectors, on January 23, 1935, as reported by Schmitt (1936). It is also known from Clipperton Island (Garth, 1965).

Pseudocryptochirus crescentus (Edmondson), seen by G. M. Wellington.

This species is included in the Galápagos fauna on the assurance of Dr. Wellington that he has seen its characteristic burrows in *Pavona* coral (letter of 28 May 1975). The species has been reported from Clipperton (Garth, 1965), and the Gulf of California (Garth and Hopkins, 1968).

1.2.2. New to Islands within the Group

To Isla Fernandina, Punta Espinosa: June 24, 1974, G. M. Wellington, collector: *Teleophrys cristulipes* Stimpson, 2♂ (B-109); *Microphrys aculeatus*

(Bell), 1♂ (B-110); *Podochela margaritaria* Rathbun, 1♂, 1♀ov (B-111): "*Micropanope*" *fraseri* Garth, 1♂ (B-112), 1♀ (B-113); *Eurytium affine* (Streets and Kingsley), 1♂ (B-114); July 25, 1938, W. L. Schmitt, collector, *Ozius verreauxii* Saussure, 1♀, Sta. 16-38.

The *Velero III* established but a single station on Isla Fernandina on January 14, 1934, at which *Mithrax nodosus*, *Xanthodius cooksoni*, *Leptodius snodgrassi*, *Epixanthus tenuidactylus*, *Geograpsus lividus*, and *Uca helleri* were encountered. In revisiting the site on July 25, 1938, as a member of the Presidential Cruise aboard the *U.S.S. Houston*, W. L. Schmitt added *Ozius verreauxii*, but failed to include it in his report of that expedition (Schmitt, 1939). The specimen came to me among unidentified Xanthidae sent to me by the Smithsonian Institution as part of its AEC/ONR Project of the early 1960's. The importance of these records, and of those of G. M. Wellington, above, is that Isla Fernandina is the most northwesterly of the island group (excluding outliers), and, together with the adjacent northwest coast of Isla Isabela, is immersed in cold water.

To Isla Isabela, Caleta Iguana: *Eupleurodon rathbunae* Garth, from branches of *Amphiroa dimorpha*, 3 m, December 6, 1974, 1♂; on *Corallina pinnatifolia*, April 15, 1975, 1 megalopa (B-148), G. M. Wellington, collector.

To Isla Genovesa: *Nanocassiope polita* (Rathbun), 20-30 fms, January 19, 1975, 3♂, 2♀ov (B-132), G. M. Wellington, collector.

To Isla Champion, Isla Santa Maria: *Eupleurodon rathbunae* Garth, February 27, 1974, from branches of *Amphiroa dimorpha*, 3 m, 1 specimen, G. M. Wellington, collector.

To Isla Española, Punta Suarez: *Planes cyaneus* Dana, June 24, 1974, 1♂ (B-115), G. M. Wellington, collector.

1.2.3. Otherwise Noteworthy

"*Actaea*" *angusta* Rathbun, Isla Floreana, NW of Isla Onslow, 20 fms, nullipore coral, January 13, 1975, 1♀ (B-117), G. M. Wellington, collector.

Known previously only from the holotype, a female collected off Isla Española in 20 fms by the *Albatross* in 1888, and from a young female collected off Gardner Bay, Isla Española, in 25–35 fms by the *Velero III* in 1935, this distinctive species is destined to remain a rarity in collections. The male of the species is unknown.

Platypodiella rotundata (Stimpson), Academy Bay, Isla Santa Cruz, November 29, 1974, 1♀ (B-128), G. M. Wellington, collector.

Although the extensive series of *Platypodiella* obtained by *Velero III* collectors was reported as the endemic *P. gemmata* by Garth (1946a), the finding by Wellington, as had Crossland before him (Finnegan, 1931), of a specimen clearly referable to the mainland *P. rotundata*, lends credence to the possibility that the two are one species, *P. gemmata* being the young of *P. rotundata*, and hence not a Galápagos endemic, as heretofore supposed.

Uca galapagensis Rathbun, Punta Cormorant, Isla Floreana, February 26, 1974, 1♀ (B-117), G. M. Wellington, collector.

Uca helleri Rathbun, Academy Bay, Isla Santa Cruz, April 7, 1974, 3♂, 2♀ (B-118), G. M. Wellington, collector.

As the Galápagos Islands become increasingly populated, particularly at Academy Bay, which houses a sizeable community, but also near Punta Cormorant, which abuts a lagoon from which salt was formerly harvested, the fiddler crabs, which take alarm at the slightest intrusion, face increasing pressure. It is therefore reassuring to know that, as of 1974, these two colonies were holding up well.

2. Distribution of Galápagos Brachyura

2.1. Introduction

Since my Distribution Studies of Galápagos Brachyura was completed (Garth, 1946b), knowledge of natural phenomena in the surrounding region has increased greatly. Plate tectonics, submerged seamounts, sub-surface currents (i.e., the Cromwell Current), then unknown, may be mentioned; in addition, the El Niño phenomenon is recognized not as a local event emanating from the Bay of Panama, but as a hemispheric event, originating in the western Pacific and influencing climate as far north as California and as far south as Chile. The oceanic transport of planktonic larvae of many invertebrates has been well documented, and the geological age of the islands is known with greater certainty. As a result of these developments, and of the finer tuning of the systematic relationships of the crabs themselves, their intra- and extra-Galápagan distributions can be reexamined now with more assurance.

2.2. Changes in Distributional Status

From time to time Galápagos material in Hancock collections has been sent to carcinologists intent on revising genera or larger groups on a world-wide basis. This permits Galápagos specimens to be compared with the same species, or with related species, from distant parts of the globe. The decisions reached by these specialists often reveal relationships unsuspected from studying the insular fauna in isolation. For example, a Galápagos specimen identified by me as *Euchirograpsus americanus* A. Milne Edwards 1880, sent to Michael Türkay for examination, was returned as the holotype of *E. pacificus* n. sp. (Türkay, 1975), when it proved possible to separate it from the west Atlantic species. Similarly, the *Tetraxanthus* reported by Garth (1946a, p. 465) as *T. rathbunae* Chace, an Atlantic species,on the basis of a single female, is likely, when males are known, to prove distinct from that species. In making critical species distinctions, the male first pleopod is essential.

Specimens of *Domecia hispida* sent to Danièle Guinot enabled her to verify the identity of eastern and western Pacific material of this species (Guinot, 1964), and to revive, for western Atlantic material formerly referred to *D. hispida*, the name *D. acanthophora* Desbonne (in Desbonne and Schramm, 1867). Previously considered to occur on both sides of the Isthmus of Panama, *D. hispida* now becomes a Pacific species with an Atlantic analogue. *Maldivia galapagensis* Garth 1939 was shown by Galil and Takeda (1986) to be identical with *M. triunguiculatus* Borradaile 1903, and the generic name *Jonesius* Sankarankutty

1962 applied. *J. triunguiculatus*, therefore, joins other coral-inhabiting species common to both western and eastern Pacific, and Galápagos consequently loses a presumed endemic species.

2.3. Intra-Galápagan Distribution

In my distribution studies (Garth, 1946b), a group of species, consisting of *Palicus fragilis*, *P. cortezi*, *Microphrys platysoma*, and *Globopilumnus xantusii*, was listed as found only at the extreme northerly islands of Wolf, Genovesa, and the northern part of Isabela (Tagus Cove, Albemarle Point) and Santiago (Sulivan Bay), while *Eurytium affine* was found on the north shore of Santa Cruz (Conway Bay). Similarly, *Eupleurodon rathbunae* and *Microphrys aculeatus* were mentioned as found only at the extreme southeasterly islands of Santa Maria and Española. The northern group was said to have affinities with the Gulf of California, while the latter group was said to be related to the Chile–Peru fauna. The implication was that the northerly group of species, after arriving at the northernmost islands from the Gulf of California–Baja California region via the California and Panama currents, were unable to disperse further southward within the Galápagos Islands, while the southerly group of species, carried from Peru and Chile by the Peru Current, were likewise unable to advance further northward. The configuration of ocean currents within the islands was considered as barring further penetration.

Additional material collected by Gerard Wellington from both northerly (Fernandina) and southerly (Española) islands provides an opportunity to test these speculations. Apparently, the one that holds up is that the northwesterly islands are inhabited by cold-water species of crabs (and shrimps), as they are by penguins and flightless cormorants. In a letter dated August 21, 1975, Wellington wrote: "[I have] just received word from Dr. [Fenner A.] Chace that the shrimp [from Caleta Iguana] is very probably *Rhynchocinetes typus*, previously known only from Peru and Chile and Juan Fernandez (a cool water species). In Galápagos it is restricted to Fernandina and western Isabela. This information adds further evidence to my contention that this western region forms a separate faunal–floral province in the islands." The *Microphrys* species found by Wellington at Punta Espinosa, Isla Fernandina, was *M. aculeatus*, the Peru–Chile species.

The Gulf of California is recognized as belonging to the warm-water Panamic province; it is named separately in Table 1. Any north-temperate species found in the Galápagos Islands would have had to come from Baja California north of Punta Eugenia, rather than from the Cabo San Lucas region, which is an island to warm-water species (see Garth, 1955). Further studies of islands off the coast of Mexico (Socorro, Clarion, Clipperton) have led to the category of insular endemic for species common to Galápagos and these offshore islands.

Finally, in a letter dated November 28, 1974, Gerard Wellington writes: "I suspect that the majority of species will eventually be found widely distributed in the archipelago. This past month I have been working in Academy Bay and have found that microhabitats play a major role in species abundance and distribution. I have found species, which were apparently absent in an area, abundant 100 m or less distance from the area I was sampling. The only difference between the two sites was somewhat looser rocks and a bit more sand. There is much work to be done in habitat descriptions before any statement can be made about inter-archipelago distributions."

2.3.1. Galápagos Endemic Species

It is obvious that the 23 species found only in the Galápagos Islands to date cannot all be endemic: that is to say, having originated there and occurring nowhere else. Some, like *Euchirograpsus pacificus* and *Anomalothir hoodensis*, undoubtedly occur widely throughout the eastern Pacific, as do their counterpart species throughout the western Atlantic, but have not been collected elsewhere. Others, like *Tyche galapagensis*, are genuinely autochthonous, being represented on the mainland by *T. lamellifrons*, to which Galápagos specimens were formerly referred, and in the Revillagigedo Islands, a northern hemispheric group, by *T. clarionensis* Garth, 1958. In the present state of our knowledge, however, we have no choice but to group them as endemics, pending clarification of their status by further collecting and taxonomic refinement.

Since 1946, when the presumed endemics were first listed, two additional species have been described that have been found in the Galápagos and nowhere else: the aforementioned *Euchirograpsus pacificus* by Türkay (1975) and *Pinnixa darwini* by Garth (1960). Both have Atlantic counterparts, the former in *E. americanus* A. Milne Edwards 1880, the latter in *P. faxoni* Rathbun 1918. This strongly suggests that the Galápagos species, or an as yet undescribed species having even closer affinities with the Atlantic one, should occur in the Bay of Panama, as the two oceans were recently confluent over the present Isthmus of Panama. One previous "endemic," the former *Maldivia galapagensis*, has been synonymized with *M. triunguiculatus* Borradaile 1903, a widely ranging Indo-west Pacific species, for which the name *Jonesius* Sankarankutty 1962 was revived by Galil and Takeda (1986), while *Platypodiella gemmata*, another presumed endemic, is now thought to represent the young of the mainland *P. rotundata*. Thus the list, although refined, still stands at 23 Galápagos endemic species (Table 1).

2.4. Extra-Galápagan Distribution

Because of their location at the convergence of three or more ocean currents, the Galápagos Islands share a portion of their brachyuran fauna with other regions, some adjacent, others remote. These are other islands in the eastern Pacific, the Gulf of California, the Bay of Panama, the western Atlantic, the western Pacific, and the Peru-Chile area. Because of the geologically recent Isthmian closure, ca. three million years ago, a special relationship exists with the Caribbean region, from which a current must formerly have flowed as well.

2.4.1. Insular Endemic Species

Other islands of the eastern Pacific share with the Galápagos a number of species that do not occur along adjacent mainland shores. Common to Clarion and Galápagos is *Clythrocerus laminatus*, to Socorro and Galápagos is *Ebalia hancocki*, and to both Clarion and Socorro is *Xanthodius cooksoni*. Common to Cocos and Galápagos are *Euprognatha granulata*, *Microphrys triangulatus*, and *Nanocassiope polita*, while common to these islands and to La Plata Island, Ecuador, is *Parthenope triangula*. These islands serve as stepping-stones in a broad arc extending from Cape San Lucas to Ecuador, touching the mainland only at its ends.

2.4.2. Panamic Species

The Panamic province, as broadly defined, stretches from the Gulf of California–Lower California to Ecuador. As might be expected from their location, on the Equator and 600 miles west of Ecuador, the Galápagos share with this faunistically rich province more species than with any other, a total of 63, or more than half of its total fauna. A few, like *Mursia gaudichaudii* and *Stenorhynchus debilis*, exceed this range, extending from California to Chile; others, 15 in number, are restricted to the Bay of Panama alone. The Panama Current, which flows from the Bay of Panama toward the Galápagos from January to April, provides a way for larval stages to reach the islands on a regular basis, with recruitment from Mexico via the California Current only in exceptional years.

2.4.3. Atlantic Species

As time and taxonomy advance, fewer species are found in common between the eastern Pacific and western Atlantic. Of the seven mentioned in 1946, only five remain: *Pachygrapsus transversus*, *Percnon gibbesi*, *Cycloes bairdii*, *Acanthonyx petiverii*, and *Cronius ruber*. *Euchirograpsus* now has a Pacific, as well as an Atlantic species; *Tetraxanthus rathbunae*, an Atlantic species reported from Galápagos, will probably prove distinct when more material becomes available. All are species that have not diverged sufficiently since the closure of the Isthmus of Panama to be recognized by taxonomists as different species, although the populations are completely separated, and it is presumed that no gene flow occurs between them.

2.4.3a. Atlantic Analogues. If it were only for the species common to the two oceans, Atlantic and Pacific, the relationship between the Galápagos fauna and that of the West Indies would be a relatively weak one. Strengthening it immeasurably are the "twin" or geminate species, occurring on each side of the Isthmus of Panama. These are species that are more closely related to each other than are other species, such as replace each other from one faunal region to another throughout the world. They are recognized, not only among crabs, but among echinoderms, fishes, and other groups, and are considered *prima facie* evidence that the two oceans were confluent across the Isthmus of Panama in comparatively recent geological times, probably late Miocene (Durham and Allison, 1960).

Some 38 species of Galápagos brachyurans were recognized as having such analogues in 1946. This number has been increased by the discovery in the Galápagos of new species like *Pinnixa darwini* Garth 1960, whose closest relative and undoubted Atlantic analogue is *P. faxoni* Rathbun 1918 from Trinidad. The number has been decreased by finding that the Galápagos species of *Tyche* is not *T. lamellifrons* Bell 1835, the Panamic species and analog of the Atlantic *T. emarginata* White 1847, but an endemic species, *T. galapagensis* Garth 1958. These changes tend to cancel each other out, leaving the present list with 38 Galápagos species with analogues.

2.4.4. Indo-Pacific Species

As developed in a series of papers (the latest of which was Garth, 1974), the brachyuran crabs that are common to both western and eastern Pacific waters are mostly those that inhabit reef-building corals. These are *Trapezia ferruginea* and *T. digitalis*, *Domecia hispida*, and *Hapalocarcinus marsupialis* in the coral *Pocillopora* and *Jonesius triunguiculatus* in the coral *Pavona*, to which may now be added the coral-burrowing species *Pseudocryptochirus crescentus* (Edmondson). The other group that transgresses the Pacific Oceanic Barrier (Hawaii to California, or Tuamotu to Chile) are the pelagic grapsoids *Plagusia immaculata*, the Pacific log rider, and *Planes cyaneus*, the turtle crab. Gone are the circum-tropical grapsoids of 1946, *Grapsus grapsus* becoming amphi-American and west African, replaced in the Indo-west Pacific by *G. tenuicrustatus* (Herbst, 1783).

To my knowledge, no comprehensive resurvey of the Brachyura has been attempted since the severe incursion of El Niño in the early 1980's. It would not be unexpected to find that there has been an invasion of western Pacific species, some of which may become permanently established in the Galápagos Islands.

2.4.5. Temperate South American Species

These are poorly represented in the Galápagos Islands, despite the proximity of Cape Blanco, the Peruvian promontory that deflects the Peru Current northwestward toward the archipelago. Some, like *Mursia gaudichaudii* and *Stenorhynchus debilis*, are Panamic species whose ranges extend south, as well as north, of the tropics, well into temperate waters. Others, like *Geograpsus lividus* and *Pachygrapsus transversus*, while limited to tropical waters in the north, apparently stray into southern temperate waters. True temperate south Pacific species are *Microphrys aculeatus* and *Pinnaxodes chilensis*, the former free-living, the latter commensal with the sea urchin, *Strongylocentrotus gibbosus*. Clearly, the Galápagos brachyuran fauna shows little effect of south-temperate invasions.

2.4.6. Species Absent from the Galápagos

Of equal importance in assessing the brachyuran fauna of the Galápagos, or of any other island group, are the species that should be there, but are not. For example, the mangrove crab, *Aratus pisonii*, found in mangrove swamps along both the Atlantic and Pacific coasts of tropical America, does not occur in the Galápagos Islands, although mangrove trees are found on many of the islands. Similarly, land crabs of the genus *Gecarcinus* do not occur in the Galápagos, although *G. planatus* occurs on Clipperton Island and *G. malpilensis* on Malpelo Island, where each is a dominant land form. Nor do land crabs of the genus *Cardisoma* occur in the Galápagos, although *C. crassum* is found at Cocos Island, where it burrows at the base of coconut palms, which are also absent from the Galápagos. One may conclude either that their larval stages never reached the archipelago, or that, having reached it, they found conditions unsuited to their survival. Thus chance has played a role in the "sweepstakes" route that leads to successful colonization of the Galápagos marine environment.

The absence of the entire family Goneplacidae, and of most of the family Xanthidae, subfamily Panopeinae, the exceptions being *Eurytium affine* and *Hexapanopeus cartagoensis*, is more readily explainable. These are the mud crabs, and, with the exception of the mangrove lagoon, mud habitats are lacking in the Galápagos. In particular, the detritus-rich mud that is carried down from the Andes by such rivers as the Guayas in Ecuador, or the Atrato in Colombia, is absent from these islands, which have no such runoff. Their oceanic waters are shared by other islands and island groups, such as the Revillagigedos (Socorro, Clarion, San Benedicto), Clipperton, Cocos, and Malpelo. In various combinations, these islands (and the tip of Baja California) share with the Galápagos a number of brachyuran species not found on the adjacent mainland, for which the term "insular endemic" has been coined.

3. Ecology of Galápagos Brachyura

3.1. Introduction

Located on the Equator, the Galápagos Islands are warm, though tempered by the cold Peru Current. Their shores, of volcanic rock and sand, break the long Pacific swells into a surf that oxygenates their coastal waters, provides a variety of habitats, and brings an abundance of food. Absent are the great rivers of the continent that dilute the salt water, muddy it for hundreds of miles, and cover the ocean bottom with gas- and bacteria-producing detritus. Near enough to the mainland to participate in the wealth of forms characteristic of continental waters, yet sufficiently remote to have evolved forms of their own, the Galápagos Islands offer the ecologist a promising field for the pursuit of his studies.

In any classification of animals according to their habitat, the Brachyura are but a small segment of the whole. Necessary to the interpretation of the complex picture are the results of studies on the coelenterates, echinoderms, mollusks, annelids, fishes, reptiles, and birds. Specialists in all these groups participated in early voyages of the *Velero III*, under the command of Captain Allan Hancock, which visited the islands in 1932, 1933, 1934, 1935, and 1938. Their results, together with mine, appear in many volumes of the Hancock Pacific Expeditions and its sequels, the Monographs, Occasional Papers, and Technical Reports of the Allan Hancock Foundation. As my contribution to these others, and to an all-embracing ecology yet to be written, I present this brief consideration of brachyuran habitats and associations within the waters of the Galápagos Islands.

3.2. Types of Brachyuran Habitats

3.2.1. Above High Tide Line (*Supra* Tidal)

3.2.1a. Spray Zone. Along the rocky shore, perpetually moistened by the spray that rises from the breaking waves, is the spray zone. It is the home of those

Brachyura that are in the process of becoming terrestrial, but have not yet become emancipated from the sea. The most evident member of the spray zone fauna is the ubiquitous "Sally Lightfoot" crab (*Grapsus grapsus*). Wherever one lands on rocky shore, one is met by a small regiment of these bright red and yellow grapsids, contesting every foot of one's advance. More secretive, *Geograpsus lividus* hides in moist crevices, from which it rarely ventures forth in daytime. These crabs have long, slender legs for rapid transit, powerful muscles for jumping from rock to rock, and nail-like dactyls for holding themselves to porous lava. Their bodies are compressed, enabling them to force their way into narrow crevices. They follow the receding tide, keeping always out of water but in the spray, which constantly moistens their gill chambers. *G. grapsus* has been observed feeding on sea lettuce, or *Ulva*, but is also known to be a scavenger. Spray zone species belong to the family Grapsidae, placed next to the family Gecarcinidae, containing the true land crabs, in systematic treatises. They are almost, but not quite, terrestrial.

3.2.2. Between Tide Levels (Tidal)

3.2.2a. Depositing Shore. The fine coral sand that alternates with rocky promontories along the Galápagos coastline supports an entirely different fauna. There is no spray zone on the sandy beach because the waves roll in smoothly, spreading a thin sheet of solid water to their uppermost reaches. In a situation roughly analogous to that occupied by the grapsids on rocky shore, namely, the highest level attained by moisture, lives the ocypodid crab, *Ocypode gaudichaudii*. He is the color of sand, and he runs over it so rapidly as to deserve the name "El Carretero," the cart driver, given him by the natives of Peru. At first sign of danger he withdraws into the mouth of his burrow, from which he surveys the world through periscopic eyes. His cylindrical body closes the aperture, and a heavy claw with truncate pincers bars access to the tunnel, which extends an arm's length or more to water level.

The pebble beach is less susceptible to wave action than the coral sand because of the larger size of the particles. It is not as stable as solid rock, however, and, because of the constant friction of pebble on pebble, no plant, sponge, coelenterate, or other sessile form can establish itself. These loose stones are inhabited by a different set of crab species from what one finds on solid rock. Known collectively as "pebble crabs," they belong to the family Xanthidae. Their most noticeable characteristic is the thickening of their carapace and chelipeds until they come to resemble the pebbles among which they hide, and they are capable of withstanding the same sort of abrasive treatment. Nor do they scuttle away if disturbed, but "freeze" in place, hoping thereby to escape detection. *Lophoxanthus lamellipes*, *Xanthodius cooksoni*, and the species of the genus *Ozius*, *O. perlatus*, *O. verreauxii*, and the former *O.* (now *Epixanthus*) *tenuidactylus*, are typical species of this habitat. The first two favor situations high upon the beach, where the only moisture is that left by the receding tide beneath the pebbles, the upper surfaces of which are dry and hot under the equatorial sun. The *Ozius* species, particularly *O. perlatus* and the former *O. tenuidactylus*, may be found by the hundreds in pockets of solid rock in which pebbles have accumulated, as at James Bay, Isla Santiago.

3.2.2b. Eroding Shore. The solid rock that is being slowly worn away by wave action is predominantly of volcanic origin. It is either basaltic, hence smooth, as at

Darwin Bay, Isla Genovesa, or scoriaceous, hence rough, as in the majority of situations. Sedimentary rock was observed at beach level in but two Galápagos localities: at Baltra Island and at Floreana Island, near Cormorant Point. Here the purple bristle crab, *Eriphides hispida*, has riddled the crumbling cliffs with its shallow burrows.

Because of its great stability, solid rock provides the required foundation for the complex communities of algae, sponges, coelenterates, tunicates, ascidians, barnacles, and other sessile forms that provide footholds, concealment, food, and adornment for crabs and other free-living creatures. Because of the great variety of habitats and associations thus provided, the crab life of the rocky shore is rich in number of species in contrast to the meager representation is shifting sand and swirling pebbles.

Some of the richest stations established by early Hancock expeditions, including station 69-33, were along the east shore of Isla Isabela, from Albemarle Point to Cartago Bay, where a series of rocky promontories alternates with sandy beaches. Here *Platypodiella gemmata*, previously known only from the type specimen and a few specimens obtained by the Crossland Expedition (Finnegan, 1931), was found abundantly when it was learned that it excavates a hollow for itself in sponges. Likewise, the grapsid crab, *Percnon gibbesi*, not previously encountered in the Galápagos Islands, was obtained through the accidental discovery of its particular habitat, the under sides of turnable boulders in knee- to hip-depth water, thoroughly aerated by the surf.

As so large a number of Galápagos Brachyura occur on rocky shore, it is not possible to discuss them individually. A typical list from the rocky shore at Sulivan Bay, Isla Santiago, station 343-35, together with the numbers of individuals observed on a single low tide, follows:

Dromidia larraburei	1	*Platyactaea dovii*	12
Dynomene ursula	9	*Cycloxanthops vittatus*	2
Herbstia pyriformis	3	*Xanthodius cooksoni*	3
Herbstia edwardsii	4	*Lophoxanthus lamellipes*	6
Mithrax nodosus	22	*Paraxanthias insculptus*	7
Teleophrys cristulipes	26	*Microcassiope xantusii*	21
Tyche galapagensis	1	*Epixanthus tenuidactylus*	5
Daldorfia garthi	1	*Eriphia granulosa*	40
Liomera cinctimana	1	*Eriphides hispida*	1
Daira americana	3	*Pachygrapsus transversus*	1

From the list just given it may be seen that the dominant species were *Mithrax nodosus*, *Eriphia granulosa*, *Teleophrys cristulipes*, and *Microcassiope xantusii*. Of lesser abundance, but common, were *Platyactaea dovii*, *Dynomene ursula*, *Paraxanthias insculptus*, *Lophoxanthus lamellipes*, and *Epixanthus tenuidactylus*. The balance, represented by three specimens or less, might be considered rare.

3.2.2c. Mangrove Association. The brackish lagoon is the habitat of the species of *Uca*, or fiddler crabs *U. galapagensis* and *U. helleri*, both Galápagos endemic species. They are confined to colonies in a few sheltered areas, as at Cormorant Point, Isla Floreana, at Academy Bay, Isla Santa Cruz, at James Bay, Isla Santiago, and at Darwin Bay, Isla Genovesa. The lagoons are fringed with mangrove trees, the roots of which form an impenetrable thicket and provide shelter for the burrows. The two species of *Uca* occupy different localities, or, if the same locality, as at Academy Bay, separate coves. The colonies remain distinct; the species do not mingle. It would seem that some slightly different requirement must be met for each in the way of habitat. At Darwin Bay, Isla Genovesa, and at Mangrove Point, Isla Fernandina, where only *U. helleri* occurs, the inlets are drained by estuarine outfall, whereas at Isla Floreana and Isla Santiago, where *U. galapagensis* is found, only seepage drainage occurs. At Academy Bay, where both species are found, *U. galapagensis* occurs at greater distance from the sea. Apparently it is more tolerant of the salinity that develops as the shallow sheet of water evaporates.

3.2.3. Just Below Low Tide (Adtidal)

3.2.3a. Corals. Exposed at extreme low tide, and seldom covered by more than 4 m of water, are the reef-building corals of the genera *Pocillopora* and *Pavona*. The former is in the shape of a head of cauliflower, with many branches. The latter is the massive "brain" coral, with convoluted ridges. In the protective labyrinth of passageways provided by the corals, and particularly by the *Pocillopora* coral, are to be found species of Brachyura that do not occur in any other habitat, as well as species that find in the coral the same type of physical shelter they would seek from any other solid object of irregular shape.

Species that inhabit *Pocillopora* coral exclusively are, in the eastern Pacific, members of the family Xanthidae. They include the species of *Trapezia*, *T. ferruginea* and *T. digitalis*, *Domecia hispida*, and *Liomera cinctimana*. Of these, the *Trapezia* species dominate, *T. ferruginea* outnumbering *T. digitalis* by about four to one. A number of other species are found commonly in *Pocillopora* coral, as well as in other sheltered crevices at low tide:

Platyactaea dovii	*Microcassiope xantusii*
Paractaea sulcata	*Daira americana*
Globopilumnus xantusii	*Paraxanthias insculptus*
Pilumnus pygmaeus	*"Micropanope" fraseri*

A few other species, such as *Cycloxanthops vittatus* and *Acidops fimbriatus*, are less often encountered in *Pocillopora* than free living. Of the exclusively coral-inhabiting species, only *Trapezia ferruginea* had been recorded from the Galápagos Islands previous to the Hancock expeditions, and that by but a single specimen (Boone, 1927).

Several heads of massive brain coral, genus *Pavona*, weighing several hundred pounds each, were grappled from a depth of 4 m off Onslow Island, and examined for commensals. In each was found a small crab with affinities to the western Pacific and Indian oceans! Believed to be an endemic species, it was first named *Maldivia galapagensis* Garth 1939, but is now known as *Jonesius triunguiculatus* (Borradaile 1903) (See Galil and Takeda, 1986). None of the brachyuran species common to *Pocillopora* was found in *Pavona*, nor was any other brachyuran species encountered. With the *Jonesius*, however, was the peculiar shrimp, *Discias serrifer* Rathbun, for which the habitat was previously unknown.

The gorgonian coral or sea fan, *Muricea miser* (Verrill), is the habitat of the xanthid crab, *Quadrella nitida*, of which a single specimen was found by the *Velero*

III in Galápagos waters. For a classic description of the remarkable association of this crab and gorgonian, see Crane (1937).

3.2.3b. Worm Tubes. From the tubes of the annelid worm, *Chaetopterus variopedatus* (Renier) were taken a male and female specimen of the pinnotherid crab, *Pinnixa transversalis*, a species common along the mainland coast of South America from Panama to Patagonia, but not previously recorded from the Galápagos Islands (Garth, 1946a). At Tagus Cove, Isla Isabela, two females of a new species of *Parapinnixa*, *P. glasselli*, were taken from a "roach" trap attached to a lobster pot suspended in 4 m of water (Garth, 1946a). No record of the host was obtained, but it was presumed to be a worm also, the carapace and first ambulatory legs of the crab being transversely cylindrical in adaptation to the tubiculous habitat.

3.2.3c. Echinoderms. The sea urchin *Caenocentrotus gibbosus* (L. Agassiz and Desor.) is the host of the pinnotherid crab *Pinnaxodes chilensis*, commensal in its periproct. This crab is commonly found in the urchin in Peruvian waters and was first noted from the Galápagos Islands by H. L. Clark (1902), in reporting the echinoderms of the Hopkins–Stanford Expedition. He did not refer to it by name, and it remained for Schmitt (1939) to record *P. chilensis* from the Galápagos Islands. The presence of the crab in the urchin can be told by external examination because of the swollen test. More recently, the sea urchin *Lytechinus semituberculatus*, was found by G. M. Wellington to host the pinnotherid crab *Pinnotheres pubescens* Holmes, or a closely related species. The crab occurs on the oral area of the urchin.

3.2.4. Deeper Water

The waters of the Galápagos Islands are crystal clear, owing to the clean nature of the bottom, which is uniformly composed of white coral sand with patches of lava rock protruding. The sand is frequently accompanied by rounded particles of stony corals in process of disintegration, and by nullipore corals, living or dead. Broken shells sometimes occur in sufficient numbers to have been noted in dredge hauls, as at Isla Wolf. Rock beneath the surface supports algal life to the limit of penetration of sunlight, and animal life to deepest levels. Algae are mainly of two kinds: red algae and pink or purplish coralline. Heavy encrustations of sponges, hydroids, and bryozoans are commonly torn loose from solid rock or brought up in the dredge on loose boulders.

Because sand and rock occur in close proximity in bays and along coastlines of the Galápagos Islands, it was difficult to make dredge hauls exclusively of one or other type of bottom. The dredge frequently surfaced with the sand completely washed out and the sand-inhabiting animals macerated by the few pieces of rock that were not their true habitat. The accompanying characterization of brachyurans according to bottom type inhabited is therefore only an approximation, as it is based upon the various types of bottom material recovered with each species.

3.2.4a. Rock Bottom. The following are predominantly rock-dwelling species, often associated with red or coralline algae. The alga is used by spider crabs as decoration; the coralline is matched by xanthid crabs for concealment:

Podochela margaritaria	*Tyche galapagensis*

Lissa aurivilliusi
Microphrys triangulatus
Lipaesthesius leeanus

3.2.4b. Sand Bottom. The following are predominantly sand-dwelling species, often with dead coral, broken shell, and nullipore:

Notosceles ecuadorensis
Ranila fornicata
Hypoconcha panamensis
Ethusa lata
Clythrocerus laminatus
Ebalia hancocki
Calappa convexa
Randallia agaricias
Mursia gaudichaudii
Cycloës bairdii
Osachila galapagensis
Osachila levis
Stenorhynchus debilis
Anomalothir hoodensis
Podochela schmitti
Euprognatha granulata
Sphenocarcinus agassizi
Mithrax mexicanus
Stenocionops ovata
Parthenope triangula
Parthenope depressiuscula
Solenolambrus arcuatus
Mesorhoea bellii
Portunus stanfordi
Portunus angustus
Portunus tuberculatus
"Lophopanopeus" maculatus
Hexapanopeus cartagoensis
Ectaesthesius bifrons
Acidops fimbriatus
Kraussia americana
Palicus cortezi
Palicus fragilis
Palicus velerae
Palicus lucasii
Euchirograpsus pacificus

3.2.4c. Sand and rock Bottom. The following may be either sand- or rock-dwelling:

Dromidia larraburei
Edwardsium lobipes
Edwardsium crosslandi
Nanocassiope polita

3.2.4d. Bottom insufficiently known. The following do not appear in one of the above lists because of inadequate sampling:

Lithadia cumingii
"Actaea" angusta
Tetraxanthus rathbunae
Microphyrys platysoma

3.2.5. Pelagic

While many, if not most species of Brachyura are pelagic in the early life stages of zoeae and megalopa, very few are pelagic as adults. In the Galápagos Islands the only species found by *Velero III* scientists at the surface as adults were members of the family Portunidae, the swimming crabs. On one occasion a dozen of the stalk-eyed *Euphylax dovii* came to the gangway light at Tagus Cove, Isla Isabela, and were dipped with a long-handled net. On another occasion, at Conway Bay, Isla Santa Cruz, four specimens of *Portunus stanfordi* were captured similarly, and at Aeolean Bay, Isla Baltra, *Portunus angustus* came to light. The latter two species were dredged abundantly, but *Euphylax* was found by *Velero III* parties only at the surface.

The Pacific log rider, *Plagusia immaculata*, was not encountered by the *Velero III* in Galápagos waters, although it has been recorded (Boone, 1927), but was taken by Gerard Wellington from the hull of a tourist vessel that operates in the Galápagos full time. Nor was the ubiquitous *Planes cyaneus* discovered in Galápagos waters by *Velero III* parties, although several green sea turtles were overturned and examined. Rathbun (1918, p. 257) records "Off Galápagos Islands, from a green turtle." Wellington discovered it in 1974 at Punta Suarez, Isla Española, but the host, if any, was not mentioned.

It is noteworthy that there are no kelps in the Galápagos Islands. Algal communities are dominated either by red or green, seldom brown algae. An exception was the *Sargassum*, from which were obtained specimens of *Acanthonyx petiverii*, a member of the family Majidae, subfamily Acanthonychinae, to which the well known *Epialtus* of temperate shores belongs.

3.2.6. Unexplored

Two habitats that were not explored by Hancock or earlier expeditions became accessible only with the advent of self-contained underwater breathing apparatus (SCUBA). These were the vertical walls of steep-sided cliffs, as at Malpelo Island, Colombia, and the newly discovered anchialine caves, described by Iliffe in Chapter 10 of this volume. The former can be treated as subtidal rocky shore, having, like the intertidal rocky shore, a solid substrate, to which corals and coralline algae attach, as do sponges, hydroids, and algae. Specimens obtained by SCUBA divers from such vertical surfaces are the same as those encountered at extreme low tide on rocky shore. The few crabs recovered to date from anchialine caves in the Galápagos are still under study, but preliminary examination suggests that this unique environment has produced modifications on its brachyuran inhabitants that should be explored further.

Note Added in Proof

Since this chapter has gone to press, the former *Micropanope fraseri* Garth, 1946, has been placed in the new genus *Garthiope* Guinot (1990), and the first of the two crabs recovered from Galápagos anchialine caves, *Garthiope anchialina* Guinot and Iliffe (1991), has been described and illustrated. The second crab is being studied.

References

Bell, T., 1835, Some accounts of the Crustacea of the coasts of South America, with descriptions of new genera and species, founded principally on collections obtained by Mr. Cuming and Mr. Miller, *Proc. Zool. Soc. London* **3**:169–173.

Bell, T., 1836, [Same Title], *Trans. Zool. Soc. London* **2**:39–66.

Bell, T., 1855, A monograph of the Leucosiadae, with observations on the relations, structure, habits, and distribution of the family; a review of the generic characters; and descriptions of new genera and species, *Trans. Linn. Soc. London* **21**:277–314.

Boone, L., 1927, The littoral crustacean fauna of the Galápagos Islands, Part I: Brachyura, *Zoologica* **8**:127–288.

Borradaile, L. A., 1903, Marine Crustaceans, III, The Xanthidae and some other crabs, in: *Fauna and Geography of the Maldive and Laccadive Archipelagoes*, Volume 1 (J. S. Gardner, ed.), University Press, Cambridge, pp. 237–271.

Castro, P., 1982, Notes on symbiotic decapod crustaceans from Gorgona Island, Colombia, with a revision of the eastern Pacific species of *Trapezia* (Brachyura, Xanthidae), symbionts of scleractinian corals, *An. Inst. Inv. Mar.* **12**:9–17.

Clark, H. L., 1902, Papers from the Hopkins–Stanford Galápagos Expedition, 1898–1899, XII, Echinodermata, *Proc. Washington Acad. Sci.* **4**:521–531.

Crane, J., 1937, The Templeton Crocker Expedition, III, Brachygnathous crabs from the Gulf of California and the west coast of Lower California, *Zoologica* **22**:47–78.

Desbonne, I., 1867, Desbonne comparé avec les Echantillons de Crustacés de sa collection et les dernières publications de MM. H. de Saussure et W. Stimpson, I. Partie, Brachyures, in: (I. Desbonne and A. Schramm) *Crustacés de la Guadeloupe*, (publisher and city unknown), Bas-Terre (The Netherlands), pp. 1–160.

DiMauro, A., 1982, Rediscovery of Professor Thomas Bell's type Crustacea (Brachyura) in the dry crustacean collection of the Zoological Collections, University Museum, Oxford, *Zool. J. Linn. Soc. London* **76**:155–182.

Durham, J. W., and Allison, E. C., 1960, The geologic history of Baja California and its marine faunas, in: The geologic history of Baja California and Adjacent Seas (Symposium), Part 1, Geologic History, *Syst. Zool.* **9**:47–91.

Finnegan, S., 1931, Report on the Brachyura collected in Central America, the Gorgona and Galápagos Islands, by Dr. Crossland on the "St. George" Expedition to the Pacific, 1924–25, *J. Linn. Soc. London, Zool.* **37**:607–693.

Forest, J., and Guinot, D., 1961, Crustacés Décapodes Brachyoures de Tahiti et des Tuamotu, in: *Expédition française sur les récifs coralliens de la Nouvelle-Calédonie*, Volume préliminaire, pp. 1–195, Editions de la Fondation Singer-Polignac, Paris.

Galil, B., and Takeda, M., 1986, Resurrection of the genus *Jonesius* and establishment of a new genus: Commensal crabs associated with corals from the Indo-Pacific Ocean, *Bull. Natl. Sci. Mus.* (Tokyo) A, **12**:163–171.

Garth, J. S., 1939, New brachyuran crabs from the Galápagos Islands, *Allan Hancock Pacif. Exped.* **5**: 9–29.

Garth, J. S., 1946a, Littoral brachyuran fauna of the Galápagos Archipelago, *Allan Hancock Pacif. Exped.* **5**:341–601.

Garth, J. S., 1946b, Distribution studies of Galápagos Brachyura, *Allan Hancock Pacif. Exped.* **5**: 603–638.

Garth, J. S., 1955, The case for a warm-temperate fauna on the west coast of North America, in: *Essays in the Natural Sciences in Honor of Captain Allan Hancock*, pp. 19–27, University of Southern California Press, Los Angeles.

Garth, J. S., 1958, Brachyura of the Pacific Coast of America, Oxyrhyncha, *Allan Hancock Pacif. Exped.* **21**:1–854.

Garth, J. S., 1960, *Pinnixa darwini*, a new species of pinnotherid crustacean from the Galápagos Islands, *Pacif. Sci.* **14**:39–42.

Garth, J. S., 1965, The brachyuran decapod crustaceans of Clipperton Island, *Proc. Calif. Acad. Sci.* (series 4)**33**:1–46.

Garth, J. S., 1968, *Globopilumnus xantusii* (Stimpson), n. comb., a stridulating crab from the west coast of tropical America, with remarks on discontinuous distribution of some West American and West African genera of brachyrhynchous crabs, *Crustaceana* **15**:312–318.

Garth, J. S., 1974, On the occurrence in the eastern tropical Pacific of decapod crustaceans commensal with reef-building corals, *Proc. 2nd Intl. Coral Reef Symp.* (Brisbane) **1**:397–404.

Garth, J. S., and Hopkins, T. S., 1968, *Pseudocryptochirus crescentus* (Edmondson), a second crab of the corallicolous family Hapalocarcinidae (Crustacea Decapoda) from the eastern Pacific, with remarks on phragmosis, host specificity, and distribution, *Bull. South. Calif. Acad. Sci.* **67**:40–48.

Garth, J. S., and Stephenson, W., 1966, Brachyura of the Pacific Coast of America, Brachyrhyncha: Portunidae, *Allan Hancock Monogr. Mar. Biol.* **1**:1–154.

Gerstaecker, K. E. A., 1857, Carcinologische Beiträge, *Arch. f. Naturg.* **22**:101–162.

Glassell, S. A., 1936, The Templeton Crocker Expeditions, 1, Six new brachyuran crabs from the Gulf of California, *Zoologica* **31**:213–218.

Goeke, G. D., 1981, Symethinae, new subfamily, and *Symethis garthi*, new species, and the transfer of *Raninoides ecuadorensis* to *Notosceles* (Raninidae; Brachyura; Gymnopleura), *Proc. Biol. Soc. Washington* **93**:971–981.

Guinot, D., 1964, Les trois espèces du genre *Domecia* (Decapoda, Brachyura): *D. hispida* Eydoux & Souleyet, *D. glabra* Alcock, et *D. acanthophora* (Desbonne and Schramm), *Crustaceana* **7**:267–283.

Guinot, D., 1966, Recherches préliminaires sur les groupements naturels chez les Crustacés, I. Les affinités des genres *Aethra, Osachila, Hepatus, Hepatella*, et *Actaeomorpha* (suite et fin), *Bull. Mus. Natl. Hist. Nat.* (Paris), sér. 2 **38**:828–845.

Guinot, D., 1967a, [Same title], II, Les anciens genres *Micropanope* Stimpson et *Medaeus* Dana, *Bull. Mus. Natl. Hist. Nat.* (Paris), sér. 2 **39**:345–374.

Guinot, D., 1967b, [Same title], III, A propos des affinités des genres *Dairoides* Stebbing et *Daira* de Haan, *Bull. Mus. Natl. Hist. Nat.* (Paris), sér. 2 **39**:540–563.

Guinot, D., 1968, [Same title], IV, Observations sur quelques genres de Xanthidae, *Bull. Mus. Natl. Hist. Nat.* (Paris), sér. 2 **39**:695–727.

Guinot, D., 1969, Sur divers Xanthidae, notamment sur *Actaea* de Haan et *Paractaea* gen. nov. (Crustacea Decapoda Brachyura), *Cah. Pacif.* **13**:222–267.

Guinot, D., 1971, Recherches préliminaires sur les groupements naturels chez les Crustacés Décapodes brachyoures, VIII, Synthèse et bibliographie, *Bull. Mus. Natl. Hist. Nat.* (Paris), sér. 2 **42**:1063–1090.

Guinot, D., 1976, Constitution de quelques groupes naturels chez les Crustacés Décapodes Brachyoures, I, La superfamille des Bellioidea et trois soufamilles des Xanthidae (Polydectinae Dana, Trichiinae de Haan, Actaeinae Alcock), *Mém. Mus. Natl. Hist. Nat.* (Paris), sér. A Zool. **97**:1–308.

Guinot, D., 1990, Établissement du genre *Garthiope* gen. nov., ses relations avec le genre *Coralliope* Guinot, 1967, et leurs affinités avec les Trapeziidae *sensu lato* (Crustacea Decapoda Brachyura), *Bull. Mus. Natl. Hist. Nat.* (Paris), sér. 4, 12, sec. A, **2**:469–487.

Guinot, D., and Iliffe, T. M., 1991, *Garthiope anchialina* sp. nov., espèce anchialine des Galápagos, île Isabela, Cueva de la Cadena, avec remarques sur la faune carcinologique anchialine (Crustacea Decapoda Brachyura), *Bull. Mus. Natl. Hist. Nat.* (Paris), sér. 4, 12 (1990), sec. A, **3–4**:607-621.

Holmes, S. J., 1900, Synopsis of the California stalk-eyed Crustacea, *Occas. Pap. Calif. Acad. Sci.* **7**: 1–262.

Holthuis, L. B., 1979, A small collection of decapod crustacea from Galápagos Islands, *Spedizione L. Mares-G.R.S.T.S.*, pp. 1–11, Firenze, Italy.

Holthuis, L. B., and Gottlieb, E., 1958, An annotated list of the Decapoda Crustacea of the Mediterranean coast of Israel, with an appendix listing the Decapoda of the Eastern Mediterranean, *Bull. Res. Council Israel* **7B**:1–126.

Menzies, R. J., 1948, A revision of the brachyuran genus *Lophopanopeus*, *Occas. Pap. Allan Hancock Found.* **4**:1–45.

Milne Edwards, A., 1880, Reports on the results of dredging under the supervision of Alexander Agassiz in the Gulf of Mexico, 8, Etudes préliminaires sur les Crustacés, pt. 1, *Bull. Mus. Comp. Zool., Harvard* **8**:1–68.

Monod, T., 1956, Hippidea et Brachyura ouest-africains, *Mém. IFAN*, **45**:1–674.

Rathbun, M. J., 1910, The stalk-eyed Crustacea of Peru and the adjacent coast, *Proc. U.S. Nat. Mus.* **38**:531–620.

Rathbun, M. J., 1911, The Percy Sladen Trust Expedition in the Indian Ocean in 1905 under the leadership of Mr. J. Stanley Gardiner, Volume III, No. XI, Marine Brachyura, *Trans. Linn. Soc. London* ser. 2, Zool. **14**:191–261.

Rathbun, M. J., 1981, The Grapsoid Crabs of America, *Bull. U.S. Nat. Mus.* **97**:1–461.

Sakai, T., 1976, *Crabs of Japan and Adjacent Seas*, Kodansha, Ltd., Tokyo.

Sankarankutty, C., 1962, On Decapoda Brachyura from the Andaman and Nicobar Islands: 2, Family Xanthidae, *J. Mar. Biol. Assoc. India* **4**:121–150.

Schmitt, W. L., 1936, Hancock Pacific Expedition, 1935, *Explorations and Fieldwork of the Smithsonian Institution in 1935*, pp. 29–36.

Schmitt, W. L., 1939, Decapod and other Crustacea collected on the Presidential Cruise of 1938 (with introduction and station data), *Smithson. Misc. Coll.* **98**:1–29.

Smith, S. I., 1869, in A. E. Verrill, On the parasitic habits of some Crustacea. *Amer. Nat.* **3**:239–250.

Stimpson, W., 1871, Notes on North American Crustacea in the Museum of the Smithsonian Institution, No. 3, *Ann. Lyc. Nat. Hist. New York* **10**:92–136.

Türkay, M., 1975, Zur Kenntnis der Gattung *Euchirograpsus* mit Bemerkungen zu *Brachygrapsus* und *Litocheira* (Crustacea: Decapoda), *Senckenbergiana Biol.* **56**:103–132.

Wellington, G. M., 1976, A prospectus: Proposal for a Galápagos Marine Park, *Noticias de Galapagos* **24**:9–13.

Chapter 6

Caridean and Stenopodid Shrimp of the Galápagos Islands

MARY K. WICKSTEN

1. Introduction

At least 65 species of caridean and stenopodid shrimp live in the waters of the Galápagos Islands. These animals can be common, and often are a food source for fishes, birds or cephalopods. Many engage in mutualistic or commensal relationships with other animals. These shrimp, however, often go unnoticed because of cryptic coloration, small size (1 cm or less in total length), nocturnal activity patterns, or inaccessible habitats.

Much information on deep-sea species of the area comes from trawling by the U.S. Fisheries Steamer *Albatross* in 1889–92 (Faxon, 1893, 1895). Both benthic and pelagic species were taken. Since 1895, there are less than 10 additional records of species living in midwater habitats or deeper than 200 m near these islands because of lack of collecting activity in deep water.

The most extensive collections of decapods from the Galápagos were made by the *Velero III* under the sponsorship of Captain G. Allan Hancock in 1931–1939. Animals were taken at 257 stations from the intertidal zone to 300 fathoms (554 m) by hand, dredge, trawl, and dip net. Sampling was conducted among algae, coral, and rocks as well as on sand and mud.

The collections of the *Velero III* were sent to the University of Southern California and the Smithsonian Institution for study. The Brachyura and Porcellanidae of the islands were identified and recorded in the works by Garth (1939, 1946; see also Chapter 5, this volume) and Haig (1960; see also Chapter 7, this volume). Other decapods, particularly the shrimp, need further study. Most of the caridean shrimp taken by the *Velero III* at the Galápagos Islands were examined by Waldo L. Schmitt of the U.S. National Museum of Natural History (Smithsonian

MARY K. WICKSTEN • Department of Biology, Texas A & M University, College Station, Texas 77843.

Institution). Dr. Schmitt, who participated in the expeditions, made color notes on living specimens for many species. Unfortunately, he had only partially sorted the specimens to species, and died before publishing anything on the collection.

L. B. Holthuis of the Rijksmuseum van Natuurlijke Historie examined the palaemonid shrimp of the Hancock Galápagos expeditions during his study of the holdings of the U.S. National Museum. Most of the records of this family in the Galápagos are contained in his works (1951, 1952a). Goy (1987) described a stenopodid shrimp from the collections. Records of some of the snapping shrimp (Alpheidae) from the collections were included in the work by Kim and Abele (1988).

Part of the collections of the *Velero III* remained at the University of Southern California, where they form part of the collection of the Allan Hancock Foundation (AHF). (Portions of the collection currently are being moved to the Los Angeles County Museum of Natural History). During preparation of a work on caridean shrimp of the Gulf of California (Wicksten, 1983) and routine cataloguing of AHF specimens, I examined all of the carideans from the Galápagos Islands. During visits in 1980 and 1987 to the USNM, I identified many of the carideans from the collections of the *Velero III*, but did not completely examine the specimens of the families Pasiphaeidae, Hippolytidae and Alpheidae. Records in this chapter include some published previously by me (Wicksten 1978, 1983, 1989a, 1989b) as well as unpublished records from the catalogues of the USNM and AHF. I also have examined specimens from the Galápagos Islands in the collections of the California Academy of Sciences (CAS).

Other accounts of carideans from the Galápagos are contained in expedition reports and species descriptions. Short accounts of collections can be found in the works of Schmitt (1924), Sivertsen (1933), Hult (1939) and Holthuis (1978). De Ridder (1980) studied the natural history of species of *Veleronia* of the Galápagos. De Ridder and Holthuis (1979) and Bruce (1978) described new species from the islands. Paul Humann of Hollywood, Florida has observed and photographed subtidal marine life of the Galápagos (Humann, 1986).

Most records of carideans and decapods come from intertidal and shallow subtidal habitats. Deeper subtidal areas have been sampled mostly by trawls and dredges. It is likely that species living in caves, under rocks, or in narrow crevices are underrepresented in collections or have not yet been collected. Almost no collections have been made in surf-swept or very steep areas, or near Islas Darwin, Wolf, Pinta, and Marchena.

2. Nearshore and Freshwater Species

A list of species living at 50 m or less is provided in Table 2. Records are taken from the published literature as well as from unpublished records of the institutions indicated in the references. Notes are provided for first records and range extensions of species at the Galápagos Islands. Station numbers refer to those of the *Velero III*, as published by Fraser (1943).

Kim and Abele (1988) described eight new species of *Alpheus* that included records from the Galápagos Islands. This work did not include comparisons to material previously reported from the Galápagos. Some of these eight species may have been identified previously as *A. leviusculus* or *A. armillatus* (Sivertsen, 1933;

Table 1. Biogeographic Affinities of Caridean and Stenopodid Shrimp from the Galápagos Islands[a]

Region	No. of species	% of total
TEP	27	46
TIP	10	17
CAR-WA	7	11
CP	4	7
G	4	7
TC	3	5
Other or unknown	4	7

[a]Abbreviations: TEP = tropical eastern Pacific, TIP = tropical Indo-Pacific, TC = tropical cosmopolitan, CP = Chilean Province, CAR-WA = Carribean-western Atlantic, G = known only from Galápagos Islands.

Wicksten, 1983). The absence of comparative information in the descriptions of Kim and Abele (1988) rendered them unusable for the present study. Species of *Alpheus* are notably polymorphic, and often have extensive ranges. (See Banner and Banner, 1982, for a discussion of variation in *Alpheus*.) I have provided records, therefore, of species of *Alpheus* examined by me prior to 1988 or which belong to species described before 1988. The task of comparing the remaining alpheids from the Galápagos with specimens from other areas of the eastern and western Pacific as well as the Atlantic side of the Panamic land mass will require extensive time and the use of numerical taxonomic methods or other techniques of species discrimination.

Most of the species taken in the Galápagos Archipelago have wide ranges. Only four (*Palaemonella asymmetrica*, *Pontonides sympathes*, *Typton crosslandi* and *Philocheras lapillus*) are known only from the islands, but these are small animals that easily could be overlooked by collectors and may occur elsewhere in the eastern Pacific. *Palaemon gladiator* is known only from the Galápagos and Clipperton Island. Kim and Abele (1988) elevated *Alpheus strenuus galapagensis* Sivertsen, 1933 to a distinct species, *A. galapagensis*, but did not compare specimens from the islands with an extensive series of specimens from the western Pacific. Whether or not the Galápagos population is sufficiently distinct to warrant designation as a separate species remains in doubt.

3. Biogeographic Affinities

The biogeographic affinities of the shallow-water caridean and stenopodidean shrimp are given in Table 1. Most of the shallow species of the Galápagos (27 species, or 46%), including two freshwater species, occur in the tropical eastern Pacific or nearby river drainages from western Mexico south to the Galápagos or the coasts of Colombia, Ecuador, or northern Peru. Ten species (*Harpiliopsis depressus*, *Fennera chacei*, *Alpheus lottini*, *A. leviusculus*, *Alpheus pacificus*, *A. strenuus*, *A. splendidus*, *Neoalpheopsis euryone*, *Synalpheus nobili* and *S. biunguiculatus*, 17%) live in both the eastern Pacific and tropical Indo-West Pacific. *Alpheus sulcatus* lives in the Indo-Pacific and eastern Pacific as well as the eastern

Table 2. Species Occurrences, Notes, and References for Stenopodid and Caridean Shrimp of the Galápagos Islands[a]

Infraorder STENOPODIDEA
  Family STENOPODIDAE
    Microprosthema emmiltum Goy, 1987; loc. 16, Goy 1987, TEP.
Infraorder CARIDEA
  Family ATYIDAE, loc. 11, near Darwin Research Station, 25 Jan. 1964; 27 Jan. 1964, D. Cavagnaro and R. Schuster, CAS, TEP?
  Family PALAEMONIDAE
    Brachycarpus biunguiculatus (Lucas, 1849); locs. 4, 5, 6, 7, 10, 13, 14, 15, 17, 21, 22; Holthuis, 1952a, TC.
    Macrobrachium hancocki Holthuis, 1952; loc. 15; Holthuis, 1952a, TEP.
    Macrobrachium americanum Bate, 1868; locs. 11, 15; Holthuis, 1952a, TEP.
    Palaemon ritteri Holmes, 1895; locs. 4, 5, 8, 10, 11, 12, 16; Holthuis, 1952a, TEP.
    Palaemon gladiator Holthuis, 1952; locs. 4, 6, 12, 22; Holthuis, 1952a, TEP.
    Palaemonella holmesi (Nobili, 1907); loc. 6; Holthuis, 1951, TEP.
    Palaemonella asymmetrica Holthuis, 1951; locs. 4, 7, 11, 14, 15, 17; Holthuis, 1951, G.
    Periclimenes infraspinis (Rathbun, 1902); locs. 10, 11, 19, 20; Holthuis, 1951, TEP.
    Periclimenes veleronis Holthuis, 1951; loc. 5; "Albermarle Is.," 22 May 1932, *Zaca*, CAS, TEP.
    Harpiliopsis depressus (Stimpson, 1860); locs. 4, 6, 11, 13, 14, 20; Holthuis, 1951, 1979, TEP.
    Periclimenaeus pacificus Holthuis, 1951; loc. 8; Holthuis, 1951, TEP.
    Pontonia margarita Smith, 1869, loc. 4; Holthuis, 1951, CAR-WA.
    Typton serratus Holthuis, 1951; loc. 6; Holthuis, 1951, TEP.
    Typton crosslandi Bruce, 1978; loc. 19; Bruce, 1978, G.
    Fennera chacei Holthuis, 1951; loc. 3; Wicksten, 1989a, TIP.
    Pseudocoutierea elegans Holthuis, 1951; loc. 3; Holthuis 1951, TEP.
    Veleronia serratifrons Holthuis, 1951; loc. 20; Holthuis, 1951, TEP.
    Veleronia laevifrons Holthuis, 1951; locs. 8, 20; Holthuis, 1951; S. James Bay, 5 Feb. 1974, G. Wellington, CAS, TEP.
    Pontonides sympathes de Ridder and Holthuis, 1979; locs. 9, 15; de Ridder and Holthuis, 1979, G.
    Gnathophyllum panamensis Faxon, 1893; loc. 4, 5, 7, 8, 10, 16, 17, 19, 20, 22; Sivertsen, 1933; Wicksten 1983, Punta Espinosa, Isla Fernandina, 18 Sept. 1974, G. Wellington, AHF, Bahía Gardner sta. 27-33, 30-33, Bahía Cartago sta. 73-33, 800-38, 76-33; Bahía James sta. 10-33; S. Seymour sta. 360-35; Bahía Sullivan sta. 795-38 USNM, TEP.
  Family BRESILIIDAE
    Discias serrifer Rathbun, 1902; locs. 5,6; Kensley, 1983, CP.
  Family PANDALIDAE
    Plesionika mexicana Chace, 1937; locs. 6, 7; Wicksten, 1983, Bahía Sullivan sta. 795-38, USNM, TEP.
  Family RHYNCHOCINETIDAE
    Rhynchocinetes typus H. Milne-Edwards, 1837; locs. 6, 15, 17, 22; Tagus Cove (juv.) Holthuis, 1979, N. of Tagus Hill, Isla Isabela sta. 154-34; off Bahía Stephens, Isla San Cristobal sta. 171-34; Post Office Bay, 5 Feb. 1933, USNM. Isla Isabela, Isla Fernandina-P. Humann, 1986; pers. comm, CP.
  Family HIPPOLYTIDAE
    Trachycaris restrictus (A. Milne-Edwards, 1878); loc. 20; Off Bahía Gardner sta. 201-34, USNM, CAR-WA.
    Latreutes antiborealis Holthuis, 1952; locs. 7, 10, 11, 14, 15, 17, 20; Wicksten, 1983; S. Seymour Is., sta. 87-83; Isla San Cristobal, sta. 41-33; Bahía Sullivan sta. 177-34, Off Post Office Bay sta. 197-34, USNM, TEP.
    Hippolyte williamsi Schmitt, 1924; locs. 5, 10, 11, 12, 14, 21; "Eden", Schmitt 1924; N. Shore Isla Española, 11 Feb. 1967, V. Walters, CAS; Bahía Cartago, sta. 187-34; S. Seymour Is. sta. 87-33, Bahía Academy, 4 Feb. 1933, USNM, TEP.
    Lysmata intermedia (Kingsley, 1878); loc. 17; Sivertsen, 1933, CAR-WA.
    Lysmata galapagensis Schmitt, 1924; locs. 4, 5, 6, 7, 10, 12, 13, 14, 16, 20; NE of Eden, Schmitt, 1924; Bahía Darwin sta. 101-33, Bahía Cartago sta. 73-33, N. Tagus Cove sta. 154-34, Bahía Sullivan sta.

Table 2. *(continued)*

343-35, S. Seymour Is. sta. 86-33, Isla Pinzón sta. 80-33, Isla Santa Fé sta. 811-38, Black Beach sta. 162-34, Bahía Gardner sta. 30-33 USNM, TEP.

Lysmata californica (Stimpson, 1866); locs. 5, 6, 11, 13, 14, 16, 17, 19, 20, 21; Bahía Cartago sta. 76-33, Tagus Cove, 15 Jan. 1934, Bahía Academy sta. 314-35, Isla Pinzón sta. 80-33, Isla Santa Fé sta. 811-38, 313-35, Black Beach sta. 162-34, Post Office Bay 5 Feb. 1933, Bahía Gardner sta. 359-35, Isla Onslow sta. 804-38, USNM, Gordon Rocks, sta. 315-35, AHF, TEP.

Family PROCESSIDAE

Ambidexter swifti Abele, 1972; locs. 4, 10, 13; Bahía Darwin sta. 94-33, S. Seymour Is. sta. 87-33, Isla Pinzon sta. 80-33, USNM, TEP.

Ambidexter panamensis Abele, 1972; loc. 20; Wicksten, 1983, TEP.

Processa peruviana Wicksten, 1983; locs. 1, 5, 6, 7, 14, 17, 20, 21; Isla Wolf sta. 143-34, Bahía Cartago sta. 185-34, Tagus Cove sta. 156-34, Bahía Sullivan sta. 178-34, S. Seymour Is. sta. 87-33, Isla Santa Fé sta. 810-38, Post Office Bay sta. 198-34, Bahía Gardner sta. 201-34, N. of Isla Española sta. 814-38, USNM, TEP.

Family CRANGONIDAE

Philocheras lapillus Wicksten, 1989; locs. 5, 6, 15, 20; Wicksten, 1989b, G.

Family ALPHEIDAE

Neoalpheopsis euryone (de Man, 1910); locs. 4, 5, 10, 11, 20; Wicksten, 1983, TIP.

Alpheopsis sp.; loc. 4; Bahía Darwin sta. 101-33 USNM, TEP?.

Salmoneus ortmanni (Rankin, 1898); locs. 17, 20; Gordon Rocks, sta. 315-34, Post Office Bay, sta. 167-34, Bahía Gardner sta. 27-33 USNM, CAR-WA.

Salmoneus serratidigitus (Coutière, 1896); locs. 5, 6, 9, 16, 19, 20; Bahía Darwin sta. 101-33, Bahía Cartago sta. 800-38, Black Beach sta. 166-34, Bahía Gardner sta. 27-33, Isla Onslow sta. 804-38 USNM, TEP.

Automate dolichognatha de Man, 1888; locs. 5, 11, 14, 20; Wicksten, 1981; Bahía Cartago sta. 76-33, Isla Sante Fe sta. 46-33, USNM, TC.

Synalpheus nobilii Coutière, 1909; loc. 11; "Eden", Schmitt, 1924; Albermarle Pt., Isla Isabela (Wicksten, 1983), TIP.

Synalpheus biunguiculatus (Stimpson, 1860); locs. 14, 20; Wicksten, 1983, TIP.

Synalpheus digueti Coutière, 1909; loc.; Gordon Rocks, Isla Santa Cruz sta. 315-34, AHF, TEP.

Pomagnathus corallinus Chace, 1937; locs, 5, 19, 20; Wicksten, 1983; Isla Onslow sta. 194-34, 804-38 USNM, TEP.

Alpheus inca Wicksten and Méndez, 1981; Gordon Rocks, Isla Santa Cruz sta. 315-35 USNM, CP.

Alpheus bellimanus Lockington, 1877; locs. 5, 7, 11, 12, 14, 16, 17; Kim and Abele 1988, Bahía Sullivan sta. 341-35 USNM, TEP.

Alpheus splendidus Coutière, 1897; loc. 10; S. Seymour Is. sta. 87-33 USNM, TIP.

Alpheus websteri Kingsley, 1880; locs. 10, 14; Wicksten, 1983, CAR-WA.

Alpheus normanni Kingsley, 1878; loc. 20; Bahía Gardner sta. 358-35 AHF, CAR-WA.

Alpheus lottini Guérin, 1830; locs. 4, 5, 10, 11, 13, 14; Holthuis, 1979, Kim and Abele 1988; Bahía Cartago sta. 800-38, Bahía Academy, no date, Karl Kubler; Isla Isabela sta. Sante Fé sta. 811-38 AHF, S. Seymour sta, 789-38, Albemarle Pt., Isla Isabela sta. 69-33, Isla Pinzón sta. 80-33, Isla Pinzón sta. 80-33, Isla Santa Fé sta. 811-38 USNM, TIP.

Alpheus malleator Dana, 1852; locs. 4, 5, 6, 7, 9, 10, 11, 14, 16, 19, 20; Wicksten, 1983; Bahía Darwin sta. 97-33, Bahía Cartago sta. 73-33; Tagus Cove sta. 152-34, SE of Isla Daphne Major sta. 789-38, Bahía Sullivan sta. 343-35; W. coast Isla Santiago sta. 333-35, Isla Bartolomé sta. 344-35, S. Seymour Is. sta. 789-38, Bahía Academy sta. 168-34, Isla Sante Fé sta. 48-33, Black Beach sta. 33-33, Isla Onslow sta. 804-38, Bahía Gardner sta. 31-33 USNM, CAR-WA.

Alpheus saxidomus Holthuis, 1980; locs. 4, 6, 7, 19, 20; Bahía Darwin, 22 Feb. 1933, Tagus Cove sta. 152-34, Albemarle Pt., Isla Isabela sta. 69-33, Tagus Cove sta. 152-34, Sullivan Bay sta. 180-34, Isla Onslow sta. 804-38, Isla Gardner sta. 357-35 USNM, TEP.

Alpheus paracrinitus Miers, 1881; locs. 4, 5, 7, 10, 11, 14, 16, 20; Wicksten, 1983; Kim and Abele 1988, TC.

Alpheus sulcatus Kingsley, 1878; locs. 4, 5, 10, 11, 16, 17, 20, 22; Sivertsen, 1933; Kim and Abele 1988, Punta Espinosa, Isla Fernadina 18 Sept. 1974, G. M. Wellington AHF; Bahía Darwin sta.

Table 2. *(continued)*

101-33, Bahía Cartago sta. 73-33, S. Seymour Is. sta. 789-38, Black Beach sta. 33-33, Bahía Gardner sta. 27-33, USNM, TC except CAR-WA.

Alpheus leviusculus Dana, 1852; locs. 12, 20; Bahía Conway sta. 82-33, AHF, Bahía Academy, no date, CAS, TIP.

Alpheus strenuus Dana, 1852; loc. 17; Silvertsen, 1933, TIP

Alpheus pacificus Dana, 1852; loc. 20; Kim and Abele, 1988, TIP.

Alpheus cf. *armillatus* Milne-Edwards, 1837; loc. 22; Punta Espinosa, Isla Ferndandina 18 Sept. 1974, G. M. Wellington, AHF, TEP?.

Alpheus chilensis Coutière, 1902; "Eden", off Isla Santa Cruz, Schmitt, 1924, CP.

[a]Key to collection sites (loc): 1, Isla Wolf; 2, Isla Pinta; 3, Isla Marchena; 4, Bahía Darwin, Isla Genovesa; 5, Bahía Cartago, Isla Isabela; 6, Tagus Cove, Isla Isabela; 7, Bahía Sullivan, Isla Santiago; 8, Bahía James, Isla Santiago; 9, Isla Daphne; 10, Isla Seymour; 11, Bahía Academy, Isla Santa Cruz; 12, Bahía Conway, Isla Santa Cruz; 13, Isla Pinzón; 14, Isla Santa Fé; 15, Isla San Cristobal; 16; Black Beach, Isla Floreana; 17, Post Office Bay, Isla Floreana; 18, Cormorant Bay, Isla Floreana; 19, Isla Onslow; 20, Isla and Bahía Gardner; 21, Isla Española; 22, Isla Ferndina.

Atlantic. *Brachycarpus biunguiculatus*, *Automate dolichognatha*, and *Alpheus paracrinitus* (5%) are cosmopolitan in tropical areas (Wicksten, 1983, 1989; Banner and Banner, 1985). Three species (*Alpheus chilensis*, *Rhynchocinetes typus*, and *Alpheus inca*) reach their northern range limits in the Galápagos, and usually are found along the coast of Peru and Chile (Méndez, 1981). *Discias serrifer* has been collected at the Galápagos and the Juan Fernandez Islands (Kensley, 1983). *Pontonia margarita*, *Lysmata intermedia*, *Trachycaris restrictus*, *Salmoneus ortmanni*, *Alpheus malleator*, *A. normanni*, and *A. websteri* (11%) are found both in the Caribbean and in the eastern Pacific, while *Latreutes antiborealis* and *Philocheras lapillus* are sibling species of the Caribbean and western Atlantic *L. parvulus* and *P. gorei* (Holthuis, 1952b; Wicksten, 1983, 1989b). The distributions of the caridean and stenopodid shrimp suggest that the Galápagos Islands have been populated by long-distance dispersal, and that insufficient time has passed for evolution of distinct insular species. The relatively young geologic age of the islands (perhaps 3.3 million years, Hall, 1983) indicates that they were populated during or after the closing of the Panamic seaway between the Caribbean and eastern Pacific (about 2–5 million years ago).

Only three species of freshwater shrimp are known from the islands. *Macrobrachium americanum* has been found in pools at Academy Bay, Isla Santa Cruz. *Macrobrachium hancocki* has been collected at Freshwater Bay, Isla San Cristobal, upstream of the bay proper. Both species of *Macrobrachium* also are widely distributed on the west coast of South America (Holthuis, 1952a). Individuals of *Macrobrachium* can disperse across salt water during floods, and have been found at other offshore islands of the eastern Pacific, including Isla Cocos (Wicksten, 1989a). Shrimp of the family Atyidae have been collected in a swamp near Academy Bay. Species of this family live in streams on the mainland coast of South and Central America.

4. Habitats

Tidepools and shallow rocky areas of the Galápagos contain many common carideans. The small translucent species of *Palaemon* are abundant. *Hippolyte williamsi* usually lives among algae, which it matches in color. *Alpheus sulcatus*, *A. malleator*, *Synalpheus digueti*, and other snapping shrimp usually live in burrows in sand under rocks, among algal holdfasts, or in cracks. *Lysmata galapagensis* tends to hide under rocks by day and forages at night.

Shallow subtidal rocky areas also have abundant carideans. *Brachycarpus biunguiculatus*, species of *Palaemonella*, *Periclimenes infraspinis*, and *Gnathophyllum panamense* can be found. *Trachycaris restrictus* and *Latreutes antiborealis* may escape notice because of their small size and cryptic coloration, which matches pieces of debris, shell chips, or sand. Species of *Alpheus* and related alpheids are common in holes, under rocks, and in crevices among algae or corals.

During daylight hours, carideans and stenopodids may hide in caves and then emerge to forage at night. Species of *Lysmata* are particularly common in caves. *Lysmata californica* and other species often associate with moray eels (family Muraenidae), which they clean in a symbiotic relationship (Limbaugh, 1961). *Rhynchocinetes typus* also often is a cave-dweller, as is the stenopodid *Microprosthema emilltum*. Patches of branched corals (*Pocillopora* spp.) harbor symbiotic carideans. *Alpheus lottini*, a colorful species with stripes of red, orange, black and/or white, is common among corals. It feeds on small invertebrates and algae among the coral branches, but also may ingest coral mucus and tissue (Castro, 1971). *Fennera chacei* and *Harpiliopsus depressa* cling to the branches. The latter species feeds on coral mucus, zooxanthellae, and algal spores (Barry, 1965; *in* Castro, 1971). These shrimp have mouthparts specially modified for feeding on mucus and fine particles, as well as short dactyls that grip the coral firmly during daylight hours. (See Patton, 1974, and Bruce, 1976, for reviews of shrimp of coral reefs.) Another commensal species, *Synalpheus charon*, has not been reported yet from the Galápagos, but is common among corals elsewhere in the tropical eastern Pacific and may also be found at the islands with further collecting (Wicksten, 1983).

Species of the family Palaemonidae include symbionts of larger invertebrates. *Pontonides galapagensis* lives on black corals (*Antipathes galapagensis*) (de Ridder and Holthuis, 1979). Species of *Typton* are found in sponges (Holthuis, 1951), as are species of *Discias* (family Bresiliidae, Kensley, 1983). *Veleronia serratifrons* and *V. laevifrons* live on a gorgonian, *Muricea appressa* (de Ridder, 1980). *Pontonia margarita* lives on pelecypods (Holthuis, 1951).

Although sandy coasts are not common in the Galápagos, these areas are rich in species of the family Penaeidae (order Decapoda, infraorder Penaeidea), which have not yet been studied in the islands. Common carideans of sandy areas include *Processa peruviana*, *Plesionika mexicana*, *Philocheras lapillus*, and species of *Ambidexter*.

5. Offshore and Deep Benthic Species

There are few records of shrimp or other crustaceans from deeper areas of the Galápagos. Two species, *Encantada spinoculata* (Caridea: Bresiliidae) and *Spongicoloides galapagensis* (Stenopodidea: Spongicolidae) probably inhabit rocky areas. Each is known only from a single specimen. The bresiliid was collected at

55–92 m east of Jervis Island (Wicksten, 1989d), and the stenopodid was trawled off Isla Santa Cruz at 717 m (Goy, 1980). Both are of phylogenetic and biogeographic interest. *Encantada spinoculata* is related to deep-sea hydrothermal vent shrimp (*Alvinocaris* and *Rimicaris* spp.), but lives at a much lesser depth and has pigmented eyes. *Spongicoloides galapagensis* is the only eastern Pacific species of this genus, whose species often live in association with deep-sea hexactinellid sponges.

It is likely that species of the families Pasiphaeidae and Oplophoridae are abundant in mesopelagic habitats near the Galápagos, as they are elsewhere in the eastern Pacific. The only species of these two caridean families reported from the islands are *Pasiphaea americana* Faxon, at 478–1017 m; and *Acanthephyra faxoni* Calman, taken at 738 m (Méndez, 1981; USNM unpublished data for *Velero III* sta. 812–38). *Nematocarcinus agassizii* Faxon (family Nematocarcinidae) has been taken at 247–1883 m (Méndez, 1981). These species range widely in the deeper parts of the tropical eastern Pacific.

Two deep benthic species have been reported from the Galápagos: *Glyphocrangon loricata* Faxon at 605–769 m, and *Pontophilus gracilis occidentalis* Faxon at approximately 1700–2100 m (Méndez, 1981; Wicksten, 1989c). The former species reaches its northern range limit in the Galápagos; the latter ranges from southern California to Chile. Both of these species probably crawl and dig in muddy bottoms.

Although they have not been reported from the islands, species of the family Pandalidae probably also live in deeper areas near there. Species of *Heterocarpus* are common on lower continental shelves and continental slopes along the coasts of central and South America, as well as in the Indo-Pacific region. Species of *Plesionika* and *Stylopandalus* also have been collected in benthic or pelagic habitats from southern California to Peru (AHF, unpublished records).

6. Faunal Recruitment

Carideans and stenopodids have planktonic larval stages which evidently have little difficulty crossing the distance between the mainland of southern and central America and offshore islands such as the Galápagos. Certain carideans, including species of *Lysmata* and *Hippolyte*, may have been transported to the islands while clinging to floating algae or debris. The presence of commensals of branched corals (*Pocillopora* spp.) also suggests recruitment from the tropical Indo-West Pacific, although whether this was directly from the western Pacific or by way of previously established populations in the eastern Pacific is uncertain. The mixture of cold and warmer currents in the Galápagos has enabled elements of the Peru–Chilean fauna to become established in the islands.

At present, it is impossible to compare the caridean and stenopodid fauna of the Galápagos with that of comparable mainland areas of central or South America. Biases in collecting, lack of sampling, uncertain identifications, or natural fluctuations in biotas prevent biologists from knowing with any certainty whether a species is consistently present or absent in a particular area. The species composition of the Galápagos on the whole is similar to that obtained by the Hancock expeditions at sites in the southern Gulf of California, western Mexico and the coasts of Panama and Ecuador. However, because the specimens have not been completely catalogued by station nor even fully identified, a station-by-station comparison is not possible.

Taxonomic problems hamper further understanding of the distributions of eastern Pacific carideans, in particular the family Alpheidae. The range of polymorphism among wide-ranging Indo-Pacific species is not known, nor have eastern Pacific populations been carefully compared with their western Pacific counterparts. Larval histories, color patterns, commensal associations and habitat requirements for most species of the area are unknown. The biology of caridean shrimp could provide many topics for study by visiting or resident scientists and their students at the islands.

References

Banner, D. M., and Banner, A. H., 1982, The alpheid shrimp of Australia. Part III: the remaining alpheids, principally the genus *Alpheus*, and the family Ogyrididae, *Rec. Aust. Mus.* **34**(1–2):1–357.

Banner, D. M., and Banner, A. H., 1985, The alpheid shrimp of Indonesia, based upon J. G. de Man's "The Decapoda of the Siboga Expedition, Part II. Family Alpheidae." (1911), *Mar. Res. Indones.* **25**:1–79.

Barry, C. K., 1965, Ecological study of the decapod crustaceans commensal with branching coral *Pocillopora meandrina* var. *nobilis* Verrill. M.S. Thesis, Univ. Hawaii, Honolulu.

Bruce, A. J., 1976, Shrimps and prawns of coral reefs, with special reference to commensalism, in: *Biology and Geology of Coral Reefs*, Volume 3, Biology 2 (O. A. Jones and R. Endean, eds.), pp. 37–94, Academic Press, New York.

Bruce, A. J., 1978, *Typton crosslandi* sp. nov., a new pontoniine shrimp from the Galápagos Islands, *Crustaceana* **35**(3):294–300.

Castro, P., 1971, The natantian shrimps (Crustacea, Decapoda) associated with invertebrates in Hawaii, *Pac. Sci.* **25**(3):395–403.

Faxon, W., 1893, Reports on the dredging operations off the west coast of central America to the Galápagos, to the west coast of Mexico, and in the Gulf of California, in charge of Alexander Agassiz, carried on by the U.S. Fish Commission steamer "Albatross" during 1891, Lieut. Commander Z. L. Tanner, U.S.N., commanding. VI. Preliminary descriptions of new species of Crustacea, *Bull. Mus. Comp. Zool. Harv. Univ.* **24**(7):149–220.

Faxon, W., 1895, Reports on an exploration off the west coasts of Mexico, central and South America, and off the Galápagos Islands, in charge of Alexander Agassiz, by the U.S. Fish Commission steamer "Albatross" during 1891, Lieut. Commander Z. L. Tanner, U.S.N., commanding. 15. The stalk-eyed Crustacea, *Mus. Comp. Zool. Harv. Univ. Mem.* **18**:1–292.

Fraser, C. M., 1943, General account of the scientific work of the *Velero III* in the eastern Pacific, 1931–41. Part III. A ten-year list of the *Velero III* collecting stations, *Allan Hancock Pac. Exped.* **1**(3):1–431.

Garth, J. S., 1939, New brachyuran crabs from the Galápagos Islands, *Allan Hancock Pac. Exped.* **5**(2): 9–29.

Garth, J. S., 1946, Littoral brachyuran fauna of the Galápagos Archipelago, *Allan Hancock Pac. Exped.* **5**(10):1–601.

Goy, J. W., 1980, *Spongicoloides galapagensis*, a new shrimp representing the first record of the genus from the Pacific Ocean (Crustacea: Decapoda: Stenopodidea), *Proc. Biol. Soc. Wash.* **93**(3):760–770.

Goy, J. W., 1987, *Microprosthema emmiltum*, new species, and other records of stenopodidean shrimps from the eastern Pacific (Crustacea: Decapoda), *Proc. Biol. Soc. Wash.* **100**(4):717–725.

Haig, J., 1960, The Porcellanidae (Crustacea Anomura) of the eastern Pacific, *Allan Hancock Pac. Exped.* **24**:1–440.

Hall, M. L., 1983, Origin of Española Island and the age of terrestrial life on the Galápagos Islands, *Science*, **221**:545–547.

Holthuis, L. B., 1951, A general revision of the Palaemonidae (Crustacea Decapoda Natantia) of the Americas. I. The subfamilies Euryrhynchinae and Pontoniinae, *Occ. Pap. Allan Hancock Found.* **11**:1–332.

Holthuis, L. B., 1952a, A general revision of the Palaemonidae (Crustacea Decapoda Natantia) of the Americas. II. The subfamily Palaemoniinae, *Occ. Pap. Allan Hancock Found.* **12**:1–396.

Holthuis, L. B., 1952b, Reports of the Lund University Chile Expedition 1948–49. 5. The Crustacea Decapoda Macrura of Chile, *Lunds Univ. Årsskrift* N.F. Avd. 2, Bd. 47 **10**:1–109.

Holthuis, L. B., 1979, A small collection of decapod Crustacea from Galápagos Islands, *Spedizione L. Mares-G.R. T. S.*, Florence, Italy. 11 pp.

Hult, J., 1939, Crustacea Decapoda from the Galápagos Islands collected by Mr. Rolf Blomberg, *Ark. Zool.* **30**A(5):1–18.

Humann, P., 1986, Galápagos Archipelago, *Ocean Realm* Winter 1986–87, pp. 62–69.

Kensley, B., 1983, New records of bresiliid shrimp from Australia, South Africa, Caribbean and Gulf of Mexico (Decapoda: Natantia: Caridea), *Smithson. Contrib. Zool.* **394**:1–31.

Kim, W., and Abele, L. G., 1988, The snapping shrimp genus *Alpheus* from the eastern Pacific (Decapoda: Caridea: Alpheidae), *Smithson. Contrib. Zool.* **454**:1–119.

Limbaugh, C., 1961, Cleaning symbiosis, *Sci. Amer.* **205**(2):42–49.

Méndez, M., 1981, Claves de identificacion y distribucion de los langostinos y camarones (Crustacea: Decapoda) del mar y ríos de la costa del Peru, *Bol. Inst. Mar Peru–Callao.* Vol. 5. 170 pp.

Patton, W. K., 1974, Community structure among the animals inhabiting the coral *Pocillopora damicornis* at Heron Island, Australia, in: *Symbiosis in the sea*, (W. B. Vernberg, ed.), University of Southern Carolina Press, Columbia, S. Carolina, pp. 219–243.

de Ridder, C., 1980, Etude des populations de deux especes de *Veleronia* (Decapodes, Pontoniinae) associees aux Gorgonacea (Cnidaria, Octocorallia) de l'archipel des Galápagos, *Cahiers Biol. Mar.* **21**(2):181–199.

de Ridder, C., and Holthuis, L. B., 1979, *Pontonides sympathes*, a new species of commensal shrimp (Crustacea, Decapoda, Pontoniinae) from Antipatharia in the Galápagos Islands, *Zool. Meded. Leiden* **54**(7):101–110.

Schmitt, W. L., 1924, The Macrura and Anomura collected by the Williams Galápagos Expedition, 1923, *Zoologica* **5**(15):161–171.

Sivertsen, E., 1933, The Norwegian Zoological Expedition to the Galápagos Islands 1925, conducted by Alf Wollebaek. VII. Littoral Crustacea Decapoda from the Galápagos Islands, *Meddel. Zool. Mus., Oslo* No. 38:1–23.

Wicksten, M. K., 1981, The species of *Automate* (Caridea: Alpheidae) in the eastern Pacific Ocean, *Proc. Biol. Soc. Wash.* **94**(4):1104–1109.

Wicksten, M. K., 1983, A monograph on the shallow water caridean shrimps of the Gulf of California, Mexico, *Allan Hancock Monogr. Mar. Biol.* No. 13 (59 pp).

Wicksten, M. K., 1989a, A key to the palaemonid shrimp of the eastern Pacific region, *Bull. South. Calif. Acad. Sci.* **88**:11–20.

Wicksten, M. K., 1989b, Two new species of caridean shrimp from the tropical eastern Pacific, *Proc. Biol. Soc. Wash.* **102**:78–83.

Wicksten, M. K., 1989c, Ranges of offshore decapod crustaceans in the eastern Pacific Ocean, *Trans. San Diego Soc. Nat. Hist.* **21**:291–316.

Wicksten, M. K., 1989d, *Encantada spinoculata*, a new genus and species of shrimp from the Galápagos Islands (Caridea: Bresiliidae), *J. Crust. Biol.* **9**:667–671.

Chapter 7

Biogeographic Patterns of the Galápagos Porcelain Crab Fauna

ALAN W. HARVEY

1. Introduction

Twelve species of porcelain crabs are currently known from the Galápagos Islands. The basic question addressed by this chapter is: Why do these twelve species, and not any of the nearly 80 other species from the eastern Pacific, or of the more than 90 species from the Indo-West Pacific, constitute the Galápagos porcellanid fauna? The geographic distribution of a species is affected by many different factors. These include biological attributes of the species, such as dispersal and competitive abilities and habitat specificity; regional characteristics, such as temperature regimes, habitat diversity and geographic isolation; geological history of the region and of the species; and random chance. Identifying the contributions of any of these factors is complicated by the generally nonexperimental nature of biogeography (see Simberloff and Wilson, 1969, for an example of experimental biogeography) and by the fact that many biogeographic processes appear to generate similar patterns (Connor and Simberloff, 1979; McCoy and Heck, 1987).

The Galápagos porcelain crabs are particularly well-suited for biogeographic analyses, because attributes of both the taxa and the geographic region serve to reduce the number of processes to consider. First, recent geological evidence strongly suggests that the Galápagos Islands are truly oceanic and were never closely connected to the mainland (Simkin, 1984). This effectively eliminates vicariance hypotheses and allows a clearer focus on dispersal-related processes.

ALAN W. HARVEY • Shannon Point Marine Center, Anacortes, Washington 98221.

Second, porcellanids are an unusually homogeneous assemblage in terms of morphology and ecology. Most eastern Pacific species are found at intertidal or shallow subtidal depths less than a few dozen meters. All porcellanids are filter feeders, and occur in habitats containing crevices or similar cover, primarily under stones or among coral branches. A significant minority are commensal to varying degrees with other organisms, such as echinoderms, cnidarians, or hermit crabs. The homogeneity within the Porcellanidae greatly reduces the 'owl and hummingbird' problem (Diamond and Gilpin, 1982) of making inappropriate comparisons among dissimilar taxa. Finally, the porcellanid faunas of potential source regions (eastern Pacific, Indo-West Pacific, and western Atlantic) have been the focus of intensive taxonomic study (Haig, 1960, 1962, 1964, 1966, 1968, 1981; Johnson, 1958; Gore and Abele, 1976; Gore, 1981).

In this chapter, I first provide basic distributional data for the twelve species of porcelain crabs reported from the Galápagos. Next, I summarize the biogeographic relationships of this fauna with those of potential source regions, specifically identifying relationships that are statistically improbable. Finally, I summarize information about certain biological attributes of the species in each region that, within the context of regional oceanographic patterns, may contribute to the observed biogeographic patterns.

2. Annotated Species List

All known localities of porcellanids within the Galápagos Islands are listed below. This information was obtained primarily from Haig (1960) and references cited therein, unless otherwise noted. Curiously, the vast majority of specimens were collected intertidally, even though many of these forms commonly occur subtidally elsewhere in the eastern Pacific. The extent to which this represents a real phenomenon or an artifact of sampling is unclear, as there were numerous shallow subtidal stations during the various Velero expeditions to the Galápagos that apparently did not locate porcellanids. On the other hand, a 1984 expedition by the Natural History Museum of Los Angeles County, which focused primarily on fishes, managed to collect several subtidal porcellanid specimens (J. Haig, pers. comm.).

Petrolisthes edwardsii (Saussere, 1853): Eden off Santa Cruz; Isabela (south of Cape Berkeley, north of Tagus Hill); Santiago (James Bay, west coast); Floreana (southeast of Cormorant Point, Black Beach); Española (Gardner Bay, Osborn in Gardner Bay); Marchena. Intertidal; under rocks, sometimes in corals [Panamic, Cortez].

Petrolisthes glasselli Haig, 1957: Española (Gardner Bay); Genovesa (Darwin Bay); Isabela (Cartago Bay); Bartolomé. Intertidal, shallow subtidal; usually associated with corals, sometimes under rocks [Panamic, Cortez].

Petrolisthes polymitus Glassell, 1937: Genovesa (Darwin Bay); Isabela (Cartago Bay); Santiago (Sullivan Bay); Baltra; Santa Cruz (opposite Gordon Rocks); Floreana (Black Beach); Española (Gardner Bay). Intertidal; under rocks, sometimes with corals or sponges [Panamic, Cortez].

Petrolisthes haigae Chace, 1962; Española (Gardner Bay); Genovesa (Darwin Bay);

Isabela (Isabela Point); Santiago (James Bay, Sullivan Bay); Seymour Norte; Santa Cruz (opposite Gordon Rocks, Academy Bay?); Santa Fé; Onslow; Culpepper; Wenman. Intertidal to 12 meters; usually associated with corals, sometimes under rocks or with sponges [Panamic, Cortez].

Petrolisthes armatus (Gibbes, 1850); Eden. Haig (1960) considers this to be an extralimital occurrence. Intertidal; under rocks, with corals and sponges [Panamic, Cortez, Peruvian, also western Atlantic, eastern Atlantic].

Petrolisthes galapagensis Haig, 1960: Española (Gardner Bay); Isabela (Black Bight, south of Cape Berkeley); Bartolomé; Santiago (James Bay); Santa Cruz (Conway Bay, Academy Bay); Floreana (Cormorant Bay, Flamingo Bay). Intertidal to 5 meters; under rocks [endemic to the Galápagos, with extralimital records in the Panamic].

Petrolisthes tonsorius Haig, 1960: Española (Gardner Bay); Isabela (south of Cape Berkeley, Black Bight); Bartolomé; San Cristobal (Stephens Bay); Floreana (southeast of Cormorant Point). Intertidal; under rocks [Panamic, Cortez].

Petrolisthes brachycarpus Sivertsen, 1933; Floreana (Post Office Bay, Cormorant Bay, Flamingo Point, southeast of Cormorant Point); Marchena; Genovesa (Darwin Bay); Fernandina (northeast point); Santiago (James Bay, Sullivan); Seymour; Baltra; Santa Cruz (Conway Bay, Academy Bay); Santa Fé; San Cristobal (Stephens Bay, Wreck Bay); Española (Gardner Bay); Osborn. Intertidal, one record from 8–20 meters; under rocks sometimes with corals [endemic to the Galápagos, with extralimital records in the Panamic].

Neopisosoma bicapillatum Haig, 1960: Isabela (north of Tagus Hill); Santiago (Sullivan Bay); Floreana (Black Beach). Intertidal; under rocks [Panamic].

Neopisosoma mexicanum (Streets, 1871): Floreana (Black Beach). Haig (1960) considered this an extralimital record. Intertidal; under rocks [Panamic].

Pachycheles biocellatus (Lockington, 1878): Santiago (Sullivan Bay); Española (Gardner Bay); Genovesa (Darwin Bay); Isabela (Isabela Point, Tagus Cove, Carrago Bay); Bartolomé; Onslow; Floreana (southeast of Cormorant Point, Gardner Bay); Wenman. Intertidal to perhaps 8 meters; usually associated with corals, sometimes with sponges or under rocks [Panamic, Cortez].

Pachycheles velerae Haig, 1960: off Marchena. Only specimen known from 40 meters.

3. Biogeographic Patterns

3.1. Endemics and Extralimitals

One of the linchpins of traditional biogeographic studies is an analysis of endemism, which yields equivocal results in the present system. Of the 12 species known from the islands, only *Petrolisthes galapagensis*, *Petrolisthes brachycarpus* and *Pachycheles velerae* have been considered endemics. The first two were reported, possibly as extralimitals, from elsewhere in the eastern Pacific: *Petrolisthes brachycarpus* from Bahia Honda, Panama (three females, one of which was ovigerous; Haig, 1960), and *Petrolisthes galapagensis* from Jasper Island, Gulf of Nicoya, Costa Rica (five males and four females; Haig, 1968). The only porcellanid strictly endemic to the Galápagos is *Pachycheles velerae*, which is known from a single soft-shelled female (Haig, 1960).

At the other extreme are *Petrolisthes armatus* and *N. mexicanum*, which are each represented in the Galápagos fauna by only a single specimen, and considered by Haig (1960, 1968) to be extralimitals. Whether this status applies to *N. mexicanum* is unclear, given that the Galápagos specimen was an ovigerous female. The record of *Petrolisthes armatus* is more likely to be an exceptional occurrence. This species (or perhaps superspecies complex, see Gore, 1973a) is for some reason rarely found on the oceanic islands within its tremendous geographic range, which includes both the eastern and western Atlantic, as well as the eastern Pacific, from the headwaters of the Gulf of California to Peru.

Thus, five of the 12 porcelain crabs recorded from the Galápagos are of questionable biogeographic status either within or outside the archipelago. To minimize the degree of uncertainty that these effects could have on biogeographic analyses, I took a conservative approach to constructing regional species pools by not distinguishing between typical and "extralimital" occurrences in the mathematical analyses, but being alert for statistical anomalies associated with these occurrences. Therefore, the following analyses include all 12 species in the Galápagos species pool, of which only *Pachycheles velerae* is considered endemic to the Galápagos.

3.2. Biogeographic Affinities

3.2.1. General Description of Galápagos Faunal Affinities

The eastern shores of the Pacific Ocean have been subdivided into biogeographic provinces by numerous researchers, with only minor variations among them. In the present analysis, I follow Maluf (1988), noting that for porcellanids there are no additional faunal changes north of the Californian province; the Peruvian and Chilean provinces are also nearly identical, and are treated as a unit in the statistical analysis.

The Galápagos porcelain crab fauna is overwhelmingly Panamic in affinity; all species except the endemic *Pachycheles velerae* have been recorded in the Panamic province. Seven of these also occur in the Cortez province. *Petrolisthes armatus* ('extralimital' in the Galápagos) barely reaches the Peruvian province, but no species are shared between the Galápagos and Chilean provinces. Likewise, the Galápagos do not share any species with the northern Californian, Oregonian, or Alaskan provinces. There are also no Indo-West Pacific porcellanids reported from the Galápagos (for that matter, the only member of the Anomura that occurs in both the eastern Pacific and the Indo-West Pacific is the hippid *Hippa pacifica* (Dana); J. Haig, pers. comm.).

According to present nomenclature, two species (*Petrolisthes tonsorius* and *Petrolisthes armatus*) are found in both the Galápagos and the western Atlantic. However, it is unlikely that the trans-Isthmian populations of either of these taxa are conspecific (Gore, 1973a, and pers. comm.). Until detailed phylogenetic relationships within the porcelain crabs are known, the biogeographic relationships of the Galápagos and the western Atlantic populations must remain somewhat ambiguous. There is clearly no direct connection, exclusive of the intervening Panamic region, between these populations (as was reported for other groups; Garth, 1946; Rosen, 1975).

3.2.2. Statistical Biogeography

3.2.2.a. General Considerations. The preceding section summarizes data in a format that should be very familiar to most biogeographers (see, for example, Garth, 1946; Brusca, 1987). This traditional approach, however, lacks any indication of statistical significance, which makes it difficult if not impossible to develop or investigate defensible hypotheses about biogeographical patterns (Connor and Simberloff, 1978; McCoy and Heck, 1987).

Virtually all statistical analysis requires an *a priori* working, or null, hypothesis. The working hypothesis of this chapter, then, is that there is nothing unusual about the composition of the Galápagos porcellanid fauna. That is, the collection of species in the Galápagos could statistically be considered a random sample of the potential source region's species pool. I test this hypothesis by asking: What is the probability that X species are shared between the Galápagos and a second area, given a common source pool containing P species? Under the working hypothesis, the appropriate probability distribution follows a hypergeometric distribution (Connor and Simberloff, 1978; Wright and Biehl, 1982):

$$p(x) = \frac{\binom{n}{x}\binom{P-n}{m-x}}{\binom{P}{m}}$$

where m and n = the number of species on the two areas being compared ($m > n$, $0 \leq x \leq m$). The expected number of shared species is simply

$$E_s = \frac{mn}{P}.$$

The distinction between this approach and the commonly made descriptive statements of the form "34% of the species found in area m were also found in area n" is the recognition of the species pool P from which the two areas may draw. Without specifying P there is no way to estimate how many species should (by whatever criterion) be shared by two areas, nor to calculate the statistical improbability of the observed data. However, determining P can be as difficult as it is essential, and careless estimates can lead to misleading results (McCoy and Heck, 1987).

It is important to realize that "expected" numbers of shared species do not necessarily correspond to what a knowledgeable biogeographer might expect, but rather to what is expected under the conditions of the working hypothesis, which does not recognize, for example, that nearby areas should share more species than distant sites, or that some species are better dispersers than others. Also, because this type of analysis takes as given the number of species present at each site, it cannot address the question of why the Galápagos has only (or as many as) twelve porcellanid species.

3.2.2.b. Global and Pacific Relationships. There are roughly 250 described species of porcelain crabs worldwide, not all of which have had an equal opportunity to colonize the Galápagos! The entire eastern Pacific holds 90 species, and I estimate a roughly similar total from the Indo-West Pacific, with no overlap between the two regions. Comparisons of expected *vs.* actual Galápagos species overlaps at this global level reveals that the Galápagos share 1) more species than can be explained by chance with the entire Pacific, and more specifically with the eastern Pacific, but 2) significantly fewer species than expected with the Indo-West Pacific (Table 1). Thus, the eastern Pacific clearly represents a more realistic, and statistically supportable, source pool than the entire world or even the entire Pacific. I present this seemingly trivial analysis as an example of the methods described in Section 3.2.2, but note that it is not always easy to make even such large scale distinctions (e.g., McCoy and Heck, 1987).

3.2.2.c. Eastern Pacific Provinces. Within the eastern Pacific region, the Galápagos share significantly more species than expected with the Panamic Province (Table 1). Differences between observed and expected values were not significant for the Cortez Province, which is essentially a subset of the Panamic Province (35 of the 41 species recorded in the Cortez Province are also found in the Panamic Province). Perhaps more surprisingly, no significant differences from random were found for the Oregonian–Alaskan and Peruvian–Chilean provinces. These cases are complicated by the small numbers of species involved, which serve to make statistically significant differences almost impossible to detect. All porcellanids within the Oregonian and Alaska provinces are also found in the Californian Province, which shares with the Galápagos significantly fewer species than expected. Thus, the three northeast Pacific provinces can be discounted as potential source areas for Galápagos porcellanids. The Peruvian–Chilean Province, although adjacent to the Panamic Province, shares far fewer species with it than expected ($O_S = 2$; $E_S = 9.49$; $p(X) = 0.00001$). Given the equally strong positive relationship of the Galápagos fauna to the Panamic fauna, it seems safe to exclude the Peruvian–Chilean region from the Galápagos source pool as well. Thus, *P* for the Galápagos is restricted to the Panamic region total of 62 species.

3.2.2.d. Panamic Islands. The porcellanid fauna of the Galápagos bears an unusually strong resemblance to the faunas of other eastern Pacific islands within the Panamic Province. Table 2 compares the Galápagos fauna to three oceanic and three continental islands for which reasonable data on porcellanid species are available. In all cases the islands share more species than expected, and in most cases the differences are significant or nearly so. The oceanic islands show noticeably stronger relationships with the Galápagos than do the continental islands.

These results suggest that the Galápagos porcellanid fauna is a nonrandom sample of the Panamic species pool, with an overrepresentation of 'insular' species. Such 'insular' Panamic species may be either particularly well-adapted to insular conditions or ill-adapted to mainland conditions. Alternatively, they simply may be excellent dispersers that are equally common on the mainland (see Sections 4.2 and 4.3.2). My preliminary studies suggest that mainland records far outnumber insular records for the two Galápagos 'extralimitals', *N. mexicanum* and *Petrolisthes armatus*, whereas the reverse is true for six of seven 'non-endemic' Galápagos species (only *N. bicapillatum* is recorded in almost equal numbers at

Table 1. Composition of the Galápagos Porcellanid Fauna Compared to Other Regions

Region	Species	P	O_s	E_s	p(x)[a]
Pacific	180	250	12	8.64	[0.01745]
Eastern Pacific	90	250	12	4.32	[0.000003]
Alaskan–Oregonian	4	90	0	0.53	0.5582[b]
Californian	25		0	3.33	0.0147
Cortez	41		7	5.47	0.8971[b]
Panamanian	61		11	8.13	[0.0506]
Peruvian–Chilean	14		1	1.87	0.4060[b]
			0[c]	1.71	0.1377[b]
		73[d]	0[c]	2.11	0.0786[b]
Indo-West Pacific	90	250	0	4.32	0.00405
		180	0	6.00	0.00017

[a] Probability of obtaining O_s shared species or *less* under the working hypothesis of no difference between areas; [brackets] indicate the probability of obtaining O_s shared species or *more*. Regions in **boldface** share significantly more species with the Galápagos than expected by chance; regions in *italics* share significantly fewer species.
[b] Not statistically significant.
[c] Excluding *Petrolisthes armatus* (as extralimital) from Galápagos fauna.
[d] Includes Panamanian plus Peruvian–Chilean species.

mainland and island sites). This pattern is at least consistent with the hypothesis of a suite of insular specialists or 'supertramps' (Diamond, 1975).

4. Biological Attributes of Porcelain Crabs as Colonists

The analyses discussed in the preceding section essentially ignore specific differences among taxa, or among regions, noting only the number of species present in each area. Thus, they can identify statistically improbable relationships among regions, but provide no insight as to the underlying forces behind these patterns. This section summarizes attributes of porcelain crabs as a group, and differences within the group, that might influence the accessibility and suitability of the Galápagos as an outpost for the fauna of potential source regions.

4.1. Phylogenetic Relationships

Only three of the 13 Panamic genera of porcellanids are found on the Galápagos: *Petrolisthes* (8/26 Panamic species shared), *Pachycheles* (2/12 species shared) and *Neopisosoma* (2/3 species shared). The latter two genera are distinguished by only a single character (the side of the carapace, which is entirely calcified in other porcellanids, is fragmented in *Pachycheles*, whereas it is completely membraneous in *Neopisosoma*), and are likely to eventually be synonomized (Haig, 1960; Gore, 1977; Harvey, unpub. data). Thus, the Galápagos porcellanid fauna draws completely from the two most speciose genera (or lineages, as presently defined), with no representation from the remaining 10 genera (Table 3). This is a statistically nonrandom distribution ($X^2 = 7.61$, $p < 0.01$), and may be biologically peculiar as well. Within higher taxa, high species diversity is often correlated with low dispersal ability and restricted species geographic ranges, because populations within poorly dispersing species are more easily

Table 2. Composition of the Galápagos Porcellanid Fauna Compared to Other Panamic Islands.

Region	Species	P	O_s	E_s	$p(x)^a$
Clipperton	**3**	**62**[b]	**3**	**0.58**	[0.0058]
Cocos	**5**		**3**	**0.97**	[0.0456]
Revillagigedos	**6**		**5**	**1.16**	[0.0007]
Tres Marias	**12**		**5**	**2.32**	[0.0443]
Isabela	13		4	2.52	[0.2132]
Gorgona	18		6	3.48	[0.0707]

[a] probability of obtaining O_s shared species or *less* under the working hypothesis of no difference between areas; [brackets] indicate the probability of obtaining O_s shared species or more. Regions in **boldface** share significantly more species with the Galápagos than expected by chance; regions in *italics* share significantly fewer species.
[b] Includes Panamic species plus *Pachycheles velerae*.

isolated or fragmented under a given set of environmental conditions than those of high-dispersal species (Hansen, 1980; Jablonski, 1982).

4.2. Dispersal Ability

One of the two great controversies that have rocked the field of biogeography through the last decade pits vicariance *vs.* long-distance dispersal as primary forces underlying the distribution of species (Nelson and Rosen, 1981; Sims, Price, and Whalley, 1983). Given that these issues are far from resolved, it is perhaps fortunate that the Galápagos are generally considered to be true oceanic islands, with no prior connection to the mainland (Cox, 1983; Simkin, 1984; Hickman and Lipps, 1985). Thus, the Galápagos must have acquired their entire biota via long-distance dispersal; even endemic species must be directly derived from prior colonizing species.

The pelagic larva is the primary dispersal phase of porcelain crabs, as in the vast majority of benthic marine invertebrates (Gurney, 1942). Oceanic current patterns are likely to define the limits of potential dispersal avenues for planktonic larvae from potential source areas, but species may vary in their ability to exploit prevailing oceanographic conditions to disperse to remote locations (Abbott, 1966; Scheltema, 1986).

4.2.1. The Oceanographic Setting

Although the oceanographic conditions influencing the Galápagos Archipelago are intricate (Houvenaghel, 1984), three main currents are probably responsible for most larval transport from mainland locations to the islands (Brusca, 1987; Chapter 1, this volume). The South Equatorial Current (SEC) brings waters from the southeast at velocities between 10 and 100 cm/s (current velocities estimated from Walton Smith, 1974), and is strongest from July to January (Houvenaghel, 1984). The main west-to-east current, the Equatorial Countercurrent (ECC), runs strongest from May to August (Houvenaghel, 1984), but rarely exceeds

Table 3. Representation of Panamic Genera in the Galápagos

	Number of species	
Genera	Galápagos	Panamic only
Petrolisthes	8	18
Pachycheles (including *Neopisosoma*)	4	11
Megalobrachium	0	7
Porcellana	0	3
Clastotoechus	0	2
Euceramus	0	2
Polyonyx	0	2
Orthochela	0	1
Minyocerus	0	1
Pisidia	0	1
Ulloaia	0	1
Heteroporcellana	0	1

50 cm/s. The Panama Current, from the northeast, includes waters from the California Current and North Equatorial Current (Brusca, 1987) along with tropical waters from offshore Central America (e.g., the Costa Rica Coastal Current). These currents are relatively slow, ranging between 10 and 25 cm/s.

Most porcellanids require about three weeks to pass through the three larval states at temperatures characteristic of tropical surface waters (see Section 4.2.2.b below). The nearest shallow water habitats west of the Galápagos, the Marshall Islands, are 7000 km distant (Brusca, 1987). Thus, a porcellanid larva hatched in the Marshalls would require a mean current speed of 385 cm/s to safely reach the Galápagos, far faster than the normal maximum velocity of the ECC.

Using this type of logic to examine why the Californian and Peruvian provinces are not represented in the Galápagos is complicated by the initially lower temperature of waters leaving these regions. Here I simply make the conservative assumption that temperatures will remain somewhere between 15° and 30°C, which roughly correspond to larval durations of 3 to 6 weeks. Larvae released from the southern limit of the Californian Province, some 4000 km distant (straight line) from the Galápagos, would require average current speeds of 110–220 cm/s to safely reach the Galápagos. This is much faster than the 5–16 cm/s current speeds that prevail in this region (Walton Smith, 1974). The northern limit of the Peruvian Province, in contrast, is only about 1100 km from the Galápagos. This is hardly further than the minimum Panamic distance of 970 km (Brusca, 1987), and, in terms of straight line distances, the Galápagos in general can hardly be considered much more isolated from the Peruvian Province than from the Panamic Province. Current speeds to transport northern Peruvian larvae to the Galápagos need only be on the order of 30–60 cm/s, and in fact appear to range between 20 and 70 cm/s (Walton Smith, 1974).

The implications from the above analyses are that 1) the "east Pacific barrier" (Ekman, 1953; Briggs, 1974) operates in full force against the porcellanids, since the ECC is not capable of transporting porcellanid larvae across the barrier; 2) the northeast Pacific provinces are too distant from the Galápagos to provide significant numbers of porcellanid larvae, and 3) the Galápagos are well within range of porcellanid larvae from the Peruvian Province, suggesting that factors other than current/dispersal interactions may be responsible for the weak faunal affinity of these two regions (keep in mind, however, that this relationship is not *statistically* improbable).

These estimates may be conservative, because they are based on straightline distances, whereas the connecting currents follow gyres and thus tend to be distinctly curved (Scheltema, 1986). On the other hand, porcellanid larvae may be able to extend their time in the plankton (thus increasing the odds of reaching suitable habitat), for example, by delaying metamorphosis to the benthic juvenile stage (Gore, 1975; Miller and Hadfield, 1990). This hypothesis is not strongly supported by published laboratory studies on porcelain crab larval development (e.g., citations listed in Fig. 1), which suggest that the duration of each larval stage, including the megalopa, is relatively inflexible; typically, all megalopae either die or molt to the first juvenile crab stage within a week of the "normal" molting peak. Alternatively, pelagic juveniles may extend their duration in the plankton by migrating to deeper, colder water where metabolic processes and development rates would be slowed (Sulkin and van Heukelem, 1986), but it is unknown whether this strategy is used by any porcellanid.

4.2.2. Porcellanids as Dispersers

Because oceanic islands are colonized strictly via long-distance dispersal, a reasonable hypothesis is that species with exceptional dispersal abilities should predominate in the faunas of oceanic islands. Some organisms have special adaptations for long-range dispersal (e.g., flight, or seeds, spores, cysts or other long-lived dormant stages) and are consistently well-represented in insular faunas (Carlquist, 1974). Compared to other decapod crustaceans, porcelain crabs produce relatively small numbers of large eggs. The planktonic larvae are also large, and undergo a brief, accelerated development that consists of only two zoeal stages plus the megalopa (Gore, 1975). Small clutches of large eggs with a short larval cycle commonly characterize species poorly suited to long-distance dispersal (Scheltema, 1971; Hansen, 1980; Jablonski, 1982). However, even within a morphologically homogeneous assemblage of probable mediocre dispersers like the Porcellanidae, there is still the potential for considerable interspecific variation in relative dispersal ability. This variation is likely to be a function of population size over the range of the species, and the duration of the dispersal stage.

4.2.2.a. Geographic Range Size. Generally, the larger the breeding population of a species, the more propagules that it will be able to disperse, and the greater the probability that at least some will successfully reach a remote site. Among closely related species, geographic range size should be positively correlated with total population size (Hanski, 1982; Bock and Ricklefs, 1983; Brown, 1984), and thus may be a useful indicator of *potential* dispersal ability. Geographic range size may also indicate *realized* dispersal ability, to the extent that the ability of a species to extend its range depends on its dispersal abilities (Brown and Gibson, 1983).

A detailed, quantitative analysis of the geographic distributions of the porcellanids is beyond the scope of the present study. Instead, I divided the 53 porcel-

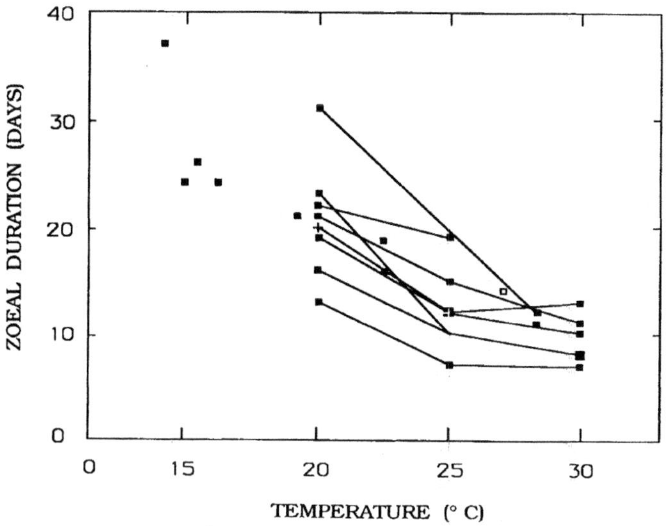

Figure 1. Zoeal duration as a function of temperatures for 18 species of eastern Pacific and western Atlantic porcelain crabs. Connected squares represent a single species raised at different temperatures. Data from Knight, 1966; Gore, 1968, 1970, 1971a–c, 1972a,b, 1973a,b, 1975, 1977; MacMillan, 1972; Pellegrini and Gamba, 1985; Saelzer, Quintana and Quiñones, 1986; also J. K. Blevins, unpublished data. ■, one species; ■, two and three species; □, *Petrolisthes tonsorius*; +, Pacific *Petrolisthes armatus*.

lanid species whose normal geographic range included the Panamic Province (that is, I excluded Panamic extralimitals, as defined by Haig, 1960) into two groups: those also present in the Galápagos ($n = 9$), and those absent ($n = 44$). Of the former group, 89% occur in one or more provinces (Californian, Cortez, or Peruvian) in addition to the Panamic, *vs.* only 52% of nonGalápagos species ($X^2 = 4.12$, $p < 0.05$). Thus, the Galápagos porcellanid fauna definitely favors widespread species (most of which seem to favor insular sites; see Section 3.2.2.d above).

4.2.2.b. Larval Duration The potential of a species to colonize distant sites also depends on the length of the dispersal phase (Scheltema, 1971). In porcelain crabs, as in most marine organisms, larval developmental rates, and thus duration of this primary dispersal phase, are closely dependent on temperature (Fig. 1). In fact, preliminary data suggest that interspecific and even intergeneric differences in larval duration are insignificant compared to temperature-related differences (Fig. 1). Temperature effects on larval duration are also independent on the geographic origins of species: cold-temperate species take the same time to develop under warm temperatures as tropical forms. Larval duration data for two species known to occur in the Galápagos (*Petrolisthes tonsorius* and *Petrolisthes armatus*, a probable extralimital) provide no indication that these species have an unusually long dispersal phase (Fig. 1). Thus, the available data, though incomplete, suggests that interspecific variation in larval duration is not an important consideration regarding the composition of the Galápagos porcellanid fauna.

4.3. Interactions with Other Species

4.3.1. Competitors

The second great controversy in contemporary island biogeography revolves around the role of competitive interactions in structuring insular communities, and whether distributional data constitutes acceptable evidence for or against competition (Diamond, 1975; Simberloff, 1978; Connor and Simberloff, 1978, 1979, 1983; Diamond and Gilpin, 1982; Wright and Biehl, 1982; Abbott, 1983; Strong *et al.*, 1984). One indication of the intractibility of this controversy is the tendency for data sets, methods, and empirical results of researchers with opposing viewpoints to converge over time, while interpretations continue to diverge.

Here I simply ask the question: Do porcellanids tend to show mutually exclusive distributions among the islands of the archipelago? Although such "checkerboard" distributions (Diamond, 1975) can arise through noncompetitive mechanisms (Connor and Simberloff, 1983), their absence should at least eliminate one possible role for competition in a given community. In the Galápagos, porcelain crabs show a strong tendency towards aggregation, as opposed to exclusivity (Table 4). Using a slightly modified version of Patterson and Atmar's (1986) N index of nestedness (Harvey, in prep.), and Monte Carlo simulation models that determine the distribution of N under random conditions, I find that the Galápagos porcellanids are significantly more nested than would be expected by chance alone ($N = 113$, expected $N = 167.44 \pm 7.04$, based on 100 replications; $p < 0.00001$). Thus, there is little indication of competitively influenced distributions within the archipelago. This analysis does not address the role competition might have on distributions on a larger scale, for example by limiting the number or combinations of Panamic species that can coexist in the archipelago (see Diamond, 1975).

4.3.2. Commensals

Species that are more or less obligately associated with other organisms might be expected to be inferior long-distance colonists if, as is the case with most marine associations, long-distance dispersal occurs separately in the host and "guest." This is because, almost by definition, very few propagules will disperse to the limit of a species' dispersal range (Brown, 1984). The odds of an obligate commensal species locating its host with only a few propagules are likely to be far less than those of a species with more generalized habitat requirements. Commensals with corals would be the most probable exceptions to this rule; in terms of "target size," coral reefs are more like "habitats" than isolated "hosts."

Nineteen of the 61 Panamic porcellanids with substratum data are considered to be characteristically commensals. Eleven species are associated with corals; the remaining species are found with gorgonians ($n = 2$), hermit crabs ($n = 2$), sponges ($n = 2$), asteroids ($n = 1$), or basket stars ($n = 1$). Three of the coral-associated Panamic species are found in the Galápagos, which is not significantly different from the "null" prediction of 2.16 species (all expected numbers of shared species are calculated using equation 2, although the specific composition of the source pool may vary as indicated). Thus, porcellanids that are commensal with corals do not appear to be underrepresented in the Galápagos porcellanid fauna. None

Table 4. Distribution of Porcelain Crabs among Islands of the Galápagos

Species	IS	ES	FL	SG	SF	CZ	GE	MA	BA	BR	CB	ON	SE	ED	FR	PI	WN	CU	Total
P. brachycarpus	0	1	1	1	1	1	1	1	1	0	1	0	1	0	1	0	0	0	11
P. biocellatus	1	1	1	1	1	1	1	0	0	1	0	1	0	0	0	1	1	0	11
P. haigae	1	1	0	1	1	1	1	0	0	0	0	1	1	0	0	0	0	1	9
P. polymitus	1	1	1	1	0	1	1	0	1	0	0	0	0	0	0	0	0	0	7
P. galapagensis	1	1	1	1	1	1	0	0	0	0	0	0	0	0	0	0	0	0	6
P. edwardsii	1	1	1	1	0	0	0	1	0	0	0	0	0	1	0	0	0	0	6
P. tonsorius	1	1	1	0	1	0	0	0	0	0	1	0	0	0	0	0	0	0	5
P. glasselli	1	1	0	0	0	0	1	0	0	1	0	0	0	0	0	0	0	0	4
N. bicapillatus	1	0	1	1	0	0	0	0	0	0	0	0	0	0	0	0	0	0	3
P. velerae	0	0	0	0	0	0	0	1	0	0	0	0	0	0	0	0	0	0	1
N. mexicanus	0	0	*	0	0	0	0	0	0	0	0	0	0	0	0	0	0	0	(1)[a]
P. armatus	0	0	0	0	0	0	0	0	0	0	0	0	0	*	0	0	0	0	(1)[a]
Total	8	8	(8)	7	5	5	5	3	2	2	2	2	2	(2)	1	1	1	1	10 + 2

[a] Island abbreviations: IS = Isabela, ES = Española, FL = Floreana, SG = Santiago, SF = Santa Fé, CZ = Santa Cruz, GE = Genovesa, MA = Marchena, BA = Baltra, BR = Bartolomé, CB = San Cristobal, ON = Onslow, SE = Seymour Norte, ED = Eden, FR = Fernandina, PI = Pinzon, WN = Wenman, CU = Culpepper.
* Probably extralimital.
[a]Recorded on one island, but as probable extralimital.

of the other commensal porcellanids are known from the Galápagos (expected = 1.57), even though suitable hosts are present. Unfortunately, the small number of noncoral commensals in the Panamic Province render it impossible to statistically determine whether they are in fact underrepresented in the Galápagos fauna.

5. Discussion

The above sections summarize distribution patterns of the porcelain crabs of the Galápagos within the context of several biological and environmental factors that may influence these patterns. Unfortunately, some of these patterns and processes are likely to be intercorrelated, so that it is often not clear which can safely be considered causal explanation. For example, do the biases favoring *Petrolisthes* and *Pachycheles* reflect superior colonizing or competitive abilities in these lineages, or simply that both genera have generalized habitat requirements (e.g., no noncoral commensalism)? The small number of species present weakens the explanatory power of most multivariate approaches (e.g., Schoener and Schoener, 1983a, 1983b).

Despite these caveats, certain generalizations can be made about the biogeographic relationships of the porcelain crabs of the Galápagos. The Galápagos porcellanid fauna is a small, nonrandom subset of the Panamic fauna, favoring wide-ranging species that are disproportionally likely to be found on, and even be largely restricted to, Panamic islands (particularly oceanic islands). Prevailing oceanographic patterns, combined with a relatively short larval dispersal phase, exert a powerful influence on distribution patterns. Potential source regions in the Indo-West Pacific and the northeastern Pacific are too distant to successfully transport porcellanid larvae before metamorphosis or death. The Galápagos Islands would appear to be within colonizing range of the Peruvian Province. The lack of shared Galápagos–Peruvian species, while not statistically unreasonable, may thus reflect environmental differences between the two regions. The Galápagos porcellanid fauna is restricted to the two lineages in the family that are most speciose, both within the Panamic region and worldwide. Interspecific differences in the dispersal ability of individual propagules appear to be small and unlikely to strongly affect porcellanid distribution patterns. Coral-associated commensals are not underrepresented in the Galápagos porcellanid fauna, but noncoral commensals may be. The distribution of porcellanids within the archipelago shows no indication of exclusive, "checkerboard" patterns commonly interpreted as an outcome of interspecific competition.

ACKNOWLEDGMENTS. Janet Haig reviewed the manuscript, provided unpublished distributional data, and offered much helpful advice; my interest in and ideas about island biogeography were inspired and improved by many lively discussions with James H. Brown and L. Yvonne Maluf.

References

Abbott, D. P., 1966, Factors influencing the zoogeographic affinities of Galápagos inshore marine fauna, in: *The Galápagos: Proceedings of the Symposia of the Galápagos International Scientific Project* (R. I. Bowman, ed.), pp. 108–122, University of California Press, Berkeley.

Abbott, I., 1983, The meaning of z in species/area regressions and the study of species turnover in island biogeography, *Oikos* **41**:385–390.

Bock, C. E., and Ricklefs, R. E., 1983, Range size and local abundance of some North American songbirds: A positive correlation, *Am. Nat.* **122**:295–299.

Briggs, J. C., 1974, *Marine Zoogeography*, McGraw-Hill, New York.

Brown, J. H., 1984, On the relationship between abundance and distribution of species, *Am. Nat.* **124**:255–279.

Brown, J. H., and Gibson, A, 1983, *Biogeography*, C. V. Mosby, St. Louis, Missouri.

Brusca, R. C., 1987, Biogeographic relationships of Galápagos marine isopod crustaceans, *Bull. Mar. Sci.* **41**:268–281.

Carlquist, S., 1974, *Island Biology*. Columbia University Press, New York.

Connor, E. F., and Simberloff, D., 1978, Species number and compositional similarity of the Galápagos flora and fauna, *Ecol. Monogr.* **48**:219–248.

Connor, E. F., and Simberloff, D., 1979, The assembly of species communities: Chance or competition? *Ecology* **60**:1132–1140.

Connor, E. F., and Simberloff, D., 1983, Interspecific competition and species co-occurrence patterns on islands: Null models and the evaluation of the evidence, *Oikos* **41**:455–465.

Cox, A., 1983, Ages of the Galápagos Islands, in: *Patterns of Evolution in Galápagos Organisms* (R. I. Bowman, M. Berson, and A. E. Leviton, eds.), pp. 11–23, Amer. Assoc. Adv. Sci., Pac. Div., San Francisco.

Diamond, J. M., 1975, Assembly of species communities, in: *Ecology and Evolution of Communities* (M. L. Cody and J. M. Diamond, eds.), pp. 342–444, Harvard University Press, Cambridge, Mass.

Diamond, J. M., and Gilpin, M. E., 1982, Examination of the 'null' model of Connor and Simberloff for species co-occurrences on islands, *Oecologia* **52**:64–74.

Ekman, S., 1953, *Zoogeography of the Sea*, Sidgwick and Jackson, London.

Garth, J. S., 1946, Distribution studies of Galápagos Brachyura, *Allan Hancock Pacific Exped.* **5**: 603–638.

Gore, R. H., 1968, The larval development of the commensal crab *Polyonyx gibbesi* Haig, 1956 (Crustacea: Decapoda), *Biol. Bull.* **135**:111–129.

Gore, R. H., 1970, *Petrolisthes armatus*: A redescription of larval development under laboratory conditions (Decapoda, Porcellanidae), *Crustaceana* **18**:75–89.

Gore, R. H., 1971a, *Megalobranchium poeyi* (Crustacea, Decapoda, Porcellanidae): Comparison between larval development in Atlantic and Pacific specimens reared in the laboratory, *Pacific Sci.* **25**: 404–425.

Gore, R. H., 1971b, The complete larval development of *Porcellana sigsbeiana* (Crustacea: Decapoda) under laboratory conditions, *Mar. Biol.* **11**:344–355.

Gore, R. H., 1971c, *Petrolisthes tridentatus*: The development of larvae from a Pacific specimen in laboratory culture with a discussion of larval characters in the genus (Crustacea: Decapoda: Porcellanidae), *Biol. Bull.* **141**:485–501.

Gore, R. H., 1972a, *Petrolisthes armatus* (Gibbes, 1850): The development under laboratory conditions of larvae from a Pacific specimen (Decapoda; Porcellanidae), *Crustaceana* **22**:67–83.

Gore, R. H., 1972b, *Petrolisthes platymerus*: The development of larvae in laboratory culture (Crustacea: Decapoda: Porcellanidae), *Bull. Mar. Sci.* **22**:336–354.

Gore, R. H., 1973a, *Pachycheles monilifer* (Dana, 1852): The development in the laboratory of larvae from an Atlantic specimen with a discussion of some larval characters in the genus (Crustacea: Decapoda: Anomura), *Biol. Bull.* **144**:132–150.

Gore, R. H., 1973b, Studies on decapod Crustacea from the Indian River region of Florida. 2. *Megalobranchium soriatum* (Say, 1818): The larval development under laboratory conditions (Crustacea: Decapoda: Porcellanidae), *Bull. Mar. Sci.* **23**:837–856.

Gore, R. H., 1975, *Petrolisthes zacae* Haig, 1968 (Crustacea, Decapoda, Porcellanidae): The development of larvae in the laboratory, *Pacific Sci.* **29**:181–196.

Gore, R. H., 1977, *Neopisosoma angustifrons* (Benedict, 1901): The complete larval development under laboratory conditions with notes on larvae of the related genus *Pachycheles* (Decapoda, Anomura, Porcellanidae), *Crustaceana* **33**:284–300.

Gore, R. H., 1981, Porcellanid crabs from the coasts of Mexico and Central America (Crustacea: Decapoda: Porcellanidae), *Smithson. Contr. Zool.* **363**:1–34.

Gore, R. H., and Abele, L. G., 1976, Shallow water porcelain crabs from the Pacific coast of Panama and adjacent Caribbean waters (Crustacea: Anomura: Porcellanidae), *Smithson. Contr. Zool.* **237**: 1–30.

Gurney, R., 1942, *Larvae of Decapod Crustacea*, Ray Society, London.

Haig, J., 1960, The Porcellanidae (Crustacea Anomura) of the eastern Pacific, *Allan Hancock Pacific Exped.* **24**:1–440.

Haig, J., 1962, Papers from Dr. Th. Mortensen's Pacific Expedition 1914–16. 79. Porcellanid crabs from eastern and western America, *Vidensk. Medd. Dansk. Naturh. Foren. Kbh.* **124**:171–192.

Haig, J., 1964, Papers from Dr. Th. Mortensen's Pacific Expedition 1914–16. 81. Porcellanid crabs from the Indo-West Pacific, Part I, *Vidensk. Medd. Dansk. Naturh. Foren. Kbh.* **126**:355–386.

Haig, J., 1966, Sur une collection de crustacés porcellanes (Anomura:Porcellanidae) de Madagascar et des Comores, Cah. ORSTOM, *Océanogr.* **3**:39–50.

Haig, J., 1968, Eastern Pacific expeditions of the New York Zoological Society. Porcellanid crabs (Crustacea: Anomura) from the west coast of tropical America, *Zoologica* **53**:57–74.

Haig, J., 1981, Porcellanid crabs from the Indo-West Pacific, part II, *Steenstrupia* **7**:269–291.

Hansen, T. A., 1980, Influence of larval dispersal and geographic distribution on species longevity in neogastropods, *Paleobiology* **6**:193–207.

Hanski, I., 1982, Dynamics of regional distribution: The core and satellite species hypothesis, *Oikos* **38**:210–221.

Hickman, C. S., and Lipps, J. H., 1985, Geologic youth of Galápagos Islands confirmed by marine stratigraphy and paleontology, *Science* **227**:1578–1580.

Houvenaghel, G. T., 1984, Oceanographic setting of the Galápagos Islands, in: *Key Environments: Galápagos* (R. Perry, ed.), pp. 43–54, Pergamon Press, New York.

Jablonski, D., 1982, Evolutionary rates and modes in late Cretaceous gastropods: Role of larval ecology, *Proc. 3rd N. Am. Paleont. Conv.* **1**:257–262.

Johnson, D. S., 1958, The Indo-West Pacific species of the genus *Polyonyx* (Crustacea, Decapoda, Porcellanidae), *Ann. of Zool.* [Agra, India] **2**:95–118.

Knight, M. D., 1966, The larval development of *Polyonyx quadriungulatus* Glassell and *Pachycheles rudis* Stimpson (Decapoda, Porcellanidae) cultured in the laboratory, *Crustaceana* **10**:75–97.

MacMillan, F. E., 1972, The larval development of northern California Porcellanidae (Decapoda, Anomura). I. *Pachycheles pubescens* Holmes in comparison to *Pachycheles rudis* Stimpson, *Biol. Bull.* **142**:57–70.

Maluf, L. Y., 1988, Biogeography of the central eastern Pacific shelf echinoderms, in: *Proceedings of the Sixth International Echinoderm Conference* (R. D. Burke, P. V. Mladenov, P. Lambert, and R. L. Parsley, eds.), pp. 389–398, A. A. Balkema, Rotterdam.
McCoy, E. D., and Heck, K. L., Jr., 1987, Some observations on the use of taxonomic similarity in large-scale biogeography, *J. Biogeo.* **14**:79–87.
Miller, S. E., and Hadfield, M. G., 1990, Developmental arrest during larval life and life-span extension in a marine mollusc, *Science* **248**:356–358.
Nelson, G., and Rosen, D. E. (eds.), 1981, *Vicariance Biogeography: a critique*, Columbia University Press, New York.
Patterson, B. D., and Atmar, W., 1986, Nested subsets and the structure of insular mammalian faunas and archipelagos, in: *Island Biogeography of Mammals* (L. R. Heaney and B. D. Patterson, eds.), pp. 65–82, Academic Press, London.
Pellegrini, N. C., and Gamba, A. L., 1985, Larval development of *Petrolisthes tonsorius* Haig, 1960 under laboratory conditions, (Decapoda, Porcellanidae), *Crustaceana* **49**:251–267.
Rosen, D. E., 1975, A vicariance model of Caribbean biogeography, *Syst. Zool.* **24**:431–464.
Saelzer, H. E., Quintana, R., and Quiñones, R., 1986, Larval development of *Petrolisthes granulosus* (Guérin, 1835) (Decapoda: Anomura: Porcellanidae) under laboratory conditions, *J. Crust. Biol.* **6**:804–819.
Scheltema, R. S., 1971, The dispersal of shoal-water benthic invertebrate species over long distances by ocean currents, in: *Fourth European Marine Biology Symposium* (D.J. Crisp, ed.), pp. 7–28, Cambridge University Press, Cambridge.
Scheltema, R. S., 1986, Long distance dispersal by planktonic larvae of shoal-water benthic invertebrates among central Pacific islands, *Bull. Mar. Sci.* **39**:241–256.
Schoener, T. W., and Schoener, A., 1983a, Distribution of vertebrates on some very small islands. I. Occurrence sequences of individual species, *J. Anim. Ecol.* **52**:209–235.
Schoener, T. W., and Schoener, A., 1983b, Distribution of vertebrates on some very small islands. II. Patterns in species number, *J. Anim. Ecol.* **52**:237–262.
Simberloff, D., 1978, Using island biogeographic distributions to determine if colonization is stochastic, *Am. Nat.* **112**:713–726.
Simberloff, D. S., and Wilson, E. O., 1969, Experimental zoogeography of islands: The colonization of empty islands, *Ecology* **50**:278–296.
Simkin, T., 1984, Geology of the Galápagos Islands, in: *Key Environments: Galápagos* (R. Perry, ed.) pp. 15–42, Pergamon Press, New York.
Sims, R. W., Price, J. H., and Whalley, P. E. S. (eds.), 1983, *Evolution, Time and Space: The Emergence of the Biosphere*, Academic Press, New York.
Strong, D. R., Jr., Simberloff, D., Abele, L. G., and Thistle, A. B. (eds.), 1984, *Ecological Communities: Conceptual Issues and the Evidence*, Princeton University Press, Princeton.
Sulkin, S. D., and van Heukelem, W. F., 1986, Variability in the length of the megalopal stage and its consequence to dispersal and recruitment in the portunid crab *Callinectes sapidus* Rathbun, *Bull. Mar. Sci.* **39**:269–278.
Walton Smith, F. G. (ed.), 1974, *Handbook of Marine Science*, Vol. 1, CRC Press, Cleveland.
Wright, S. J., and Biehl, C. C., 1982, Island biogeographic distributions: Testing for random, regular, and aggregated patterns of species occurrence, *Am. Nat.* **119**:345–357.

Chapter 8

Zoogeography of the Shallow-Water Cirriped Fauna of the Galápagos Islands and Adjacent Regions in the Tropical Eastern Pacific

VICTOR A. ZULLO

1. Introduction

The association of the Galápagos Islands with the birth of Darwinian evolutionary theory has inspired many detailed studies of their fauna and flora. Yet the cirripeds, the only invertebrates to receive Darwin's personal attention as a systematist, are virtually unknown for the region. His monographs on the Cirripedia (Darwin, 1851, 1854) contain references to three species collected during the voyage of the *Beagle: Balanus tintinnabulum* (Linnaeus) and *Tetraclita porosa* (Gmelin) from rocky intertidal areas, and *Platylepas decorata* Darwin from marine turtles.

Pilsbry (1907, 1916) provided additional information on the fauna in his monographs on American species. The two intertidal barnacles were reevaluated from material collected by the *Albatross* in 1888 and described as the new subspecies *Balanus tintinnabulum galapaganus* Pilsbry and *Tetraclita squamosa milleporosa* Pilsbry. Another turtle barnacle, *Chelonibia testudinaria* (Linnaeus)

VICTOR A. ZULLO • Department of Earth Sciences, University of North Carolina at Wilmington, Wilmington, North Carolina 28403-3297.

was reported from Isabela Island, and two new species of *Scalpellum*, *S. galapaganus* Pilsbry and *S. soror* Pilsbry, were described from an 1891 *Albatross* station at 1100 m depth off the Galápagos Islands. More recently Newman (Zullo and Newman, 1964) reported *Heteralepas quadrata* (Aurivillius) from the spiny lobster *Panulirus pencillatus* (Oliver), Zullo and Beach (1973) described the new species *Hexacreusia straeleni* Zullo and Beach from a balanophyllid coral at 60 m depth and *Membranobalanus nebrias* (Zullo and Beach) from a clionid sponge at subtidal depths, Henry and McLaughlin (1986) reported *Megabalanus coccopoma* (Darwin), and Zullo (1986) discussed the Quaternary fossil barnacle fauna.

These records represent all of the published systematic data on Galápagos cirripeds. However, various expeditions to the Galápagos made general collections that contain barnacles. Notable are those of the Allan Hancock Foundation and the California Academy of Sciences. In addition, the archipelago was visited by Henry A. Pilsbry as a member of the 1929 Pinchot South Sea Expedition. His collection is housed at the Academy of Natural Sciences of Philadelphia.

This study is based on personal observations and collections made as a member of the 1964 Galápagos International Scientific Project (GISP), and is augmented by the unpublished collections at the museums listed below. During the six weeks spent in Galápagos, I collected barnacles from the following islands: Baltra (South Seymour), Bartholomé, Eden, Genovesa (Tower), Islas Plaza, San Salvador (James), Santa Cruz (Indefatigable) and Santa Fe (Barrington). Additional samples were provided by expedition members from Isabela (Albemarle), Marchena (Bindloe), San Cristobal (Chatham) and Santa Maria (Charles).

Museum collections used in this study include: Allan Hancock Foundation (AHF), University of Southern California, courtesy of John S. Garth; Academy of Natural Sciences of Philadelphia (ANSP), Pennsylvania, courtesy of R. Tucker Abbott; California Academy of Sciences, San Francisco (CAS); 1953–1954, Galáagos Expedition of the International Institute for Submarine Research, Vaduz, Liechtenstein (IISR), courtesy of J. Wyatt Durham; Los Angeles County Museum of Natural History, Invertebrate Paleontology (LACMIP), courtesy of Edward C. Wilson; Museum of Paleontology, University of California, Berkeley (UCMP), courtesy of J. Wyatt Durham; and National Museum of Natural History, Smithsonian Institution (USNM), Washington, D.C., courtesy of Raymond B. Manning.

2. The Galápagan Cirriped Fauna

The Galápagan cirriped fauna includes 18 species-group taxa, all of which belong to the order Thoracica (Table 1). No Verrucomorpha or Chthamaloidea were observed. The absence of chthamaloids is particularly surprising considering their abundance on intertidal rocky shores of the Panamic province mainland. Although a determined search was made for Acrothoracica, none was found. Ascothoracica and Rhizocephala were not sought specifically, and their absence in the Galápagan fauna is probably the result of collection bias.

The Galápagos fauna did not include any introduced species as of 1965. Henry and McLaughlin (1986), however, recorded *Megabalanus coccopoma*, a Panamic province mainland species, from Baltra and Santa Cruz Islands. As these islands were explored rather thoroughly during the 1964 GISP expedition, and have since

Table 1. Composition and Biogeographic Distribution of the Fossil and Extant Galápagan Cirriped Fauna

Taxon	Distribution[a]		
	Mainland Panamic province	Cocos Island	Other
Lepadomorpha			
Conchoderma virgatum			cosmopolitan
Heteralepas quadrata	X	X	western Pacific
Lepas anatifera			cosmopolitan
Lepas pectinata			cosmopolitan
Oxynaspis n. sp.			
Balanomorpha			
Balanus calidus[b]	X		western Atlantic
Balanus poecilus[b]	X		
Balanus trigonus[b]	X		pan-tropical
Chelonibia testudinaria			pan-tropical
Concavus panamensis (cf. *eyerdami*[c])	X	*	
Conopea galeata	X		western Atlantic
Coronula diadema[c]			cosmopolitan
Cylindrolepas darwiniana			pan-tropical
Hexacreusia straeleni			
Megabalanus coccopoma	X		
Megabalanus peninsularis[b]	**	**	
Membranobalanus nebrias			
Platylepas decorata			western Pacific
Stephanolepas muricata			pan-tropical
Tetraclita milleporosa[b]			
T. rubescens rubescens?[c]			Californian province

[a] Distribution: X indicates presence of species; * indicates related subspecies, *C. panamensis panamensis*, found; **, Henry and McLaughlin (1986) synonymized *M. galapaganus* from the Galápagos and Cocos Island with *M. peninsularis* of the mainland Panamic province (see Appendix for discussion).
[b] Also known from Pleistocene deposits in the Galápagos.
[c] In the Galápagos known only from Pleistocene deposits.

become primary to the tourist trade, I would suggest that this species was introduced by ships. Such additional species as *Balanus amphitrite amphitrite* Darwin, *B. eburneus* Gould, *B. improvisus* Darwin, and *B. reticulatus* Utinomi, that have achieved wide distribution through ship transport (Carlton and Zullo, 1969; Henry and McLaughlin, 1975; Laguna, 1985) are known from mainland tropical eastern Pacific localities, and could be present in the islands as a result of vastly increased tourism over the past 20 years.

3. Cirriped Fauna of the Tropical Eastern Pacific

3.1. Other Tropical Eastern Pacific Islands

Limited information is available concerning the cirriped fauna of other islands in the eastern Pacific. Cirripeds collected at Cocos Island off the coast of Costa Rica during the 1964 GISP expedition and on Cruise 18 of the *Te Vega* included

Heteralepas quadrata, *Concavus (Arossia) panamensis panamensis* (Rogers), *Megabalanus peninsularis* (Pilsbry) and *Tetraclita* sp. aff. *T. panamensis* Pilsbry. A single barnacle, *Megabalanus clippertonensis* (Zullo), is known from Clipperton Island, an atoll about 1000 km west of Mexico (Zullo, 1969a). Hertlein (1932) reported *Tetraclita milleporosa* from Cocos Island, Costa Rica and Malpelo Island, Panama. These specimens were reexamined during the course of this study, and are reidentified as *Tetraclita* sp. aff. *T. panamensis* and *T. panamensis*, respectively.

3.2. Tropical Eastern Pacific Mainland

In an earlier paper prepared before the 1964 GISP expedition (Zullo, 1966), an attempt was made to evaluate the balanomorph fauna of the tropical eastern Pacific (Panamic province) from the limited data available in the literature. The Panamic province cirriped fauna was considered to extend from Cape San Lucas, Baja California, south to the vicinity of Paita, Peru. The Gulf of California fauna was essentially excluded from the Panamic fauna and was viewed rather as a mixture of Panamic and north temperate (Californian province) taxa. From the present more detailed study, and from recent works (notably the publications of Henry and McLaughlin, 1975, 1986; and the unpublished master's thesis of Laguna, 1985), several species are added to the fauna, and the distribution of previously known species is more precisely determined. The revised species distribution list for the tropical eastern Pacific is presented in Table 2.

The Gulf of California cirriped fauna includes 23 taxa whose distribution is self-determined.* These taxa, arranged in distributional categories, are presented in Table 3. The Gulf fauna is now seen to be dominated by Panamic species or their derivatives, and is here considered as a northern extension of the Panamic fauna.

Garth (1960, p. 119) in his summary of the distribution of Brachyura of the Gulf, divided the fauna into three parts: (1) a large Panamic element occurring throughout, but attenuating northward; (2) a warm temperate element consisting of endemic and southern California–northern Baja California species that occurs in the northern Gulf; and (3) an "insular" littoral element of the Cape San Lucas region related to the faunas of such offshore islands as Socorro, Clarion, and the Galápagos. Walker (1960, p. 129–133) recognized a similar distributional pattern in the marine fish fauna, and described four faunal assemblages related to ecologic conditions: (1) the upper Gulf, north of the southern tip of Tiburon Island, which is characterized by a depauperate fauna containing moderate depth endemics and species from more northern parts of the outer Pacific coast; (2) the central Gulf, from San Francisquito Bay–Kino Bay south to La Paz–Guaymas, characterized by a rich fauna containing a large endemic element in shallow waters, and a number of north-ranging species; (3) the southeastern Gulf, south of Guaymas, having a fauna similar to that of southern Mexico; and (4) the Cape San Lucas region, from La Paz

*The term "self-determined" is used here to distinguish distribution patterns that result primarily from larval dispersal, although rafting of adults (including transport as fouling organisms) may also occur. This category includes sessile benthonic taxa and obligate commensals of sessile or sedentary benthonic organisms, and excludes those species normally attached to drifting substrata (e.g., the species of *Lepas*), or to actively swimming hosts such as turtles and whales (e.g., the species of *Coronula* and *Chelonibia*).

Table 2. Composition and Distribution of the Thoracican Cirriped Fauna of the Tropical Eastern Pacific Mainland Coast

Taxon	Distribution[a]							
	1	2	3	4	5	6	7	Authority
Lepadomorpha								
Heteralepas quadrata (Aurivillius)	X			X				Laguna (1985)
Lepas anatifera Linnaeus				X				Pilsbry (1907)
L. anserifera Linnaeus	X							Pilsbry (1907)
L. denticulata Gruvel		X						MacDonald (1929); ? = *L. anserifera*
L. hilli (Leach)						X		Pilsbry (1907)
Octolasmis californiana Newman		X		X		X		Newman (1960b); Laguna (1985)
Oxynaspis rossi Newman	X							Newman (1972)
Pollicipes elegans Lesson	X	X	X			X	X	Henry(1960); Laguna (1985)
Balanomorpha								
Balanus amphitrite amphitrite Darwin[b]	X	X						Henry and McLaughlin (1975)
B. calidus Pilsbry					X			This chapter
B. eburneus Gould[b]	X	X	X	X				Henry and McLaughlin (1975); Laguna (1985)
B. improvisus Darwin[b]	X	X				X	X	Henry and McLaughlin (1975)
B. inexpectatus Pilsbry	X	X	X	X		X		Henry and McLaughlin (1975); Laguna (1985)
B. mexicanus Henry	X	X	X	X		X	X	Henry and McLaughlin (1975)
B. parkeri Zullo	X							Zullo (1967)
B. peruvianus Pilsbry			X	X			X	Henry and McLaughlin (1975)
B. poecilus Darwin	X			X	X	X	X	Laguna (1985); this chapter
B. reticulatus Utinomi[b]		X	X	X				Laguna (1985)
B. trigonus Darwin	X	X	X	X		X		This chapter
Catophragmus pilsbryi Broch			X	X				Stanley and Newman (1980)
Chelonibia patula dentata Henry	X							Henry (1960)
C. testudinaria (Linnaeus)	X				X			Henry (1960); this chapter
Chthamalus anisopoma Pilsbry	X							Henry (1960)
C. fissus Darwin	X							Henry (1960)
C. panamensis Pilsbry s.l.		X	X	X		X		Pilsbry (1916); Henry and McLaughlin (1975); this chapter
Concavus panamensis eyerdami (Henry)	X							Henry (1960)
Concavus panamensis panamensis (Rogers)				X				Newman (1982)
Conopea galeata (Linnaeus)	X			X	X			Laguna (1985); this chapter
C. masignota (Henry and McLaughlin)		X	X	X		X		Henry and McLaughlin (1967)
Euraphia imperatrix (Pilsbry)		X	X	X	X			Pilsbry (1916); Laguna (1985); this chapter
E. eastropacensis Laguna		X	X	X	X			Laguna (1988)
Fistulobalanus dentivarians (Henry)		X		X		X		Henry and McLaughlin (1975)
F. suturaltus (Henry)			X					Henry and McLaughlin (1975)
Hexacreusia durhami (Zullo)	X	X		X				Zullo and others (1972)
Megabalanus californicus (Pilsbry)	X							Henry and McLaughlin (1986)
M. coccopoma (Darwin)	X	X	X	X	X	X	X	Henry and McLaughlin (1986); this chapter
M. peninsularis (Pilsbry)	X	X	X	X	X	X		Henry and McLaughlin (1986)
M. vinaceus (Darwin)			X			X		Henry and McLaughlin (1986)
Membranobalanus orcutti (Pilsbry)	X							Zullo and Beach (1973)

continued

Table 2. *(continued)*

Taxon	Distribution[a] 1	2	3	4	5	6	7	Authority
M. robinae Van Syoc	X							Van Syoc (1988)
Notomegabalanus concinnus (Darwin)							X	Henry and McLaughlin (1986); this chapter
Tetrabalanus polygenus Cornwall			X	X		X		Zullo (1969b); Laguna (1985)
Tetraclita confinis Pilsbry	X	?						This chapter
T. panamensis Pilsbry			X	X	X	X		Laguna (1985); this chapter
T. rubescens Darwin	X							Henry (1960)
T. stalactifera (Lamarck)	X	X	X					This chapter

[a]Distribution: 1 = Gulf of California; 2 = southern Mexico; 3 = Guatemala to Costa Rica; 4 = Panama; 5 = Colombia; 6 = Ecuador; 7 = northern Peru.
[b]= Introduced species.

south, characterized by an "insular" fauna that lacks northern elements and contains a number of western Pacific species and species ranging south from Mazatlán. Walker (1960, p. 132) concluded that "the fish fauna of The Gulf of California is clearly part of the Panamic fauna, but it has certain peculiarities and distinctive elements which characterize it."

Too little is known of the distribution of cirriped species within the Gulf to provide such an analysis. However, it does appear that cirriped distribution and faunal composition is similar in gross aspect to that of the Brachyura and fish in having a strong Panamic emphasis, a small, but distinct endemic element, some mixture of north- and south-ranging species, and at least one species, *Megabalanus peninsularis*, in the Cape San Lucas region with possible "insular" affinities.

The cirriped fauna of the Panamic province from Mazatlán south to Paita, Peru is distinctive and relatively homogeneous. As more localities are investigated in this extensive region, the more apparent it becomes that most of the known

Table 3. Biogeographic Distribution of the Gulf of California Cirriped Fauna

Endemic species (5)	Californian province species (5)
Balanus parkeri	*Chthamalus fissus*
Chthamalus anisopoma	*Megabalanus californicus*
Concavus panamensis eyerdami	*Oxynaspis rossi*
Membranobalanus robinae	*Membranobalanus orcutti*
Tetraclita confinis	*Tetraclita rubescens*
Panamic province species (9)	Californian and Panamic species (2)
Balanus inexpectatus	*Balanus trigonus*
Balanus poecilus	*Conopea galeata*
Balanus mexicanus	Introduced species (3)
Concavus panamensis panamensis	*Balanus amphitrite amphitrite*
Hexacreusia durhami	*Balanus eburneus*
Megabalanus coccopoma	*Balanus improvisus*
Megabalanus peninsularis	
Pollicipes elegans	
Tetraclita stalactifera	

Panamic cirripeds range throughout the province. This is especially true of lower intertidal and subtidal species. The southern boundary of the Panamic province in the vicinity of Paita is striking in its abruptness. Only one species, *Pollicipes elegans* Lesson, is widespread both in the Panamic and south temperate (Peruvian) provinces. A Peruvian species, *Notomegabalanus concinnus* (Darwin) extends barely into the Panamic province, reaching as far north as Talara, Peru. This definite faunal break reflects the abrupt change in marine hydroclimates at Point Agujas, Peru, where major upwelling occurs and the cold, north-flowing Peru Current swings westward, transporting inshore waters into the central Pacific basin. The northern boundary of the Panamic province is not as sharply defined, but still distinct. Only 11 of the 38 species with self-determined distributions found in the Panamic province are also found to the north in the warm temperate Californian province, and of these 11 species, five are Californian province species that occur in the Gulf of California. The biogeographic distribution of the mainland Panamic province cirriped fauna is summarized in Table 4.

4. Ecology and Local Distribution of Galápagos Cirripeds

The Galápagos are oceanic islands of volcanic origin that lie on the equator about 1000 km west of Ecuador. There are 16 major islands and numerous islets and

Table 4. Biogeographic Distribution of the Mainland Panamic Province Cirriped Fauna, Gulf of California to Paita, Peru

Endemic (24)	Present in Peruvian province (2)
Balanus inexpectatus	*Notomegabalanus concinnus*
Balanus mexicanus	*Pollicipes elegans*
Balanus parkeri	Californian province species (6)
Balanus peruvianus	*Chthamalus fissus*
Balanus poecilus	*Megabalanus californicus*
Catophragmus pilsbryi	*Membranobalanus orcutti*
Chelonibia patula dentata	*Octolasmis californiana*
Concavus panamensis eyerdami	*Oxynaspis rossi*
Concavus panamensis panamensis	*Tetraclita rubescens*
Conopea masignota	Tropical amphi-American (3)
Chthamalus anisopoma	*Balanus calidus*
Chthamalus eastropacensis	*Conopea galeata*
Chthamalus panamensis	*Tetraclita stalactifera*
Euraphia imperatrix	Pan-tropical (3)
Fistulobalanus dentivarians	*Balanus trigonus*
Fistulobalanus suturaltus	*Chelonibia testudinaria*
Hexacreusia durhami	*Heteralepas quadrata*
Megabalanus coccopoma	Cosmopolitan (3)
Megabalanus peninsularis	*Lepas anatifera*
Megabalanus vinaceus	*Leaps anserifera*
Membranobalanus robinae	*Lepas hilli*
Tetrabalanus polygenus	Introduced (4)
Tetraclita confinis	*Balanus amphitrite*
Tetraclita panamensis	*Balanus eburneus*
	Balanus improvisus
	Balanus reticulatus

rocks that cover a sea area from 1° 40′ N to 1° 25′ S latitude and 89° 15′ W to 92° W longitude. The largest island, Isabela, is about 140 km in length and has the greatest elevations, with Volcan Wolf reaching 1700 m. Santa Cruz Island, near the center of the main island group, has a maximum diameter of 41 km and an elevation of over 850 m. The smallest of the major islands, Genovesa, has a diameter of about 5 km and a height of 64 m.

The marine hydroclimate of the Galápagos is complex and atypical for an equatorial location. This complexity is the direct result of the interaction of the major oceanic current systems influencing the regions and the episodic, short-term effects of El Niño (Abbott, 1966; Philander, 1990). The Peru Oceanic Current traveling northward off the coast of Peru carries subtropical water at its surface and subantarctic water at 200–300 m depth. This current turns westward between 10° and 20° south latitude to form the South Equatorial Current that passes through the Galápagos. The South Equatorial Current is supported throughout the year by prevailing southeasterly trade winds, but decreases in strength during February and March when these winds are weakest. From February through April strong currents associated with upwelling in the Gulf of Panama move southward along the coasts of Colombia and Ecuador and merge with the South Equatorial Current to pass through the Galápagos. Cooler waters of the eastward-flowing Equatorial Undercurrent upwell along the western wall of the Galápagos Platform enhancing warm temperate conditions along the western margins of the archipelago. During normal years, the strength of the westward-flowing equatorial currents prevent warmer, eastward-flowing waters of the Equatorial Counter Current from reaching the Galápagos Islands. During years of El Niño, however, warmer waters from the equatorial regions of the western Pacific spread far eastward, bathing the Galápagos and the mainland of North and South America during the early months of the year. During particularly severe El Niños, coastal water temperatures from California to southern Peru are measurably increased (Philander, 1990).

The influence of these interacting water masses is reflected by yearly variations in surface water temperatures in various parts of the archipelago (Abbott, 1966). At Wreck Bay, San Cristobal Island, sea surface temperatures reach a maximum of 28°C in March and a minimum of 15°C in November, with a mean annual range of 25° to 18.5°C. Overall, sea temperatures north of the equator achieve a maximum of 30°C in February and March, and a minimum of 20°C in July, with a mean annual range of 27° to 24°C. South of the equator maximum sea temperatures of 28°C occur from March through May, and minima of less than 18°C in August, with a mean annual range of less than 27° to 21°C. If, as proposed by Ekman (1953), a tropical hydroclimate is defined by minimum monthly temperatures of 20°C, and a subtropical hydroclimate by minimum monthly temperatures between 16° and 18°C, then the Galápagos hydroclimate can be roughly divided into northern tropical and southern subtropical zones.

The spring tide range in the archipelago is 1.5 to 2 m, and the mean range is 1.2 to 1.5 m. As indicated by Hedgpeth (1969), this is an unusually large range for an equatorial island. The prevailing oceanic swell is damped for many of the eastern islands by the barrier provided by Isabela Island on the western side of the archipelago. The regions of greatest wave shock are along southern shores exposed to the southeasterly trade winds. Salinities measured at Wreck Bay in 1960 ranged from 32.9 to 37.1 ppt, with a mean of 34.7 ppt (U.S. Department of Commerce, 1962).

The coasts of the Galápagos Islands consist of volcanic headlands and cliffs, rock reefs, and occasional stretches of coral and volcanic sand beaches. In some areas small embayments are present, formed either by breachment of volcanic craters (e.g., Darwin Bay, Genovesa Island) or by graben faulting (e.g, Academy Bay, Santa Cruz Island).

4.1. The Intertidal Environment

Hedgpeth (1969) remarked on the apparent dearth of exposed marine life in the rocky intertidal zone, and barnacles are no exception. *Tetraclita milleporosa* is the only true intertidal barnacles in the Galápagos. This species occurs in the mid-intertidal about 1 m above mean sea level in a narrow band whose thickness averages 30 cm and was not observed to exceed 45 cm. The lateral distribution of *Tetraclita* is quite irregular. In many areas it is absent, in others scattered at low densities, and in a few areas dense populations up to three individuals per square cm carpet the rocks. This type of distribution was observed for *T. stalactifera* (Lamarck) in the Florida Keys (Stephenson and Stephenson, 1972) and for *T. japonica* Pilsbry in Japan (Suzuki and Syuiti, 1963). The highest densities of *T. milleporosa* in the Galápagos were found to coincide with areas in which slope of the coast was quite gentle and, conversely, *Tetraclita* was most often absent on vertical surfaces. In addition, the best developed populations were found in areas of strong wave action, generally along exposed southern coasts.

The importance of herbivorous and corallivorous rasping fish (e.g., wrasses, parrotfish, butterflyfish, surgeonfish) as a limiting factor in the distribution of intertidal barnacles and other sessile or sedentary invertebrates in tropical regions was discussed by Newman (1960a) and Bakus (1964). The damage caused by these fish is clearly evident in Galápagos *Tetraclita* populations. Each outgoing tide uncovers fresh evidence of rasping. In several instances, rasping is so severe as to penetrate the body chamber and kill adult barnacles. It is possible that local differences in population density are a function of the activity of rasping fish. At high tide, steep intertidal slopes are adjacent to deeper water where fish can readily maneuver and feed, whereas gentler slopes have a shallower, more turbulent water cover in which fish may find feeding difficult. These observations on the distribution of Galápagos *Tetraclita* populations contrast with those of Vermeij (1978, p. 104), who indicated that barnacles on open tropical shores are restricted to topographic highs or steep slopes.

Vermeij (1978, p. 105) also suggested that the relatively low outline and thickened walls of *Tetraclita* were adaptations to hinder predation by muricid gastropods. No evidence of muricid (or dorid nudibranch) predation of Galápagan *Tetraclita* populations was observed during the 1964 GISP expedition. Furthermore, it is unlikely that gastropod predation could account for the observed variations in distribution of *Tetraclita* in the Galápagos. My observations of intertidal barnacle populations in temperate regions of the Pacific and Atlantic coasts of North America suggest that, even in areas of severe gastropod predation, only the largest (i.e., usually the oldest year class) individuals are preyed upon. High barnacle densities in areas of severe predation are maintained by yearly recruitment and by the smaller individuals of more recent year classes. Thus, predation appears to be an effective means of removing older individuals from the substratum, making space available for settlement of future generations.

The three barnacles of the lower intertidal, *Megabalanus peninsularis*,* *Balanus calidus* Pilsbry and *B. trigonus* Darwin, are not truly intertidal organisms, but rather can be regarded as subtidal barnacles that extend into the lower reaches of the tidal zone. However, it is in high energy lower intertidal areas (between 0 and +0.3 m above MSL) that *Megabalanus peninsularis* reaches its maximum size and density in the Galápagos. *Balanus trigonus*, on the other hand, prefers waters well protected from surf, and is equally abundant in the subtidal and lower intertidal. In the intertidal zone, *B. trigonus* is usually found on the undersides of rocks in tide pools and flats. *Balanus calidus* was collected only once in the lowermost intertidal on a rock reef in Academy Bay. Henry and McLaughlin (1986) reported *Megabalanus coccopoma* from Baltra and Santa Cruz Islands. Considering my failure to find any specimens of *M. coccopoma* during extensive surveys of these islands in 1964, and considering the major increases in shipborne tourist traffic to these two islands during the past two decades, it is more likely that *M. coccopoma* was introduced by ships, rather than having arrived through larval dispersal or natural rafting.

4.2. The Subtidal Environment

Benthic substrata are predominantly rock and coral, with only occasional flats of medium-grained coral sand in protected bays. Coral growth is commonly encountered, and is locally widespread and massive (see Glynn and Wellington, 1983). *Megabalanus peninsularis* and *Balanus trigonus* are the most common barnacles of the subtidal, and are found on a variety of substrata, including rocks, dead coral, mollusk shells, and crustacean carapaces. *Balanus calidus* has similar habitat preferences, but is rare in the Galápagos. Another rare species, *B. poecilus* Darwin, appears to prefer attachment to the spines of living cidaroid echinoids and the shells of living epibenthic bivalves.

The remaining benthonic species occurring subtidally are commensals. *Heteralepas quadrata* is primarily an epizoite of the spiny lobster *Panulirus pencillatus* in the Galápagos, although one group of individuals were found attached to the columella of the gastropod *Malea* inhabited by a hermit crab. An undescribed species of *Oxynaspis* Darwin on an antipatharian host is known from a single locality below wave base. *Conopea galeata* (Linnaeus) is found in association with gorgonians. The Galápagos specimens occurred with *Pacifigorgia* Bayer, and were obtained wherever the gorgonian was encountered. *Membranobalanus nebrias* lives within a small, bright-red clionid sponge that bores into the nongrowing parts of coral heads and, in one instance was found occupying vesicles in basaltic rock. *Hexacreusia* Zullo, an obligate commensal of corals, is represented by *H. straeleni* Zullo and Beach found in a balanophyllid coral from single, deeper water locality.

*Henry and McLaughlin (1986) synonymized *Megabalanus galapaganus* with the Panamic province mainland species *M. peninsularis*. See Appendix, this chapter, for discussion.

4.3. Pelagic Species and Commensals of Marine Turtles

Species of the genus *Lepas* and *Conchoderma virgatum* Spengler are usually found in oceanic waters attached to floating weed and debris. The two species of *Lepas*, *L. anatifera* Linnaeus, and *L. pectinata* Spengler, found during the 1964 GISP expedition, were represented by ovigerous adults attached to beached boards. The specimens of *Conchoderma virgatum* were associated with juveniles of an undetermined species of *Lepas* on a floating wooden box.

Green and hawksbill turtles in Galápagan waters host a variety of barnacles. Four species of four genera are known from the Galápagos, including *Chelonibia testudinaria* (Linnaeus), *Cylindrolepas darwiniana* Pilsbry, *Platylepas decorata* Darwin, and *Stephanolepas muricata* Fischer. Each of these turtle barnacles has a preferred site on the turtle. *Chelonibia testudinaria* is found attached to the carapace, whereas *Platylepas decorata* is usually found embedded in the skin of the neck and the upper surface of the front flippers. *Cylindrolepas darwiniana* is deeply embedded to its orifice in and between the plates of the plastron, and *Stephanolepas muricata* is deeply embedded in skin of the leading edge of the front flippers.

5. Zoogeography

Discussion of the affinities of the Galápagos cirriped fauna is best considered in three parts corresponding to the environmental divisions described in Section 4. In terms of distribution, the species of the intertidal exhibit the greatest provincialism, whereas those of the subtidal and the pelagic or motile-host environments have much broader geographic ranges.

Tetraclita milleporosa is apparently endemic to the Galápagos Islands. It is morphologically similar to the amphi-American species *T. stalactifera* and the Panamic species *T. panamensis*. However, little phenotypic variation was observed in the large sampling made from the islands throughout the archipelago, and no individuals resembling any of the eastern Pacific mainland species were encountered. The morphologic distinctiveness of *T. milleporosa* suggests that it has been a geographic isolate for some period of time, and that dispersal and gene flow from the Panamic mainland is of exceedingly rare occurrence. The antiquity of *T. milleporosa* is further supported by its presence in Quaternary deposits on the islands (Zullo, 1986). Although brooding of larvae to the cyprid stage has not been verified for *T. milleporosa*, a few cyprids and numerous late stage nauplii were observed in the mantle cavities of a number of individuals. A tendency for an abbreviated larval history in this and, possibly, other species of *Tetraclita* could account for the high degree of morphological integrity exhibited by island populations in the eastern Pacific, as dispersal would depend almost entirely on accidental introduction of adults through mechanisms such as rafting.

The three species found in the lower intertidal zone exhibit broader distributions. *Megabalanus peninsularis*, as redefined by Henry and McLaughlin (1986), occurs at lower intertidal and subtidal depths at Cocos Island off Costa Rica, Malpelo Island in the Bay of Panama, and along the mainland coast of the Panamic province from Cape San Lucas; Baja California to Ecuador. This species is not closely related to other mainland species. Rather, *M. peninsularis*, together with *M. clippertonensis* from Clipperton Island and *M. tanagrae* (Pilsbry) from the Hawaiian Islands, form a complex characterized by a flat to nearly flat, narrow scutum. *Megabalanus peninsularis* is also present in Quaternary deposits in the Galápagos (Zullo, 1986).

The discovery of *Balanus calidus* in the Galápagos and on the Pacific coast of Colombia adds another species to the list of amphi-American cirripeds that have retained their identity in the western Atlantic and eastern Pacific since the closing of Central American seaways in the Neogene. A related pan-tropical species, *B. trigonus*, is a ubiquitous element of the Galápagan lower intertidal and subtidal zones. This species is generally considered to have been present in the Atlantic and Pacific prior to the formation of the Central American land barrier, but recent evidence suggests that its modern distribution in the Atlantic basin may be the result of introduction. There are no verifiable fossil records of *B. trigonus* in the Atlantic basin (Zullo, 1986), although it is a common element in the Pleistocene of Hawaii (Newman and Abbott, 1980), the Gulf of California, and the Galápagos (Zullo, 1986).

In addition to the aforementioned species, the subtidal environment includes two endemic obligate commensals, *Membranobalanus nebrias* in a clionid sponge, and *Hexacreusia straeleni* in a balanophyllid coral. The presumed endemism of *Membranobalanus nebrias* may be an artifact of sampling, as I examined few ahermatypic corals from other eastern Pacific localities. On the other hand, clionid sponges and reef-forming corals from numerous eastern Pacific localities were examined by me without success, both for *Membranobalanus* and *Hexacreusia*. *Membranobalanus nebrias* is a member of the amphi-American *Membranobalanus* species-group complex, which is differentiated from the Indo-West Pacific complex by the armature of Cirrus IV. Unlike amphi-American species that have retained their identity on both sides of the Central American barrier, the species of *Membranobalanus* have diversified, at least partially, as a result of sponge-host specificity. *Membranobalanus nebrias*, *M. orcutti* (Pilsbry) of the Californian province and the Gulf of California, *M. robinae* Van Syoc from the Gulf of California, and *M. declivis* (Darwin) and *M. costatus* Zullo and Standing from the western Atlantic form a close-knit group. The rather subtle, but consistent differences that distinguish the species of this group suggest that their diversification was quite recent. *Hexacreusia straeleni* is related to the mainland Panamic species *H. durhami*, the type and only other known species of *Hexacreusia*. It is unlikely that the Galápagan species was derived directly from its mainland counterpart. *Hexacreusia durhami* is a commensal of *Porites*, and at least two species of *Porites* occur in the Galápagos. As indicated previously, *H. straeleni* is found on ahermatypic balanophyllid corals, and it does not seem probable that larvae emigrating from the mainland would pass over potential poritid or related coral hosts in favor of settlement on deeper water balanophyllids.

Although *Balanus poecilus* was reported only twice previously, the number of museum lots now available indicates that it is a common, widespread element of the Panamic shelf fauna. *Balanus poecilus* is endemic to the Panamic province. It is not closely related to any other species of *Balanus*, but has affinities with *B. calidus* and *B. trigonus*. The amphi-American gorgonian commensal *Conopea galeata* has a distribution similar to that of *B. trigonus* in the Americas, ranging from central California to Ecuador in the eastern Pacific, and from North Carolina to Venezuela in the western Atlantic.

The lepadomorph *Oxynaspis* has been reported only once previously from the eastern Pacific (Newman, 1972). The new Galápagan species of *Oxynaspis*, like *O. rossi* Newman from the Gulf of California and the Californian province, is a member of the *O. celata* Darwin group, which occurs world-wide in warmer waters. *Heteralepas quadrata* is a tropical Pacific species that is widespread both in the eastern and western parts of the basin. The crustacean hosts of this barnacle, primarily spiny lobsters, are not considered highly motile, suggesting that long-distance larval dispersal is a major factor in the geographic distribution of *H. quadrata*. The two recorded species of *Lepas* and the single species of *Conchoderma* are virtually cosmopolitan. The turtle barnacle *Chelonibia testudinaria* has been reported from all of the world's temperate and tropical regions that are frequented by marine turtles. *Cylindrolepas darwiniana*, *Platylepas decorata*, and *Stephanolepas muricata* are known from turtles in the tropical Pacific and Caribbean.

6. Derivation and Age of the Galápagos Cirriped Fauna

The Galápagan cirriped fauna can be characterized as a depauperate outlier of the Panamic province fauna. Of the 18 extant taxa presently known from the islands, 11 are benthic forms whose distribution is self-determined, and seven are pelagic forms or commensals of marine turtles. The pelagic taxa and turtle barnacles are either pan-tropical or cosmopolitan and, although some have yet to be recorded from the mainland parts of the Panamic province, all are probably represented in the region. Of the species with self-determined distribution, four are endemic to the Galápagos, six are Panamic province species and one, *Heteralepas quadrata*, is found throughout the tropical Pacific. Three of the four Galápagan endemic species, *Tetraclita milleporosa*, *Membranobalanus nebrias*, and *Hexacreusia straeleni*, are closely allied to Panamic mainland species. The fourth, *Oxynaspis* n. sp. is allied to a pan-tropical group with representatives both in the western Pacific and the Californian province of the eastern Pacific.

The Quaternary Galápagan cirriped fauna is essentially similar to the extant fauna (Zullo, 1986). The most common fossils are specimens of *Tetraclita milleporosa*, *Megabalanus peninsularis*, and *Balanus trigonus*. *Balanus poecilus* and a questionable specimen of *B. calidus* were identified from one locality, and a single example of *Chelonibia testudinaria* was found at another. Two species found in Quaternary deposits are not present in the extant fauna. One, *Concavus panamensis*, is an extant mainland Panamic species,and the other, tentatively identified with *Tetraclita rubescens*, is an extant Californian province species.

Table 5 summarizes the distributional components of the Galápagos barnacle fauna. Only 22 percent of the total fauna is endemic to the Galápagos Islands. However, if only those 11 species exhibiting self-determined distribution are considered, the percentage of endemics increases to 36 percent. The percentage of endemism for the total cirriped fauna agrees well with estimates ranging from 26 percent (Finet, 1985) to 23.2 percent (Shasky, 1989) determined for the Galápagan molluscan fauna, but is considerably less than that for isopods (51%) and pycnogonids (75%) (see Brusca, 1987). The percentage of endemic species for those shallow marine groups that have been studied is usually no more than one-third of the total fauna, thus supporting the conclusion by James (1984) that endemism in

Table 5. Distributional Analysis of the Galápagos Cirriped Fauna

Group (no.)	No. (%) endemic Galápagan species	No. (%) endemic Panamic species in the Galápagos	No. (%) endemic amphi-American species in the Galápagos
Total fauna (18)	4 (22%)	3 (17%)	2 (11%)
Species with self-determined distribution (11)	4 (36%)	3 (27%)	2 (18%)

the shallow-marine environment of the Galápagos is considerably less than that in the terrestrial environment.

Based on radiometric and paleontologic evidence, Hall (1983) and Hickman and Lipps (1985) concluded that the archipelago has been emergent for only the past three to four million years, and that no known fossiliferous marine deposit in the islands was older than two million years. These age determinations are in marked contrast to the Miocene and Pliocene ages proposed earlier for some Galápagan fossil deposits (e.g., Hertlein and Strong, 1939; Durham and McBirney, 1975). The fossil barnacles from localities cited by Hickman and Lipps (1985) are attributable to extant species, and support arguments for a Pleistocene or Holocene age for the marine deposits (Zullo, 1986). On the other hand, one of the more common fossil barnacles is the mid-intertidal endemic species *Tetraclita milleporosa*. Its presence in the Pleistocene fauna indicates that its origins must be sought in the Neogene.

One possibility is that *Tetraclita milleporosa* originated elsewhere in the eastern Pacific, colonized the Galápagos in the Pleistocene, and became extinct in this area of origin. Newman and Foster (1983) suggested that the shallow-marine endemics of Easter and Sala y Gomez Islands in the subtropical southeastern Pacific were derived from nearby, now submerged, volcanic islands. The ocean floor to the east and northeast of the Galápagos is dotted with numerous, small volcanic seamounts, a few of which are less than 500 m from the ocean surface (Batiza, 1989). None of these, however, appear to have ever been above sea level or within the wave base zone (McBirney and Williams, 1969; Batiza, 1989). The Galápagos Islands are situated on the Galápagos Platform which forms the western end of the east–west-trending Carnegie Ridge. Cocos Island is situated on the northeast–southwest-trending Cocos Ridge to the north of Carnegie Ridge. These two ridges are separated by the east–west-trending Galápagos Spreading Center, which forms the boundary between the Nazca Plate to the south (on which Carnegie Ridge is located), and the Cocos Plate to the north (on which Cocos Ridge is located). Both of these ridges are considered to have been formed by the Galápagos Hot Spot beginning some 12 to 15 million years ago, and have been moving apart since that time (Batiza, 1989). It is possible that Cocos Island, when it was located closer to the present site of the Galápagos in the late Neogene, could have provided the necessary stepping-stone for migration of *T. milleporosa*, but Cocos Island is also considered to be no more than three million years old (Batiza, 1989). More importantly, analysis of magnetic anomaly patterns in the area suggest that the distance separating Cocos Island from the Galápagos two million years ago was not significantly less than at present (see Brusca, 1987; Atwater, 1989).

It is more likely that at least part of the Galápagos Archipelago was emergent for some, possibly considerable, period of time prior to the Quaternary to allow for migration and subsequent divergence of the Panamic precursor of *Tetraclita milleporosa*. Older fossiliferous marine deposits may be exposed in some parts of the Galápagos, but most of this earlier geologic record has been buried and subsequently submerged under younger volcanic rocks.

ACKNOWLEDGMENTS. I thank the organizers and members of the Galápagos International Scientific Project for making this study possible. The project was sponsored by the National Science Foundation (grant no. GE-2370) and the University of California, Berkeley, in cooperation with the government of Ecuador and the Charles Darwin Foundation for the Galápagos Islands. The field work could not have been completed without the aid of Fritz and Carmen Angermeyer and Carl Angermeyer, the late Allan Cox, J. Wyatt Durham, the late Carl B. Kofoid, Richard N. Mariscal, Theodore Pappenfuss, Robert O. Schuster, and the late Allyn G. Smith in the Galápagos and Cocos, and Amy Jean Gilmartin in Ecuador. Thanks also to Dora P. Henry and A. Ross Kiester for their advice and the loan of specimens, and to William A. Newman for our many discussions on cirriped zoogeography and systematics and for his review of this manuscript. The following individuals provided access to museum collections: John S. Garth, R. Tucker Abbott, J. Wyatt Durham, and Raymond B. Manning.

This study is dedicated to the memory of my fellow expedition members and friends Allan Cox and Allyn G. Smith.

7. Appendix: Annotated Systematic List of Galápagan Cirripeds, and Cirripeds from New Localities in the Panamic Province

Suborder LEPADOMORPHA

Family Heteralepadidae

Heteralepas quadrata (Aurivilliuis, 1894). **Galápagos, Recent:** CAS 40825, CAS 40831, on the lobster *Panulirus pencillatus*, Academy Bay, Santa Cruz Island; CAS 40799, on columella of *Malea* shell inhabited by hermit crab, Barrington Bay, Santa Fe Island. **Cocos Island, Recent:** CAS 40829, on the lobster *Panulirus pencillatus*, Chatham Bay.

Family Oxynaspididae

Oxynaspis n. sp. **Galápagos, Recent:** CAS collection, 17 individuals on antipatharian, 25 m depth off Academy Bay, Santa Cruz Island February 18, 1971, M. Christian, collector.

Family Lepadidae

Conchoderma virgatum (Spengler, 1790). **Galápagos, Recent:** CAS 40800, on floating wooden box with Lepas sp. off Santa Cruz Island near Islas Plaza.

Lepas anatifera Linnaeus, 1767. **Galápagos, Recent:** CAS 40777, several hundred specimens on beached boards, Academy Bay, Santa Cruz Island.

Lepas pectinata Spengler, 1793. **Galápagos, Recent:** CAS 40781, numerous specimens on beached board, Academy Bay, Santa Cruz Island.

Lepas sp. **Galápagos, Recent:** CAS 40800, cyprids and recently metamorphosed juveniles on floating wooden box with *Conchoderma virgatum*, off Santa Cruz Island near Islas Plaza.

Suborder BALANOMORPHA
Superfamily Chthamaloidea
Family Chthamalidae
Subfamily Euraphiinae
Euraphia imperatrix (Pilsbry, 1916). **Colombia, Recent:** CAS collection, Cruise 18, *Te Vega*
Superfamily Coronuloidea
Family Coronulidae
Subfamily Chelonibiinae
Chelonibia testudinaria (Linnaeus, 1767). **Galápagos, Recent:** Pilsbry (1916). **Galápagos, Quaternary:** CAS 61281, north of Punta Miedo, Santa Fe Island (Zullo, 1986).
Subfamily Coronulidae
Coronula diadema (Linnaeus, 1767). **Galápagos, Quaternary:** CAS 61229, Villamil, Isabela Island (Zullo, 1986).
Subfamily Platylepadinae
Cylindrolepas darwiniana Pilsbry, 1916. **Galápagos, Recent:** CAS 40816, between and in plates of the plastron of female hawksbill turtle, Port Nuñez, Santa Cruz Island; CAS 40818, between plates of plastron of juvenile green turtle, Porto Nuñez, Santa Cruz Island.
Platylepas decorata Darwin, 1854. **Galápagos, Recent:** Darwin (1854); CAS 40817, on female green turtle, Porto Nuñez, Santa Cruz Island.
Stephanolepas muricata Fischer, 1886. **Galápagos, Recent:** CAS 40819, on leading edge of front flippers of male green turtle, Porto Nuñez, Santa Cruz Island.
Family Tetraclitidae
Subfamily Tetraclitinae
Tetraclita milleporosa Pilsbry, 1916. **Galápagos, Recent:** UNCW Z-319, Baltra Harbor, Baltra Island; ANSP unnumbered locality, Daphne Island; CAS 40813, Eden Island; CAS 40834, Post Office Bay, Floreana Island; CAS 40785, CAS 40786, CAS 40792, ANSP unnumbered locality, Darwin Bay, Genovesa Island; CAS 40833, Marchena Island; ANSP unnumbered locality, Kicker Rock, off San Cristobal Island; CAS 40780, Wreck Bay, San Cristobal Island; CAS 40807, CAS 40810, Sullivan Bay, San Salvador Island; CAS 40776, CAS 40778, CAS 40779, Academy Bay, Santa Cruz Island; ANSP unnumbered locality, Seymour Bay, Santa Cruz Island; CAS 40339, South Channel, Santa Cruz Island; CAS 40796, Barrington Bay, Santa Fe Island; CAS 40801, South Plaza Island. **Galápagos: Quaternary:** CAS 61281, CAS 61282, CAS 61286, near Punta Miedo, Santa Fe Island; CAS 61386, James Bay, San Salvador Island; CAS 61387, Rabida Island; CAS 61391, Baltra Island (Zullo, 1986).
Tetraclita panamensis Pilsbry, 1916. **Ecuador, Recent:** CAS 40823, Punta Centinela, Santa Elena Bay.
Tetraclita sp., aff. *T. panamensis*. **Cocos Island, Recent:** CAS 40826, CASS 40827, Chatham Bay.

Superfamily Balanoidea

Family Archaeobalanidae

Subfamily Archaeobalaninae

Conopea galeata (Linnaeus, 1771). **Galápagos, Recent:** AHF 306-35, Marchena Island; CAS 40797, Barrington Bay, Santa Fe Island; CAS 40811, off Bartolomé Island, Sullivan Bay, San Salvador Island; CAS 40821, Academy Bay, Santa Cruz Island. **Ecuador, Recent:** AHF 850-38, Cape San Francisco; **Gulf of California, Recent:** AHF 1056-40, Puerto Refugio Bay, Mejia Island; **California, Recent:** CAS unnumbered locality, Monterey Bay.

Membranobalanus nebrias Zullo and Beach, 1973. **Galápagos, Recent:** CAS 40803, South Channel, Baltra island; CAS 40808, Sullivan Bay, San Salvador Island; CAS 40814, Conway Bay, Santa Cruz Island; CAS 40822, Academy Bay, Santa Cruz Island (Zullo and Beach, 1973).

Family Pyrgomatidae

Hexacreusia straeleni Zullo and Beach, 1973. **Galápagos, Recent:** LACMIP unnumbered locality, 55-90 m depth, off Beagle Island (Zullo and Beach, 1973).

Family Balanidae

Subfamily Balaninae

Balanus calidus Pilsbry, 1916. **Galápagos, Recent:** CAS 40795, Academy Bay, Santa Cruz Island; CAS 40797, Barrington Bay, Santa Fe Island; CAS 40812, Sullivan Bay, San Salvador Island. **Galápagos, Quaternary:** CAS 61932(cf.), Cerro Colorado, Santa Cruz Island (Zullo, 1986).

Balanus poecilus Darwin, 1854. **Galápagos, Recent:** CAS 40820, Academy Bay, Santa Cruz Island; CAS collection, 25 m depth off Academy Bay, Santa Cruz Island on antipatharian, February 18, 1971, M. Christian, collector. AHF 355-35, off Gardner Island, near Española Island. **Galápagos, Quaternary:** CAS 61392, Cerro Colorado, Santa Cruz Island (Zullo, 1986). **Ecuador, Recent:** AHF 209-34, off La Libertad; **Colombia, Recent:** AHF 851-38, off Gorgona Island. **Panama, Recent:** AHF 240-34, off Jicarita Island; AHF 863-38, Bahia Honda. **Gulf of California, Recent:** AHF 574-36, off San Pedro Nolasco Island; CAS 27587, off Punta San Lucas; CAS 40771, off Espiritu Santo Island; CAS collection, 50 m depth, off Las Animas Islet, on antipatharian.

Balanus trigonus Darwin, 1854. **Galápagos, Recent:** CAS 40802, CAS 40804, CAS 40806, South Channel, Santa Cruz Island; CAS 40815, Conway Bay, Santa Cruz Island; ANSP 6516, Iguana Cove, Isabela Island. **Galápagos, Quaternary:** CAS 61388, CAS 61389, CAS 61390, Baltra Island; CAS 61392, Cerro Colorado, Santa Cruz Island (Zullo, 1986). **Panama, Recent:** UCMP 2416. **Gulf of California, Pleistocene:** UCMP B-1395, Puerto Peñasco.

Concavus (Arossia) panamensis panamensis (Rogers, 1948). **Cocos Island, Recent:** CAS collection, Cruise 18, *Te Vega*, Wafer Bay.

Concavus (Arossia) panamensis eyerdami (Henry, 1960). **Galápagos, Quaternary:** CAS 61390(cf.), Baltra Island (Zullo, 1986).

Subfamily Megabalaninae

Megabalanus coccopoma (Darwin, 1854). **Peru, Recent:** CAS 41293, Talara, associated with *Notomegabalanus concinnus*. **Ecuador, Recent:** CAS

40823, Santa Elena Bay. **Costa Rica, Recent:** UCMP B-7818, Matalimon. **Mexico, Recent:** UCMP 1660, Mazatlán. The Pliocene specimens from Turtle Bay, Baja California assigned to this species by Jordan and Hertlein (1926) are of *Balanus pacificus* Pilsbry.

Megabalanus peninsularis (Pilsbry, 1916). Henry and McLaughlin (1986, p. 36, 37) included *M. galapaganus* (Pilsbry) in the synonymy of *M. peninsularis*, noting that: "Local populations [of *M. peninsularis*] exhibit striking differences from the type series: however, a continuum in morphological characters has been observed. *M. galapaganus* falls well within this range, and we have not been able to find any characters by which this taxon can be distinguished from *M. peninsularis*." My examination of several hundred specimens from localities throughout the Galápagos Island and on Cocos Island indicates that the species present in these areas is quite conservative throughout its adult life, both in coloration and opercular plate morphology. Furthermore, the Galapagan-Cocos species is readily distinguished on the basis of these features from *M. peninsularis* from the type locality in the Cape San Lucas, Baja California region. I have, however, examined only two lots of specimens referable to either of these species from other mainland Panamic province localities. I would assign the specimens from Port Utria, Columbia (AHF 419-35) to *M. galapaganus*, whereas the Costa Rican specimens collected during *Te Vega* Cruise 18 appear to represent typical *M. peninsularis*. As an alternative to Henry and McLaughlin's conclusion that the two species are one variable species, I would tentatively suggest that both *M. peninsularis* and *M. galapaganus* are represented in the mainland Panamic fauna. For the present, however, I am following their synonymy. **Galápagos, Recent:** unnumbered ANSP locality, Daphne Island; CAS 40813, Eden Island; ANSP paratypes (of *M. galapaganus*) 6538, 6539, Española Island; CAS 40835, Punta Daylight, Floreana Island; unnumbered ANSP locality and IISR 79, Gardner Island off Española Island; CAS 40784, CAS 40787, CAS 40788, CAS 40789, CAS 40790, CAS 40791, CAS 40793, AHF 782-38, Darwin Bay, Genovesa Island; CAS 40832, Punta Berkeley, Isabela Island; unnumbered ANSP locality, Kicker Rock off San Cristobal Island; ANSP 170390, CAS 40780, Wreck Bay, San Cristobal Island; AHF 12-32, James Bay, San Salvador Island; CAS 40809, CAS 40810, Sullivan Bay, San Salvador Island; CAS 40782, CAS 40783, Academy Bay, Santa Cruz Island; CAS 40804, CAS 40805, South Channel, Santa Cruz Island; CAS 40797, Barrington Bay, Santa Fe Island; CAS 40798, Melville Bay, Santa Fe Island; CAS 40775, South Plaza Island. **Galápagos, Quaternary:** CAS 61281, CAS 61282, CAS 61286, near Punta Miedo, Santa Fe Island; CAS 61392, Cerro Colorado, Santa Cruz Island (Zullo, 1986). **Cocos Island, Recent:** CAS 40827, CAS 40830, Chatham Bay; IISR 107, Ulloa Island; IISR 114, Wafer Bay. **Colombia, Recent:** AHF 419-35, Port Utria. **Costa Rica, Recent:** CAS collection, Cruise 18, *Te Vega*.

Notomegablanus concinnus (Darwin, 1854). **Peru, Recent:** CAS 41293, Talara, in association with *Megabalanus coccopoma*; CAS collection, Cruise 18, *Te Vega*, Paita.

References

Abbott, D. P., 1966, Factors influencing the zoogeographic affinities of the Galápagos inshore marine fauna, in: *The Galápagos* (R. I. Bowman, ed.), pp. 108–122, University of California Press, Berkeley.

Atwater, T., 1989, Plate tectonic history of the northeast Pacific and western North America, in: *The Geology of North America, Vol. N, The Eastern Pacific Ocean and Hawaii* (E. L. Winterer, D. M. Hussong and R. W. Decker, eds.), Geological Society of America, Boulder, Colorado, pp. 21–72.

Bakus, G. J., 1964, The effects of fish-grazing on invertebrate evolution in shallow tropical waters, *Allan Hancock Found. Pub.* **27**:1–29.

Batiza, R., 1989, Seamounts and seamount chains of the eastern Pacific, in: *The Geology of North America, Vol. N, The Eastern Pacific Ocean and Hawaii* (E. L. Winterer, D. M. Hussong and R. W. Decker, eds.), Geological Society of America, Boulder, Colorado, pp. 289–306.

Brusca, R. C., 1987, Biogeographic relationships of Galápagos marine isopod crustaceans, *Bull. Mar. Sci.* **41**(1):268–281.

Carlton, J., and Zullo, V. A., 1969, Early records of the barnacle *Balanus improvisus* Darwin from the Pacific coast of North America, *Occas. Pap. Calif. Acad. Sci.* **75**:1–6.

Darwin C., 1851, *A monograph on the sub-class Cirripedia with figures of all the species. The Lepadidae; or, pedunculated cirripedes*, Ray Society, London, 400 pp.

Darwin, C., 1854, *A monograph on the sub-class Cirripedia with figures of all the species. The Balanidae, the Verrucidae*, Ray Society, London, 684 pp.

Durham, J. W., and McBirney, A. R., 1975, Galápagos Islands, in: *The Encyclopedia of World Regional Geology, Part 1* (R. W. Fairbridge, ed.), Dowden, Hutchinson and Ross, Inc., Stroudsburg, Pennsylvania, pp. 285–290.

Ekman, S., 1953, *Zoogeography of the Sea*, Sidgwick and Jackson, London, xiv + 417 pp.

Finet, Y., 1985, Preliminary faunal list of the marine mollusks of the Galápagos Islands, *Documents de Travail No. 20, Inst. roy. Sci. natur. Belgique (Brussels)*, 50 pp.

Garth, J. S., 1960, Distribution and affinities of the brachyuran Crustacea, *Syst. Zool.* **9**:105–123.

Glynn, P. W., and Wellington, G. M., 1983, *Corals and Coral Reefs of the Galápagos Islands. With an Annotated List of the Scleractinian Corals of the Galápagos, by J. W. Wells*, Univ. California Press, Berkeley, xvi + 330 pp.

Hall, M. L., 1983, Origin of Española Island and the age of terrestrial life on the Galápagos Islands, *Science* **221**:545–547.

Hedgpeth, J. W., 1969, An intertidal reconnaissance of rocky shores of the Galápagos, *Wasmann Jour. Biol.* **27**:1–24.

Henry, D. P., 1960, Thoracic Cirripedia of the Gulf of California, *Univ. Wash. Pub. Oceanogr.* **4**:135–158.

Henry, D. P., and McLaughlin, P. A., 1967, A revision of the subgenus *Solidobalanus* Hoek (Cirripedia, Thoracica) including a description of a new species from complemental males, *Crustaceana* **12**: 43–58.

Henry, D. P., and McLaughlin, P. A., 1975, The barnacles of the *Balanus amphitrite* complex (Cirripedia, Thoracica), *Zool. Verhandel.* **141**, 254 pp.

Henry, D. P., and McLaughlin, P. A., 1986, The recent species of *Megabalanus* (Cirripedia: Balanomorpha) with special emphasis on *Balanus tintinnabulum* (Linnaeus) sensu lato, *Zool. Verhandel.* **235**, 69 pp.

Hertlein, L. G., 1932, Mollusks and barnacles from Malpelo and Cocos islands, *Nautilus* **46**:43–45.

Hertlein, L. G., and Strong, A. M., 1939, Marine Pleistocene mollusks from the Galápagos Islands, *Proc. Calif. Acad. Sci.*, series 4, **23**:367–380.

Hickman, C. S., and Lipps, J. H., 1985, Geologic youth of Galápagos Islands confirmed by marine stratigraphy and paleontology, *Science* **227**:1578–1580.

James, M. J., 1984, A new look at evolution in the Galápagos: evidence from the late Cenozoic marine molluscan fauna, *Biol. J. Linn. Soc.* **21**:77–95.

Jordan, E. K., and Hertlein, L. G., 1926, Expedition to the Revillagigedo Islands, Mexico, in 1925, VII. Contributions to the geology and paleontology of the Tertiary of Cedros Island and adjacent parts of Lower California, *Proc. Calif. Acad. Sci.* **15**:409–464.

Laguna, G. J., 1985, Systematics, ecology and distribution of barnacles (Cirripedia; Thoracica) of Panama, *Unpublished Master's thesis, University of California*, San Diego, 234 pp.

Laguna, G. J., 1988, *Euraphia eastropacensis* (Cirripedia, Chthamaloidea), a new species of barnacle from the tropical eastern Pacific: Morphological and electrophoretic comparisons with *Euraphia rhizophorae* (deOliveira) from the tropical western Atlantic and molecular evolutionary implications, *Pac. Sci.* **41**:132–140.

McBirney, A. R., and Williams, H., 1969, Geology and petrology of the Galápagos Islands, *Geol. Soc. Amer. Mem.* **118**, xiii + 197 pp.

McDonald, R., 1929, A report on some cirripeds collected by the S.S. "Albatross" in the eastern Pacific during 1891 and 1904, *Bull. Mus. Comp. Zool. Harv. Univ.* **69**:527–538.

Newman, W. A., 1960a, The paucity of intertidal barnacles in the western Pacific, *Veliger* **2**:89–94.

Newman, W. A., 1960b, *Octolasmis californiana*, spec. nov., a pedunculate barnacle from the gills of the California spiny lobster, *Veliger* **3**:9–11.

Newman, W. A., 1972, An oxynaspid (Cirripedia, Thoracica) from the eastern Pacific, *Crustaceana* **23**:202–208.

Newman, W. A., 1982, A review of the extant taxa of the "Group of *Balanus concavus*" (Cirripedia, Thoracica) and a proposal for genus-group ranks, *Crustaceana* **43**:25–36.

Newman, W. A., and Abbott, D. P., 1980, Cirripedia: the barnacles, in *Intertidal Invertebrates of California* (R. H. Morris, D. P. Abbott and E. C. Haderlie, eds.), pp. 504–535, Stanford University Press, Stanford, California.

Newman, W. A., and Foster, B. A., 1983, The Rapanuian faunal district (Easter and Sala y Gomez): In search of ancient archipelagos, *Bull. Mar. Sci.* **33**(3):633–644.

Philander, S. G. H., 1990, *El Niño, La Niña and the Southern Oscillation*, Academic Press, New York, 293 pp.

Pilsbry, H. A., 1907, The barnacles (Cirripedia) contained in the collections of the U.S. National Museum, *U.S. Nat. Mus. Bull.* **60**, x + 122 pp.

Pilsbry, H. A., 1916, The sessile barnacles (Cirripedia) contained in the collections of the U.S. National Museum; including a monograph of the American species, *U.S. Nat. Mus. Bull.* **93**, xi + 366 pp.

Shasky, D. R., 1989, An update on the molluscs from the Galápagos Islands as listed in the preliminary faunal list of the marine mollusks of the Galápagos Island by Yves Finet, *Annu. Rep. West. Soc. Malacol.* **21**:8–10.

Stanley, S. M., and Newman, W. A., 1980, Competitive exclusion in evolutionary time: The case of the acorn barnacles, *Paleobiology* **6**:173–183.

Stephenson, T. A., and Stephenson, A., 1972, *Life between tidemarks on rocky shores*, W. H. Freeman and Co., San Francisco, California, 425 pp.

Suzuki, N., and Syuiti, M., 1963, Adaptation phenomena concerning the water content of barnacles, *Jap. J. Ecol.* **13**:1–9.

U.S. Department of Commerce, 1962, Surface water temperature and salinity, Pacific coast, North and South America and Pacific Ocean islands. *Coast and Geodetic Surv. Pub.* **31–3**, 71 pp.

Van Syoc, R. J., 1988, Description of *Membranobalanus robinae*, a new species of sponge barnacle (Cirripedia, Archaeobalanidae) from Baja California, with a key to the genus, *Proc. Biol. Soc. Wash.* **101**:832–837.

Vermeij, G. J., 1978, *Biogeography and Adaptation, Patterns of Marine Life*, Harvard Univ. Press, Cambridge, Massachusetts, xi + 332 pp.

Walker, B. W., 1960, The distribution and affinities of the marine fish fauna of the Gulf of California, *Syst. Zool.* **9**:123–133.

Zullo, V. A., 1966, Zoogeographic affinities of the Balanomorpha (Cirripedia: Thoracica) of the eastern Pacific, in *The Galápagos* (R. I. Bowman, ed.), pp. 139–144, University of California Press, Berkeley.

Zullo, V. A., 1967, A description of *Balanus parkeri* (Cirripedia, Thoracica), a new species from the Gulf of California, *Notulae Naturae* 406, 5 pp.

Zullo, V. A., 1969a, A new subspecies of *Balanus tintinnabulum* (Linnaeus, 1758) (Cirripedia, Thoracica) from Clipperton Island, eastern Pacific, *Proc. California Acad. Sci.*, series 4, **36**:501–510.

Zullo, V. A., 1969b, Additional records of *Tetrabalanus polygenus* Cornwall, 1941 (Cirripedia, Thoracica), *Occas. Papers California Acad. Sci.* **74**, 8 pp.

Zullo, V. A., 1986, Quaternary barnacles from the Galápagos Islands, *Proc. California Acad. Sci.* **44**:55–66.

Zullo, V. A., and Beach, D. B., 1973, New species of *Membranobalanus* Hoek and *Hexacreusia* Zullo (Cirripedia, Balanidae) from the Galápagos Archipelago, *Contrib. in Sci., Natur. Hist. Mus. Los Angeles County* **249**, 16 pp.

Zullo, V. A., Beach, D. B., and Carlton, J., 1972, New barnacle records (Cirripedia, Thoracica), *Proc. Calif. Acad. Sci.* **39**:65–74.

Zullo, V. A., and Newman, W. A., 1964, Thoracic Cirripedia from a southeast Pacific guyot, *Pac. Sci.* **18**:355–372.

Chapter 9

Amphipoda of the Galápagos Islands

J. L. BARNARD

1. Introduction

Amphipoda are small crustaceans about 5–10 mm long which look either like shrimps, but lacking the carapace, or like isopods (pill bugs), but mostly flattened from side to side. They are easily identified because they are crustaceans with only 3 pairs of pleopods (swimmerets). Two body extremes are shown in Fig. 1; the rarely expressed flattened form resembles an isopod.

The status of amphipods in the intertidal zone of the Galápagos Islands can be summarized as 'an assemblage of weedy invaders or their descendants.' The word 'weedy' implies that the species (or closely related twin species) are common in other regions of the world, and tend to be species that disperse widely in opportunistic fashion. In cases of speciation resulting from insular isolation, the descendent species retain a clear relationship to these same weedy species. We will now turn to the questions of where they came from, how they got to the islands, and how normal the resultant faunal composition is in the Galápagos. We will even discuss why anybody would bother to go to the Galápagos to study just amphipods when there are so many other classic biotic groups to investigate. There's a little story to tell about it all.

Of all the animals on earth visible to the naked eye, amphipods are one of the most plentiful, and for some unknown reason one of the most ignored. Amphipods occur in almost every body of fresh and salt waters on earth except in the east African rift lakes (perhaps because of predation by special fishes). When the Russians sampled the deepest ocean bottom on earth a few years ago they brought up an amphipod from a depth of over 10,000 meters. Some amphipods are even terrestrial and one species has been found living high in a tree in moss in the rainforest of New Guinea. Why then, have they been so little studied? Perhaps, because of the commonly held belief among invertebrate zoologists, that they all

J. L. BARNARD • Department of Invertebrate Zoology, National Museum of Natural History, Smithsonian Institution, Washington, D.C. 20560.

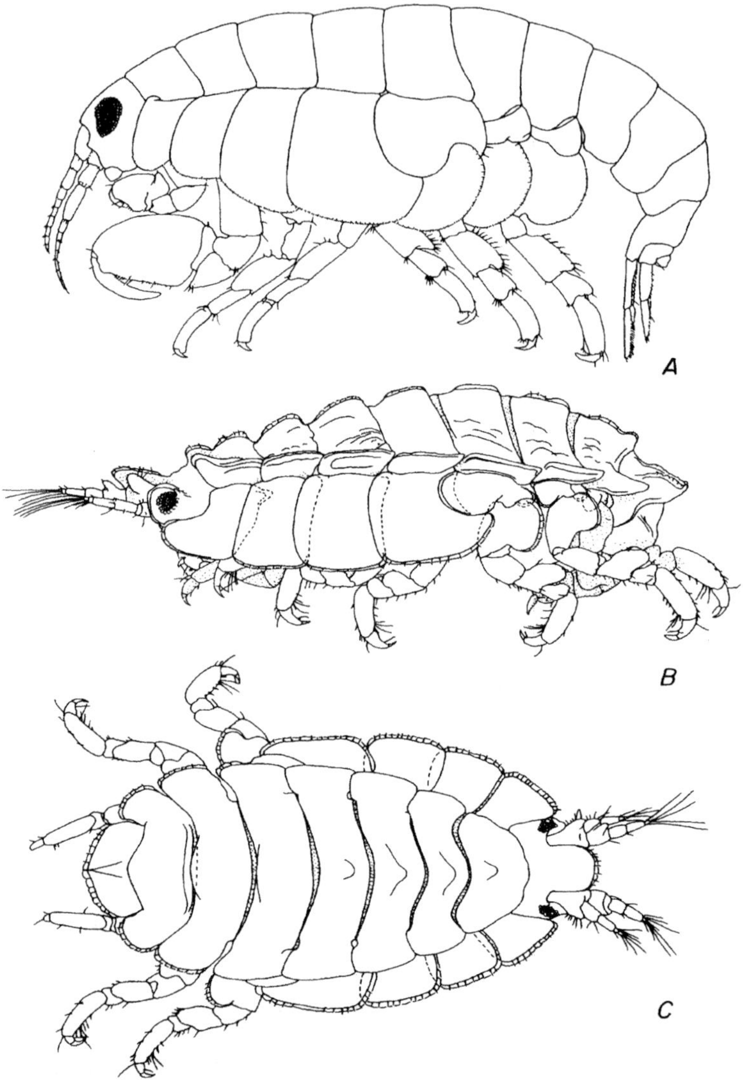

Figure 1. Extremes of amphipod shape in Galápagos Islands; A, *Hyachelia tortugae*, left lateral view; B, *Heterophlias galapagoanus*, left lateral view; C, same, dorsal view.

look alike and are difficult to identify. This neglect has, however, led to some wonderful results, because the study of amphipods has suddenly caught on in the last two decades. New students entering the field have some excellent fresh biological material available that is unencumbered with complicated taxonomic problems regarding names. Modern students can jump right in and investigate species that are either new to science or well described in modern times and they can spend their valuable time working with advanced techniques not available to the 19th century taxonomists, who looked at more obvious members of the fauna such as birds, mollusks, or other crustaceans.

2. Historical Perspective

The Galápagos Islands were in my blood long before I knew what an amphipod was. As a 12 year old in 1940 I had the good fortune to see a movie made by biologists at the Allan Hancock Foundation of the University of Southern California (U.S.C.). They brought their movies to schools in the Los Angeles area. The philanthropist Allan Hancock had established as one of his missions in life the exploration of eastern Pacific shores (aboard his palatial research vessel, the *Velero III*). Beautiful color movies were made; several of the trips in the 1930s concentrated on Galápagan seashores. The movies were enthralling to a child who, 7 years later, out of the blue, was offered a scholarship to this institution. What a thrill to come aboard and meet so many of the biologists who years earlier had participated in those wonderful expeditions—to see the vast collections of yet unworked animals waiting for somebody to tackle.

The progressive system at U.S.C. allowed students to start research in their junior year, and I commenced work on corals—but soon ran out of material because the eastern Pacific is depauperate of corals. By this time a young graduate student in isopods (R. J. Menzies) had focused my attention on crustaceans. One day he and the well-known authority on eastern Pacific Crustacea, Professor Garth, gave me a tour of the crustacean collections. Dr. Garth wanted me to study large crustaceans such as mole crabs but Mr. Menzies wanted me to work on small crustaceans. Crabs weren't appealing (I had no talent in illustrating or photography) but I knew how to operate a camera lucida, used for drawing small crustaceans. I asked them which group that was well-represented in their collections might be the most difficult to study; they both replied: "Amphipods." I was on my way the next day—but six months later I had grave doubts because I still hadn't been able to identify the first amphipod they gave me. I had, however, been able to identify the second and third ones, so I was not totally discouraged.

In my peregrinations in the collections I occasionally noticed an amphipod labeled "Galápagos," but there were very few; not enough to undertake a significant scientific contribution. I soon realized that nobody ever collected amphipods except in a cursory fashion. When biologists on expeditions went ashore they collected large things they could easily see. They also had a propensity to collect isopods because of their attractive flattened shapes. As a predoctoral student at U.S.C. my major assignment was to produce published monographs on amphipods; I was beset by a lot of taxonomic trouble, and soon realized most amphipod students were in the same boat: they were dealing with collections mostly of the giant and therefore aged amphipod specimens not fully typical of a species.

By that time Menzies had reached the same conclusion and had 'invented' what we called a "formalin wash" for collecting amphipods and isopods. On our subsequent expeditions we would pull up seaweed, rocks, sand, and other substrata, and put them in buckets with seawater and a bit of formalin. Tiny animals were quickly killed and fell to the bottom of the bucket. The large items in the bucket were removed and the rich soup of tiny animals in the bottom was preserved and sorted out in the laboratory. Suddenly we were inundated with hundreds of kinds of amphipods, almost all of them new to science. My life's work was starting.

The U.S. Navy appropriated the old *Velero III* for submarine patrolling during World War II. After the war, Allan Hancock built a new serviceable research vessel, the *Velero IV*. Unfortunately his new vessel was kept so busy in Californian and Mexican waters that no more trips to the Galápagos were undertaken. Finally, in 1964, I had the good fortune to be invited to participate in the Galápagos International Scientific Expedition (GISP) (Bowman, 1966). This expedition would make a modern exploration of the islands, call attention to the need for conservation in the islands, and deliver heavy equipment for the newly established Charles Darwin Research Station (CDRS). By 1964 the shallow-water amphipod fauna of southern California and northern Mexico had been described, but nothing was known about the amphipod fauna of the Galápagos nor about that of any of the coastline from middle Mexico to middle Chile.

In preparation for the expedition and subsequent laboratory workup of material I reviewed Garth's (1946) classic paper on origin, evolution, and speciation of the crab fauna in the Galápagos Islands. I was certain that amphipods, because they bear live young from brooded eggs, would not have the same evolutionary history as brachyuran crabs which emit planktonic larvae. An organism releasing larvae has the chance to be dispersed widely by ocean currents, yet the Galápagos fauna harbored many endemic crabs. Garth was able to match up several Galápagan crabs with American species that might have had common ancestors. If amphipods cannot disperse larvae then there might be a chance that those that reached the Galápagos, by whatever means, would have diverged more strongly than would have crabs. If crab larvae are commonly dispersed widely every year, then amphipods, which may be dispersed only occasionally by some unknown means, would rarely reach the Galápagos. Amphipods would therefore be subject to longer periods of isolation and might therefore exhibit more adaptive radiation. One might predict, *a priori*, that amphipods would exhibit a higher degree of endemism.

Determining the endemic status of Galápagan amphipods without knowing anything about the amphipods of the mainland between Mexico and middle Chile was a serious concern. My reckoning was that if I found a large number of unfamiliar new species I would have to devote many subsequent years to a study of Panamic amphipods in order to make any satisfactory taxonomic judgments. I had already collected extensively on Panamanian shores but had never had time to analyze the specimens. During our cruise to the Galápagos, other members of the expedition and I made plans to explore those shores at some point in the future. Owing to the decline of interest in taxonomy in the United States these plans never matured.

Upon arrival at our first collecting locality in the Galápagos, I was dismayed at the moonlike barrenness of the intertidal region. Nothing from the book by Beebe (pp. 126 and 128) and others, who had extolled the wonderful life of the Galápagos, had prepared me for this amazing wasteland. But soon, my fears of coming up empty handed were allayed by J. Wyatt Durham and E. Yale Dawson. Durham, the giant of geology and paleontology, instructed us to lift up huge slabs of lava with crowbars and look underneath; to our surprise the cracks a foot or so below the surface were teeming with life. And Dawson, the fearless algologist, plunged into the mightiest of breakers where marine iguanas feared to tread and came up with great masses of algae, full of amphipods.

The problem with the Galápagan intertidal is the low range of tidal amplitudes, the burning sun, the relentless activities of Sally Lightfoot Crabs scraping away most of the surface scum, and the marine iguanas consuming most of the large fleshy algae except in the heaviest surf. We had therefore to seek out the cracks, the inner pools not frequented by scavengers, and the substrata occupied by coelenterates that were unpalatable to iguanas. Between Dawson and our small party we came back with a healthy collection of amphipods. During the collecting process we were careful to reset all of our lava slabs back in their normal positions and to take only the minimum samples we felt would adequately survey the fauna.

On my return to California I found myself suddenly out of work and scrambling for a new situation. Fortunately, the Smithsonian Institution was hiring and I went aboard, carrying my collections from Galápagos, Panama, Mexico, and California. But I soon found myself confronted with so many opportunities to look at the larger world picture on amphipods by traveling all over the world for several years that the Galápagan collections stayed on the shelf until the mid 1970s. Meanwhile, I'd had the opportunity to publish studies on Ifaluk, Hawaii, Australia, New Zealand, the Gulf of California, and other regions (Barnard, 1965, 1969a,b, 1970, 1972a, 1972b, 1974, 1977, 1979).

When, in 1976, I opened the first, long-preserved sample of Galápagan amphipods I was sure I had made a great mistake in labeling. I thought I was looking at a sample collected at Corona del Mar, California, a place I had sampled intensively before 1950 and before its closure for conservation reasons. The similarity of the Galápagan samples to parts of California and Mexico was astounding. Sample after sample confirmed this faunal similarity.

Ultimately, my impression of compositional similarity was determined to be a congruence in the dominant genera and how closely their species resembled each other. Most of the new species differed only minutely from species in the Mexican–Californian region.

3. Composition of the Fauna

The data presented in Table 1 show 50 Galápagan species, of which 19 are endemic (38%), 5 are cosmopolitan (10%), 24 occur widely in the eastern Pacific (48%), one occurs also only in Cocos Island (2%), and 2 are undescribed (4%).

The data presented in Table 2 show 13 island groups, of which, for example, the Hawaiian Chain has a level of endemicity in amphipods of 46%. When Barnard (1970) published the original data, the endemicity level was 66%, but since that time many species, known originally only from Hawaii, have been found widely distributed in the Pacific and Indiana Oceans. The data for the Galápagos come from 1979 and no further work has been done in the eastern Pacific. I suspect that several of the now apparent endemics will be found more widely distributed and the Galápagan endemics might be reduced downward to about 25%, perhaps lower.

The data in Table 2 show that an island group similar to the Galápagos, the Canary Islands, has almost no endemic species; the Canaries are even closer to a

Table 1. Galápagan Amphipods and Their Relationships[a]

Factor[a,b]	Name of species	Affinity
1176	*Hyale darwini* Barnard, 1979	Panama
384	*Ericthonius brasiliensis* (Dana, 1853)	Cosmopolitan
(234)	*Ampithoe pollex* Kunkel, 1910	Pan-American
(234)	*Ampithoe tahue* Barnard, 1979	Endemic sibling of *pollex*
(234)	*Ampithoe vacoregue* Barnard, 1979	Endemic sibling of *pollex*
212	*Elasmopus tubar* Barnard, 1979	Gulf of Callifornia
197	*Elasmopus mayo* Barnard, 1979	Eastern Pacific
130	*Elasmopus serricatus* Barnard, 1969a	Eastern Pacific
93	*Microjassa chinipa* Barnard, 1979	Panama
82	*Posophotis seri* Barnard, 1979	Gulf of California
79	*Elasmopus zoanthidea* Barnard, 1979	Endemic sibling of Californian *E. holgurus*
75	*Lembos schire* Barnard, 1979	Cocos Island
63	*Ampithoe plumulosa* Shoemaker, 1938	Eastern Pacific
41	*Hyale zuaque* Barnard, 1979	Eastern Pacific
37	*Elasmopus ocoroni* Barnard, 1979	Endemic sibling of Californian *E. mutatus*
36	*Bemlos macromanus* Shoemaker, 1925	Eastern Pacific
36	*Amphilochus neapolitanus* Della Valle, 1893	Cosmopolitan on coelenterates
27	*Photis elephantis* Barnard, 1962	California
16	*Chevalia* sp. (as *C. aviculae* in Barnard, 1979)	Endemic, now realized as a new species
16	*Maera chinarra* Barnard, 1979	Endemic sibling of worldwide species complex
15	*Meximaera diffidentia* Barnard, 1969b	Gulf of California
14	*Hyale humboldti* Barnard, 1979	Endemic sibling of worldwide complex
11	*Microdeutopus hancocki* Myers, 1968	Eastern Pacific
7	*Elasmopus ecuadorensis* Schellenberg, 1936	Sibling of Hawaiian *E. hawaiensis* but is a known "drifter"
6	*Ampelisca lobata* Holmes, 1908	Eastern Pacific
5	*Ampithoe guaspare* Barnard, 1979	Endemic sibling of cosmopolitan *A. ramondi*
	Maera reishi Barnard, 1979	Endemic sibling of worldwide species complex, like *M. chinarra*
	Hyale guasave Barnard, 1979	Gulf of California
	Varohios topianus Barnard, 1979	Gulf of California
	Heterophlias galapagoanus Barnard, 1979	Endemic sibling of American complex
	Gitanopsis baciroa Barnard, 1979	Gulf of California
	Leucothoe spinicarpa (Abildgaard, 1789)	Cosmopolitan
	Podocerus n. sp.	Sibling of circumtropical complex
	Corophium baconi Shoemaker, 1934	East Pacific enormously adaptable
	Parhyale hawaiensis (Dana, 1853)	Cosmopolitan
	Polycheria osborni Calman, 1898	East Pacific
	Anamixis sp. (as *Leucothoides pottsi* in Barnard, 1979	Endemic new species
	Melita sp.	Endemic new species

continued

Table 1. (*continued*)

Factor[a,b]	Name of species	Affinity
	Elasmopus temori Barnard, 1979	Endemic sibling of Californian *E. holgurus*
	Colomastix sp.	Unknown
	Dulzura gal Barnard, 1979	Endemic sibling of American complex
	Microdeutopus trichopus Myers, 1968	Pan-American
	Neomegamphopus roosevelti Shoemaker, 1942	Eastern Pacific
	Ampelisca milleri Barnard, 1960	Eastern Pacific
	Ampelisca pugetica (Stimpson, 1854)	Pan-American
	Eyakia calcarata (Gurjanova, 1938)	Pan-American
	Proharpinia tropicana Barnard, 1979	Endemic, unusual Antarctic immigrant
	Hyachelia tortugae Barnard, 1967	Cosmopolitan, in sea-turtle mouths
	Anchialella vulcanella Barnard, 1979	Endemic in brackish lens, Indo-Pacific
	Galapsiellus leleuporum (Monod, 1970)	Endemic in brackish lens, Indo-Pacific

[a] Listed in order from dominant to rare based on a factor calculated from abundance in the samples (from Barnard, 1979).
[b] The factor (234) for three species groups them together, as their juveniles cannot be separated.

continent than are the Galápagos. The Azores are much farther away from a continent than the Galápagos, and at the moment have a 19% level of endemism.

Sweepstakes rafting or chance dispersal (*sensu* Ekman, 1953, or Zimmerman, 1948) is a subject that has been sub-rosa for most of the past three decades. Zoologists who bring it up in meetings or try to write about it are often 'laughed out of court' because the 'fad' in the past three decades has been tectonics and vicariance. Zoologists have spent their time discussing how species are distributed because they were actually carried about on tectonic plates long ago or were easily dispersed between closely proximate places that today may be far apart because of geological drift or change in sea levels (for example Azores and Europe). On the other hand Springer (1982;p. 5) pointed out that "dispersal is not the material of [vicariography]" but becomes interesting to a vicariographer "when the dispersal is interrupted." Vicariance therefore seems to be involved more with explaining speciation and endemism than accounting for the mechanism of dispersal.

A large number of deep-sea seamounts occurs to the north and east of the Galápagos (Heezen and Tharp, 1977) but there is little or no evidence that the youthful Galápagos have ever been connected by shallow-water stepping-stones nor enjoyed physical proximity to the continent or to island chains (Marquesas) 1600 km to the west.

The amphipods known from the Galápagos are presumed to have arrived mostly by rafting on debris and flotsam brought to the islands on the California and Humboldt currents. Although the California current does not strike the Galápagos directly as does the Humboldt current, the El Niño events do bring water outward to the Galápagos from the Gulf of Panama.

Most of the species in Table 1 would appear to an amphipod taxonomist to be

Table 2. Number of Amphipod Species in Island Groups[a]

South Georgia 160 (62 intertidal) (33 endemic but artificial)
Falkland 63 (intertidal) (14 endemic but artificial)
Kerguelen 81 (42 endemic but artificial)
Tristan da Cunha 30 (6 endemic)
Hawaii 120 (55 endemic, = 46%)
Bermuda 45 (17 endemic but reducible)
Galápagos 50 (19 endemic, = 38%)
Canary 75 (1 endemic, = 1%)
Azores 61 (5 endemic, = 8%)
Cape Verde 17 (no endemics)
New Zealand 133 (100 endemic, = 75%)
Fiji 80 (many endemic but contiguous exploration weak)
Madagascar 287 (a giant subcontinental island)

[a] Data and ideas in Table 2 compiled from numerous sources, principally Afonso, 1976; Barnard, 1965, 1969c, 1970, 1972, 1976, 1977, 1979; Barnard and Barnard, 1983; Barnard and Clark, 1985; Barrois, 1988; Bellan-Santini and Ledoyer, 1974; Chevreux, 1888, 1925, 1926; Kunkel, 1918; Ledoyer, 1982, 1986; MacArthur and Wilson, 1957; Myers, 1985; Schellenberg, 1931; Stephensen, 1949).

good candidates for rafting because they are either nestlers in intertidal plants or they occupy tubes glued to plants and objects with might float. The exceptions are species in the genera *Amphilochus*, *Ampelisca*, *Anamixis*, *Colomastix*, *Eyakia*, *Gitanopsis*, *Heterophlias*, *Leucothoe*, and *Polycheria*. However, *Anamixis*, *Colomastix*, *Leucothoe*, and *Polycheria* live inside tunicates or sponges which might also float or be rafted in entanglements of other materials. *Amphilochus* and *Gitanopsis* probably occur as commensals with coelenterates, particularly hydroids, which might also be rafted. The greatest mysteries are *Ampelisca* and *Eyakia*. *Ampelisca* is a genus living in tubes it constructs on the sea bottom but the three species found in the Galápagos also have been found on the American continent mixed into intertidal or very shallow-water habitats (Barnard, 1954). *Eyakia* is a genus that burrows into moderately coarse sediment in habitats that, one imagines, would have little or no chance of being rafted.

Species imported by humans have not been identified in the Galápagos although *Ericthonius brasiliensis* and *Corophium baconi* might be good candidates for study. These two species belong to genera which are commonly transported by shipping or drifting logs or by imports of fishery materials (such as oysters).

4. Evolutionary Affinities of the Fauna

The geographic affinity of the Galápagan amphipod fauna with the American continent is partly supported by the presence of genera found only in the eastern Pacific: *Eyakia*, *Dulzura*, *Heterophlias*, *Meximaera*, *Posophotis* and *Varohios* (Table 1).

In the matter of species evolved from repeated immigrations by a widespread parent species or ancestral 'type,' the Galápagan fauna of Table 1 shows some possible cases, but these are only tentative because so little is known of continental tropical amphipods. For example, *Elasmopus temori* and *E. zoanthidea* appear to be siblings of the mainland *E. holgurus*; *Maera reishi* and *M. chinarra* appear to be siblings of a circumtropical complex of species known as the *M. tenuimana* complex; *Ampithoe tahue* and *A. vacoregue* appear to be siblings of *A. pollex*, a widely distributed species also occurring in the Galápagos (it would be intriguing to try to demonstrate that the two endemic species are results of early repeated immigrations of *A. pollex* each of which diverged specifically before the parent species arrived a third time to remain unchanged); *Elasmopus ocoroni* "looks" like a sibling of the mainland *E. mutatus*; *Hyale humboldti* "looks" like a sibling of an Indo-Pacific group of species named the *Hyale nigra* complex; *Ampithoe guaspare* "looks" like a sibling of the Indo-Pacific *Ampithoe ramondi* complex, and indeed, it seems strange that *A. ramondi*, a widely distributed tropical species, as yet has not been found in the Galápagos. *Elasmopus ecuadorensis* appears to be a sibling of a complex of species which is poorly understood but which has some members collected far out at sea (between Hawaii and Galápagos) in flotsam; and finally, *Chevalia* sp. appears in retrospect (Barnard and Thomas, 1987b) to be a new species in the circumtropical *Chevalia aviculae* complex.

There is virtually no way to determine if any amphipods have their origins solely in the "classic" Indo-Pacific fauna to the west. There are several well-speciated genera of that Indo-Pacific fauna that are missing from the Galápagos (for example, *Mallacoota*, *Ceradocus*, *Parelasmopus*, *Paradexamine*). All of the species have a recognizable degree of connection to mainland American species. One must presume that these Indo-Pacific species, if indeed ancestral to any Galápagan amphipods, would have reached the American mainland or sunken seamounts long before the birth of the Galápagos Islands.

Because the Galápagos Islands have been explored to about the same degree (= effort × size) as other tropical island systems listed in Table 2, one may safely compare the taxonomic composition to determine degrees of depauperation at various taxonomic levels. In addition to what is known in the literature I can add some experiences from preliminary looks at collections I have made (in some cases jointly with colleagues) in 22 island systems in the Pacific, Atlantic, and Indian Oceans. These experiences (and those of my colleagues) are poorly documented in the literature and might be termed "lore." What they provide is the rudimentary capability to look at the taxonomic names of the species in various island groups and make either conclusions or predictions on the richness or depauperateness of certain taxa, taxal groups, and ecological groups in various island chains. We have built up a poorly documented knowledge about the general ecological factors and behavior of many taxa. For example, we know by name the majority of the tube-dwellers, nestlers, inquilines (semi-parasites), and burrowers, and we know many generalities about where members of each genus live in the environment and to some extent what they "do" for a living.

In composing Table 3, I have therefore pooled into groups various taxa which are believed to have similar kinds of habitat and behavior, and then drawn comparisons among several island groups. A 'factor' at the top of the table is the multiplier of the 50 species in the Galápagos so that when one sees more than twice as many nestlers in Hawaii as in the Galápagos, one may attribute part of this to the fact that there are 2.4 times as many species in Hawaii as in the Galápagos.

The term 'nestler' refers to unspecialized amphipods, 'tubicolous' to those amphipods that build tube-like houses, 'inquilinous' to those that are semi-parasites on sessile organisms.

By ignoring all discrepancies such as difference in size of environment

Table 3. Cohort and Taxonomic Comparisons of Island Groups[a,b]

Item	Galápagos	Hawaii	Azores	Bermuda	Canary	Kerg	Fiji
Factor	1	2.4	1.1	0.9	1.5	1.6	1.6
Domicolous	17	45	21	13	23	14	31
Nestlers	17	39	26	22	32	42	26
Inquilines	5	19	7	5	8	16	10
Ampelisca	2	1	2	0	1	0	0
Ampithoidae	4	10	5	6	11	1	12
Elasmopus	7	10	5	3	2	0	6
Hyalidae	5	9	12*	5	9	1	6
Aorids	2	8	4	4	5	5	5
Maerids	5	9	3	8	4	1	10
Photids	2	3	0*	0	0	1	2
Gammaropsids	0	7	1	0	4	5	6
Leucothoids	1	4	1	1	1	1	4
Amphilochids	3	6	1	2	2	2	5
Colomastix	1	4	0	0	0	1	1
Phliantids	1	1	0	1	2	1	1
Stenothoids	0	5	4	2	1	5	1
Lysianassids	0	1	0	1	5	21	2

[a] The 'factor' is the number of species divided by 50 (= Galápagos total).
[b] * = Anomalous situation.

between Hawaii and the Galápagos, the much greater isolation of Hawaii, the stronger development of corals in Hawaii, and a myriad of other obvious differences, there remains a remarkable consistency in the presence of various amphipod groups. Bermuda, the Azores, and the Canaries are also considered in Table 3, and to some extent their minor differences can be explained by variance in size and isolation, but also can be explained because they are in a different ocean, and their sources of amphipods from Atlantic-basin continents are somewhat different; for example, stenothoids are few in the Atlantic basin compared to the Pacific.

Kerguelen, being a remote cold subantarctic island, is a very different environment from the other islands; it has a typically cool-water amphipod fauna with strong influence of Antarctica, as seen in the high number of lysianassids, for example. The contrast with the tropical islands is an extreme one but is demonstrative of the particularity of the tropical taxa expected to inhabit tropical islands as opposed to cool southern island groups.

None of the Caribbean island systems with which I am now familiar is of the same biogeomorphologic and oceanographic structure we see in the island groups of Table 2. The Caribbean island chains are integrated, are only weakly isolated, and are in much more complex hydrographic systems than the Galápagos and similar islands. However, the anchialine parts of the fauna have been subject to heavy vicariance (Stock, 1981). Our examination of Caribbean islands (Barnard and Thomas, 1983, 1987a, 1987b) shows little or no insular endemism in intertidal habitats.

I will refrain from discussing discrepancies in comparisons between these island groups and the Galápagos, since future SCUBA collecting should alter these figures in all the island groups. But by tracing all the taxa individually in the island groups of Tables 2 and 3, examining the size, degree of isolation, and kind of nearest neighbor (island or continent) for each group, one can make a preliminary encapsulated judgment for each.

Kerguelen is clearly in an entirely distinctive province from the tropical islands but contains species of the so-called "weedy" genera that are non-vicariant (they overcome the isolation factor by repeated dispersal) and are probably dispersed by surface currents. Fiji is a way-station in the midstream between the heart of the Indo-Pacific fauna and Hawaii (for example) where many more species will probably be found when SCUBA work is undertaken. Because of its size, neighbors, and geological history, Fiji supports several taxa of vicariant origin (for example *Parambasia*). The Canaries are but a satellite in the east Atlantic warm water provinces, too close to the continent for any significant endemism to have developed. The Azores are sufficiently removed from the same system as the Canaries to support a surprisingly low, but significant, degree of endemism. The Azores have at least one anchialine species in *Pseudoniphargus*, which may be present because the Azores were at one time very close to Eurafrica. This genus is a vicariant in Springer's sense (1982) only in that its dispersal is obviously interrupted but it remains unchanged despite this isolation. The low degree of endemism is surprising because of the isolation of the Azores. The much higher degree of endemism (37%) in Bermuda seems to be appropriate and the taxonomic composition suggests that Bermuda should be called an 'outpost' of the Gulf Stream. Unfortunately, the faunal source, the Caribbean amphipods, are not well studied, and the degree of endemism in Bermuda may therefore be reduced when more Bermuda species are found to occur in the Caribbean. Bermuda is very small and would be expected to have a lower number of species than the other island groups. Hawaii can be encapsulated as the 'ultimate but very isolated outpost of the Indo-Pacific.' Hawaii is said to be the most isolated archipelago in the world, but geologists and biologists have said that one must be careful about making conclusions on Hawaii based on isolation because evidence may develop that Hawaii was connected to the Indo-Pacific through seamounts. When Hawaiian amphipods were first examined by Barnard (1970), a high degree of endemism was found. However, in the past two decades, as more work is being completed in the main part of the Indo-Pacific, more and more of Hawaiian amphipod species are being found in the west. This does not apply, however, to the small species "swarms," for example the endemic Hawaiian species of the genus *Elasmopus* presumably derived from "weedy" Indo-Pacific ancestors. Several of the taxa that 'looked' in 1970 to be connected directly to American evolutionary centers, such as *Konatopus*, have lately been found to be of Indo-Pacific origins and therefore the American components must be considered as Indo-Pacific rather than American.

The capsule statement for the Galápagos reflects 'a close satellite of middle America, with a surprisingly high endemism at species level, but populated by weedy genera.' The high endemism at the specific level will probably be diminished as more middle American studies are made and the now endemic species are found to occur in the Panamic province. Little influence from the Magellanic fauna via the Humboldt Current can be discerned.

5. Prospectus

Many research problems remain unexplored with Galápagan amphipods. Despite many weeks of collecting in the islands, the surface has scarcely been scratched; most of what I had collected came from only 2 islands (Santa Cruz and Genovesa), and only 2–3 poorly collected SCUBA samples were available outside of those made by Dawson, who had collected all of his samples from the heaviest surf zones. I was able to describe only half of the total collections owing to time constraints, although most of the samples were examined cursorily for conspicuous differences from the half on which I did work. If more species were present in those collections they could easily have been overlooked. A few more dredged samples from greater depths are left in the old Allan Hancock collections now housed at the Los Angeles County Museum (LACM). These are mostly composed of one specimen per lot and would be tedious to identify because of their tiny size and complicated taxonomic problems.

At present, to better understand the amphipod fauna, the following needs must be met: (1) further SCUBA samples from collectors, such as those who have recently found so many unexplored habitats and taxa in the Caribbean sea and on the Great Barrier Reef; (2) the minute combing of many habitats, especially substrata where commensals often occur; (3) exploration of the intertidal of many additional islands; and finally, (4) a major effort is needed for comparative purposes on the mainland from middle Mexico to middle Chile. Unfortunately, the attention of taxonomists has been riveted on far more distant places, for example, Australia, with much more potential for the discovery of evolutionary information at higher systematic levels such as family and tribe. Because a few taxonomists have taken cursory looks at widely scattered samples in the Mexico to middle Chile coasts, they have reached the provisional conclusion that this area contains largely a "weedy" fauna with little evolutionary input of taxa at levels higher than species. The Pan-American area has only one endemic family out of 91 known families. Few endemic genera are present. There will be many new species and a few new genera to describe, and there will be many interesting inner regional comparisons to make at the species level. The Galápagos Islands will be a major focus because they remain the most isolated area within the eastern Pacific Ocean.

References

Abildgaard, P. C., 1789, Zoologia Danica seu animalium Daniae et Norvegiae rariorum ac minus notorum, *Descriptiones et Historia* **3**:71 pp. Havniae: N. Molleri et Filii.

Afonso, O., 1976, Amphipoda des Açores cueillis par scaphandrier autonome (avec la description d'une nouvelle espece), *Publs. Inst. Zool. "Dr. Augusto Nobre", Fac. Cienc. Porto* **130**:9–38.

Audouin, V., 1826, Explication sommaire des planches de crustaces de l' Egypte et de la Syrie, publiées par Jules-César Savigny, membre de l'Institut; offrant un exposé des caractéres naturels des génres, avec la distinction des espéces, *Descr. l'Egypte, Hist. Nat.* **1**:77–98.

Barnard, J. L., 1954, Amphipoda of the family Ampeliscidae collected by the *Velero III* in the Caribbean Sea, *Allan Hancock Atl. Exped. Rep.* **7**:13 pp.

Barnard, J. L., 1960, The amphipod family Phoxocephalidae in the eastern Pacific Ocean, with analyses of other species and notes for a revision of the family, *Allan Hancock Pac. Expeds.* **18**:175–368.

Barnard, J. L., 1962, Benthic marine Amphipoda of southern California: families Aoridae, Photidae, Ischyroceridae, Corophiidae, Podoceridae, *Pac. Nat.* **3**:1–72.

Barnard, J. L., 1965, Marine Amphipoda of atolls in Micronesia, *Proc. U.S. Nat. Mus.* **117**:459–552, 35 figures.

Barnard, J. L., 1967, A new genus of Galápagan amphipod inhabiting the buccal cavity of the sea-turtle, *Chelonia mydas*, *Proc. Symp. Crustacea Ernakulum, India* **1**:119–125.

Barnard, J. L., 1969a, Gammaridean Amphipoda of the rocky intertidal of California: Monterey Bay to La Jolla, *U.S. Nat. Mus. Bull.* **258**:230 pp.

Barnard, J. L., 1969b, A biological survey of Bahia de Los Angeles, Gulf of California, Mexico, IV. Benthic Amphipoda (Crustacea), *Trans. San Diego Soc. Nat. Hist.* **15**:175–228.

Barnard, J. L., 1969c, The families and genera of marine gammaridean Amphipoda, *U.S. Nat. Mus. Bull.* **271**:535 pp.

Barnard, J. L., 1970, Sublittoral Gammaridea (Amphipoda) of the Hawaiian Islands, *Smithson. Contr. Zool.* **34**:286 pp.

Barnard, J. L., 1972a, Gammaridean Amphipoda of Australia, Part I, *Smithson. Contr. Zool.* **103**:333 pp.

Barnard, J. L., 1972b, The marine fauna of New Zealand: Algae-living littoral Gammaridea (Crustacea Amphipoda), *N. Z. Oceanogr. Inst. Mem.* **62**:7–216.

Barnard, J. L., 1976, Amphipoda (Crustacea) from the Indo-Pacific tropics: A review, *Micronesica* **12**:169–181.

Barnard, J. L., 1977, The cavernicolous fauna of Hawaiian lava tubes 9. Amphipoda (Crustacea) from brackish lava ponds on Hawaii and Maui, *Pac. Insects* **17**:267–299.

Barnard, J. L., 1979, Littoral Gammaridean Amphipoda from the Gulf of California and the Galápagos Islands, *Smithson. Contr. Zool.* **271**:vi and 149 pp.

Barnard, J. L., and Barnard, C. M., 1983, *Freshwater Amphipoda of the World, I. Evolutionary Patterns and II. Handbook and Bibliography*, xix and 830 pp. Hayfield Associates, Mt. Vernon, Virginia.

Barnard, J. L., and Clark, J., 1985. A new sea-cave amphipod from Bermuda (Dulichiidae), *Proc. Biol. Soc. Wash.* **98**:1048–1053.

Barnard, J. L., and Thomas, J. D., 1983. A new species of *Amphilochus* from the gorgonian *Pterogorgia anceps* in the Caribbean Sea, in *Selected Papers on Crustacea*, (P. A. John, ed.) pp. 179–187. Trivandrum: The Aquarium.

Barnard, J. L., and Thomas, J. D., 1987a, New species of *Neomegamphopus* from tropical America (Crustacea: marine Amphipoda), *Proc. Biol. Soc. Wash.* **100**:147–163.

Barnard, J. L., and Thomas, J. D., 1987b, A new species of *Chevalia* from the Caribbean Sea (Crustacea: Amphipoda), *Proc. Biol. Soc. Wash.* **100**:532–542.

Barrois, T., 1888, *Catalogue des crustaces marins recueillis aux Acores durant les mois d'Aout et de Septembre 1887*, 110 pp. Lille: le Bigot.

Beebe, W., 1924, *Galápagos: world's end*, G. P. Putnam's Sons, New York and London, 443 pp.

Bellan-Santini, D., and Ledoyer, M., 1974, Gammariens (Crustacea-Amphipoda) des Îles Kerguelen et Crozet, *Tethys* **5**:635–707.

Bowman, R. I. [ed.], 1966, The Galápagos; Proceedings of the symposia of the Galápagos international scientific project, University of California Press, Berkeley, 318 pp.

Calman, W. T., 1898, On a collection of Crustacea from Puget Sound, *Ann. N. Y. Acad. Sci.* **11**:259–292.

Chevreux, E., 1888, Troisiéme campagne de l'*Hirondelle*, 1887. Sur quelques crustacés amphipodes du littoral de Açores, *Bull. Soc. Zool. France* **13**:31–35.

Chevreux, E., 1925, Amphipodes I.—Gammariens. Voyage de la goelette *Melita* aux Canaries et au Sénégal (1889–1890), *Bull. Soc. Zool. France* **50**:278–311.

Chevreux, E., 1926, Amphipodes I.—Gammariens (*suite*). Voyage de la goelette *Melita* aux Canaries et au Sénégal 1889–1890, *Bull. Soc. Zool. France* **50**:365–398.

Dana, J. D., 1853, Crustacea. Part II, *U.S. Expl. Exped.* **14**:689–1618.

Della-Valle, A., 1893, Gammarini del Golfo di Napoli, *Fauna und Flora der Golfes Neapel und der angrenzenden Meeres-Abschnitte, Monog.* **20**:xi and 948 pp.

Ekman, S., 1953, Zoogeography of the Sea. Sidgwick and Jackson, Ltd., London.

Garth, J. S., 1946, Distribution studies of Galápagos Brachyura, *Allan Hancock Pac. Exped*s. **5**:603–638.

Gurjanova, E., 1938, Amphipoda, Gammaroidea of Siaukhu Bay and Sudzukhe Bay (Japan Sea), *Repts. Japan Sea Hydrobiol. Exped. Zool. Inst. Acad. Sci. USSR in 1934* **1**:241–404. [In Russian.]

Heezen, B. C., and Tharp, M., 1977, *World ocean floor* [chart], U.S. Navy, Office of Naval Research, Washington, D.C.

Holmes, S. J., 1908, The Amphipoda collected by the United States Bureau of Fisheries Steamer, "Albatross", off the west coast of North America, in 1903 and 1904, with descriptions of a new family and several new genera and species, *Proc. U.S. Nat. Mus.* **35**:489–543.

Kunkel, B. W., 1910, The Amphipoda of Bermuda, *Trans. Conn. Acad. Arts Sci.* **16**:1–116.
Ledoyer, M., 1982, Crustacés amphipodes gammariens familles des Acanthonotozomatidae á Gammaridae, *Faune de Madagascar* **59**(1):1–598.
Ledoyer, M., 1986, Crustacés amphipodes gammariens, *Faune de Madagascar* **59**(2):599–1112.
MacArthur, R. H., and Wilson, E. O., 1967, *The Theory of Island Biogeography*, ix + 203 pp.
Monod, T., 1970, Sur quelques crustacés malacostraces des Îles Galápagos recoltes par N. et J. Leleup (1964–1965), *Miss. Zool. Belge Galápagos Ecuador (N. et J. Leleup, 1964–1965)* **2**:11–53.
Myers, A. A., 1968, Some Aoridae (Amphipoda: Gammaridea) collected by the Hancock Expeditions to the eastern Pacific, 1931–1941, *Pac. Sci.* **22**:497–506.
Myers, A. A., 1985, Shallow-water, coral reef and mangrove Amphipoda (Gammaridea) of Fiji, *Rec. Australian Mus. Suppl.* **5**:144 pp.
Schellenberg, A., 1931, Gammariden und Caprelliden des Magellangebietes, Sudgeorgiens und der Westantarktis, *Further Zool. Res. Swedish Antarc. Exped. 1901–1903* **2**(6):290 pp.
Schellenberg, A., 1936, Zwei neu Amphipoden des Stillen Ozeans und zwei Berichtigungen, *Zool. Anz.* **116**:153–156.
Shoemaker, C. R., 1925, The Amphipoda collected by the United States Fisheries Steamer 'Albatross' in 1911, chiefly in the Gulf of California, *Bull. Amer. Mus. Nat. Hist.* **52**:21–61.
Shoemaker, C. R., 1934, Two new species of *Corophium* from the west coast of America, *J. Wash. Acad. Sci.* **24**:356–360.
Shoemaker, C. R., 1938, Three new species of the amphipod genus *Ampithoe* from the west coast of America, *J. Wash. Acad. Sci.* **28**:15–25.
Shoemaker, C. R., 1942, Amphipod crustaceans collected on the Presidential Cruise of 1938, *Smithson. Misc. Colls.* **101**(11):52 pp.
Springer, V. G., 1982, Pacific plate biogeography, with special reference to shorefishes, *Smithson. Contr. Zool.* **367**:182 pp.
Stephensen, K., 1949, The Amphipoda of Tristan da Cunha, *Res. Norwegian Sci. Exped. Tristan da Cunha 1937–1938* **19**:61 pp.
Stimpson, W., 1864, Descriptions of new species of marine invertebrata from Puget Sound, collected by the naturalists of the North-West Boundary Commission, A. H. Campbell, Esq., Commissioner, *Proc. Acad. Nat. Sci., Philadelphia* **16**:153–165.
Stock, J., 1981, L'origine geologique des iles des Indes occidentales en relation avec dispersion de quelques malacostraces stygobiontes, *Geobios* **14**:219–227.
Zimmerman, E. C., 1948, *Insects of Hawaii* **1**:i–xx, 1–206.

V

Marine Cave Fauna

Chapter 10

Anchialine Fauna of the Galápagos Islands

THOMAS M. ILIFFE

1. Introduction

From the biogeographic viewpoint, marine caves on oceanic islands represent 'islands within islands.' Due to the isolation and environmental stability of this habitat, such caves act as refuges, preserving primitive relict taxa. Investigations of marine caves in the Atlantic have resulted in the discovery of many higher taxa including the new crustacean class Remipedia (Yager, 1981), the new peracarid order Mictacea (Bowman and Iliffe, 1985), the new copepod order Platycopioida (Fosshagen and Iliffe, 1985), the new isopod family Atlantasellidae (Sket, 1979) and the new caridean family Agostocaridae (Hart and Manning, 1986). Unexpectedly however, the troglobitic (i.e., cave-limited) fauna of Atlantic marine caves show highly anomalous biogeographical distributions, as well as close taxonomic affinities to deep sea species. Particularly noteworthy in this regard is the fauna of the Jameos del Agua, a marine lava tube cave in the Canary Islands. Members of six

THOMAS M. ILIFFE • Department of Marine Biology, Texas A & M University at Galveston, Galveston, Texas 77553.

crustacean genera inhabiting this cave, including the remipede *Speleonectes*, the anthurid isopod *Curassanthura*, the amphipod *Spelaeonicippe*, the mysid *Heteromysoides*, the ostracod *Danielopolina*, and the thermosbaenacean *Halosbaena*, also have species inhabiting caves on the western side of the Atlantic. This amphi-Atlantic distribution suggests that these taxa are Tethyan relics having an origin in caves early in the history of the Atlantic and subsequently dispersed by plate tectonics and sea floor spreading (Iliffe *et al.*, 1984; Wilkens *et al.*, 1986). Also from the same cave, the galatheid crab *Munidopsis polymorpha* and the polynoid polychaete *Giesiella jameensis* belong to groups known primarily from the deep sea.

Occupying an analogous position to the Canary Islands of the Atlantic, the Galápagos Islands are situated in an isolated location on the eastern edge of a major ocean near the meeting point of several major current systems. Geologically, as in the case of the Canaries, the Galápagos Islands are relatively young, semi-arid volcanic peaks with varying eruptive histories. Marine lava tube caves and similar groundwater habitats in the Galápagos represent the only known subterranean ecosystem of this type in the East Pacific. Located midway between the cave areas of the South Pacific and the Caribbean, the marine caves of the Galápagos Islands hold the potential to provide yet another type of natural laboratory for the testing of biogeographical and evolutionary hypotheses.

2. Anchialine Habitats

The term 'anchialine' refers to "pools with no surface connection to the sea, containing salt or brackish water, which fluctuates with the tides" (Holthuis, 1973). It was originally proposed to describe open pools near the sea in both limestone and volcanic terrains which were found to contain unusual hypogean shrimp and other crustaceans. With the discovery of similar pools inside caves, a more rigorous definition of the term was suggested: "bodies of haline waters, usually with a restricted exposure to open air, always with more or less extensive subterranean connections to the sea, and showing noticeable marine as well as terrestrial influences" (Stock *et al.*, 1986). In contrast to anchialine caves, submarine caves are those wholly below sea level lacking significant terrestrial influences. The term 'marine cave' is used here to include both anchialine and submarine caves, i.e., any cave containing haline waters of marine origin.

2.1. Grietas

Three types of anchialine habitats are found in the Galápagos Islands. The first type consists of anchialine pools in tectonic faults, locally referred to as 'grietas'. Extensive recent faulting has occurred in the Galápagos Islands, most notably along the south coast of Santa Cruz Island. In this area, faults as much as 20 km long generally trend in an east–west direction parallel to the coast (Fig. 1). Some of these faults may have been formed within the last few thousand years (McBirney and Williams, 1969). Apparently all of the faulting is normal faulting in which one crustal block has moved up or down with respect to its neighbor along a nearly vertical fracture or fault plane (Simkin, 1984). Such faulting has produced an

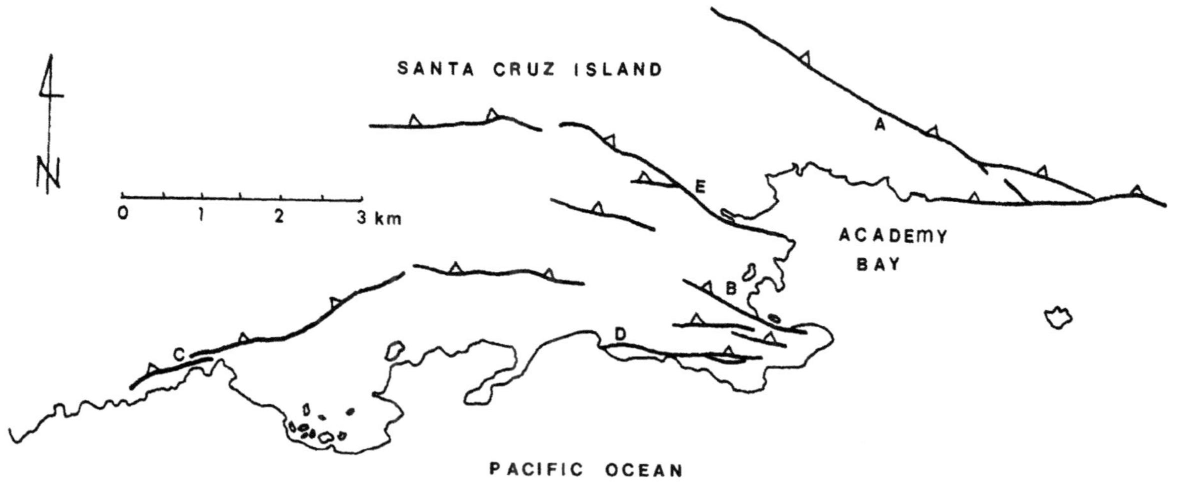

Figure 1. Position of major faults along the southern coast of Santa Cruz, Galápagos Islands inferred from aerial photographs. A = Cueva de Iguana, B = Grieta de Delfine, C = Grieta de Caleta la Torta, D = Deep Grieta, E = Grieta north of trail to Tortuga Bay.

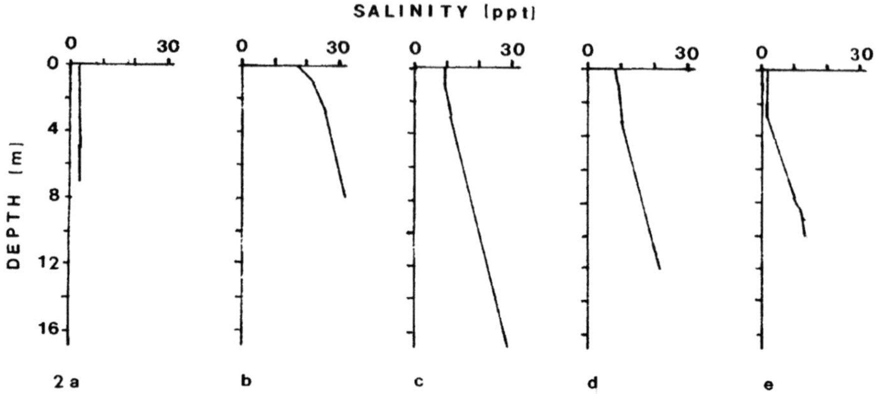

Figure 2. Salinity/depth profiles from five grieta pools on Santa Cruz measured between 25 April and 18 June 1987. Water samples collected by diving and measured with a calibrated refractometer. a = Cueva de Iguana, b = Grieta de Delfine, c = Grieta de Caleta la Torta, d = Deep Grieta, e = Grieta north of trail to Tortuga Bay.

impressive series of high cliffs and deep fissures. Typically, fissures consist of opposite-facing sheer rock walls, 1–10 m apart and as much as 30 m high. Those fissures nearest to the coast may extend down below sea level and contain anchialine pools. Large collapse breakdown blocks, which were probably generated by the earthquakes forming the fissure, make up the floor. Occasionally, talus caves occur, both above and below water, where it is possible to enter totally dark areas under wedged breakdown blocks and between fissure walls.

Several grietas are here described in order to illustrate the range of environmental parameters existing in this habitat. Figures 2 and 3 present depth profiles of salinity and temperature in five grietas on Santa Cruz. Distinct haloclines, producing much steeper salinity gradients than those depicted, occur in most of the deeper grieta pools. Noticeable tidal currents exist in those pools closer to the sea,

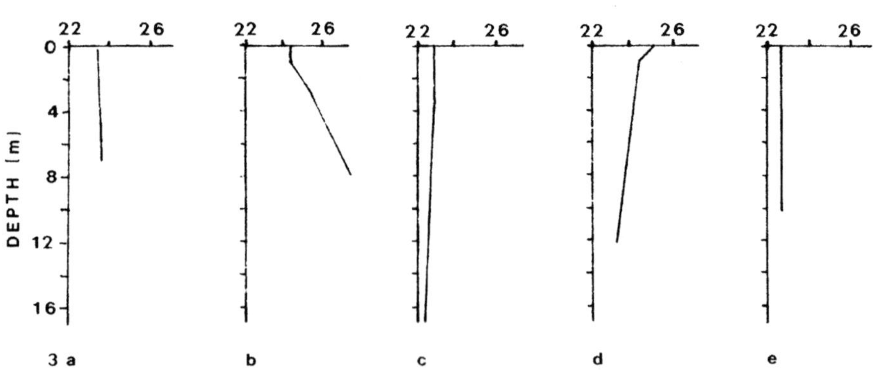

Figure 3. Temperature/depth profiles from the same grieta pools as in Fig. 2.

while in more remote pools, gently oscillating water-level fluctuations are the only evidence of tidal connections.

Cueva de Iguana, also known as Grieta de Estación Darwin, is located near Puerto Ayora on the right side of the trail leading past the iguana enclosures and heading toward a high fault cliff (Peck and Peck, 1986). This 'earthquake fissure,' situated about 1 km inland from the coast, is about 12 m deep, 1–3 m wide and 100–150 m long. The main pool from which the Darwin Station pumps brackish water is 7 m deep and ends in a sediment and collapse plug. The pool is well shaded and is in almost total darkness at the bottom. Salinity and temperature were uniform at 3 ppt and 23.6°C (Figs. 2a and 3a). The water was very clear.

Grieta de Delfine is located on the west side of Puerto Ayora past a series of shallow lagoons behind the Hotel Delfine. This grieta runs parallel to the coastline and is about 300 m inland. It is about 5 m wide, several hundred meters long and is bordered by 10 m high cliffs. The pool is 10 m deep at its deepest point and is quite open so that it receives considerable sunlight. Tidal currents are evident, especially at deeper depths near the west end where it is apparently connected to the sea through breakdown. Salinity and temperature both increased with depth in the pool so that at 8 m, values of 31.5 ppt and 27.0°C approached those of the open sea (Figs. 2b and 3b). Underwater visibility was 5 m or less.

Grieta de Caleta la Torta (Fig. 4) is located 7 km west of Puerto Ayora and 100 m inland from a small, but well protected bay. The grieta is at the western side of a prominent, 15 to 20 m high cliff line and consists of 3 large pools separated by collapse barriers. The easternmost pool is 3–5 m wide by 100 m long. By passing through a narrow hole between breakdown blocks at 12 m depth, a large chamber was reached which extended to 29 m depths. This chamber contained clear water, with visibilities of 20 m or more, in total darkness. Water temperatures decreased slightly at the deepest depths to 22.3° C, the lowest values found in any of the grietas. Salinity increased with depth from 9 ppt at the surface to 29 ppt at 17 m (Figs. 2c and 3c).

Deep Grieta is located about 250 m east of the beach at Tortuga Bay and 80 m inland, behind a conspicuous round lava mound (Peck and Peck, 1986). This large grieta pool runs parallel to the coast. It is situated between 10 m high cliffs and is about 8 m wide by 40 m long. Maximum depth in the pool is 12 m. The rear part of the pool is roofed over with wedged breakdown blocks and is thus in darkness. Masses of roots hang down into the water in this section of the pool. Water temperature decreased slightly with depth, while salinity increased from 8 ppt at the surface to 22 ppt at 12 m (Figs. 2d and 3d). Underwater visibility was about 5 m.

The grieta north of the trail to Tortuga Bay is located about 100 m past the last house in Puerto Ayora (Peck and Peck, 1986). Immediately on the north side of the trail is a sheer-walled fissure, 1 to 2 m wide by 10 m deep. About 30 m along this fissure is a 30 m long pool divided by breakdown into two sections. Water is pumped from near the surface of the pool, probably for domestic purposes. In the second part of the pool, a deep crack descends underwater past a marked halocline at 9 m depth before ending in collapse at 10 m. Moderate tidal currents are present at the surface. The deepest part of the pool is in nearly total darkness as are narrow sections of the fissure extending through the breakdown at either end. Water temperatures were uniform with depth, while salinity increased markedly at the halocline from 1.5 to 12.5 ppt (Figs. 2e and 3e). Visibility was 10 m or more.

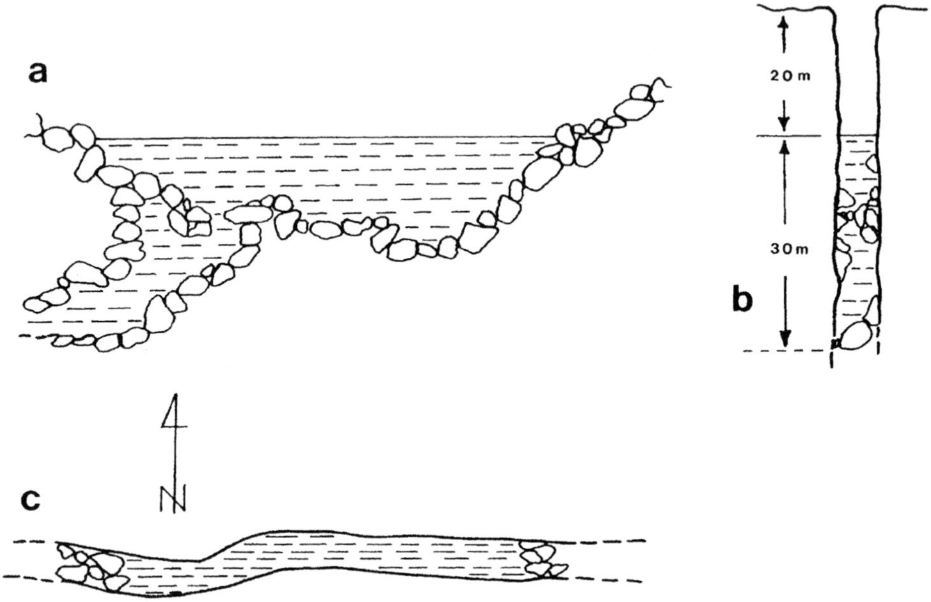

Figure 4. Profile (a), section (b), and plan (c) maps of Grieta de Caleta la Torta, Santa Cruz.

2.2. Lava Tubes

A second type of anchialine habitat occurs in coastal lava tubes. Lava tubes are conduits, formed during volcanic eruptions and serving in the transport of molten lava for considerable distances away from the eruptive centers. After the eruption ends, molten lava flows out of the tube leaving behind these volcanic caves. Lava tubes in the Galápagos are most often found at higher elevations on the islands where caves in excess of 2250 m length have been explored (Balazs, 1975; Montoriol-Pous and De Mier, 1977).

Although many holes along the coastline superficially resemble the entrances to lava tubes, most of these are short, erosional sea caves, not of biospeleological interest. Only two anchialine lava tubes are known from the Galápagos, though many other must exist. Post Office Bay Cave (Fig. 5) is a lava tube located 300 m inland from the north coast of Floreana at Post Office Bay (Montoriol-Pous and Escolá, 1975; Steadman, 1986; Peck and Peck, 1986). The tube consists of two segments with separate entrances isolated from each other by an area of massive roof collapse. The lower (Inferior) section of the cave is 202 m long and ends in a 60 m long, tidal brackish water pool. The pool begins at a point where the elevation of the downward sloping tube passes below sea level. The pool gradually increases in depth to reach a maximum of 2 m at the far end. At this point, the ceiling of the cave dips below water level and the underwater passage becomes much lower. A final small, submerged chamber exists only a few meters past the end of the lake, but it pinches out as lava bedrock fills the passage. Water in the pool is exceptionally clear with no evident particulate matter being present. A distinct halocline is

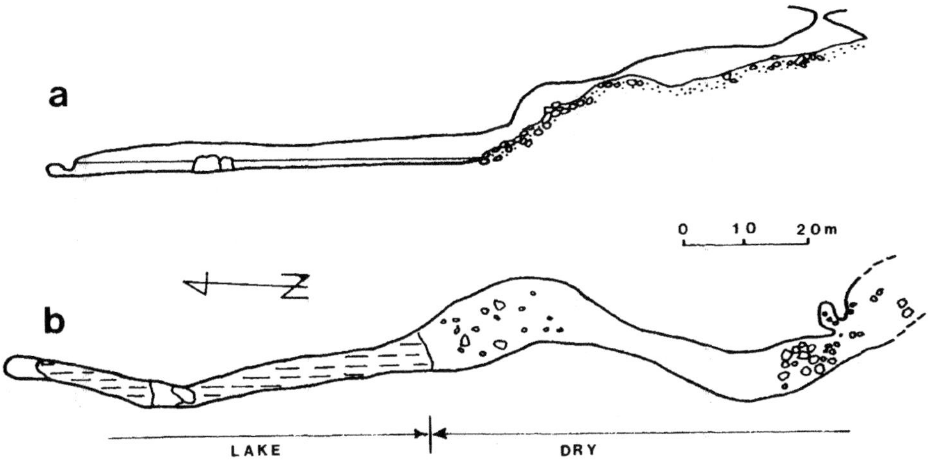

Figure 5. Profile (a) and plan (b) maps of Post Office Bay Cave, Floreana; modified from Montoriol-Pous and Escolá (1975).

evident at 1.5 m depth. Water temperatures are uniform at a relatively cool 20.3°C, while salinity increases from 17 to 22 ppt across the halocline. A particularly unusual feature of this pool is the presence of a crystalline carbonate submarine cement coating rocks and even tortoise bones on the floor of the underwater cave.

Cueva de la Cadena (Fig. 6) is a lava tube located 2 km west of Puerto Villamil on the southwest coast of Isabela (Montoriol-Pous and Escolà, 1978; Peck and Peck, 1986). The cave is in a pahoehoe lava ridge which extends out into the sea. Depending on the level of the tides, seawater floods the cave to at least its overhead

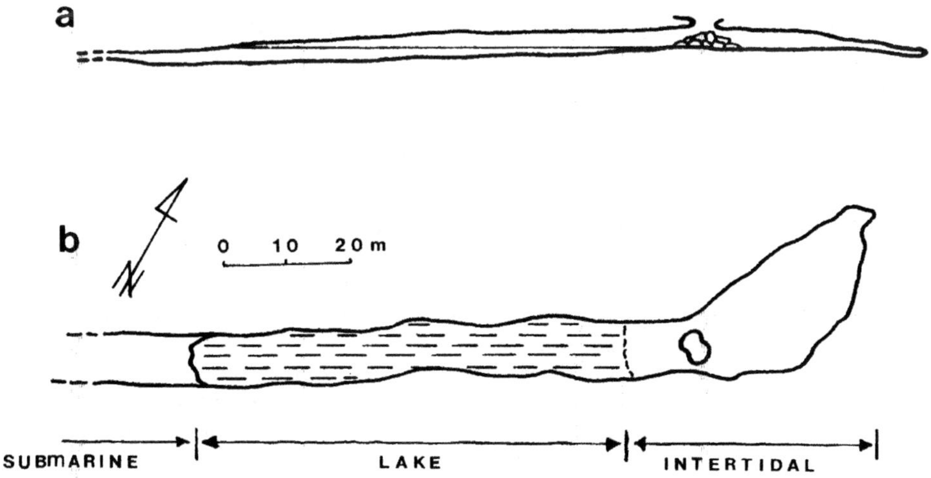

Figure 6. Profile (a) and plan (b) maps of Cueva de la Cadena, Isabela; modified from Montoriol-Pous and Escolá (1978).

collapse entrance or beyond. At low tide, the lake within the tube is 98 m long with an additional 16 m of dry gallery behind the entrance. Maximum depth within the pool is 1.5 m. The tube continues seaward for about 15 m past the last air bell in the ceiling before becoming too low and being plugged with sediment. Underwater visibility was about 5 m. A slightly depressed surface salinity of 32 ppt indicates the presence of some brackish runoff through the tube. At 1 m depth, salinity and temperature were identical with the open sea (33.5 ppt and 27.1°C). The presence of open water fish in the pool indicates that a direct connection with the sea must exist. Tidal period and range in the cave appear to be indistinguishable from those of open waters.

2.3. Lava Rock Pools

In low-lying terrain near the coast, highly uneven and irregular lava flows have left behind low spots, cracks, and gas pockets which extend below sea level and are now filled with anchialine groundwater. Along the west coast of Isabela in the area immediately north of Punta Tortuga, numerous small anchialine pools exist in a relatively recent lava flow. These pools range from small depressions less than a meter in diameter and 30 cm deep to partially rubble filled fissures up to 15 m long and 1.5 m deep. Water in these pools is clearly tidal with salinities from 19.5 to 22.5 ppt. Similar pools located half a kilometer east of Puerto Villamil on Isabela have even lower salinities to 1.5 ppt. Many of the pools are at least partially roofed over limiting light penetration. Cracks and fissures within the submerged bedrock are sometimes evident, greatly extending the volume of these voids.

3. Composition of the Anchialine Fauna

Previous studies of the cave fauna of the Galápagos Islands include those of Leleup (1968, 1970, 1976), Peck and Peck (1986) and Peck and Kulalova-Peck (1986). Material collected by the author during a 2-month expedition in 1987 is also reported below (as Iliffe).

3.1. Caridea

3.1.1. *Typhlatya galapagensis* Monod and Cals, 1970

Characteristic features.-Family Atyidae. Eyes without pigment, lacking faceted cornea. Body white, without dark pigment. Conspicuous setal tufts on fingers of chelae of first and second pereiopods.
Size.-Postorbital carapace length of juveniles 2 to 3 mm (Monod and Cals, 1970).
Species range.-Galápagos Islands. Santa Cruz: grietas on the south and southeast coast including Leleup's station G4, G14, G22 and G23 (Monod and Cals, 1970); Cueva de Iguana, Grieta del Pozo de Puerto Ayora, and Grietas de Tortuga Bay (Peck and Peck, 1986); Cueva de Iguana at 0 to 7 m water depths, Grieta de Caleta la Torta

at 24 m depth, Deep Grieta at 6 to 12 m depths, and grieta north of trail to Tortuga Bay at 9 to 10 m depths (Iliffe). Isabela: Leleup's station G14 (Monod and Cals, 1970); Cueva de la Cadena (Peck and Peck, 1986); lava rock pools at Punta Tortuga and Puerto Villamil at 0 to 0.7 m water depths (Iliffe).
Number of species in genus-Nine, all troglobitic.
Genus range.-Ascension Island, Bermuda, West Indies (Barbuda, Caicos Islands, Cuba, Dominican Republic, Mona, Puerto Rico), Mexico (Campeche, Yucatan) and Galápagos Islands (Hobbs *et al.*, 1977).
Ecological notes.-Of the nine species in the genus, four (*T. galapagensis*, *T. iliffei* from Bermuda, *T. rogersi* from Ascension Island, and *T. garciai* from the Bahamas) are known to inhabit brackish or marine waters, while the remainder are found in freshwater habitats. *T. galapagensis* is often present in extremely large numbers especially at Cueva de Iguana, Deep Grieta and lava rock pool at Puerto Villamil. Individuals were observed both walking on sediment surfaces and rock walls as well as actively swimming in water column. Salinities at which this shrimp was found ranged from 1.5 to 29 ppt. Collection depths ranged from the surface to 24 m.
Life history.-Specimens reported by Monod and Cals (1970) and collected in October, November, December and February were all juveniles.
Closest related species.-*T. galapagensis* and *T. rogersi* (from Ascension Island) are considered to be the most divergent species in the genus (Hobbs and Hobbs, 1976). *T. galapagensis*, together with 3 other Caribbean congeners, belong to a group in which the rostrum does not extend anteriorly beyond the eyes (Hobbs *et al.*, 1977).

3.2. Amphipoda

3.2.1. *Galapsiellus leleuporum* (Monod, 1970) (see also Barnard, 1976 and Stock and Iliffe, 1990)

Characteristic features.-Family Melitidae. Eyeless.
Size.-Body length, 1 to 4.3 mm. Largest female, 4.3 mm; largest male, 3.3 mm (Stock and Iliffe, 1990).
Species range.-Galápagos Islands. Santa Cruz: Leleup's station G6 (taken with meat baits in deep freshwater at 5 m elevation) and G22 (brackish water accessible by digging) (Monod, 1970); anchialine mangrove pool 300 m from the sea near the Darwin Station (Barnard, 1976); Cueva de Iguana at 7 m water depth, Grieta de Caleta la Torta at 17 to 29 m depths, Grieta Cave at 0 to 1 m depths, and grieta north of trail to Tortuga Bay at 0 to 10 m depths (Stock and Iliffe, 1990). Isabela: lava rock pools near Punta Tortuga at 0 to 0.3 m depths (Stock and Iliffe, 1990).
Number of species in genus.-One.
Ecological notes.-A maximum of 7 specimens were collected at one location, Grieta Cave, but more typically, only 1 to 2 specimens were found. Specimens collected by Leleup at his station G6 were taken with meat baits (Monod, 1970). Samples studied by Stock and Iliffe (1990) were collected with a suction bottle from the surface of rocks and bottom sediments, and with a plankton net from the water column and cave walls. Salinities at collection sites ranged from 3 to 29 ppt.

Life history.-Monod's description is apparently based on juveniles or young females (1–3 mm). A 4.3 mm female studied by Stock and Iliffe (1990) had well developed oostegites. Marked sexual dimorphism.
Closest related genera.-Barnard (1976) believed this genus to be derived from *Eriopisa*, a wide-spread marine genus, while similar morphologically to *Paraniphargus*, a genus consisting of two freshwater hypogean species on western Pacific and Indian Ocean islands.

3.2.2. *Anchialella vulcanella* Barnard, 1979

Characteristic features.-Family Gammaridae. Eyes absent.
Size.-Body length 4.06 mm.
Species range.-Galápagos Islands. Santa Cruz: intertidal mangrove tidepool, 305 m from sea near lower bodega of Academy Bay (Barnard, 1979).
Number of species in genus.-One.
Genus range.–Galápagos Islands.
Ecological notes.-No specimens were found by the author in his 1987 collections from anchialine habitats.
Life history.-One 2.10 mm juvenile collected by Barnard from the type locality. Since only the female has been found, sexual dimorphism is unknown.
Closest related genera.-*Anchialella* appears to be derived from an Indo-Pacific *Eriopisa* and is transitional between this genus and *Galapsiellus* (Barnard, 1979).

3.2.3. *Valettietta cavernicola* Stock and Iliffe, 1990

Characteristic features.-Family Lysianassidae. No eyes visible.
Size.-Body length, 2.5 mm.
Species range.-Galápagos Islands. Santa Cruz: Grieta de Caleta la Torta at 17 to 29 m water depths (Stock and Iliffe, 1990).
Number of species in genus.-Four, other three from deeper waters to far over 4000 m.
Genus range.-North Atlantic abyssal, Pacific collected in vertical haul between 0 and 5,300 m, and Galápagos anchialine.
Ecological notes.-Single specimen collected from the water column in 17–29 m depths of totally dark section of grieta. Salinity was at least 29 ppt. Only two other lysianassids have so far been reported from anchialine caves: *Idunella sketi* from Bermuda and *Lucayarina catacumba* from the Bahamas. This specimen was smaller than other members of the genus such as *V. castellana* which reaches 4 mm. Larger size is generally related to increasing water depth.
Life history.-Not known.
Closest related genera.-Similar to the genus *Valettiopsis*, another deep sea amphipod.

3.2.4. *Antronicippe serrata* Stock and Iliffe, 1990

Characteristic features.-Family Pardaliscidae. No eyes visible.

Size.-Body length, 4.5 mm.
Species range.-Galápagos Islands. Santa Cruz: Grieta de Caleta la Torta at 24 m water depth (Stock and Iliffe, 1990).
Number of species in genus.-One.
Ecological notes.-Single specimen collected from the water column in 20 m depths of totally dark section of grieta. Salinity was at least 29 ppt.
Life history.-Not known.
Closest related genera.-*Anteronicippe*, together with *Nicippe* and *Spelaeonicippe*, form a monophyletic group. *Nicippe* and *Spelaeonicippe* differ in 12 character states; *Anteronicippe* agrees with *Spelaeonicippe* in all 12 (Stock and Iliffe, 1990). *Spelaeonicippe* contains two species inhabiting anchialine caves in the Canary Islands and the Caicos Islands (West Indies).

3.3. Ostracoda

3.3.1. *Skogsbergia galapagensis* Kornicker and Iliffe, 1989

Characteristic features.-Family Cypridinidae. Medial eye small and unpigmented; lateral eye indistinct, unpigmented and minute with globular cells but no ommatidia.
Size.-Adult female: length, 2.02 to 2.14 mm; height, 1.35 to 1.41 mm (n = 6). Adult male: length, 1.76 mm; height, 1.10 mm (n = 1).
Species range.-Galápagos Islands. Santa Cruz: Deep Grieta at 6 to 12 m water depths, Grieta de Caleta la Torta at 10 to 18 m depths, and grieta north of trail to Tortuga Bay at 9 to 10 m depths.
Number of species in genus.-Ten, of which only *S. galapagensis* is anchialine.
Genus range.-Widespread between latitudes 60°N and 34°S in the open sea, mostly at shelf depths.
Ecological notes.-Specimens of *S. galapagensis* were collected with both plankton net and suction bottle by divers and with plastic bottle traps baited with crushed crab and left overnight in the grietas. Collections came from 6 to 18 m depths where salinities ranged from 13 to 29 ppt. The gut of adult specimens was gorged with fine grained, light amber colored particles. A few fragments bearing claws, suggesting an arthropod origin, were present in the gut of a 1st instar.
Life history.-Five juvenile instars are present for *S. galapagensis*, the same as for the open water species *S. lerneri* and for species in three other genera of Cypridinini. Thus, the number of juvenile instars appears to be constant in the Cypridinini, and the number of growth stages is not affected by the troglobitic habit of *S. galapagensis* (Kornicker and Iliffe, 1989). The shape of the juvenile carapace of *S. galapagensis* is similar to that of the adult, while the order of appearance of appendages is similar to that reported for many myodocopids. The lateral eye is vestigial in all instars. The growth curves are parallel for *S. galapagensis* and *S. lerneri*.
Closest related species.-*S. galapagensis* differs from all previously described species of *Skogsbergia* in lacking well developed lateral eyes. Except for this, the morphology and ontogeny of *S. galapagensis* are quite similar to that of *S. lerneri* collected in the Gulf of Mexico and Caribbean Sea (Kornicker, 1984) and shown to be a scavenger of dead animals (Cohen, 1983).

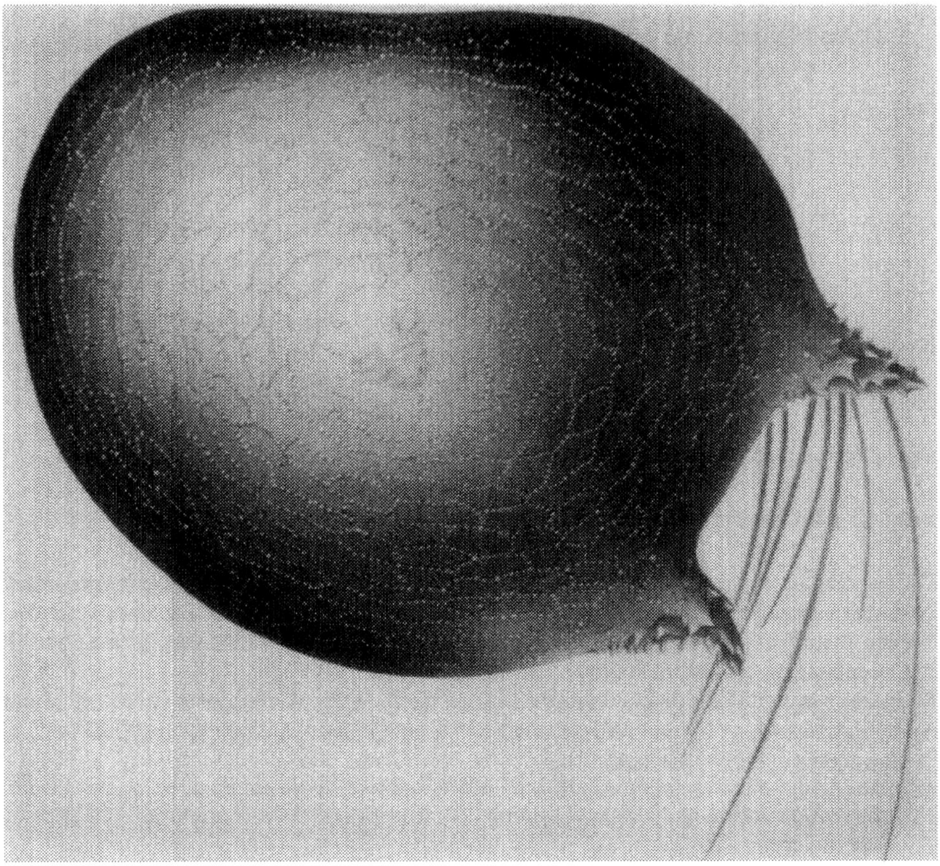

Figure 7. *Danielopolina styx*, a troglobitic halocyprid ostracod from Santa Cruz, length, including anterior process, 0.57 mm.

3.3.2. *Danielopolina styx* Kornicker and Iliffe, 1989 (Fig. 7)

Characteristic features.-Family Thaumatocyprididae. Furca of *D. styx* has only one articulated anterior claw, differing from all known species of Thaumatocyprididae which possess 2 claws.
Size.-Instar IV: length with anterior process, 0.68 mm; height, 0.51 mm.
Species range:Galápagos Islands. Santa Cruz: Deep Grieta at 6 to 12 m water depths, Grieta de Caleta la Torta at 17 to 29 m depths (Kornicker and Iliffe, 1989).
Number of species in genus.-Six species, five of which are troglobitic and one from the deep sea.
Genus range.-Cuba, Bahamas, Canary Islands, Mexico (Yucatan), Galápagos—all anchialine; South Atlantic from near the Equator at a depth of 3459 m.
Ecological notes.-Of the 5 genera comprising the family Thaumatocyprididae, 2 are known only as fossils (from the Permian and Upper and Lower Jurassic) and 2

known only from the deep sea. The genus *Danielopolina* is the only member of the family inhabiting marine caves. Salinities at depths where *D. styx* was found ranged from 13 to 29 ppt. Unrecognized amber to brown particles were present in the gut.

Life history.-As in *Thaumatoconcha radiata* (from the deep sea) and *D. bahamensis* (from anchialine caves in the Bahamas), the 1st and 2nd antennae, mandible, maxilla, 5th limb and furca are already present in the 1st instar. The 6th limb appears in the 3rd instar, while the 7th limb appears in the 4th. The total number of growth stages for *D. styx* is unknown. Other species in the Thaumatocyprididae have either 5 or 7 growth stages.

Closest related species.-The carapace of *D. styx* resembles that of *D. orghidani* (from Cuba) and *D. carolynae* (from the South Atlantic), but differs from both species in having a small posterodorsal process on each valve (Kornicker and Iliffe, 1989).

3.3.3. Podocopid and Platycopid Ostracoda

Eleven species of podocopid and platycopid Ostracoda have been collected from anchialine habitats in the Galápagos (Maddocks and Iliffe, 1991; Maddocks, 1991a&b). None of these species are derived from present-day Pacific coastal faunas of North, Central or South America, with most having Atlantic and West Indian affinities. Two species, *Iliffeoecia iliffei* and *Paracypris crispa* previously known only from anchialine caves and coral reefs in Bermuda (Maddocks and Iliffe, 1986), are present in Cueva de la Cadena. *Kareloecia woutersi*, from Cueva de la Cadena and the grieta north of the trail to Tortuga Bay, is most closely related to two anchialine congeners from Bermuda. *Strandesia (Neocypris) stocki*, found in large numbers in a small freshwater lava rock pool at Puerto Villamil, has been previously reported from wells, rivers, springs and volcanic and limestone pools in the Caribbean. *Occultocythereis angusta*, a widely dispersed species ranging from Madeira and Bermuda in the Atlantic to Midway and the Marshall Islands in the Pacific, is present in Cueva de la Cadena. The Galápagos endemic *Cyprideis stenopora* inhabits Deep Grieta, a lava rock pool at Puna Tortuga and the anchialine Crater Lake Arcturus on Genovesa. Three new endemic species (*Dolerocypria ensigera*, *Mungava recta* and *Hansacypris galapagosensis*) belonging to circumtropically distributed genera have been described from the grieta north of the trail to Tortuga Bay. *Anchisrocheles darwini* from Deep Grieta is a member of a phytal, cryptic or interstitial genus having a circumtropical distribution in reef environments. While all the preceding are members of the Order Podocopida, a *Cytherella* sp. from the Order Platycopida occurs in Cueva de la Cadena.

3.4. Copepoda

3.4.1. *Expansophria galapagensis* Boxshall and Iliffe, 1990 (Fig. 8)

Characteristic features.-Family Misophriidae. Highly distensible gut modified for an opportunistic gorging strategy. Exoskeletal modifications at cephalosome—first pedigerous somite boundary permits swelling of prosome to accommodate distension of gut; this region of body exposed, not covered by carapace.

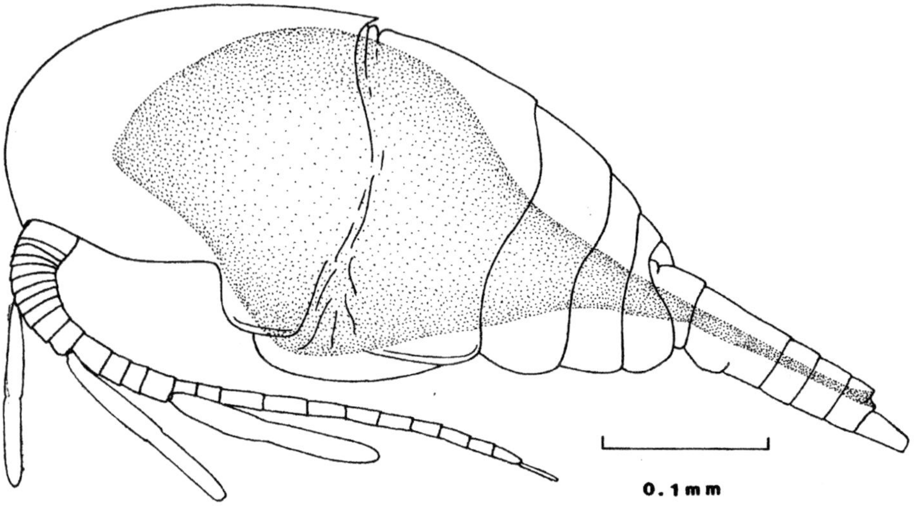

Figure 8. *Expansophria galapagensis*, a troglobitic misophrioid copepod from Santa Cruz, length to 0.66 mm. Note the highly distensible gut.

Size.-Female, 0.66 mm; male, 0.50 mm.

Species range.-Galápagos Islands. Santa Cruz: Deep Grieta at 6 to 12 m water depths, and grieta north of trail to Tortuga Bay at 9 to 10 m depths (Boxshall and Iliffe, 1990).

Number of species in genus.-Three, all troglobitic.

Genus range.-Lanzarote (Canary Islands), Palau (Micronesia) and Galápagos Islands.

Ecological notes.-*Expansophria* is an opportunistic feeder with an enormously distensible gut and probably scavenges material transported into the cave. Specimens were collected with plankton tows in 6 to 12 m depths where salinities ranged from 13 to 22 ppt. During the tows, the walls and rocks floors of the grieta were scraped with the net. Prior to the discovery of misophrioids in anchialine caves (Boxshall and Iliffe, 1986, 1987), this group had been regarded as essentially a deep sea taxon.

Life history.-Life history probably comprising 1 nauplius stage, 5 copepodid stages and the adult, as in other misophrioids. Copepodids II through V and both sexes of adults collected at grieta north of trail to Tortuga Bay.

Closest related genera.-*Expansophria* is the most primitive genus in a group of five anchialine genera comprising, in addition, *Speleophria*, *Boxshallia*, *Dimisophria* and *Palpophria*.

3.4.2. *Enantiosis* sp. Fosshagen and Iliffe, in prep.

Characteristic features.-Family Epacteriscidae. Colorless with a whitish eyespot.

Size.-Male, 1.25–1.45 mm; Female, 1.21–1.39 mm.

Species range.-Galápagos Islands. Santa Cruz: Deep Grieta at 6 to 12 m water depths, Grieta de Caleta la Torta at 17 to 29 m depths, grieta north of the trail to Tortuga Bay at 0 to 10 m depths. Floreana: Post Office Bay Cave at 0 to 2 m depths.

Number of species in genus.-Six, five of which are new and currently being described; all anchialine or crevicular.

Genus range.-Bahamas, Bermuda, Galápagos, Fiji, Palau, Philippines.

Ecological notes.-Specimens were collected with a plankton net towed in open waters. Salinities at collection sites ranged from 1.5 to 29 ppt. The *Enantiosis* sp. from Bermuda is present in crevicular habitats outside of caves.

Life history.-Not known. The females of the known *Enantiosis* spp. are nearly indistinguishable from each other; only the males differ in the 1st antennae and the 5th legs.

Closest related species.-Morphologically, the genus *Enantiosis* is closely related to a new calanoid genus being described from caves in the Canary Islands. The new Galápagos species of *Enantiosis* is most closely related to its Bermuda congener.

3.5. Pisces

3.5.1. *Ogilbia galapagosensis* (Poll and Leleup, 1965)

Characteristic features.-Family Bythitidae. Variable reduction of eyes and pigment. Color, white to pink beige; eyes, blind to reduced but functional. Numerous sensory papillae on head.

Size.-44 to 63 mm.

Species range.-Galápagos Islands. Santa Cruz: grietas near lagoon of the pink flamingos located 2 km west of Tortuga Bay, a grieta close to the seismological station of the Charles Darwin Station (Poll and Leleup, 1965), grietas 1 km inland from Tortuga Bay, and a grieta 500 m inland from Punta Las Palmas at the southwestern corner of the island (Van Mol, 1967).

Number of species in genus.-Six, plus numerous undescribed ones (Cohen and Nielsen, 1978).

Genus range.-From tropical American reefs, freshwater caves in Yucatan, and brackish water in the Galápagos. Probably occurring in the Indo-Pacific as well (Cohen and Nielsen, 1978).

Ecological notes.-*O. galapagosensis*, rare and difficult to observe, inhabits shallow water in or around cracks or gaps between rocks of anchialine pools. Colonies are small, typically 4 to 20 individuals, with each fish having its own rock cavity. It flees into this refuge when disturbed. It is slow moving with an undulating swimming motion. It prefers staying close to rocks or root masses in darker sections of the pools, rarely venturing into more open waters. Salinity in the pools ranged from 5.3 to 15.7 ppt. All pools in which this species is found are situated between 500 and 1000 m from the coast. *O. galapagosensis* apparently feeds on the shrimp *Typhlatya galapagensis* and larvae of dragonflies and mosquitos. In turn, it may be preyed upon by the large shrimp *Macrobrachium americanum* (Van Mol, 1967).

Life history.-Ovoviviparous. Embryos of the closely related *O. deroyi* stay in the uterus to an advanced stage, up to 25 mm. A female of *O. deroyi* has been found with five embryos, while another contained 20 unfertilized eggs. Annual rainfall cycles, which may bring about salinity changes in the grietas, could influence reproductive cycles (Poll and Leleup, 1965).

Closest related species.-Derived from the endemic marine crevicular species *O. deroyi* which inhabits cracks and gaps between rocks near the lower limit of the intertidal zone on Santa Cruz and Santa Fe islands (Poll and Van Mol, 1966). Salinity barriers between coastal and inland populations of the two species may have produced the genetic isolation necessary for speciation.

3.6. Troglophilic and Other Taxa

All of the species listed above are troglobitic/anchialine. However, a brachyuran crab, *Garthiope anchialina* from Cueva de la Cadena, is not morphologically modified and is considered to be a troglophile (Guinot and Iliffe, 1991). *G. anchialina* is most closely related to *G. fraseri*, a species endemic to the Galápagos and present along the coast of many islands including Isabela. *G. anchialina* also has affinities to Caribbean and West Atlantic taxa, namely *G. spinipes* from Bermuda and the Bahamas and *G. barbadensis* from the Caribbean. The tanaidacean *Tanais stanfordi* and the shrimp *Macrobrachium americanum*, present in grietas on Santa Cruz, are also regarded as troglophiles (Peck and Kulalova-Peck, 1986).

Some specimens collected during the 1987 expedition are still under study. A calanoid copepod from the genus *Ridgewayia* was found in anchialine pools on Isabela (A. Fosshagen, pers. comm.). This genus has a world wide distribution in tropical and subtropical waters, but one species in Bermuda is common in caves, even in their most remote sections. A caridean shrimp collected from the deepest sections of Grieta de Caleta la Torta belongs to the genus *Parhippolyte*, an anchialine group widely distributed in the tropical Indo-Pacific (A. Bruce, pers. comm.). A new genus of xanthid crab is present in grietas on Santa Cruz (D. Guinot and J. Garth, pers. comm.).

4. Discussion

4.1. Anchialine Habitat

Three types of anchialine habitats are present in the Galápagos Islands: pools in tectonic faults, lava tubes, and depressions in coastal lava flows. These, however, represent only a small part of the much more extensive crevicular anchialine habitat, inaccessible to humans. It has been observed in the Canary Islands that many troglobitic species first discovered from anchialine lava tubes also inhabit groundwater-filled cracks and fissures in bedrock, accessible through wells (Wilkens *et al.*, 1986). The existence of anchialine fauna in such a broadly integrated environment demonstrates their potential for mobility, enabling them to colonize suitable biotopes in lava flows of recent origin. In addition, their mobility may be vertical as well as horizontal, in this manner directly linking shallow water anchialine habitats with the deep sea.

Salinity ranges in the grietas are considerable. At the deepest points reached by diving, nearly 30 m below sea level, the water was only beginning to approach oceanic salinities. While some anchialine species are euryhaline, other species occur only below the halocline. Still deeper, fully marine groundwaters in the Galápagos are likely to contain further new taxa.

Unfortunately, no long underwater cave systems have been discovered in the Galápagos comparable to the 1.5 km Jameos del Agua lava tube in the Canary Islands or the 3 km Green Bay Cave developed in Bermuda's Pleistocene age eolian limestone. Such large anchialine cave systems provide ample opportunity for study of a larger part of the habitat with an accompanying broader range of environmental variables.

4.2. Biogeographical Affinities

Most of the obligate anchialine fauna in the Galápagos consists of endemic species closely related to Caribbean and/or to deep sea taxa and belong to groups that have other members inhabiting caves (Table 1). Surprisingly, relatively few species are derived from the fauna of South Pacific islands or the west coast of Central or South America. The distant affinities evidenced by the Galápagos fauna are comparable to the situation in the Canary Islands where relationships of anchialine taxa are again to the Caribbean–West Atlantic region rather than to the nearby coast of West Africa.

The relatively well studied and widely distributed genera *Typhlatya* and *Danielopolina* are particularly noteworthy in their biogeographical significance. The anomalous distribution of the cave-limited genus *Typhlatya* over much of the Western Atlantic and the Galápagos has been a topic of speculation by a number of biologists. Chace and Hobbs (1969) suggest that members of this genus "represent relicts of a once much more widespread epigean stock of which no other trace is known to exist. . . . On the basis of the present distribution of the members of *Typhlatya*, there seems no reason to doubt that the Antillean members have been derived from stocks that reached the islands from the Central American–Mexican region, probably in Miocene or Pliocene times." Chace and Manning (1972:5) projected that *T. rogersi* on relatively young, 1–2 million years old, Ascension Island might be derived from "an amphi-Atlantic marine shrimp, perhaps pelagic, which existed into the Pleistocene but did not survive in the oceanic habitat into the present. . . . The close similarities of the known species of *Typhlatya* suggest a common origin, but their relationship may be obscured by convergence." Hobbs *et al.* (1977) concluded that "the presence of *T. galapagensis* on an eastern Pacific island suggests an extension of the range of the ancestral *Typhlatya* stock prior to the closing of the Middle American isthmus."

The two species of *Typhlatya* occupying marine habitats on mid-ocean sea mounts in the Atlantic, *T. iliffei* from Bermuda and *T. rogersi* from Ascension, are closely related (Hart and Manning, 1981), thus suggesting a similar derivation. Iliffe *et al.* (1983) proposed that "the closely related *T. iliffei* (on Bermuda) and *T. rogersi* (on Ascension) represent an atyid stock that originated at about the same time and place as the genus *Atya* (perhaps as early as the middle Mesozoic), surviving on submerged and emergent sea mounts along or associated with the Mid-Atlantic Ridge." It follows that the ancestral form of *Typhlatya* may have originally entered the cave environment directly from deep waters via cracks and fissures (micro-caves) on submerged sea mount slopes during the Mesozoic (Hart *et al.*, 1985;

Table 1. Composition of Troglobitic Anchialine Fauna of the Galápagos.

	Endemic	Eyes/ pigment reduced	Other anchialine affinities	Caribbean affinity	Deep sea affinity
Caridea					
Typhlatya galapagensis	+	+	+	+	
Amphipoda					
Galapsiellus leleuporum	+	+	+		
Anchialella vulcanella	+	+	+		
Valettieta cavernicola	+	+			+
Antronicippe serrata	+	+	+	+	+
Ostracoda					
Skogsbergia galapagensis	+	+		+	
Danielopolina styx	+	+	+	+	+
Podocopida spp.	+		+	+	
Copepoda					
Expansophria galapagensis	+	+	+		+
Enantiosis sp.	+	+	+	+	
Pisces					
Ogilbia galapagosensis	+	+	+	+	

Manning *et al.*, 1986). Following island emergence, subsequent dispersal into lower salinity and or even fresh groundwaters could have occurred (Iliffe, 1986).

The ostracod *Danielopolina* has a distribution reminiscent of *Typhlatya*. Both genera have species from Cuba, the Yucatan Peninsula, the Bahamas, and the Galápagos, but additionally, *Danielopolina* is represented in the Canary Islands and even includes a deep sea member. Ancestors of *Danielopolina* are represented in the fossil record from the Permian in Greece and the Lower and Upper Jurassic in Germany. According to Kornicker and Iliffe (1989), "*Danielopolina* is an apparently ancient form having its origins on the Tethys Sea, and dispersal of its ancestors was greatly influenced by plate movement. The present distribution of the genus in marine caves and in the deep sea suggests that the descendants of an initially shelf form in the Mesozoic later found refuge in caves and the deep sea; possibly shelf forms could not survive increased competition from abundant new post-tethyan shelf taxa."

Other anchialine taxa from the Galápagos are closely related to West Atlantic forms. Two species (*Iliffeoecia iliffei* and *Paracypris crispa*) and one genus (*Kareloecia*) of podocopid ostracods from the Galápagos also inhabit anchialine caves in Bermuda. In addition, *Strandesia* (*Neocypris*) *stocki* had been previously known only from a number of Caribbean islands. Galápagos podocopids, including *Anchisrocheles darwini*, *Occultocythereis angusta*, *Dolerocypria ensigera*, *Mungava recta*, and *Hansacypris galapagosensis*, are members of circumtropically distributed genera. The myodocopid ostracod *Skogsbergia galapagensis* could be derived from the open water species *S. lerneri*, still inhabiting the Gulf of Mexico and the Caribbean. The only examples of species of Indo-Pacific origin are the amphipods *Galapsiellus leleuporum* and *Antronicippe serrata*, both derived from the widespread Indo-Pacific genus *Eriopisa*, and the shrimp *Parhippolyte*, a broadly distributed anchialine genus.

4.3. Additional Deep-Sea Affinities

Even more remarkable biogeographical affinities exist for other species. Prior to discoveries in the Galápagos, the misophrioid copepod genus *Expansophria* consisted of only two species—one inhabiting a marine lava tube in the Canary Islands and the other from an anchialine limestone cave in Palau, on almost opposite sides of the Earth (Boxshall and Iliffe, 1987). Until their discovery from marine caves in Bermuda, misophrioid copepods had been regarded as a deep sea group (Boxshall and Iliffe, 1986). Based on the complex set of characters exhibited by deep sea misophrioid genera, these taxa are viewed as having a bathypelagic origin (Boxshall, 1983). Boxshall and Iliffe (1986) have suggested that the misophrioid copepods inhabiting anchialine caves in Bermuda are derived from deep sea taxa that colonized crevicular microhabitats or even lava tubes on the slopes of this sea mount. With the discoveries of 5 additional species of troglobitic misophrioids, Boxshall and Iliffe (1987) stated that these anchialine misophrioids "represent descendants of deep-sea forms which became separated by vertical vicariance events. . . . The distribution of *Expansophria*, with one Atlantic and one Indo-West Pacific species, is difficult to explain by the deep-sea emergence route unless the genus was widespread in the deep sea." The discovery of a third species of *Expansophria* in the Galápagos further documents the highly anomalous distribution of the genus.

In addition to the bathyal/abyssal affinities of *Expansophria* and *Danielopolina*, the amphipods *Valettietta cavernicola* and *Antronicippe serrata* belong to primarily, though not exclusively, deep-sea families. Such close affinities to deep-sea species have led to the obvious suggestion of a direct deep-water origin for anchialine fauna (Iliffe *et al.*, 1983, 1984). Several lines of circumstantial evidence provide additional support for this novel hypothesis. Similarities between cave and deep sea environments (e.g., lightless, stable conditions, limited food supply) could result in a preadaptation of deep-sea taxa for invasion of crevicular habitats. Submersible and other deep-sea investigations have revealed the presence of not only deep-water crevicular environments, but both limestone caves and lava tubes as well. The rare occurrence of some troglobitic taxa in well studied anchialine caves suggests that they are originating from some adjacent and interconnected habitat. During low sea stands of the Pleistocene when most of today's anchialine caves were subaerial, old species must have survived in suitable refugial habitats that now are at considerable depths.

In contradiction to the above conclusions, Stock (1986) claims that a direct origin of anchialine species from deep-water ancestors is unlikely. Citing as evidence, the extinction of Tethyan deep-sea biotas during the Oligocene and cladistic analyses, Stock instead suggests that both cave and deep-sea species are derived from infra- or circa-littoral forms.

In order to settle this controversy, investigations of deep-sea caves and crevicular habitats need to be performed in order to learn what species, if any, are present and where their affinities lie. If indeed deep-water crevicular species do exist and can migrate into shallow water anchialine habitats, then the groundwater fauna of oceanic islands such as the Galápagos may be substantially older than the islands themselves.

Corroborating evidence supports this latter assumption. The date for the initial emergence of the Galápagos Islands has been placed at about 4–5 million years BP (Hall, 1983). Yet, the original entry of the ancestors of the Galápagos anchialine fauna into the Pacific must have preceded the closing of the Panama land bridge approximately 5 million but possibly as early as 10 million years BP (Smith, 1985). Since no other known intermediary anchialine habitats exist in this area of the Pacific today or are likely to have existed during the period in question, it can be presumed that the Galápagos sea mount was the original colonization site for these taxa. In any event, the oldest inhabitants of the Galápagos Islands are likely to be its anchialine fauna.

4.4. Comparisons with Other Oceanic Islands

A number of remarkable similarities exist among the fauna of oceanic sea amount islands including the Galápagos, Canary Islands, and Bermuda. Two genera (*Danielopolina* and *Expansophria*) are common to both Galápagos and Canary Island anchialine caves, while four other genera (*Typhlatya*, *Enantiosis*, *Ridgewayia*, and *Kareloecia*) and two species (*Iliffeoecia iliffei* and *Paracypris crispa*) inhabit both the Galápagos and Bermuda anchialine habitats. Such biogeographical affinities between remote oceanic sea mounts points to a common origin for their fauna at a period when closer ties existed between them.

Despite the fact that a shallow-water connection has never existed between the various islands in the Galápagos, at least five obligate anchialine species are found on more than one island. *Typhlatya galapagensis*, *Galapsiellus leleuporum*, and *Kareloecia woutersi* are present on both Santa Cruz and Isabela, while *Enantiosis galapagensis* inhabits both Santa Cruz and Floreana. Aside from *Cyprideis stenopora* present in anchialine habitats on Santa Cruz, Isabela, and Genovesa and possibly dispersed by migratory birds (Maddocks and Iliffe, 1991), these distributions further suggest the existence of deep-water crevicular connections between the islands. Considering the limited extent of present studies, further explorations and collections would surely broaden the range of many anchialine species within these islands.

5. Conclusions

A rich and diverse fauna inhabits anchialine habitats, particularly deep pools in tectonic faults, in the Galápagos. The highly fissured, fractured nature of this solidified lava has produced an integrated groundwater biotope capable of permitting local dispersal of anchialine fauna. A number of taxa, including *Danielopolina*, *Typhlatya* and *Expansophria*, can be considered as relicts. Close taxonomic relationships exist with the fauna of Caribbean–West Atlantic caves, with deep-sea fauna or with both. Were it not for this island's position in the Pacific, the anchialine fauna of the Galápagos could be said to be typical of that of a mid-Atlantic oceanic sea mount. All of the Galápagos' Caribbean-derived fauna must have entered the Pacific prior to the closing of the Panama land bridge at the end of the Miocene. However, since a Mesozoic colonization of North Atlantic caves has been proposed for many of the same genera that inhabit the Galápagos groundwaters, a much older origin is possible. If this is indeed the case, then initial colonization of the Galápagos sea mount must have taken place prior to the emergence of the island above the surface of the sea. A deep-sea origin for at least some of the taxa inhabiting anchialine environments in the Galápagos cannot be ruled out.

ACKNOWLEDGMENTS. Cave collections in the Galápagos Islands by T. M. Iliffe in 1987 were supported by National Science Foundation grants BSR-8215672 and BSR-8417494. We especially thank the Director and staff of the Galápagos National Park and the Charles Darwin Research Station, without whose invaluable assistance this study would not have been possible. Drs. Geoffrey Boxshall, A. J. Bruce, Audun Fosshagen, J. S. Garth, Danièle Guinot, Louis Kornicker, Rosalie Maddocks, and Jan Stock identified specimens collected during the 1987 expedition and provided additional information for the preparation of this paper. Mr. Serban Sarbu assisted with the preparation of cave maps, while Ms. Aline Wardle translated papers from French. Dr. Louis Kornicker and Ms. Molly Ryan generously provided the illustration of *Danielopolina* and Dr. Geoffrey Boxshall contributed that of *Expansophria*.

References

Balazs, D., 1975, Lava tubes on the Galápagos Islands, *NSS Bull.* **37**:1–4.

Barnard, J. L., 1976, Affinities of *Paraniphargus lelouparum* Monod, a blind anchialine amphipod (Crustacea) from the Galápagos Islands, *Proc. Biol. Soc. Wash.* **89**:421–432.

Barnard, J. L., 1979, Littoral gammaridean Amphipoda from the Gulf of California and the Galápagos Islands, *Smithson. Contrib. Zool.* **271**:1–149.

Bowman, T. E., and Iliffe, T. M., 1985, *Mictocaris halope*, a new unusual peracaridan crustacean from marine caves on Bermuda, *J. Crust. Biol.* **5**:58–73.

Boxshall, G. A., 1983, Three new genera of misophrioid copepods from the near-bottom plankton community in the North Atlantic Ocean, *Bull. Br. Mus. (Nat. Hist.), Zool.* **44**:103–124.

Boxshall, G. A., and Iliffe, T. M., 1986, New cave-dwelling misophrioids (Crustacea: Copepoda) from Bermuda, *Sarsia* **70**:54–64.

Boxshall, G. A., and Iliffe, T. M., 1987, Three new genera and five new species of misophrioid copepods (Crustacea) from anchialine caves on Indo-West Pacific and North Atlantic Islands, *Zool. J. Linn. Soc., Lond.* **91**:223–252.

Boxshall, G. A., and Iliffe, T. M., 1990, Three new species of misophrioid copepods from oceanic islands, *J. Nat. Hist.* **24**:595–613.

Chace, F. A., Jr., and Hobbs, H. H., Jr., 1969, The freshwater and terrestrial decapod crustaceans of the West Indies with special reference to Dominica, *U.S. Natl. Mus. Bull.* **292**:1–258.

Chace, F. A., Jr., and Manning, R. B., 1972, Two new caridean shrimps, one representing a new family, from marine pools on Ascension Island (Crustacea: Decapoda: Natantia), *Smithson. Contrib. Zool.* **131**:1–18.

Cohen, A. C., 1983, Rearing and postembryonic development of the myodocopid ostracod *Skogsbergia lerneri* from coral reefs of Belize and the Bahamas, *J. Crust. Biol.* **3**:235–256.

Cohen, D. M., and Nielsen, J. G., 1978, Guide to the identification of genera of the fish order Ophidiiformes with a tentative classification of the order, *NOAA Technical Report NMFS Circular 417* (December 1978), p. 72.

Fosshagen, A., and Iliffe, T. M., 1985, Two new genera of Calanoida and a new order of Copepoda, Platycopioida, from marine caves on Bermuda, *Sarsia* **70**:345–358.

Guinot, D., and T. M. Iliffe, 1991, *Garthiope anchialina* sp. nov., espèce anchialine des Galápagos, île Isabela, Cueva de la Cadena, avec des remarques sur la fauna carcinologique anchialine (Crustacea Decapoda Brachyura), *Bull. Mus. Natl. Hist. Nat., Paris* 4e sér., *12*, sect. A, **(3–4)**:607–621.

Hall, M. L., 1983, Origin of Española Island and the age of terrestrial life on the Galápagos Islands, *Science* **221**:545–547.
Hart, C. W., Jr., and Manning, R. B., 1981, The cavernicolous caridean shrimps of Bermuda (Alpheidae, Hippolytidae, and Atyidae), *J. Crust. Biol.* **1**:441–456.
Hart, C. W., Jr., and Manning, R. B., 1986, Two new shrimps (Procaridae and Agostocaridae, new family) from marine caves of the western North Atlantic, *J. Crust. Biol.* **6**:408–416.
Hart, C. W., Jr., Manning, R. B., and Iliffe, T. M., 1985, The fauna of Atlantic marine caves: Evidence of dispersal by sea floor spreading while maintaining ties to deep waters, *Proc. Biol. Soc. Wash.* **98**:288–292.
Hobbs, H. H., III, and Hobbs, H. H., Jr., 1976, On the troglobitic shrimps of the Yucatan Peninsula, Mexico (Decapoda: Atyidae and Palaemonidae), *Smithson. Contrib. Zool.* **240**:1–23.
Hobbs, H. H., Jr., Hobbs, H. H., III, and Daniels, M. A., 1977, A review of the troglobitic decapod crustaceans of the Americans, *Smithson. Contrib. Zool.* **244**:1–183.
Holthuis, L. B., 1973, Caridean shrimps found in land-locked saltwater pools at four Indo-West Pacific localities (Sinai Peninsula, Funafuti Atoll, Maui and Hawaii Islands), with the description of one new genus and four new species, *Zool. Verh. (Leiden)* **128**:1–48, pls. 1–7.
Iliffe, T. M., 1986, The Zonation Model for the evolution of aquatic faunas in anchialine caves, *Stygologia* **2**:2–9.
Iliffe, T. M., Hart, C. W., Jr., and Manning, R. B., 1983, Biogeography and the caves of Bermuda, *Nature* **302**:141–142.
Iliffe, T. M., Wilkens, H., Parzefall, J., and Williams, D., 1984, Marine lava cave fauna: Composition, biogeography and origins, *Science* **225**:309–311.
Kornicker, L. S., 1984, Cypridinidae of the continental shelves of southeastern North America, the northern Gulf of Mexico, and the West Indies (Ostracoda: Myodocopina), *Smithson. Contrib. Zool.* **401**:1–37.
Kornicker, L. S., and Iliffe, T. M., 1989, Troglobitic Ostracoda (Myodocopa: Cypridinidae, Thaumatocyprididae) from anchialine pools on Santa Cruz Island, Galápagos Islands, *Smithson. Contrib. Zool.* **483**:1–38.
Leleup, N., 1968, 1970, 1976, *Mission Zoologique Belge aux Iles Galápagos et en Ecuador (N. et J. Leleup, 1964–1965)*, Resultata Scientifiques Koninklijk Museum voor Midden-Afrika, Premier partie, 1968, 272 pp.; Deuxieme partie, 1970, 237 pp; Troisième partie, 1976, 355 pp.
Maddocks, R. F., 1991a, Revision of the Family Pontocyprididae (Ostracoda), with new anchialine species and genera from the Galápagos Islands, *Zool. J. Linn. Soc., Lond.*
Maddocks, R. F., 1991b, Anchialine Cyprididae (Ostracoda) from the Galápagos Islands, with a review of the Subfamily Paracypridinae, *Zool. J. Linn. Soc., Lond.*
Maddocks, R. F., and Iliffe, T. M., 1986, Podocopid Ostracoda of Bermudian caves, *Stygologia* **2**:26–76.
Maddocks, R. F., and Iliffe, T. M., 1991, Anchialine podocopid Ostracoda of the Galápagos Islands, *Zool. J. Linn. Soc., Lond.*
Manning, R. B., Hart, C. W., Jr., and Iliffe, T. M., 1986, Mesozoic relicts in marine caves of Bermuda, *Stygologia* **2**:156–165.
McBirney, A. R., and Williams, H., 1969, Geology and petrology of the Galápagos Islands, *Geol. Soc. Am. Mem.* **118**:1–197.
Monod, T., 1970, Sur quelques crustacés malacostracés des Iles Galápagos, in: *Mission Zoologique Belge aux Iles Galápagos et en Ecuador (N. et J. Leleup, 1964–1965)* **2**:11–53.
Monod, T., and Cals, P., 1970, Sur une espéce nouvelle de crevette cavernicole: *Typhlatya galapagensis* (Decapoda Natantia; Atyidae), in: *Mission Zoologique Belge aux Iles Galápagos et en Ecuador (N. et J. Leleup, 1964–1965)* **2**:57–103.
Montoriol-Pous, J., and De Mier, J., 1977, Contribución al conocimiento vulcanoespeleológico de la isla Santa Cruz (Galápagos, Ecuador), *Speleon* **23**:75–91.
Montoriol-Pous, J., and Escolà, O., 1975, Contribución al conocimiento vulcanoespeleológico de la isla Floreana (Galápagos, Ecuador), *Speleon*, monographía **1**:57–67.
Montoriol-Pous, J., and Escolà, O., 1978, Contribución al conocimiento vulcanoespeleológico de la isla Isabela (Galápagos, Ecuador), *Speleon* **24**:101–110.
Peck, S. B., and Kulalova-Peck, J., 1986, Preliminary summary of the subterranean fauna of the Galápagos Islands, Ecuador, *Proc. 9th Congr. Int. Espeleol., Barcelona, Spain* **2**:164–169.
Peck, S. B., and Peck, J., 1986, The Galápagos Islands: Volcanic caves and cave fauna of the Galápagos Islands, *Can. Caver* **18**:42–52.

Poll, M., and Leleup, N., 1965, Un poisson aveugle nouveau de la famille des Brotulidae provenant des îles Galápagos. *Bull. Acad. R. Belg. Cl. Sci. Ser.* 5 **51**:461–474.
Poll, M., and Van Mol, J. J., 1966, Au sujet d'une espéce inconnue de Brotulidae littoral des îles Galápagos, apparentée à l'espèce aveugle *Caecogilbia galapagosensis* Poll et Leleup, *Bull. Acad. R. Belg. Cl. Sci. Ser.* 5 **52**:1444–1461.
Simkin, T., 1984, Geology of the Galápagos Islands, in: *Key Environments: Galápagos*, (R. Perry, ed.), pp. 15–41, Pergamon Press Ltd., Oxford, England.
Sket, B., 1979, *Atlantasellus cavernicolus* n.gen. n.sp. Isopoda Asellota, Atlantasellidae n.fam. from Bermuda, *Biol. Vestn. (Ljubljana)* **27**:175–183.
Smith, D. L., 1985, Caribbean Plate relative motions, in: *The Great American Biotic Interface* (F. G. Stehli and S. D. Webb, eds.), pp. 17–48, Plenum Press, New York.
Steadman, D. W., 1986, Holocene vertebrate fossils from Isla Floreana, Galápagos, *Smithson. Contrib. Zool.* **413**:1–103.
Stock, J., 1986, Deep sea origin of cave faunas, an unlikely supposition, *Stygologia* **2**:105–111.
Stock, J., and Iliffe, T. M., 1990, Amphipod crustaceans from anchihaline cave waters of the Galápagos Islands, *Zool. J. Linn. Soc., Lond.* **98**:141–160.
Stock, J., Iliffe, T. M., and Williams, D., 1986, The concept "anchialine" reconsidered, *Stygologia* **2**: 90–92.
Van Mol, J. J., 1967, Écologie comparée de deux espéces de Brotilidae (Pisces) des îles Galápagos: *Caecogilbia deroyi* Poll et Van Mol 1967 et *C. galapagosensis* Poll et Leleup 1965, *Bull. Acad. R. Belg. Cl. Sci. Ser.* 5 **53**:232–248.
Wilkens, H., Parzefall, J., and Iliffe, T. M., 1986, Origin and age of the marine stygofauna of Lanzarote, Canary Islands, *Mitt. Hamb. Zool. Mus. Inst.* **83**:223–230.
Yager, J., 1981, Remipedia, a new class of Crustacea from a marine cave in the Bahamas, *J. Crust. Biol.* **1**:328–333.

VI

Mollusca

Chapter 11

The Marine Mollusks of the Galápagos

Determinants of Insular Marine Faunas

E. ALISON KAY

1. Introduction

Charles Darwin, on his visit to the Galápagos Islands, celebrated the "eminently curious" creatures of the land, but paid scant attention to marine organisms. Two comments, however, indicate that Darwin recognized that the marine mollusks were also unique: 1) ". . . of the ninety shells no less than forty-seven are unknown elsewhere—a wonderful fact, considering how widely distributed sea-shells generally are" (Darwin, 1906, p. 376); and 2) "We have . . . in this quarter of the world, three great conchological sea-provinces, quite distinct . . ." (Darwin, 1906, p. 376).

Darwin, in the course of that brief visit to the Galápagos had noticed the relatively high proportion of endemism, now frequently cited for the Galápagos marine fauna (Brusca, 1987; McCosker and Rosenblatt, 1984; Wellington, 1984), and he had also recognized that the marine molluscan fauna represented a complex of biogeographic elements from the Panamic and Indo-Pacific faunas. Subsequent observers (Hedgepeth, 1969; Nybakken 1979; Wellington, 1984) noted a third feature of the Galápagos marine molluscan biota: some prominent mainland species are absent, others are more abundant in the Galápagos than they are on mainland coasts; some species are present on one island but not on another. And at least one species, *Cassis tenuis*, "here attains a vigorous growth and frequently an extraordinary size . . ." (Stearns, 1893).

The molluscan fauna of the Galápagos invites comparison with that of the

E. ALISON KAY • Department of Zoology, University of Hawaii, Honolulu, Hawaii 96822.

Hawaiian Islands which is also characterized by a relatively high proportion of endemism, a mosaic of biogeographical elements, and peculiarities in occurrence, size, and abundance (Kay, 1967; 1979; Kay and Palumbi, 1987).

Like the Galápagos, the Hawaiian Islands are of volcanic origin, rising thousands of meters from the crustal platform of the sea floor. The Galápagos, lying between 1°N and 1°S of the equator, consist of 13 major islands and numerous lesser islands and rocks; the western islands are dated at < 2–3 my, the eastern islands are <1 my (Hickman and Lipps, 1985). The Hawaiian Islands stretch as a chain west-northwest between 20°N and 29°N, and include eight high volcanic islands, and more than 30 lesser volcanic islets, atolls, and drowned banks. The youngest volcanic island (Hawaii) is <1 my, the oldest (Gardner) perhaps 12–14 my. Each of the archipelagoes is separated by thousands of kilometers from its neighbors: the Galápagos are about 1050 km from the west American coast; the Hawaiian Islands are 4000 km from the North American coast; and, to the southwest, 800 km from Johnston Island and 3200 km from the Marshall Islands. The Galápagos lie in a convergence of three water masses: the cool subtropical waters of the Equatorial Undercurrent (Cromwell Current); the temperate waters of the South Equatorial Current in which the islands lie throughout the year; and the warm waters from the region of the Panama Bight that are especially intense during El Niño years. Two subtropical water masses obtain in the Hawaiian Archipelago, the North Pacific Equatorial Current which flows from east to west just south of the islands, and the northeasterly flowing Subtropical Countercurrent which intermittently intersects the islands, especially during summer and fall (Grigg, 1981).

In this chapter I explore the similarities and differences between the marine molluscan faunas of these two Pacific archipelagoes. The comparison of the two insular faunas may help to test several theories of Pacific island biogeography such as those of Springer (1982), Grigg (1983), and Vermeij (1987), and help develop a unifying theory of the biogeography of marine organisms in the tropical Pacific.

2. Materials and Methods

In this assessment of the Galápagos marine molluscan fauna I have relied heavily on the faunal records and lists of Keen (1971), Finet (1985), Bernard (1983, 1988), Shasky (1988), and Emerson (1978, 1983). These records have been augmented by my own study of marine mollusks *in situ* in the uplift area and shallow waters of Urvina Bay, Isabela Island in July–August 1987, and of the collections of mollusks made at that time. I have also worked with the collections of Galápagos mollusks at the Charles Darwin Research Station and in the Angermeyer collection, both in Puerto Ayora, Santa Cruz Island, and in the collections in the California Academy of Science, San Francisco. Data for the Hawaiian marine mollusks are drawn from Kay (1979).

Three groups of marine mollusks in the Galápagos fauna are omitted in the analyses. Mollusks recorded from deep water (> 100 m) are not included because data on their distributional ranges are not sufficiently authoritative to establish the ranges of these species. The bivalve leptonaceans, except for *Kellia suborbicularis*, are excluded because many of them are of uncertain taxonomic status. The Opisthobranchia are treated by Gosliner (Chapter 13, this volume).

Indices of selectivity used in the comparisons of the Galápagos and Hawaiian faunas follow Vermeij's (1987) analysis: each fauna was divided into categories (systematic: gastropods, bivalves, *etcetera*; ecological: sand dwellers, rock dwellers, *etcetera*) and percentages of species in each category that do not reach either the Galápagos or the Hawaiian Islands from a source area were calculated.

Terminology for the biogeographical relationships includes the following distinctions. Panamic refers to the warm water Eastern Pacific faunal region (Briggs, 1974; Brusca, 1987). Indo-Pacific is used *sensu* Springer (1982): "The Indian Ocean including contiguous seas, and the Pacific Ocean as far east as Easter Island but excluding the area occupied by the coast and offshore islands . . . of the Western Hemisphere." Cosmopolitan species are regarded as widespread species not limited to tropical or temperate regions. Circumtropical species are defined *sensu* Gosliner (1987) as those recorded from at least some portion of each of the Atlantic, Indian, and Pacific oceans.

3. Results

3.1. Faunal Composition of the Galápagos

The marine molluscan fauna of the Galápagos Islands probably numbers at least 800 species. The current list (summarized in Table 1) is composed of about 700 species, of which 640 are recorded from depths of less than 100 m, and the remainder are apparently known only from depths of from 200 to 1500 m. New records are continually being added to the faunal list (see, for example, Shasky, 1988), and if they continue at the present rate of about five species per year (including the unnamed species now in collections), an estimate of 800 species does not seem unreasonable.

Gastropod species outnumber bivalves by about three-to-one (Table 2). The greater proportion of gastropods, about 80%, are caenogastropods, with the neotaenioglossans and neogastropods about equally divided (47% and 53% respectively) in numbers of species. Four gastropod families are recorded with more than 20 species each: Turridae (58), Muricidae (25), Columbellidae (23), and Epitoniidae (23). Thirty-six families in the Bivalvia are represented. The Veneridae (22), Arcidae (14), Pectinidae (13), and Mytilidae (12) are represented by the largest numbers of species.

The Polyplacophora, Scaphopoda, and Cephalopoda account for less than 4% of the marine mollusks (Table 1). Thirteen of the 14 chiton species are found in shallow water where they form a conspicuous component of the intertidal biota. Of the five scaphopod species, three are recorded only from deep water. There are four cephalopod species, three species of *Octopus* recorded from relatively shallow water, and *Argonauta pacifica* Dall, 1869, which is pelagic.

3.2. Faunal Characteristics

It has long been noticed that the Galápagos marine fauna has fewer species than are found on the west American coast: Stearns (1893) wrote that "the olives, so

Table 1. Family Composition of the Galápagos Marine Molluscan Fauna

Class–Order Family	Estimated no. species	No. endemic
Polyplacophora		
Lepidopleuridae	2	1
Ischnochitonidae	2	1
Chaetopleuridae	3	1
Chitonidae	4	
Ancanthochitonidae	4	2
Archaeogastropoda		
Scissurellidae	2	1
Fissurellidae	12	6
Acmaeidae	6	4
Trochidae	13	6
Skeneidae	3	3
Liotiidae	3	1
Turbinidae	1	1
Phasianellidae	2	1
Neritidae		
Neritidae	2	
Caenogastropoda Discopoda		
Littorinidae	9	1
Barleeidae	5	4
Rissoidae	17	6
Vitrinellidae	12	
Turritellidae	2	
Caecidae	6	
Modulidae	1	
Vermetidae	2	
Planaxidae	2	
Cerithiidae	6	
Potamidae	1	1
Strombidae	2	
Hipponicidae	3	
Fossaridae	7	1
Vanikoridae	2	1
Calyptraeidae	7	
Capulidae	2	
Naticidae	17	4
Lamellariidae	1	
Triviidae	7	2
Cypraeidae	8	
Ovulidae	4	
Tonnidae	2	
Cassididae	4	
Ranellidae	11	1
Bursidae	3	
Neogastropoda		
Muricidae	25	5
Thaididae	13	1
Coralliophilidae	10	2

continued

Table 1. (*continued*)

Class–Order family	Estimated no. species	No. endemic
Colubrariidae	3	1
Buccinidae	12	2
Columbellidae	23	2
Nassariidae	9	4
Fasciolariidae	9	2
Turbinellidae	1	1
Volutidae	1	1
Olividae	6	
Marginellidae	14	5
Harpidae	2	
Cancellariidae	9	2
Conidae	19	1
Terebridae	11	1
Turridae	58	30
Ptenoglossa		
Cerithiopsidae	10	2
Triphoridae	15	4
Epitoniidae	23	1
Janthinidae	3	
Eulimidae	10	4
Heterogastropoda		
Architectonicidae	4	
Pyramidellidae	12	4
Rissoellidae	2	
Opisthobranchia	(not listed)	
Pulmonata		
Onchididiidae	2	2
Melampidae	7	1
Trimusculidae	1	
Siphonariidae	1	
Bivalvia		
Anomiidae	1	
Arcidae	14	1
Cardiidae	5	
Carditidae	5	
Chamidae	5	1
Condylocardiidae	1	
Corbulidae	2	
Crassitellidae	3	
Galeommatidae	4	
Gastrochaenidae	2	1
Glycymerididae	3	
Isognomonidae	1	
Lasaeidae	2	
Lepitonidae	1	
Limidae	4 (1*d*)	
Lucinidae	7	

continued

Table 1. (*continued*)

Class–Order family	Estimated no. species	No. endemic
Mactridae	3	
Malleidae	1	
Mytilidae	12	
Nuculanidae	3 (1d)	
Ostreidae	3	3
Pectinidae	13 (4*d*)	
Petricolidae	1	
Pholadidae	2	
Pinnidae	3	
Plicatulidae	2	1
Psammobiidae	2	
Pteriidae	2	
Semelidae	6	2
Sportellidae	1	
Tellinidae	11 (1d)	1
Thraciidae	1	1
Ungulinidae	1	
Veneridae	22	3
Verticordidae	8	
Scaphopoda		
Siphonodentaliidae	1 (*d*)	
Dentaliidae	4 (2*d*)	
Cephalopoda		
Octopodidae	3	
Argonautidae	1	

[a]*d* represents deep–water (>100m) species.

common on the mainland . . ., seem to be of rare occurrence . . ."; Hedgepeth (1969) commented on the lack of littorines, limpets, and *Siphonaria* on intertidal shores; Nybakken (1979) noted that common mainland *Conus* species are absent in the Galápagos; and Nesbitt and Pitt (1986) remarked that "far fewer *Nassarius* species are on islands than on the mainland." Indeed, of the 3000 estimated Panamic species of mollusks (Keen, 1971) less than one-third occur in the Galápagos, and 14% of the families have not been reported from the Galápagos.

The Galápagos mollusks compared with two mainland coastal molluscan faunal assemblages (Table 3) show that in the Galápagos there are proportionately more gastropod than bivalve species and nearly equal numbers of epifaunal and infaunal bivalve species. Analysis of the Galápagos mollusks in terms of Vermeij's (1987) selectivity index (Table 4), which compares percentages of species in the source faunas with those in the "target" faunas, indicate that selectivity between the mainland and the islands is more effective against bivalves than gastropods, against archaeogastropods than neotaenioglossans or neogastropods, and against gastropods and bivalves that are burrowers or that live in sand than those which have other modes of life.

Other characteristics of the marine mollusks of the Galápagos which differ from those of the west American coast include population densities, size, and

Table 2. Composition of the Galápagos Marine Molluscan Fauna by Class and Order

	Taxonomic group	No. of species 100 m	(> 100 m)	Percent of biota	Percent of gastropoda
Class	Polyplacophora	13	(1)	2	
Class	Cephalopoda	4		< 1	
Class	Scaphopoda	2	(3)	< 1	
Class	Bivalvia	153	(8)	24	
Class	Gastropoda	468		73	
Subclass	Prosobranchia				
Superorder	Archaeogastropoda	31	(10)		7
Superorder	Neritomorpha	3			< 1
Superorder	Caenogastropoda				
Order	Neotaenioglossa	219	(5)	34	47
Suborder	Discopoda	136	(4)		29
Suborder	Heteropoda	20			4
Suborder	Ptenoglossa	63	(1)		14
Order	Neogastropoda	206	(31)	32	44
Subclass	Opisthobranchia		(see Chapter XX, this volume)		
Subclass	Pulmonata	9			1
Total		640	(58)		

patchy distribution. Several species are more common in the Galápagos than on the mainland, among them *Muricopsis princeps* (Broderip, 1833), *Purpura columellaris* (Lamarck, 1822), *Thais planospira* (Lamarck, 1822), *Conus tiaratus* Sowerby, 1833, and *C. lucidus* Wood, 1828. Some species appear to reach a larger size in the Galápagos than elsewhere: mean length of Galápagos *Thais planospira* is 45 mm, that of *T. planospira* on the mainland is 33 mm. In still other species, the shells are smaller in the Galápagos than they are on the mainland: mean length of Galápagos *Modiolus capax* (Conrad, 1837) is 62 mm, mean length of mainland shells is 109 mm. Nybakken (1979) has also found significantly fewer *Conus* species per habitat in the Galápagos compared with other Indo-Pacific areas although there is no indication of dietary specialization.

A well known difficulty in assembling a faunal list of insular marine mollusks is that records often consist of single specimens or reports of single sightings. Of the six species of Littorinidae listed by Finet (1985), for example, only three species are regularly reported, *Nodilittorina porcata* (Philippi, 1845) from mangroves, and *Nodilittorina galapagiensis* (Stearns, 1892) and *Nodilittorina modesta* (Philippi, 1846) in the rocky intertidal. Hedgepeth (1969) recorded *Nerita scabricostata* Lamarck, 1822 in the rocky intertidal of Academy Bay, Santa Cruz; Darwin Bay, Tower; Punta Espinosa, Fernandina; and Bahia Elizabeth, Isabela, but did not find it at Isla Plazas or Santa Maria. There are at least three explanations for these singular observations (apart from that of possible taxonomic misidentification): 1) the records are of species restricted in their distribution to particular islands or shorelines (i.e., patchy distribution); 2) the records are of chance occurrences or sporadic influx by transient water masses; and 3) the records are correct but the community or assemblage had subsequently become extinct.

There is some evidence of inter-island regionalism. Darwin and Wolf, the two

Table 3. Characteristics of West American Mainland, Galápagos, and Hawaiian Island Marine Molluscan Assemblages[a,b]

Category	Sonora	Bahia de Los Angeles	Galápagos	Hawaiian Islands
Number of species	559	586	666	826
Taxonomic Groups:				
Gastropod: bivalve ratio	62:31	61:39	76:24	86:14
% Archaeogastropoda	6	10	8	5
% Neotaenioglossa	34	47	43	43
% Neogastropoda	57	41	46	47
Habitat Groups:				
% Epifaunal bivalves	31	27	45	52
% Infaunal bivalves	69	70	54	40
% Epifaunal gastropods	82	82	84	74
% Infaunal gastropods	17	16	12	26
% Predatory gastropods	61	42	48	55
Endemism:				
% Endemism	–	–	18	20
% Archaeogastropod endemism	–	–	45	39
% High shoreline endemism	–	–	21	26
Number of biographic regions	1	1	4	4

[a]Sonora data from DuShane and Poorman, 1967; Bahia de Los Angeles data from Coan, 1968; McLean, 1961; Poorman, 1978; Hawaiian data from Kay, 1979.
[b]Opisthobranchs are omitted.

northernmost islands, are subject to more tropical water temperatures than are the other islands (Brennan, 1973) and the fish fauna in the region is dominated by western Pacific and pelagic species (Wellington, 1984). The western shores of Isabela and Fernandina and the southern islands of Floreana and Española are influenced by upwelling which does not affect other islands in the archipelago, but which does apparently affect both fishes and invertebrates on those coasts (Wellington, 1984).

Evidence to support the view of continuing but sporadic immigration and extinction of marine mollusks in the Galápagos comes from both Recent and fossil records. That Indo-Pacific species are found sporadically in time and space, but are apparently more commonly recorded during or just after periods of El Niño, when both temperature and current patterns are favorable for tropical species, is noted below. Nor are extinctions unknown. During the El Niño of 1982, several extensive algal, sea urchin, and coral die-offs were reported, including the almost complete extinction of the barnacle *Tetraclita* at Urvina Bay, Isabela (T. L. Smalley, personal communication). Evidence for continuing immigration and extinction also comes from the fossil record, which, although not extensive, clearly indicates that species present and abundant in the Pleistocene became extinct. Wellington (1984) calculated that 8% of the mollusks of the Pleistocene fossil record (cited as Pliocene but Pleistocene *fide* Hickman and Lipps, 1984) are now extinct, although some persist

Table 4. Selectivity Indices between the West American Coast and the Galápagos, and between Enewetak, Marshall Islands and the Hawaiian Islands[a,b]

Category	Number on west American coast	% not in Galápagos	Number at Enewetak	% not in Hawaii
Gastropoda	504	53	687	41
Archaeogastropoda	45	64	63	81
Neotaenioglossa	111	43	476	40
Neogastropoda	348	55	374	57
> 50 mm	56	55	101	49
Sand dwelling	73	71	105	67
Bivalvia	231	61	115	61
Borers	24	79	4	75
Burrowers	250	68	54	67
Epifauna	71	39	45	53
Low-shore rock-dwellers	61	47	38	47

[a]For each category, the total number of species on the west American coast and at Enewetak are given, and the percentage of species not found in the Galápagos and Hawaii. Designations for bivalve habits follow Stanley (1986) and calculations follow Vermeij's (1987) method of determining selectivity. Leptonaceans are not included in the counts of the Bivalvia.

[b]Data for the west American and Galápagos based on Keen (1971), DuShane and Poorman (1967), and T. Smalley (personal communication). Data for Enewetak and the Hawaiian Islands based on Kay (1979) and Kay and Johnson (1987).

elsewhere in the eastern Pacific. Two species which were predominant in Pleistocene benthic assemblages, the bivalve *Pegophysema spherica* (Dall and Ochsner, 1928) and the gastropod *Terebra albemarlensis* (Dall and Ochsner, 1928), are now extinct.

3.3. Biogeography

The biogeographical relationships of the Galápagos marine mollusks are shown in Table 5. As was pointed out by Darwin (1906) and confirmed many times over (cf. Abbott, 1966; Brusca, 1987; Hedgepeth, 1969), the Galápagos marine mollusks (and other marine invertebrates) are a mixture of species occurring in neighboring Pacific regions but dominated by Panamanic forms: an estimated 430 species or 70% of the molluscan fauna are found in the Panamic province (including extralimital occurrence north into the Californian province and south toward Peru). The remaining mollusks are divided among three components, a prominent endemic component (about 18%), an often-mentioned Indo-Pacific and circumtropical component (5%), and a small and somewhat curious Peruvian element (< 1%).

Despite the fact that the Peruvian province borders on the Panamic, and that a northward flowing current—the Peru Oceanic–South Equatorial Current—flows through the Galápagos (Abbott, 1966; Brennan, 1973), Peruvian representation among the Galápagos marine mollusks is negligible. Dall (1909) listed seven species collected in the Galápagos which are otherwise restricted to the Peruvian

Table 5. Indo-Pacific Mollusk Species Recorded from the Galápagos Islands

Species	Indo-Pacific	Source[a]
Polyplacophora		
Ischnochiton petaloides (Gld. 1846)	X	Finet, 1985
Gastropoda		
Neotaenioglossa		
Charonia tritonis (Linn. 1767)	X	CDRS
Cymatium nicobaricum (Roed. 1798)	X	LACM
Cymatium succincta (Linn. 1751)	X	Shasky, 1988
Cypraea annulus Linn. 1758	X	Gal. colln.
Cypraea caputserpentis Linn. 1758.	X	Gal. colln.
Cypraea moneta Linn. 1758	X	CDRS
Cypraea teres Gmelin, 1791	X	Gal. colln.
Hipponix foliaceus (Q.&G. 1835)	X	CDRS
Hipponix imbricatus Gld. 1846	X	CDRS
Hipponix pilosus (Desh. 1832)	X	CDRS
Pseudocypraea adamsoni (Sby. 1832)	X	CDRS
Tonna perdix (Linn. 1758)	X	CDRS LACM
Triphora triticea (Pease 1861)	X	Shasky, 1988
Neogastropoda		
Cantharus fumosus (Dillw. 1817)	X	Shasky, 1988
Conus chaldaeus (Roeding 1798)	X	CAS LACM
Conus ebraeus Linn. 1758	X	CAS LACM
Coralliobia cumingi (H.&A. Ad. 1864)	X	CDRS LACM
Coralliophila violacea (Kiener 1836)	X	CDRS LACM
Drupa ricina (Linn. 1758)	X	Finet, 1985
Microdaphne trichodes (Dall 1919)	X	Keen, 1971
Mitra mitra (Linn. 1758)	X	CDRS
Phyllocoma scalariformis (Brod. 1833)	X	LACM
Quoyula madreporarum (Sby. 1834)	X	CAS LACM

[a]Sources are: California Academy of Sciences (CAS), San Francisco, Charles Darwin Research Station (CDRS), Galapagos Island Collection, private collections in the Galapagos, Los Angeles County Museum of Natural History (LACM).

province but none of these records has been verified (Brennan 1973), and Bernard (1983) lists only two bivalves. I find six species possibly referable to the Peruvian province.

Twenty-four Indo-Pacific species (Table 5) have been recorded in the Galápagos. An additional 10 species are circumtropical or cosmopolitan in distribution, and are all recorded from the western Pacific. Most of the Indo-Pacific species are exceedingly widespread tropical species, and except for one species (*Ischnochiton petaloides* (Gould, 1846) from Hawaii), are not restricted in their distribution even to the Pacific Plate. The Indo-Pacific species recorded from the eastern Pacific have been described as "architecturally sturdy" (Vermeij, 1978); and primarily forms with teleplanic larvae (Emerson, 1967a; Scheltema, 1988). They are also noticeably patchy in distribution in time and in space. Virtually all the records are for one or two specimens collected at one time, and they are usually reported as dead or beachworn. Records of living Indo-Pacific specimens include those for *Charonia tritonis* (Linnaeus, 1767) in 1981 (Emerson, 1989), 1983 (Shasky, 1983), and 1987 (G. Kendrick, personal communication), *Cypraea moneta* Linnaeus, 1758, in 1984

(Finet, 1986), and *Cantharus fumosus* (Dillwyn, 1817), *Triphora triticea* Pease, 1861, and *Linatella succincta* (Linnaeus, 1771), in 1988 (Shasky, 1988). There is some evidence that Indo-Pacific mollusks are seen more frequently during periods of El Niño than at other times (C. Angermeyer, J. DeRoy, personal communication). Sixteen of the 20 records with dates of Indo-Pacific species occur in or within a year after El Niño years.

Of the 10 circumtropical and cosmopolitan species, four are limpet-like in form, *Cheilea cepacea* (Broderip, 1834), *C. corrugata* (Broderip, 1834), *Crepidula aculeata* (Gmelin, 1791), and *Crucibulum scutellatum* (Wood, 1828); one is a ranellid, *C. muricinum* (Roeding, 1798) known for its teleplanic larva, three are pelagic (*Janthina*), and two are bivalves (the leptonacean *Kellia suborbicularis* (Montagu, 1803) and the mytilid borer *Lithophaga aristata* (Dillwyn, 1817).

3.4. Endemism

Just over one hundred species in 52 families of the marine mollusks, about 18% of the shallow water mollusks, are reported as endemic to the Galápagos. The characteristics of the endemics are shown in Table 6. Endemism is at the species level only, and the species:genus ratio is about 1.9:1. Levels of endemism vary markedly, both among and within taxonomic groups: 6% of the bivalve species and 20% of the gastropods are reported only from the Galápagos. Among the gastropods, 45% of the archaeogastropods but only 15% of the neotaenioglossans are endemic. Endemic species co-occur with their congeners which are widespread in the Panamic region in 36% of the instances of endemism (Table 5).

The present distributional pattern of the Galápagos marine mollusks is one of a mosaic of representatives of perhaps five biogeographical components. Analysis of probable or inferred sister species and/or congeneric relationships of the endemic mollusk species shows a similar pattern (Table 7). Of the 32 endemics for which at least speculative relationships can be identified, 17 appear to be of Panamic origin, 11 are related to Caribbean and/or western Atlantic forms, two resemble species in the European fossil record, and two may be of Indo-Pacific derivation.

3.5. Galápagos and Hawaii Compared

Data comparing the molluscan faunas of the Galápagos and the Hawaiian Islands are set out in Table 3. As in the Galápagos biota, there are proportionately more gastropod than bivalve species, and approximately equal numbers of epifaunal and infaunal bivalve species. The Vermeij (1987) selectivity indices for marine mollusks between the Galápagos and the west American coastline, and those of the Hawaiian Islands, are remarkably similar (Table 4). The index is higher for bivalves than for gastropods, for archaeogastropods than for neotaenioglossa, and more effective against soft-sediment dwellers than for those mollusks with other modes of life.

Other parallels between the two faunas include anomalies in size and occurrence, inter-island differences, and a fossil record which is interpreted as involving

Table 6. Biogeographical Characteristics of the Shallow-Water Marine Mollusks

Taxon	No. species	No. endemic species	(%)	No. genera with endemics	Species: genus ratio	No. genera with endemics and congeners	Panamic	% Species also found: Pan–Cal–Peru	Peru	Indo–Pacific	Circum-tropical
Polyplacophora	13	9	(69)	7	1.3:1	3	7	15		7	
Scaphopoda	2	0					50	50			
Cephalopoda	4	2	(50)	1	2.0:1	1	25				
Gastropoda											
Archaeogastropoda	31	14	(45)	8	1.75:1	4	42	13			
Neritomorpha	3	0					33	67			
Neotaenioglossa	219	36	(16)	13	2.7:1	9	56	25	<1	6	4
Neogastropoda	206	42	(20)	33	1.3:1	19	60	14	<1	5	
Opisthobranchia		NA									
Pulmonata	9	2	(22)	1	2.0:1		67	11			
Bivalvia	153	10	(6)	7	1.9:1	5	40	47	<1		1
Totals	640	115	(18)	70		41					
Mean					1.9:1		42	30	<1	6.0	2.5

Table 7. Possible and Inferred Sister Species of Some Galápagos Endemics

Galápagos endemic	Inferred sister species	Sister species/ congeners range	Reference
Calloplax duncanus (Dall)		Congeners: Chile; Peru	Ferreira, 1982
Chiton goodalli Broderip	*C. marquesanas*	Marquesas Islands	Bullock, 1982
Chiton sulcatus Wood	*C. rossi*	Upper Cretaceous	Bullock, 1988
Tonicia forbesii arnheimi Dall	*T. f. forbesii*	Nicaragua	Smith and Ferreira, 1977
Acanthochitona jacquelinae Smith and Ferreira	*A. arragonites*	Gulf of California	Smith and Ferreira, 1977
Scissurella (Anatoma) epicharis McLean		Congener: Baja California	McLean, 1978
Nesta galapagaensis McLean		Congeners: Red Sea: Caribbean	McLean, 1970b
Tegula cooksoni (E.A. Smith)	*T. fascaita*	West Indies	Keen, 1971
Turbo situlus Dall	*T. squamiger*	Panamic–Peruvian	Keen, 1971
Batillaria mutata Pilsbry and Vanatta	*B. minima*	Western Atlantic	Keen, 1971
Cymatium lineatum (Broderip)	*C. krebsii*	Western Atlantic	Keen, 1971
Maxwellia angermayerae Emerson and d'Attilio	*M. gemma*	Californian	Keen, 1971
Murexiella radwini Emerson and d'Attilio	*M. hidalgoi*	Western Atlantic	Emerson and d'Attilio, 1970
Pteropurpura deroyana Berry	*P. centrifuga*	Panamic	Keen, 1971
Latiaxis santacruzensis Emerson and d'Attilio	*L. dalli*	Western Atlantic	Emerson and d'Attilio, 1970
Aspella hastula Reeve		European Miocene	Vokes, 1985
Favartia purdyae Vokes and d'Attilio	*F. erosa*	Panamic	Vokes, 1980
Nassarius caelolineatus Nesbitt and Pitt	*N. versicolor* and *N. nodicinctus*	Panamic	Nesbitt and Pitt, 1986
Phos (Metaphos) laevigatus A. Adams		1 Panamic congener	Emerson, 1967
Strombina deroyae Emerson and d'Attilio	*S. esmeraldensis*	Ecuadorian Neogene	Emerson and d'Attilio, 1969
Cancellaria deroyae Petit		Congeners: Californian	Keen, 1971
Volvarina innexa Roth	cf. *V. veliei*, *V. avenacea*	Caribbean	Roth, 1978
Volvarina nyssa Roth and Coan	cf. *V. albolineata*	Caribbean	Roth and Coan, 1971
Mitra gausapata Reeve	*M. lens*; *M. muricata*	Panamic	Cernohorsky, 1976
Conus kohni Mclean and Nybakken	*C. xanthicus* or *C. recurvus*	Panamic	McLean and Nybakken, 1979
Cerodrilla asymmetrica McLean and Poorman	*C. cybele*	Congeners Panamic	McLean and Poorman, 1970
Fusiturricula andrei McLean and Poorman		Congeners Panamic	McLean and Poorman, 1970
Mitrolumna keenae McLean and Poorman	cf. *M. olivacea*	European Miocene	McLean and Poorman, 1970
Veprecula tornipila McLean and Poorman		*Veprecula* is an Indo–Pacific genus	McLean and Poorman, 1970
Onchidella steindacheri (Semper)	*O. floridiana*	Caribbean	Keen, 1971
Semele sowerbyi Tryon	*S. decisa* Gould	California	Coan, 1988
Protothaca pertincta Dall		Congeners: Californian: Peruvian	Keen, 1971

both extinction and immigration through the Pliocene and Pleistocene (Kay and Palumbi, 1987; Grigg, 1988). Size differences between Hawaiian marine mollusks and those from elsewhere in the Indo-Pacific have been well documented: *Cypraea mauritiana* Linnaeus, 1758, *C. tigris* Linnaeus, 1758, *C. scurra* Gmelin, 1791, and other cowries are larger in the Hawaiian Islands than they are elsewhere (Kay, 1961). Several marine mollusks are common in Hawaii but uncommon elsewhere in their range: *Cymatium occidentale* (Mörch, 1877) and *C. intermedium* (Pease, 1869) are examples (Beu, 1986; Beu and Kay, 1988). Presence/absence records among islands in the Hawaiian chain are most conspicuous between the atolls of the Northwestern Hawaiian Islands and the volcanic islands to the southeast: *Cellana* (3 spp.), *Littoraria pintado* (Wood, 1828), and *Neothais harpa* (Conrad, 1837) are found only on shorelines of the volcanic islands; *Nerita plicata* Linnaeus, 1758 and *Drupa grossularia* (Roding, 1798) are common at Midway and others of the atolls, but rarely seen on basalt shorelines (Kay, 1979). In the Hawaiian fossil record it is apparent that species in several groups colonized the Hawaiian Islands only to become locally extinct and replaced by other taxa (Kay and Palumbi, 1987). The common Indo-Pacific cone, *Conus chaldaeus* (Roeding, 1798) (which is also recorded from the Galápagos), is not known from Hawaii during the Pleistocene, but a similar endemic species, *C. kahiko* Kohn, 1980, now extinct, was present on intertidal benches, the habitat of *C. chaldaeus* today (Kohn, 1980).

The Hawaiian marine mollusks, like the Galápagos mollusks represent a mix of biogeographical elements, but the Hawaiian fauna exhibits a dominant Indo-Pacific component, a lesser Pacific plate component, and a circumtropical/cosmopolitan component of only about 5%. Endemism is about 20%, with higher rates in the archaeogastropods and bivalves (the latter in contrast to endemism in the Galápagos bivalves). As in the Galápagos, endemism is at the species level only, and the species:genus ratio is 1.3:1, slightly lower than it is in the Galápagos.

4. Discussion

Although separated by more than 4800 km of ocean, and with faunal lists with an overlap of less than 10%, the marine molluscan faunas of the Galápagos Islands and the Hawaiian Islands share characteristics that are sufficiently distinctive as to characterize them as isolated, oceanic, insular biotas. The suite of characteristics that these molluscan biotas hold in common indicates that at least three factors acting in concert determine the faunal characteristics of these isolated oceanic islands: habitat substratum, oceanographic conditions, and isolation.

In both the Galápagos and the Hawaiian Islands, the marine mollusks represent assemblages drawn from substantial faunas occurring thousands of kilometers from their shores. If distance were the sole determinant of the composition of the island biotas by functioning as an unselective filter, then the assemblages should mirror the source faunas. That neither biota is even a miniature reflection of the source area faunas indicates that factors other than distance are operative in assembling the marine biotas of these islands. The emphasis on gastropods rather than bivalves and epifaunal species rather than infaunal species suggests that substrate is a key factor: islands such as the Hawaiian Islands and the Galápagos lack the vast expanses of silty, sandy ocean bottom characteristic of continental offshore areas and ancient archipelagoes which support large numbers of bivalves and infaunal gastropods thereby restricting the infaunal component (Kay, 1967). Vermeij (1987) attributes gastropod and epifaunal dominance to oceanic waters of low productivity. In the Galápagos, however, where waters at least around some of the islands are highly productive, the proportions of gastropod and bivalve species and epifaunal and infaunal gastropods are similar to those in Hawaii, and would seem to provide evidence that habitat substratum is a determining factor in faunal settlement. Organisms may arrive, but if the required habitats are not present they cannot settle and reproduce.

Changes in oceanographic conditions may result in sporadic gene flow, which must also influence faunal makeup both in terms of the ecological assemblages of organisms and the biogeographical components of the faunas. Eggs and larvae arrive from different directions, and settle in unique assemblages of species with population densities and morphological features (for example, size) different from their parent populations. Long term changes in oceanographic conditions may also result in extinctions, which in time are cancelled by the arrival of more larvae, although not necessarily of the same species.

The pattern of endemism of the marine molluscan faunas of the Galápagos and the Hawaiian islands differs in almost every aspect from that of their terrestrial counterparts, in which endemism is as high as 90%, and explosive evolution and adaptive radiation are hallmarks. As James (1984) suggested, adaptive radiation and speciation do not explain the evolutionary history of all marine invertebrates in the Galápagos. However, what does apply to the marine molluscan biotas of the Galápagos and the Hawaiian Islands is that they are faunas which have differentiated but not diversified. Darwin (1909), in a perceptive illustration of the observation that not all oceanic island species are "modified," noted that "in the Galápagos Islands there are 26 land birds; of which 21 (or perhaps 23) are peculiar, whereas of the 11 marine birds, only 2 are peculiar; and it is obvious that marine birds could arrive at these islands much more easily and frequently than land birds." The point was well made. Marine ecosystems of oceanic islands are more open to colonization than are terrestrial, as Grigg (1983) has suggested, hence distance restricts but does not prevent gene flow, and in this way curtails the opportunity for speciation in marine habitats.

5. Summary

The marine molluscan fauna of the Galápagos Islands consists of more than 700 species. It is dominated by gastropods (73%) and bivalves (24%), and is noticeably different in other respects from that on the tropical west American mainland: there are more epifaunal than infaunal gastropod species, and there are peculiarities in abundance and size, as well as sporadic records of species. The molluscan biota includes a mixture of species occurring in neighboring Pacific regions such as the Indo-Pacific, but is principally represented by Panamic forms (70%). There is an endemic component of more than 100 species (18%), with endemism varying among taxa. Many endemics co-occur with their Panamic congeners. Inferred sister species reflect a mix of biogeographical elements. Virtually all features of the Galápagos molluscan assemblages are also characteristic of the marine molluscan fauna of the Hawaiian Islands and provide the basis for a unifying theory: several factors acting in concert—habitat substratum, long term oceanographic change, and the degree of isolation of oceanic islands—appear to be the main determinants of marine faunal composition of oceanic islands.

ACKNOWLEDGMENTS. The field work at Urvina Bay, Isabela Island was supported by N.S.F. Grant EAR 8508966 to Mitchell Colgan, to whom I am grateful for permitting me to join the field team. I also thank Thomas Smalley for his help in the field and critique of the manuscript, and Isabella Abbott, Richard W. Grigg, Stephen R. Palumbi, and David Reid for critically reading the manuscript. I acknowledge, with thanks, the support of the Director and staff of the Charles Darwin Research Station, and the hospitality of Terrence Gosliner, and the staff of the Invertebrate Zoology Department, California Academy of Science, San Francisco, California, and of Jacqueline and André DeRoy and Carmen and Fritz Angermeyer of Santa Cruz, Galápagos.

References

Abbott, D. P., 1966, Factors influencing the zoogeographic affinities of the Galápagos inshore marine fauna, in: *The Galápagos Islands* (R. E. Bowman, ed.), pp. 108–122, Univ. Calif. Press, Berkeley.

Bernard, F. R., 1983, Catalogue of the living Bivalvia of the eastern Pacific Ocean: Bering Strait to Cape Horn, *Can. Spec. Publ. Fish. Aquat. Sci.* **61**:1–102.

Bernard, F. R., 1988. Living Bivalvia of the Galápagos Islands, *Abstr. Ann. Rpt. West. Soc. Malacol.* **21**:10.

Beu, A. G., 1986, Taxonomy of gastropods of the families Ranellidae (=Cymatiidae) and Bursidae. Part 2. Descriptions of 14 new modern Indo-West Pacific species and subspecies, with revisions of related taxa, *N.Z. J. Zool.* **13**:273–355.

Beu, A. G., and Kay, E. A., 1988, Taxonomy of gastropods of the families Ranellidae (=Cymatiidae) and Bursidae, Part IV. The *Cymatium pileare* complex. *J. Roy. Soc. N.Z.* **18**:185–223.

Brennan, E., 1973, Factors contributing to the unique environment of Galápagos marine mollusks, *Echo* **6**:45–57.

Briggs, J. C., 1974, *Marine Zoogeography*, McGraw-Hill, New York.

Brusca, R. C., 1987, Biogeographic relationships of Galápagos marine isopod crustaceans, *Bull. Mar. Sci.* **41**:268–281.

Brusca, R. C., and Wallerstein, B. R., 1979, Zoogeographic patterns of idoteid isopods in the northeast Pacific, with a review of shallow water zoogeography of the area, *Bull. Biol. Soc. Wash.* **3**:67–105.

Bullock, R. C., 1988, The genus *Chiton* in the New World (Polyplacophora: Chitonidae), *Veliger* **31**: 141–191.

Cernohorsky, W. O., 1976, The Mitridae of the world. Part 1. The subfamily Mitrinae, *Indo-Pacific Mollusca* **3**:273–528.

Coan, E. V., 1968, A biological survey of Bahia de los Angeles, Gulf of California, Mexico. III. Benthic Mollusca, *Trans. San Diego Soc. Nat. Hist.* **15**:107–132.

Coan, E. V., 1988, Recent eastern Pacific species of the bivalve genus *Semele*, *Veliger* **31**:1–42.

Dall, W. H., 1909, Report on a collection of shells from Peru, with a summary of the littoral marine Mollusca of the Peruvian zoological province, Proc. *U.S. Natl. Mus.* **37**:147–294.

Darwin, C. R., 1906, *Journal of Researchers into Geology and Natural History*, J. M. Dent, New York.

Darwin, C. R., 1909, *The Origin of Species*, P. F. Collier & Son, New York.

DuShane, H., and Poorman, R., A checklist of Mollusks for Guyamas, Sonora, Mexico, *Veliger* **9**:413–431.

Emerson, W. K., 1967a, Indo-Pacific faunal elements in the tropical eastern Pacific with special reference to the mollusks, *Venus* **25**:85–93.

Emerson, W. K., 1967b, On the identity of *Phos laevigatus* A. Adams, 1851 (Mollusca: Gastropoda), *Veliger* **10**:99–100.

Emerson, W. K., 1978, Mollusks with Indo-Pacific affinities in the eastern Pacific Ocean, *Nautilus* **92**: 91–96.

Emerson, W. K., 1983, New records of prosobranch gastropods from Pacific Panama, *Nautilus* **97**: 119–123.
Emerson, W. K., 1989, On the occurrence of *Charonia tritonis* (Linnaeus) in the eastern Pacific (Ranellidae: Cymatiinae). *Festivus* **21**:13–19.
Emerson, W. K., and D'Attilio, A., 1969, A new species of *Strombina* from the Galápagos Islands, *Veliger* **11**:195–197.
Emerson, W. K., and D'Attilio, A., 1970, Three new species of muricacean gastropods from the eastern Pacific, *Veliger* **12**:270–274.
Ferreira, A. J., 1982, *Calloplax hanselmani* (Mollusca: Polyplacophora) in the eastern Pacific, *Veliger* **24**:321–324.
Finet, Y., 1985, Preliminary faunal list of the marine Mollusca of the Galápagos Islands, *Inst. Roy. des Sci. Nat. de Belgique* **20**:1–50.
Finet, Y. 1986. Living *Cypraea moneta* L. in the Galápagos Islands. *La Conchiglia* **19(222–223)**:22–24.
Gosliner, T. M., 1987, Biogeography of the opisthobranch gastropod fauna of southern Africa, *Amer. Mal. Bull.* **5**:243–258.
Grigg, R. W., 1981, *Acropora* in Hawaii. Part 2. Zoogeography, *Pac. Sci.* **35**:15–24.
Grigg, R. W., 1983, Community structure, succession and development of coral reefs in Hawaii, *Mar. Ecol. Prog.* (Ser. 11), pp. 1–14.
Grigg, R. W., 1988, Paleoceanography of coral reefs in the Hawaiian-Emperor chain, *Science* **240**: 1737–1743.
Hedgepeth, J. W., 1969, An intertidal reconnaissance of rocky shores of the Galápagos, *Wasmann J. of Biol.* **27**:1–24.
Hickman, C. S., and Lipps, J. H., 1985, Geologic youth of Galápagos Islands confirmed by marine stratigraphy and paleontology, *Science* **227**:1578–1586.
James, M. J., 1984, A new look at evolution in the Galápagos: Evidence from the late Cenozoic marine molluscan fauna, *Biol. J. Linn. Soc.* **21**:77–95.
Kay, E. A., 1961, A zoogeographical analysis of species of *Cypraea* in the Hawaii Islands, *Proc. Malac. Soc. Lond.* **34**:185–198.
Kay, E. A., 1967, The composition and relationships of marine molluscan fauna of the Hawaiian Islands, *Venus* **25**:96–104.
Kay, E. A., 1979, *Hawaiian marine shells*, Bishop Mus. Press, Honolulu, Hawaii.
Kay, E. A., and Johnson, S., 1987, Mollusca of Enewetak Atoll, in: *The Natural History of Enewetak Atoll*, vol. 2 (D. M. Devaney, E. S. Reese, B. L. Burch, and P. Helfrich, eds.), pp. 105–146, Department of Energy, Washington, D.C.
Kay, E. A., and Palumbi, S. R., 1987, Endemism and evolution in Hawaiian marine invertebrates, *TREE* **2**:181–191.
Kay, E. A., and Switzer, M. F., 1974, Molluscan distribution patterns in Fanning Island lagoon and a comparison of mollusks of the lagoon and seaward reefs, *Pac. Sci.* **28**:275–295.
Keen, A. M., 1971, *Sea shells of Tropical West America: Marine Mollusks from Baja California to Peru*, 2nd ed., Stanford Univ. Press, Stanford, California.
Kohn, A. J., 1980, *Conus kahiko*, a new Pleistocene gastropod from Oahu, Hawaii, *Journ. Paleont.* **54**:534–541.
McCosker, J. E., and Rosenblatt, R. H., 1984, The inshore fish fauna of the Galápagos Islands, in: *Key environments Galápagos* (R. Perry, ed.), pp. 133–143, Pergamon Press, New York.
McLean, J. H., 1961, Marine mollusks from Los Angeles Bay, Gulf of California, *San Diego Soc. Nat. Hist., Trans.* **12**:499–476.
McLean, J. H., [1969] 1970a, New species of tropical eastern Pacific Gastropoda. *Malac. Rev.* **2**:115–130.
McLean, J. H., 1970, Descriptions of a new genus and eight new species of eastern Pacific Fissurellidae, with notes on other species, *Veliger* **12**:362–367.
McLean, J. H., and Nybakken, J., 1979, On the growth stages of *Conus fergusoni* Sowerby, 1873, the reinstatement of *Conus xanthicus* Dall, 1910, and a new species of *Conus* from the Galápagos Islands, *Veliger* **22**:135–143.
McLean, J. H., and Poorman, R., 1971, New species of tropical eastern Pacific Turridae, *Veliger* **14**:89–113.
Nesbitt, E. A., and Pitt, W. D., 1986, *Nassarius* (Gastropoda: Neogastropoda) from the Galápagos Islands, *Veliger* **28**:294–301.
Nybakken, J., 1979, Population characteristics and food resource utilization of *Conus* in the Galápagos Islands, *Pac. Sci.* **32**:271–280.

Poorman, F. L., and Poorman L. H., 1978, Additional molluscan records from Bahia de Los Angeles, Baja California Norte, *Veliger* **20**:369–374.

Roth, B., 1978, New species and records of tropical west American Marginellidae (Molusca: Neogastropoda), *Contr. Sci. Nat. Mus. L.A. County* **292**:1–18.

Roth, B., and Coan, E. V., 1971, Marginellidae (Mollusca: Neogastropoda) from the Galápagos Islands and Cocos Island, *Proc. Calif. Acad. Sci.* (**4**)37:197–255.

Scheltema, R. S.,1988, Initial evidence for the transport of teleplanic larvae of benthic invertebrates across the east Pacific barrier, *Biol. Bull.* **174**:145–152.

Shasky, D. R., 1983, New records of Indo-Pacific Mollusca from Cocos Island, Costa Rica, *Nautilus* **97**:144–145.

Shasky, D. R., 1988, An update on the molluscs from the Galápagos Islands as listed in the preliminary faunal list of the marine mollusks of the Galápagos Islands by Yves Finet, *Ann. Rpt. West. Soc. Malacol.* **21**:8–10.

Smith, A. G., and Ferreira, A., 1977, Chiton fauna of the Galápagos Islands, *Veliger* **20**:82–97.

Springer, V., 1982, Pacific Plate biogeography, with special reference to shorefishes, *Smithsonian Contrib. to Zool.* **367**:1–182.

Stearns, R. E. C., 1983, XXV. Report on the mollusk fauna of the Galápagos Islands with descriptions of new species, *Proc. U.S.N.M.* **16**:353–450.

Vermeij, G. J., 1978, *Biogeography and Adaptation: Patterns of Marine Life*, Harvard Univ. Press, Cambridge, Massachusetts.

Vermeij, G. J., 1987, The dispersal barrier in the tropical Pacific: Implications for molluscan speciation and extinction, *Evol.* **4**:1046–1058.

Vermeij, G. J., E. A. Kay, and L. G. Eldredge, 1984, Molluscs of the northern Mariana Islands, with special reference to the selectivity of oceanic dispersal barriers, *Micronesica* **19**:27–55.

Vokes, E. H., 1985, Review of the west coast aspelloids *Aspella* and *Dermomurex* (Gastropoda: Muricidae), with the descriptions of two new species, *Veliger* **27**:430–439.

Wellington, G. M., 1984, Marine environment and protection, in: *Key environments in the Galápagos* (R. Perry, ed.), pp. 247–263, Pergamon Press, Oxford.

Chapter 12

The Marine Mollusks of the Galápagos Islands

YVES FINET

1. Introduction

The marine mollusks of the Galápagos have been of great interest to malacologists since the early collecting expeditions in the nineteenth century.

Adaptive radiation and speciation among land snails in the unique terrestrial biota of the Galápagos have been studied by evolutionary biologists and malacologists (Breure and Coppois, 1978; Coppois, 1985; Coppois and De Vos, 1986; Coppois and Glowacki, 1982; Odhner, 1950; Smith, 1966; Vagvolgyi, 1977).

But it has been pointed out (James, 1984) that the evolutionary scenario resulting from Darwin's work and observations generally does not apply in the same way to marine invertebrate organisms. Many marine animals have the ability to exchange genetic material between widely separated populations through transport of planktonic larvae. The rate of marine invertebrate dispersal, which is higher than for terrestrial animals, could account for a smaller degree of evolutionary divergence from other island or mainland source populations (James, 1984).

Although speciation among Galápagos marine mollusks is not spectacular as in terrestrial vertebrates, such as reptiles or birds, there are several aspects of the malacofauna that make it interesting from a biogeographic and evolutionary perspective. These aspects include a unique geologic/geographic setting (volcanic, oceanic islands on the equator influenced by cold rather than warm ocean currents), the presence of many vagrant elements from other biogeographic provinces, and a fairly well-documented fossil record.

YVES FINET • Department of Invertebrates, Museum of Natural History, CH1211 Geneva 6, Switzerland.

In addition to the mere descriptions of new species, several groups or families of Galápagan or West American marine mollusks have formed the subject of numerous systematic accounts or revisions (Baker, 1926; Baker, Hanna and Strong, 1928, 1930a,b, 1938a,b; Bartsch, 1907, 1910, 1911a,b, 1915; Bertsch *et al.*, 1973; Bratcher, 1979; Bratcher and Burch, 1970a,b, 1971; Bullock, 1988; Cate, 1969a,b, 1973, 1979; Cernohorsky, 1975; Coan, 1979, 1988; DuShane, 1974, 1979; Emerson and D'Attilio, 1970; Ferreira, 1979; Grau, 1959; Hanna and Hertlein, 1961; Hoagland, 1977; Keen and Smith, 1961; Lindberg and McLean, 1981; Marincovich, 1977; McFadien, 1973; McLean, 1967, 1970a,b,c,d, 1971; McLean and Nybakken, 1979; McLean and Poorman, 1971; Morrison, 1963; Nesbitt and Pitt, 1986; Nesis, 1973; Nybakken, 1970; Old, 1975; Olsson, 1961; Petit, 1970; Pilsbry and Lowe, 1932; Radwin, 1974; Rost, 1955; Roth and Coan, 1968, 1971; Smith and Ferreira, 1977; Soot-Ryen, 1955; Sphon, 1969, 1976; Sphon and Mulliner, 1972).

Faunal lists for the Galápagos Islands were published beginning in the late nineteenth century (Cauquoin, 1969; Dall, 1908; Emerson and Old, 1965; Hertlein and Strong, 1955; Métivier, 1968; Sabelli and Tommasini, 1980; Schwengel, 1938; Smith *in* Günther, 1877; Soot-Ryen, 1932; Stearns, 1893; Strong and Hertlein, 1937, 1939; Taviani, 1979; Tomlin, 1927; Wimmer, 1879). But several of these faunal lists, especially the early ones, gave information that has to be very critically reviewed. They often reported species not actually living in the Galápagos. These errors were due to misidentifications, inaccurate taxonomic decisions, or wrong localities for some records, particularly those based upon specimens from old private collections or in material collected by Hugh Cuming. One of the most recent and complete compilations of data was that of Keen (1971), which was devoted to the entire Panamic Province molluscan fauna.

Zoogeographic affinities and taxonomic relationships of the fauna, as well as evolutionary patterns, were recently examined by several authors (Abbott, 1966; Brennan, 1973; Emerson, 1967, 1978, 1982, 1983; Fechter, 1981; Finet, 1987b, 1988, 1989; Hertlein, 1937; Hertlein and Emerson, 1953; James, 1984; James and Pitt, 1984; Robertson, 1975, 1976, 1979; Vermeij, 1987; Wellington, 1984). They were often enlightened by paleontological data (Dall and Ochsner, 1928; Emerson *et al.*, 1981; Hertlein, 1972; Hertlein and Strong, 1939; Hoffstetter, 1952; James, 1984; James and Pitt, 1984; Pitt, 1981; Pitt *et al.*, 1986), by ecological studies (Cinelli and Colantoni, 1974; Hedgpeth, 1969; Houvenaghel and Houvenaghel, 1974; Kohn, 1971; Miller *in* Bowman, 1966; Nybakken, 1978, 1979; Wellington, 1984) or by new climatic and oceanographic interpretations (Brennan, 1973; Houvenaghel, 1978, 1984; Laursen, 1981; Ramage, 1986; Scheltema, 1966, 1968, 1971a,b, 1972, 1986, 1988; Thorson, 1961; Wooster and Hedgpeth *in* Bowman, 1966; Wyrtki, 1965, 1967; Zinsmeister and Emerson, 1979).

In this chapter I have provided an updated account of the faunal composition and biogeographic setting of the Galápagos marine malacofauna. A complete taxonomic revision of the fauna is beyond the scope of the present work and is not presented herein. However, such a revision is in progress and will be published at a later time.

2. Taxonomic Composition and Biogeographic Affinities of the Fauna

The actual number of species of marine mollusks known to live in the Galápagos Archipelago is not well established. Data from Keen (1971), with additional information from several other publications, lead to a number of 323 species from the Archipelago (James, 1984). Using field-collected data, museum specimens, and literature records, Finet (1985) listed 593 species of verified occurrence in the Galápagos. Taking into consideration new records (Coan, 1988, 1989; Finet, in preparation; Shasky, 1989), it now appears that the fauna consists of approximately 652 species, and additional data and records given by other authors (Bernard, 1989) may increase this number.

The total of 652 includes both deep-water and shallow-water species, as both are generally quoted in the literature or the museum records as "from the Galápagos", though the deep-water species, i.e., the ones found at more than 200 m depth, are not to be considered properly as a component of the benthic littoral malacological fauna of the Archipelago. As 58 of the 652 "Galápagan" species are deep-water species, this would reduce the shallow-water component of the islands to 594 species.

An overview of the biogeographic affinities of the Galápagos indicates that it is a depauperate outpost of the Panamic marine zoogeographical Province that includes more than 3000 species (Keen, 1971). But the fauna of the Galápagos Archipelago also includes many vagrant elements from other biogeographical Provinces, as well as a significant endemic component. A compilation of the data given in Keen (1971) lead to a high 42% percentage of species endemic to the islands (James, 1984). Using more precise and extensive data from museum collections, literature, and other findings, an actual lower percentage of 22.7% is obtained. If only the shallow-water species are considered, this percentage is reduced to 18.7%. The systematic composition of the fauna and the "rates of endemicity" for each group are presented in Table 1. Other biogeographic source elements found in the fauna are Californian, Peruvian, Indo-Pacific, or Caribbean in affinity. Additionally, a few worldwide or tropical wide ranging species are also encountered in the Archipelago. The biogeographic distribution of the Galápagos species is detailed quantitatively in Table 2.

The specific composition of the fauna is detailed in the list below; this faunal list does not include information about the sources of species occurrences, because enumerating such data would be beyond the scope of this chapter; however, such a documented list is needed and will be published separately (Finet, in preparation).

3. Factors Influencing the Biogeographic Affinities of the Galápagos Malacofauna

The varying degrees of faunal affinity shown in Table 2 are related to several factors, such as land and sea barriers, ocean currents, timing and duration of spawning seasons, incidence of larvae with a long planktonic life, and thermal conditions of the seawater. The main barriers to east–west distribution are the Central American land bridge and Ekman's East Pacific Barrier between the Indo-Pacific Polynesia and the eastern Pacific. Barriers to North–South distribution are mainly thermal barriers, plus a discontinuity in shore surface currents off Guatemala.

Since the late Tertiary, the Central American land bridge isolated the Caribbean Sea from the tropical eastern Pacific. Many Panamic species are closely

Table 1. Systematic Composition of the Galápagos Islands Malacofauna, with Number and Percentage of Endemic Species[a]

	Total number of species			Endemic species			Non-endemic Species			Percentage endemic		
Taxa	sw	dw	Total	sw	dw	Total	sw	dw	Total	sw	dw	Global
Gastropoda	483	47	530	92	34	126	391	13	404	19.0%	72.3%	23.8%
(Prosobranchia)	(424)	(44)	(468)	(82)	(31)	(113)	(342)	(13)	(355)	(19.3%)	(70.5%)	(24.1%)
(Archaeogastropoda)	(41)	(8)	(49)	(16)	(5)	(21)	(25)	(3)	(28)	(39.0%)	(62.5%)	(42.9%)
(Mesogastropoda)	(183)	(12)	(195)	(25)	(8)	(33)	(158)	(4)	(162)	(13.7%)	(66.7%)	(16.9%)
(Neogastropoda)	(200)	(24)	(224)	(41)	(18)	(59)	(159)	(6)	(165)	(20.5%)	(75.0%)	(26.3%)
(Opisthobranchia, Pyramidelloidea, Gymnomorpha, and marine Pulmonata)	(59)	(3)	(62)	(10)	(3)	(13)	(49)	(-)	(49)	(16.9%)	(100%)	(21.0%)
Bivalvia	92	8	100	8	3	11	84	5	89	8.7%	37.5%	11.0%
Polyplacophora	13	—	13	9	—	9	4	—	4	69.2%	—	69.2%
Scaphopoda	2	3	5	—	—	—	2	3	5	0.0%	0.0%	0.0%
Cephalopoda	4	—	4	2	—	2	2	—	2	50.0%	—	50.0%
Total	594	58	652	111	37	148	483	21	504	18.7%	63.8%	22.7%

[a]sw = shallow water; dw = deep water.

Table 2. Biogeographic Distribution of the Species of Marine Mollusks Occuring in the Galápagos[a]

Affinity	No. of species	% of total
Endemic		
shallow-water (max. 200 m)	111	17.0
deep-water	37	5.7
Total endemic	148	22.7
Pure Panamic	337	51.7
With Californian affinities		
Panamic + Californian	46	7.05
pure Californian	0	0.0
With Peruvian affinities		
Panamic + Peruvian (+ Abyssal Magellanic)	42	6.44
Panamic + Californian + Peruvian	17	2.61
pure Peruvian	3	0.46
With Caribbean affinities		
Panamic + Caribbean (+ Patagonian)	10	1.53
With Indo-Pacific affinities		
Panamic + Indo-Pacific	8	1.22
pure Indo-Pacific	13	1.99
Western Pacific wide-ranging (incl. Japonic and/or S. Australian)	5	0.77
With a wide-ranging distribution		
circumtropical or tropical wide-ranging	14	2.15
worldwide	2	0.31
pelagic (verified occurrences only)	7	1.07

[a]Percentages are based on a total of 652 shallow-water and deep-water species.

related to corresponding species in the Caribbean. A few of them appear to be identical in gross morphology in population from both sides of the Continent. Four species recorded in the Galápagos have been identified as Caribbean species. Because the populations of the two marine Provinces are now completely separated, the species should also be separate, even if the shells look identical. Further studies using electrophoresis and other techniques will probably clarify the evolutionary relationships of these species.

The primary mechanism for dispersal of benthic invertebrates like mollusks appears to be transport of pelagic larvae by ocean currents, or very rarely drifting adults attached to debris (Abbott, 1966; Emerson, 1967; Scheltema, 1966, 1968, 1971a,b, 1972; Wyrtki, 1965, 1967; Zinsmeister and Emerson, 1979).

3.1. Factors Influencing the Faunal Affinities between the Galápagos and Peru (Fig. 1)

The primary factors influencing the marine zoogeographical affinities of the Galápagos have been discussed by Abbott (1966).

The main current flowing through the Galápagos is the South Equatorial Current (S.E.C.) originating from the Peru Oceanic Current (P.O.C.) (Fig. 1). The Peru Oceanic Current flows northward but turns westward at about 10° S. Its northern limit crosses the equator just east of the Galápagos, where it becomes the South

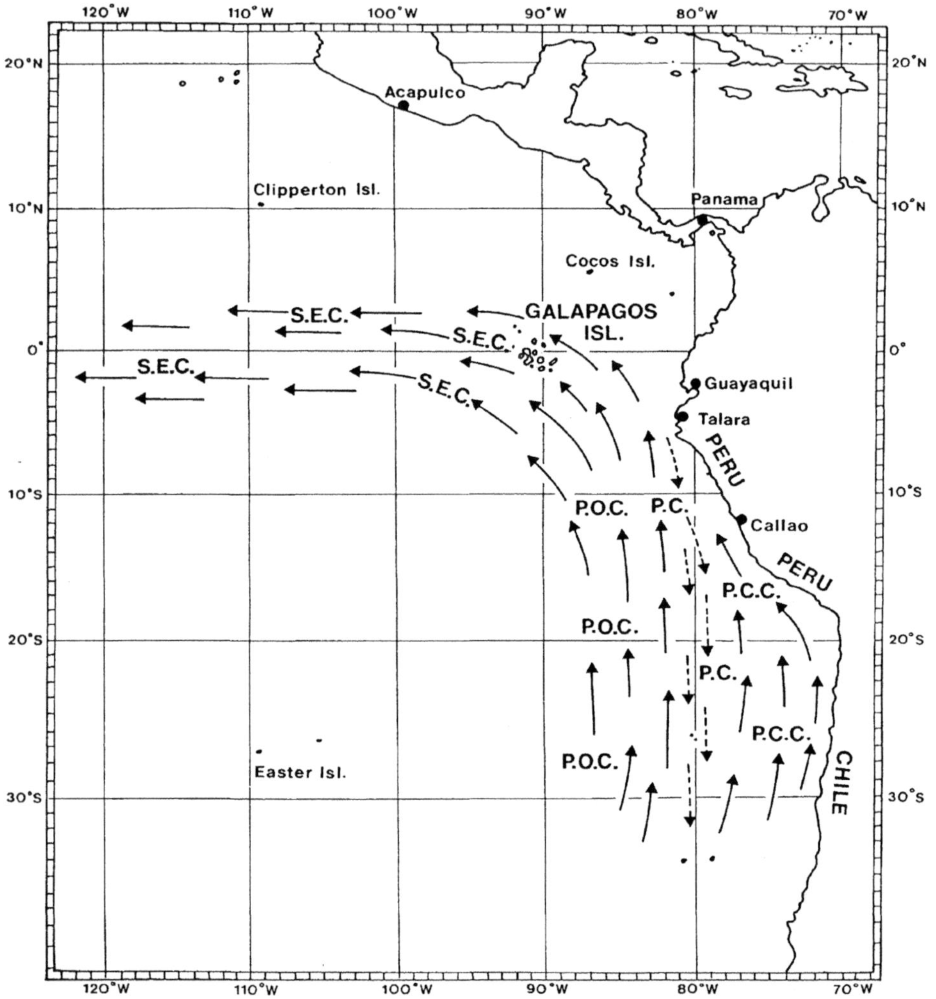

Figure 1. Ocean currents related to the faunal affinities between the Galápagos and the Peruvian biogeographical Province.

Equatorial Current. These currents carry offshore waters from off Chile and Peru up through the Galápagos. They are created by winds throughout the year, although there is some annual variation in intensity; the South Equatorial Current through the Galápagos is stronger during the southern winter (August–September) and weaker during the southern summer, principally during the February–April period.

The Peru Oceanic Current is distinct from the Peru Coastal Current or Humboldt Current (P.C.C.), flowing northward but close to the shores of Chile and Peru, providing strong upwelling along these coasts, and not extending north of Callao in Central Peru.

The waters of the Peru Oceanic Current and of the Peru Coastal Current are separated by the Peru Countercurrent (P.C.), that flows southward between them, off the Peruvian and Chilean coasts (Fig. 1).

Because the main surface current flowing through the Galápagos transports waters from the South, one would predict a fairly high biogeographical affinity between Peru and the Galápagos. But on the contrary, there are not many Peruvian species occurring in the Galápagos. Table 2 shows that there are only 3 strictly Peruvian species living in the Galápagos Archipelago (cf. Finet, 1985). Nevertheless, there may be more species with Peruvian affinities, but these species also occur along the South American coast of the Panamic Province and might be eurythermal.

Why are there so few Peruvian species in the Galápagos? First, the Peru Oceanic Current is an offshore current, distinct from the Peru Coastal Current, separated from the latter by the Peru Countercurrent, and there is little chance for larvae of coastal benthic forms to be transported by the P.O.C. The Peru Countercurrent functions as a barrier to dispersal of Peruvian species.

Second, a narrow continental shelf and strong currents sweeping offshore throughout the year are very unfavorable conditions that increase the hazards of larval existence (Thorson, 1950); it is hazardous for a larva of an adult benthic form to be swept offshore beyond a favorable site for settlement, and to be subjected to unfavorable conditions of food or temperature. In such conditions, selection favors a more direct type of development, with a reduction of the plankton-feeding habit and a shortening or even an omission of the pelagic phase. As those kinds of conditions prevail along the northern Peruvian coast, one may expect a fauna that includes species having relatively direct development, with only a few forms able to be distributed as far as the Galápagos.

Another important factor influencing zoogeographic affinities between well-separated areas is the coincidence of seasonal variations in climate and current velocity with the spawning and maximum production of larvae. In the Peruvian Province, spawning is expected to occur mainly in the austral summer, between January and April, with a fairly short breeding period as is generally the case in cold waters. But during that season, the Peru Oceanic Current and the South Equatorial Current are weaker, and it would take longer for larvae to reach the Galápagos. This further reduces the chances for species with a short larval existence to be dispersed from Peru up to the Galápagos.

3.2. Factors Influencing the Faunal Affinities between the Galápagos and the Central Pacific (Fig. 2)

The only current flowing eastward through the Galápagos is the Equatorial Undercurrent or Cromwell Current (E.U.C.). It is not a surface current, but rather an underwater stream originating in the central Pacific area of the Line Islands, remaining submerged at depths between 50 and 300 m, and upwelling at its eastern end in the Galápagos (Emerson, 1967; Wyrtki, 1967). It is not yet well established if this undercurrent is likely to play a role in transporting pelagic stages of organisms across the Pacific directly to the Galápagos.

The North Equatorial Countercurrent (N.E.C.C.), a surface current almost

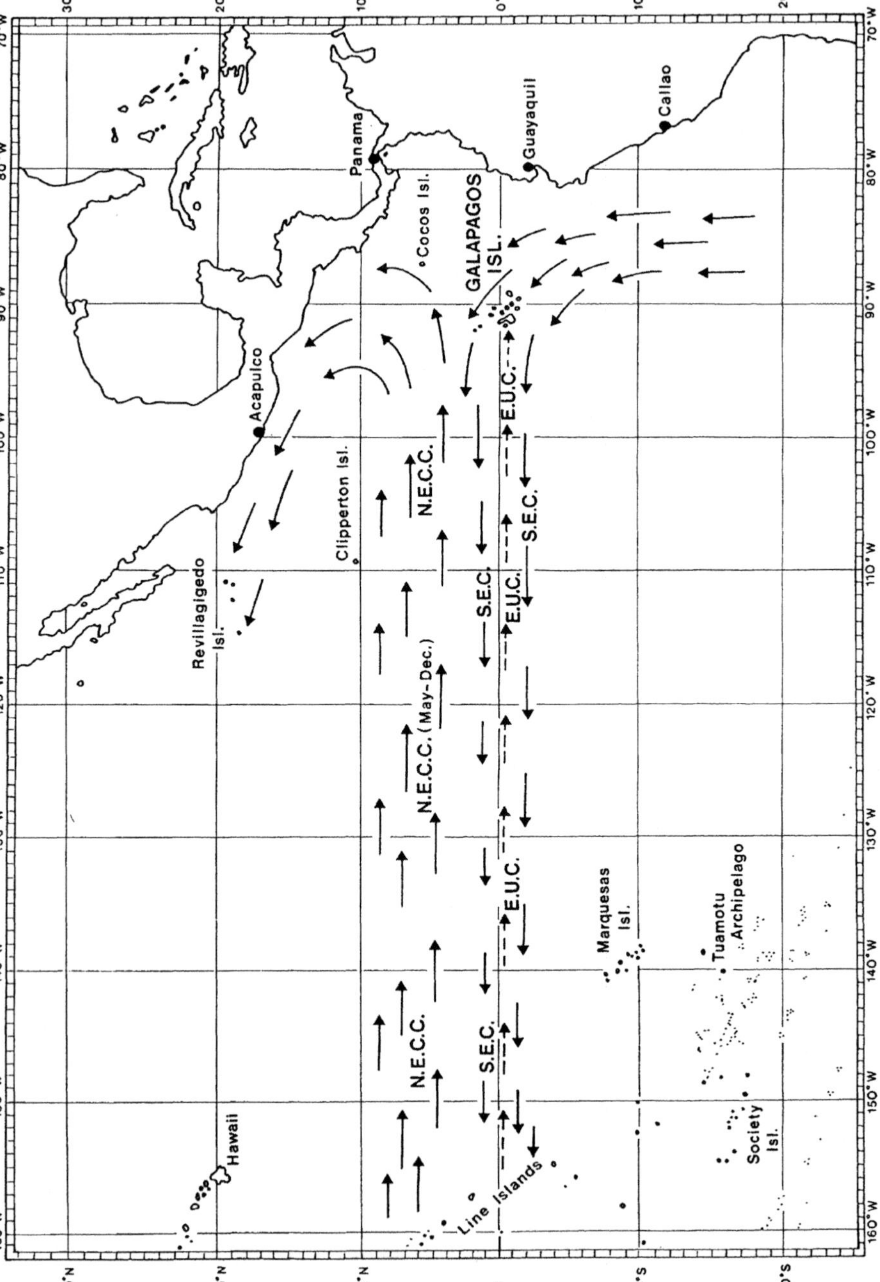

Figure 2. Ocean currents related to the faunal affinities between the Galápagos and the Indo-Pacific Province.

completely concentrated in the upper 200 m (Wyrtki, 1967), is a more likely way for transport of molluscan larvae from the western Pacific to the eastern Pacific waters around Clipperton Atoll, Cocos Island, and the Central American coast.

This current, originating in the western Pacific, flows eastward in a path between 4° and 11° north of the equator. Its size and velocity vary markedly with the season, at least in the eastern Pacific. During the northern winter, especially between February and April, it may disappear entirely. During the rest of the year, and particularly between July and November, it transports water at a velocity of 0.75 up to 2 knots from the West toward the East, terminating north of the Galápagos at a longitude of about 90° (Abbott, 1966; Wyrtki, 1965). The main residual flow flows toward the northwest along the coast of Central America (see Fig. 2).

Such ocean surface currents may be an effective factor of passive dispersal, particularly for shallow water benthic invertebrates having long pelagic larval stages (= teleplanic larvae, Scheltema, 1971b). The influence of planktotrophic larvae from the central Pacific on the occurrence of Indo-Pacific elements in the eastern Pacific may be predicted high because the favorable conditions in tropical regions like the western or central Pacific allow some 80 to 85 percent of the shallow-water benthic invertebrates to have pelagic and planktotrophic larvae, rather than relatively direct development like on the Peruvian coast (Thorson, 1950; Abbott, 1966).

Moreover, spawning and producing larvae in the tropical central Pacific islands is most important in the April–October period, precisely when the North Equatorial Countercurrent is stronger and carrying larvae more efficiently. Despite these favorable factors, the affinity between the Indo-Pacific and the Galápagos is fairly low, due to the extremely long distance and travel time for the drifting larvae (Abbott, 1966).

Using an average velocity of 1 knot for calculating duration of an eastward drift by the North Equatorial Countercurrent, a reasonable estimate of this duration would be 4 to 5 months for a Pacific crossing from the Line Islands to Clipperton Island in the eastern Pacific and 6 to 7 months for the same crossing up to the waters off Costa Rica around Cocos Island. Only organisms having teleplanic larvae with such a long pelagic life in the upper 100 m would then eventually succeed in crossing Ekman's East Pacific Barrier (Finet, 1987b). The majority of the mollusks having such long larval stages are prosobranchs of the families Cypraeidae, Cymatiidae, Tonnidae, Coralliophilidae, Mitridae, and Conidae (Laursen, 1981; Scheltema, 1986), and the majority of the species with Indo-Pacific affinities occurring in the eastern Pacific belong to these families (Emerson, 1967, 1978, 1983; Finet, 1985, 1987a,b; Shasky, 1983).

However, we cannot totally exclude the possible contribution of water exchange through eddies between the North Equatorial Countercurrent and the South Equatorial Current in the area north of the Galápagos. The North Equatorial Countercurrent does not seem to account for direct transport from the Marshall or Line Islands to the Galápagos. It seems more likely that the Indo-Pacific elements occurring in the Galápagos have reached the archipelago using other areas in the eastern Pacific as relays for dispersal. These other areas being more in the direct path of the current flowing from the west. These elements may then be transported from the southern part of Central America and from the Panama region by currents flowing from the northeast up to the Galápagos between January and April, especially during "El Niño" years, when strong currents from the Bay of Panama bring large quantities of tropical waters to the Galápagos.

There could be a correlation between the sporadic occurrences of the "Niño" climatic event in the eastern Pacific and many individual recorded occurrences of Indo-Pacific species in the Galápagos (Grove, 1989). Indeed, the occurrence of some of these Indo-Pacific species may also appear to be only temporary (Finet, 1987a), as most of them may encounter environments unsuitable for permanent colonization. In that regard, the quasi-status of stony corals in the Galápagos limits the available ecological niches for Indo-Pacific faunal elements (Emerson, 1967). Such unfavorable environmental conditions are probably another factor accounting for a low faunal affinity between the Galápagos and the Indo-Pacific.

On the other hand, species that are typically Panamic but that may also be found living in the Indo-Pacific are very rare. These species are listed in Table 2 as Panamic + Indo-Pacific or as western Pacific wide-ranging or tropical wide-ranging. They might be dispersed from the eastern Pacific and the Galápagos by way of the South Equatorial Current, flowing westward. The South Equatorial Current is the only current sweeping through the Galápagos before flowing directly toward the open Pacific Ocean (Fig. 1). The South Equatorial Current could theoretically bring some Galápagan and other Panamic elements into Indo-Pacific islands despite the following: 1) the temperature conditions are very different at the two ends of the pathway, 2) spawning occurs mainly between January and April in the Galápagos when the South Equatorial Current is weaker and less effective, and 3) the incidence of species with long planktonic larval stages is expected to be low in the Galápagos due to fairly unfavorable conditions such as those prevailing along the Peruvian coast. This mechanism could explain the presence throughout the Indo-Pacific of some Panamic prosobranch species of the families Architectonicidae, Hipponicidae, Calyptraeidae, Cymatiidae, Bursidae, Muricidae, Buccinidae, and Columbellidae (cf. Finet, 1985).

3.3. Factors Related to the Panamic Affinity of the Galápagos

During the northern winter, especially between February and April, strong upwelling occurs in the Gulf of Panama, and associated strong currents originate in this region. They flow toward the southwest, to the Galápagos where they merge with the South Equatorial Current, which is weaker during this season (Schaeffer, 1962; Abbott, 1966) (Fig. 3). At this time of year, there is thus a direct pathway for waters from Panama, Ecuador, and probably Colombia to be carried to the Galápagos. These currents are almost absent during the rest of the year. In the fairly warm tropical or subtropical waters of the Bay of Panama or of the Ecuadorian coast, the fauna is expected to be rich in species with pelagic, planktotrophic larvae, particularly because, unlike on the Peruvian coasts, the danger of these larvae being swept out to sea by strong currents is low during most of the year. The incidence of planktotrophic larvae is thus expected to be high. Maximum spawning occurs between May and September in the Bay of Panama and between January and April in Ecuador. As the relevant currents are effective during the January–April period, the strongest faunal affinity occurs between the Galápagos and the Ecuadorian coasts. This also holds true because the latter has the closest proximity

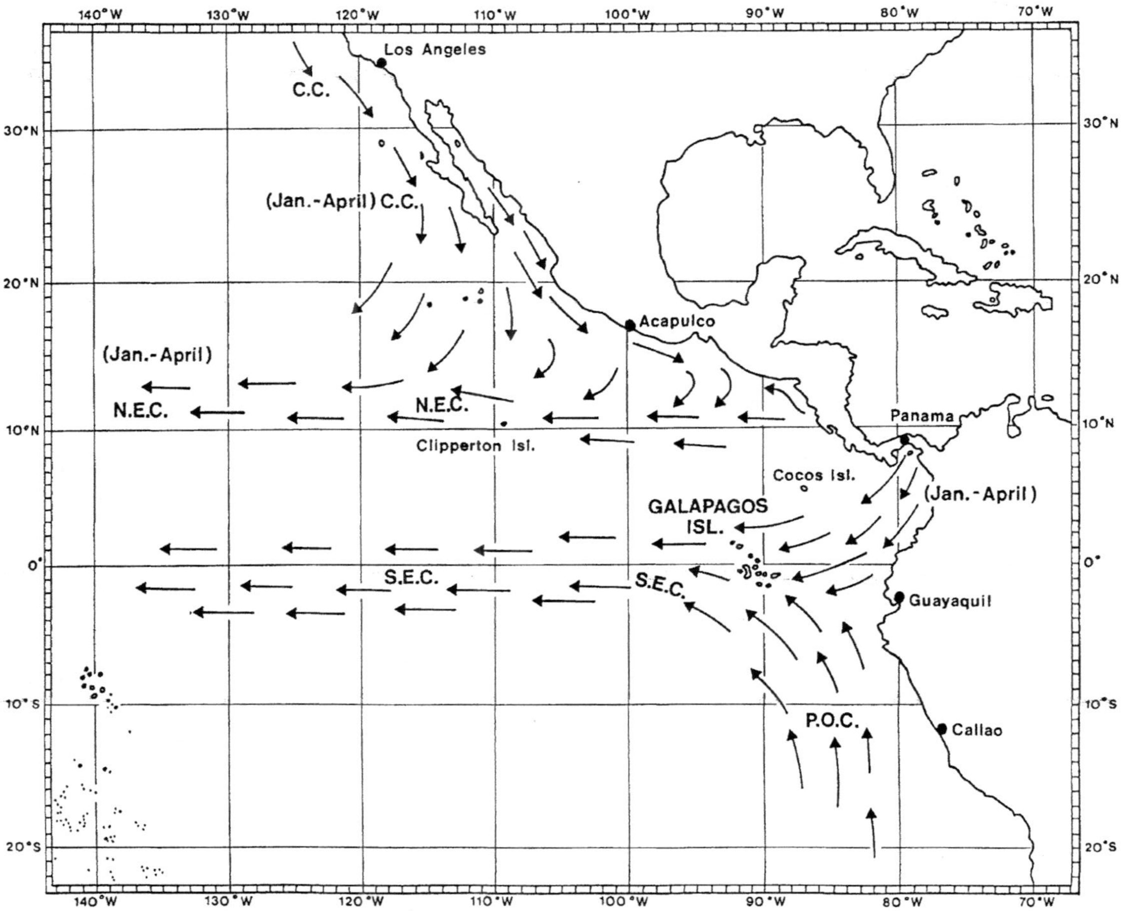

Figure 3. Ocean currents in the eastern Pacific during the January–April period, relating to the faunal affinities between the Galápagos and the Panamic and Californian Provinces.

to the Galápagos and because its ocean temperature range and annual temperature cycle are similar to those prevailing in the northern and central Galápagos Islands.

3.4. Factors Influencing the Faunal Affinities between the Galápagos and California (Figs. 3 and 4)

The California Current (C.C.), parallel to the North American coast, flows southward off the California coast and turns westward off Baja California at about 25°N, where it merges with the North Equatorial Current (N.E.C.) flowing westward across the Pacific, well to the north of the North Equatorial Countercurrent (Abbott, 1966) (Fig. 3).

During the northern winter, especially in the February–April period, these currents are stronger. During the same period, the North Equatorial Countercurrent is weak or absent. The California Current may then split south of the tip of Baja California: a western branch turns westward as a cut-off to join the North Equatorial Current, while an eastern branch joins an outflow from the Gulf of California and continues flowing along the Mexican coast toward the southeast and the area off Guatemala and El Salvador, where it also turns westward to join the North Equatorial Current. During this season, larvae of southern Californian faunal elements may be transported into Central American coastal waters. But because this season is the northern winter, reproduction (and thus larval availability) is expected to be minimal.

The southeastern extremity of that Central American coastal current does not seem to ever reach the Panama region; the few larvae transported at that time would seem most likely to be picked up and sent westward off the coast of Costa Rica by the North Equatorial Current at its peak during the same season. It thus seems almost impossible for Californian elements to be transported by currents as far as the Galápagan region and the faunal affinity between southern California and the Galápagos is expected to be low.

Consequently, no pure Californian species are known to occur in the Galápagos (see Table 2). However, there are some 46 species in the Galápagos that have a Panamic + Californian distribution, as well as 17 species having a Panamic + Californian + Peruvian distribution (Table 2). We may explain this by oceanographic events during the rest of the year. In the northern summer (the May–November period), the North Equatorial Current is weak or nearly absent, while the North Equatorial Countercurrent flowing toward Central America is stronger. At its eastern end, in the region off Costa Rica around Cocos Island, its main branch turns northward and produces currents flowing along the coasts toward the northwest (Fig. 4). The coastal currents off Mexico thus flow toward the north and because the North Equatorial Current that would send the waters back toward the west is almost absent, the coastal currents may extend as far north as off the outer coast of Baja California (Fig. 4). Indeed, during the same period, the California Current, which normally flows toward the south, is also weak or almost absent.

During the northern summer, the coastal currents flowing northward along the coasts of Central America and Mexico may thus transport Panamic elements up to the southern Californian region; this seems not unlikely because it is also the warm season when larvae are most likely to abound. Species occurring in the Galápagos

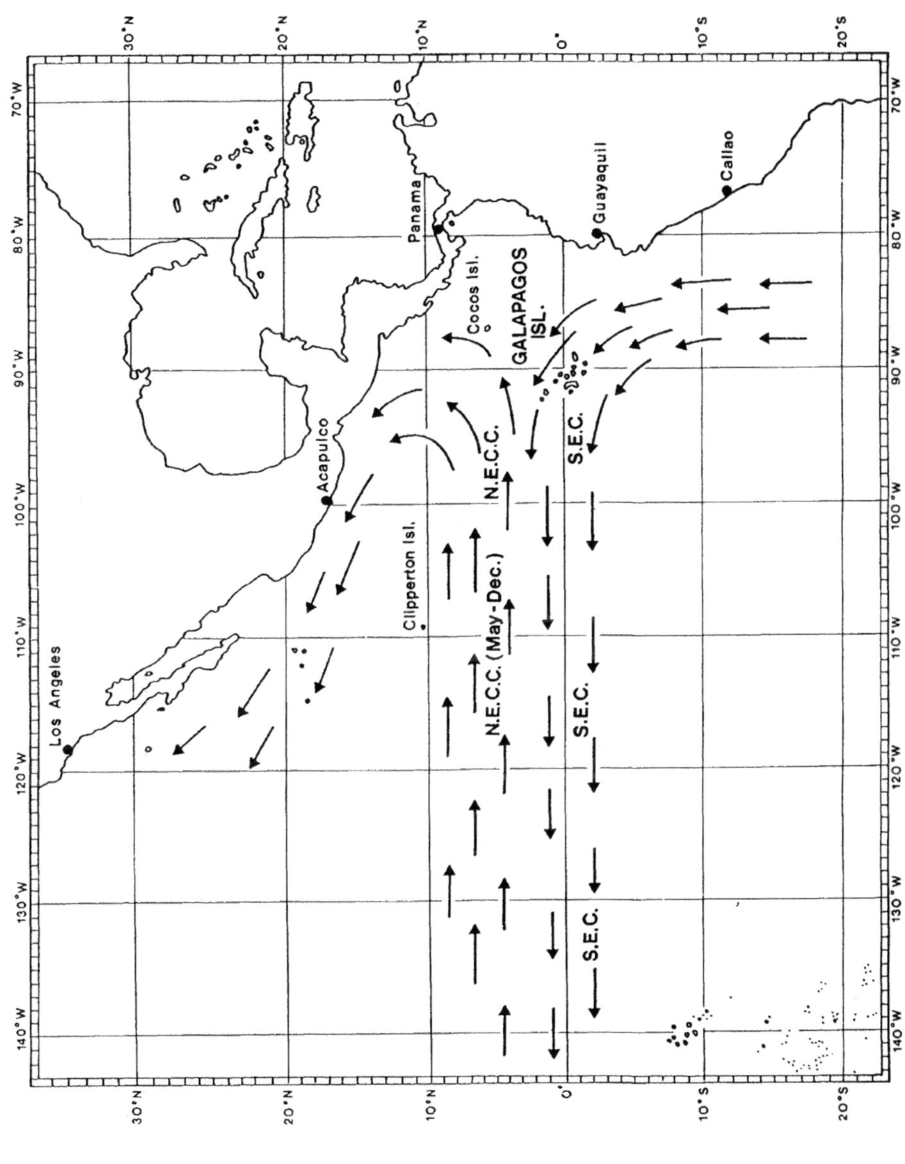

Figure 4. Ocean currents in the eastern Pacific during the May–December period, relating to the faunal affinities between the Galápagos and the Californian Province.

that have Californian affinities are thus more likely to be Panamic species ranging widely along the Central American coasts, that may extend as far north as into the Californian Province, and not the contrary.

4. Evolutionary Patterns and Endemism

James (1984) emphasized the statement that different mechanisms account for the evolution of marine and terrestrial organisms in the Galápagos. Isolation, low rate of dispersal and a fairly high rate of evolutionary change among terrestrial vertebrates such as mammals has led to spectacular evolution and endemism through adaptive radiation and speciation in the terrestrial fauna and flora of the Galápagos Islands. Though it has long been assumed, sometimes unconsciously, that the situation might be comparable for the marine biota, James (1984) clearly pointed out that an alternative evolutionary scenario more likely applies to the marine organisms, especially invertebrates. First, shallow-water marine invertebrates like the mollusks might have a slower intrinsic rate of evolutionary change and speciation (Stanley, 1978, 1979). Second, a lower degree of isolation due to higher dispersal abilities and gene flow also leads to smaller evolutionary divergence from mainland source populations.

As discussed above, the main dispersal mechanism for some shallow-water marine mollusks or invertebrates not pertaining to the meiofauna is the transport of planktonic larvae by ocean currents. This mechanism results in a continuous exchange of genetic material between island and mainland populations. It is important to point out that, unlike what applies to terrestrial organisms, dispersal generally occurs at the larval stage in marine invertebrates, i.e., at a very early stage of development in the life history of the organism (James, 1984; Lutz and Jablonski, 1980).

This genetic exchange has important implications in terms of evolution of populations of marine mollusks. Adaptive radiation and speciation into unoccupied ecological niches are expected to be less effective than they are among the terrestrial organisms, leading to a lower degree of endemism. Not only are the evolutionary changes reduced in magnitude, but they rarely concern taxa beyond the species level. The magnitude of change is generally too small to provide endemicity that would be justified at higher taxonomic levels. Although a few tropical genera seem to be represented in the Panamic Province by endemic species from the Galápagos, there are no marine molluscan genera nor families known to be endemic to the Galápagos. On the contrary, there is a fairly close similarity in faunal diversity at high taxonomic levels between the archipelago and mainland populations. This similarity is presented in Table 3, in which the number of taxa of the shallow water molluscan fauna of the Galápagos is compared to the one of the mainland Panamic Province for different taxonomic levels.

As noted above, the high value of 42% of Galápagan endemic species obtained by compilation of the data from Keen (1971) and reported by James (1984) definitely has to be ruled out. The percentage of about 22.7% reported here (or even of 18.7% if we consider only the shallow-water mollusks) is thus more reasonable and it is closer to the prediction of Briggs (1966). The real rate of endemicity may still be lower. Several findings will reduce the eventual number of endemic species from

Table 3. Richness of the Shallow-Water Malacofauna of the Galápagos Compared to the Entire Panamic Fauna for Different Taxonomic Levels[a]

Galápagos/total Panamic	Species	Genera	Families	Superfamilies
Gastropoda	20%	45%	58%	75%
(Prosobranchia)	(23.5)	(49)	(72)	(90)
(Archaeogastropoda)	(25)	(41)	(60)	(83)
(Mesogastropoda)	(23.5)	(53)	(72)	(89)
(Neogastropoda)	(23.5)	(48)	(80)	(100)
(Opisthobranchia + Pyramidell.)	(8.5)	(37)	(46)	(50)
(Gymnomorpha + mar. Pulmonata)	(24)	(54)	(100)	(100)
Bivalvia	13	33	47	64
Polyplacophora	26.5	53	75	not calculated
Scaphopoda	12	50	50	not calculated
Cephalopoda	31	100	100	100
Total Mollusca	19	43	58	70

[a]Approximate round percentages.

the Galápagos, including: 1) confirming the occurrence of additional Panamic species in the Galápagos, and 2) discovery of species originally described as endemic to the Galápagos on the mainland coast of South America.

A low degree of endemism, restricted to the species level, may not arise in the same way for deep-water organisms. Dispersal and genetic exchange are generally not affected by surface ocean currents, except for some deep-water molluscan species having planktotrophic larvae migrating vertically up to surface waters, as revealed by the study of their protoconchs (Bouchet, 1976; Bouchet and Fontes, 1981; Bouchet and Warén, 1979).

At the species level, Table 1 shows that about 63.8% of the deep-water mollusks known to the Galápagos are endemic. Relationships with adjacent abyssal communities however are not yet very well established, because of the lack of available data and material (Rokop, 1972).

A spectacular endemism at high taxonomic levels has been recently discovered and studied among the benthic hydrothermal vent communities of the Galápagos Rift, not very far from the Archipelago itself (Arp and Childress, 1981; Boss and Turner, 1980; Cavanaugh *et al.*, 1981; Felbeck, 1981; Filatova, 1980; Fretter *et al.*, 1981; Jones, 1981; Kenk and Wilson, 1985; McLean, 1981; Rau, 1981; Rhoads *et al.*, 1981; Turekian and Cochran, 1981; Wittenberg *et al.*, 1981). Lutz *et al.* (1980) indicated the presence of a planktotrophic larval stage in a mytilid bivalve from the Galápagos Rift that could account for long-range dispersal abilities through transport by currents.

In conclusion, similar neo-darwinian mechanisms probably underlie evolution among both terrestrial and shallow-water or deep-water marine organisms, namely, genes undergo mutations, individuals undergo adaptations, populations evolve and species change. However, because of dissimilar ecological, biogeographical, and life history features among various Galápagos organisms, different macroevolutionary scenarios can be observed in these two components of the biota. These differing scenarios have produced separate degrees of evolutionary divergence or rates of speciation among their respective populations.

5. Appendix: List of Marine Molluscan Fauna*

Class: POLYPLACOPHORA

Acanthochitona cf. avicula (Carpenter, 1866)
Acanthochitona hirudiniformis (Sowerby, 1832)
Acanthochitona imperatrix Watters, 1981
Acanthochitona jacquelinae Smith and Ferreira, 1977
Callistochiton carmenae Smith and Ferreira, 1977
Calloplax duncanus (Dall, 1919)
Chaetopleura cf. mixta (Dall, 1919)
Ishnochiton petaloïdes (Gould, 1846)
Ischnochiton macleani Ferreira, 1978
Leptochiton albemarlensis Smith and Ferreira, 1977
Chiton goodallii Broderip, 1832
Chiton sulcatus Wood, 1815
Tonicia forbesii arnheimi Dall, 1903

Class: GASTROPODA
Subclass: Prosobranchia
Order: Archaeogastropoda

Haliotis dalli Henderson, 1915
Scissurella epicharis McLean, 1970
Sinezona rimuloïdes (Carpenter, 1865)
Diodora alta (C. B. Adams, 1852)
Diodora inaequalis (Sowerby, 1835)
Diodora pica (Sowerby, 1835)
Diodora punctifissa McLean, 1970
Diodora saturnalis (Carpenter, 1864)
Emarginula tuberculosa Libassi, 1859
Fissurella deroyae McLean, 1970
Fissurella macrotrema Sowerby, 1835
Fissurella microtrema Sowerby, 1835
Fissurella obscura Sowerby, 1835
Fissurella virescens Sowerby, 1835
Lucapinella crenifera (Sowerby, 1835)
Nesta galapagensis McLean, 1970
Puncturella expansa Dall, 1896
Collisella mitella (Menke, 1847)
Lottia mimica Lindberg and McLean, 1981
Lottia smithi Lindberg and McLean, 1981
Tectura filosa (Carpenter, 1865)
Tectura immaculata (Lindberg and McLean, 1981)
Tectura rothi (Lindberg and McLean, 1981)
Bathybembix galapagana (Dall, 1908)
Calliostoma jacquelinae McLean, 1970
Calliostoma leanum (C. B. Adams, 1852)
Calliostoma mcleani Shasky and Campbell, 1964
Calliostoma santacruzanum McLean, 1970
Gaza rathbuni Dall, 1890
Mirachelus galapagensis McLean, 1970
Solariella diomedea Dall, 1919
Solariella tavernia Dall, 1919
Tegula cooksoni (E. A. Smith, 1877)
Tegula snodgrassi (Pilsbry and Vanatta, 1902)
Ganesa filosa Dall, 1919
Ganesa piona Dall, 1919
Cyclostremiscus cosmius (Bartsch, 1907)
Cyclostremiscus glyptomphalus Pilsbry and Olsson, 1952
Cyclostremiscus pachynepion Pilsbry and Olsson, 1945
Cyclostremiscus planospiratus (Carpenter, 1857)
Cyclostremiscus valvatoïdes (C. B. Adams, 1852)
Brookula galapagana (Dall, 1913)
Arene echinata McLean, 1970
Arene ferruginosa McLean, 1970
Arene guttata McLean, 1970
Turbo scitulus (Dall, 1919)
Tricolia diantha McLean, 1970
Nerita funiculata Menke, 1851
Nerita scabricosta Lamarck, 1822

*The six following species mentioned previously (Finet, 1985) have been left out from the list as their occurrence in the Galápagos could not be verified and appears dubious: *Simnia avena* (Sowerby, 1832); *Vitularia salebrosa* (King and Broderip, 1832); *Subcancilla sulcata* (Swainson in Sowerby, 1825) (records from the Galápagos in the collections appear to be based upon misidentifications of *Mitra sphoni* Shasky and Campbell, 1964); *Bulla gouldiana* Pilsbry, 1895; *Mactra fonsecana* Hertlein and Strong, 1950; *Solecurtus guaymasensis* Lowe, 1935.

Order: Mesogastropoda

Littorina aspera Philippi, 1846
Littorina atrata (C. B. Adams, 1852)
Littorina excavata (C. B. Adams, 1852)
Littorina modesta Philippi, 1846
Littorina paytensis Philippi, 1847
Littorina peruviana Lamarck, 1822
Littorina porcata (Philippi, 1846)
Nodilittorina galapagiensis (Stearns, 1892)
Paesiella roosevelti Bartsch and Rehder, 1939
?*Alvania chathamensis* Bartsch, MS
?*Alvania duncani* Bartsch, MS
?*Alvania oldroydae* Bartsch, 1911
Alvinia ima (Bartsch, 1911)
Alvinia lara (Bartsch, 1911)
Alvinia montserratensis (Baker, Hanna and Strong, 1930)
Alvinia profundicola (Bartsch, 1911)
Alvinia tumida (Carpenter, 1857)
Lirobarleeia galapagensis (Bartsch, 1911)
Lirobarleeia halia (Bartsch, 1911)
Lirobarleeia hoodensis (Bartsch, 1911)
Lirobarleeia nemo (Bartsch, 1911)
Amphithalamus inclusus Carpenter, 1865
Rissoella johnstoni Baker, Hanna and Strong, 1930
Rissoella tumens (Carpenter, 1857)
Rissoina dina Bartsch, 1915
Rissoina ericana Hertlein and Strong, 1951
Rissoina fortis (C. B. Adams, 1852)
Rissoina gisna Bartsch, 1915
Rissoina inca d'Orbigny, 1840
Rissoina insignis (De Folin, 1867)
Rissoina io Bartsch, 1915
Rissoina laurae De Folin, 1870
Rissoina stricta Menke, 1850
Assiminea compacta (Carpenter, 1864)
Caecum eburneum C. B. Adams, 1852
Caecum richthofeni Strong and Hertlein, 1939
Caecum semilaeve Carpenter, 1857
Elephantanellum mirificum De Folin, 1867
Fartulum dextroversum (Carpenter, 1857)
Episcynia bolivari Pilsbry and Olsson, 1946
Episcynia medialis Keen, 1971
Episcynia nicholsoni (Strong and Hertlein, 1939)
Parviturboïdes copiosus (Pilsbry and Olsson, 1945)
Parviturboïdes decussatus (Carpenter, 1857)
Solariorbis pellucidus Pilsbry and Olsson, 1952
Teinostoma pallidulum (Carpenter, 1857)
Macromphalina souverbiei (De Folin, 1867)
Heliacus bicanaliculatus (Valenciennes, 1832)
Heliacus caelatus (Hinds, 1844)
Heliacus mazatlanicus Pilsbry and Lowe, 1932
Vermicularia pellucida eburnea (Reeve, 1842)
Modulus cerodes (A. Adams, 1851)
Petaloconchus complicatus Dall, 1908
Serpulorbis margaritaceus (Chenu, 1844, ex Rousseau, MS)
Planaxis planicostatus Sowerby, 1825
Batillaria mutata (Pilsbry and Vanatta, 1902)
Cerithium adustum Kiener, 1841
Cerithium gallapaginis Sowerby, 1855
Cerithium gemmatum Hinds, 1844
Cerithium maculosum Kiener, 1841
Cerithium stercusmuscarum Valenciennes, 1833
Cerithium uncinatum (Gmelin, 1791)
Cerithiopsis bicolor Bartsch, 1911
Cerithiopsis curtata Bartsch, 1911
Cerithiopsis eiseni Strong and Hertlein, 1939
Cerithiopsis galapagensis Bartsch, 1911
Cerithiopsis guanacastensis Hertlein and Strong, 1951
Cerithiopsis neglecta (C. B. Adams, 1852)
Eumetula eucosmia Bartsch, 1911
Seila assimilata (C. B. Adams, 1852)
Alaba supralirata Carpenter, 1857
Metaxia convexa (Carpenter, 1857)
Triphora adamsi Bartsch, 1907
Triphora alternata C. B. Adams, 1852
Triphora chathamensis Bartsch, 1907
Triphora contrerasi Baker, 1926
Triphora cookeana Baker and Spicer, 1935
Triphora dalli Bartsch, 1907
Triphora escondidensis Baker, 1926
Triphora evermanni Baker, 1926
Triphora excolpa Bartsch, 1907
Triphora galapagensis Bartsch, 1907
Triphora hannai Baker, 1926
Triphora oweni Baker, 1926
Triphora postalba Bartsch, 1907
Triphora triticea (Pease, 1861)
Triphora unicolor Bartsch, 1907
Fossarus abjectus (C. B. Adams, 1852)
Fossarus angiostoma (C. B. Adams, 1852)
Fossarus angulatus Carpenter, 1857
Fossarus megasoma (C. B. Adams, 1852)
Strombus galeatus Swainson, 1823
Strombus granulatus Swainson, 1822
Epitonium acapulcanum Dall, 1917
Epitonium billeeanum (DuShane and Bratcher, 1965)
Epitonium emydonesus Dall, 1917
Epitonium eutaenium Dall, 1917
Epitonium habeli Dall, 1917
Epitonium huffmani DuShane and McLean, 1968
Epitonium indistinctum (Sowerby, 1844)
Epitonium minuticosta (De Boury, 1912)
Epitonium rhytidum Dall, 1917

Epitonium replicatum (Sowerby, 1844)
Cirsotrema togata Hertlein and Strong, 1951
Nitidiscala cumingii (Carpenter, 1856)
Nitidiscala hancocki DuShane, 1970
Nitidiscala willetti Strong and Hertlein, 1937
Sthenorhytis turbinum Dall, 1908
Amaea brunneopicta (Dall, 1908)
Amaea deroyae DuShane, 1970
Amaea pompholyx (Dall, 1890)
Opalia crenatoïdes (Carpenter, 1864)
Opalia diadema (Sowerby, 1832)
Opalia funiculata (Carpenter, 1857)
Opalia spongiosa Carpenter, 1864
Janthina janthina (Linnaeus, 1758)
Balcis falcata (Carpenter, 1865)
Balcis ochsneri (Bartsch, 1917)
Balcis panamensis (Bartsch, 1917)
Eulima elegantissima De Folin, 1867
Eulimostraca galapagensis Bartsch, 1917
Niso cf. aeglees Bush, 1895
Niso imbricata (Sowerby, 1834)
Sabinella chathamensis Bartsch, 1917
Sabinella meridionalis Bartsch, 1917
Stilifer astericola Broderip, 1832
Hipponix grayanus Menke, 1853
Hipponix panamensis C. B. Adams, 1852
Hipponix pilosus (Deshayes, 1832)
Vanikoro galapagana Hertlein and Strong, 1951
Cheilea cepacea (Broderip, 1834)
Cheilea corrugata (Broderip, 1834)
Crepidula aculeata (Gmelin, 1791)
Crepidula excavata (Broderip, 1834)
Crepidula onyx Sowerby, 1824
Crepidula striolata Menke, 1851
Crucibulum scutellatum (Wood, 1828)
Capulus sericeus J. and R. Burch, 1961
Thyca callista Berry, 1959
Eunaticina insculpta (Carpenter, 1865)
Natica brunneolinea McLean, 1970
Natica chemnitzii Pfeiffer, 1840
Natica elenae Récluz, 1844
Natica grayi Philippi, 1852
Natica idiopoma Pilsbry and Lowe, 1932
Natica othello Dall, 1908
Natica sigillata McLean, 1970
Polinices caprae (Philippi, 1850)
Polinices crawfordianus Dall, 1908
Polinices hacketti Marincovich, 1975
Polinices intemeratus (Philippi, 1851)
Polinices litorinus Dall, 1908
Polinices pardoanus Dall, 1908
Polinices uber (Valenciennes, 1832)
Lamellaria inflata (C. B. Adams, 1852)
Hespererato galapagensis Schilder, 1933
Niveria maugeriae (Sowerby, 1832, ex Gray, MS)
Niveria occidentalis (Schilder, 1822)
Niveria pacifica (Sowerby, 1832, ex Gray, MS)
Pusula fusca (Sowerby, 1832, ex Gray, MS).
Pusula rubescens (Gray, 1833)
Pusula sanguinea (Sowerby, 1832, ex Gray, MS)
Pusula solanderi (Sowerby, 1832, ex Gray, MS)
Trivia citeria (Cate, 1979)
Cypraea albuginosa Gray, 1825
Cypraea arabicula (Lamarck, 1811)
Cypraea cervinetta Kiener, 1843
Cypraea isabellamexicana Stearns, 1893
Cypraea moneta Linnaeus, 1758
Cypraea nigropunctata Gray, 1828
Cypraea robertsi (Hidalgo, 1906)
Cypraea teres Gmelin, 1791
Jenneria pustulata (Lightfoot, 1786)
Pseudocypraea adamsonii (Sowerby, 1832)
Simnia aequalis (Sowerby, 1832)
Simnia rufa (Sowerby, 1832)
Malea ringens (Swainson, 1822)
Tonna perdix (Linnaeus, 1758)
Casmaria vibexmexicana (Stearns, 1894)
Cypraecassis coarctata (Sowerby, 1825)
Cypraecassis tenuis (Wood, 1828)
Semicassis centiquadrata (Valenciennes, 1832)
Charonia tritonis (Linnaeus, 1758)
Cymatium amictoideum Keen, 1971
Cymatium gemmatum Reeve, 1844
Cymatium gibbosum (Broderip, 1833)
Cymatium lineatum (Broderip, 1833)
Cymatium muricinum (Röding, 1798)
Cymatium parthenopeum (Von Salis, 1793)
Cymatium pileare (Linnaeus, 1758)
Cymatium vestitum (Hinds, 1844)
Distorsio constricta (Broderip, 1833)
Linatella succincta (Linnaeus, 1771)
Bursa caelata Broderip, 1833)
Bursa calcipicta Dall, 1908

Order: NEOGASTROPODA

Acanthina brevidentata (Wood, 1828)
Acanthotrophon sentus (Berry, 1969)
Aspella hastula (Reeve, 1844)
Aspella pollux Radwin and D'Attilio, 1976
Aspella pyramidalis (Broderip, 1833)
Bizetiella shaskyi Radwin and D'Attilio, 1973
Drupa ricinus (Linnaeus, 1758)
Evokesia rufonotata (Carpenter, 1864)
Favartia incisa (Broderip, 1833)
Favartia purdyae Vokes and D'Attilio, 1980
Maxwellia angermayerae (Emerson and D'Attilio, 1965)

Murexiella humilis (Broderip, 1833)
Murexiella lappa (Broderip, 1833)
Murexiella perita (Hinds, 1844)
Murexiella radwini Emerson and D'Attilio, 1970
Murexiella venustula Poorman, 1983
Murexsul jacquelinae Emerson and D'Attilio, 1969
Muricanthus princeps (Broderip, 1833)
Muricopsis zeteki Hertlein and Strong, 1951
Neorapana grandis (Sowerby, 1835, ex Gray, MS)
Neorapana muricata (Broderip, 1832)
Nipponotrophon galapaganus Emerson and D'Attilio, 1970)
Phyllocoma scalariformis (Broderip, 1833)
Pteropurpura deroyana Berry, 1968
Purpura columellaris (Lamarck, 1822)
Purpura pansa Gould, 1853
Thaïs biserialis (Blainville, 1832)
Thaïs callaoensis (Gray, 1828)
Thaïs melones (Duclos, 1832)
Thaïs planospira (Lamarck, 1822)
Thaïs speciosa (Valenciennes, 1832)
Thaïs triangularis (Blainville, 1832)
Tripterotyphis lowei (Pilsbry, 1931)
Typhis clarki Keen and Campbell, 1964
Babelomurex deroyorum (D'Attilio and Myers, 1984)
Babelomurex hindsi (Carpenter, 1857)
Babelomurex santacruzensis (Emerson and D'Attilio, 1970)
Coralliophila costata (Blainville, 1832)
Coralliophila cumingii (H. and A. Adams, 1864)
Coralliophila neritoïdes (Lamarck, 1816)
Coralliophila nux (Reeve, 1846)
Coralliophila orcuttiana Dall, 1919
Coralliophila parva (E. A. Smith, 1877)
Quoyula madreporarum (Sowerby, 1834)
Caducifer biliratus (Reeve, 1846)
Caducifer cinis (Reeve, 1846)
Caducifer pictus (Reeve, 1844)
Cantharus fumosus (Dillwyn, 1817)
Cantharus janellii (Kiener, 1835–36)
Cantharus ringens (Reeve, 1846)
Cantharus sanguinolentus (Duclos, 1833)
Engina maura (Sowerby, 1832)
Engina pyrostoma (Sowerby, 1832)
Engina tabogaensis Bartsch, 1931
Phos laevigatus (A. Adams, 1851)
Anachis atramentaria (Sowerby, 1844)
Anachis fusidens (Dall, 1908)
Anachis nigricans (Sowerby, 1844)
Anachis rugulosa (Sowerby, 1844)
Anachis teevani Hertlein and Strong, 1951
Anachis uncinata (Sowerby, 1832)
Anachis varia (Sowerby, 1832)
Columbella castanea Sowerby, 1832
Columbella festiva Kiener, 1841
Columbella fuscata Sowerby, 1832
Columbella haemastoma Sowerby, 1832
? *Mitrella elegans* (Dall, 1871)
Mitrella guttata (Sowerby, 1832)
Nassarina hancocki Hertlein and Strong, 1939
Nassarina melanosticta (Pilsbry and Lowe, 1932)
Strombina bicanalifera (Sowerby, 1832)
Strombina deroyae Emerson and D'Attilio, 1969
Strombina lanceolata (Sowerby, 1832)
Strombina maculosa (Sowerby, 1832)
Strombina recurva (Sowerby, 1832)
Zafrona incerta (Stearns, 1892)
Nassarius caelolineatus Nesbitt and Pitt, 1986
Nassarius catallus (Dall, 1908)
Nassarius corpulentus (C. B. Adams, 1852)
Nassarius exsarcus (Dall, 1908)
Nassarius goniopleura (Dall, 1908)
Nassarius shaskyi McLean, 1970
Nassarius townsendi (Dall, 1890)
Nassarius taeniolatus (Philippi, 1845)
Nassarius versicolor (C. B. Adams, 1852)
Colubraria lucasensis Strong and Hertlein, 1937
Colubraria ochsneri Hertlein and Allison, 1968
Fusinus allyni McLean, 1970
Fusinus dupetitthouarsi (Kiener, 1840)
Fusinus humboldti Poorman, 1981
Latirus centrifugus (Dall, 1915)
Latirus concentricus (Reeve, 1847)
Latirus sanguineus (Wood, 1828)
Leucozonia tuberculata (Broderip, 1833)
Pleuroploca princeps Sowerby, 1825
Calliotectum vernicosum Dall, 1890
Morum tuberculosum (Reeve, 1842, ex Sowerby, MS)
Morum veleroae Emerson, 1968
Surculina galapagana (Dall, 1919)
Oliva kaleontina Duclos, 1835
Oliva porphyria (Linnaeus, 1758)
Olivella anazora (Duclos, 1835)
Olivella bitleri Olsson, 1956
Olivella fletcherae Berry, 1958
Olivella gracilis (Broderip and Sowerby, 1829)
Dentimargo eremus (Dall, 1919)
Granula insularum Roth and Coan, 1971
Granula minor (C. B. Adams, 1852)
Granula polita (Carpenter, 1857)
Granulina margaritula (Carpenter, 1857)
Persicula imbricata (Hinds, 1844)
Persicula phrygia (Sowerby, 1846)
Volvarina nyssa Roth and Coan, 1971
Volvarina taeniolata rosa (Schwengel, 1938)
Volvarina taeniolata taeniolata Mörch, 1860
Mitra crenata Broderip, 1836
Mitra effusa Broderip, 1836
Mitra gausapata Reeve, 1845

Mitra lens Wood, 1828
Mitra mitra (Linnaeus, 1758)
Mitra sphoni (Shasky and Campbell, 1964)
Mitra tristis Broderip, 1836
Subcancilla edithrexae Sphon, 1976
Thala gratiosa (Reeve, 1845)
Thala jeancatae Sphon, 1969
Thala solitaria (C. B. Adams, 1852)
Agatrix deroyae (Petit, 1970)
Cancellaria chrysostoma Sowerby, 1832
Cancellaria darwini Petit, 1970
Cancellaria gemmulata Sowerby, 1832
Cancellaria haemastoma Sowerby, 1832
Cancellaria obesa Sowerby, 1832
Cancellaria pulchra Sowerby, 1832
Cancellaria urceolata Hinds, 1843
Sveltia gladiator (Petit, 1976)
Trigonostoma milleri Burch, 1949
Tritonoharpa vexillata Dall, 1908
Conus brunneus Wood, 1828
Conus chaldeus (Röding, 1798)
Conus dalli Stearns, 1873
Conus diadema Sowerby, 1834
Conus ebraeus Linnaeus, 1758
Conus fergusoni Sowerby, 1873
Conus gladiator Broderip, 1833
Conus kohni McLean and Nybakken, 1979
Conus lucidus Wood, 1828
Conus mahogani Reeve, 1843
Conus nux Broderip, 1833
Conus perplexus Sowerby, 1857
Conus purpurascens Sowerby, 1833, ex Broderip, MS
Conus recurvus Broderip, 1833
Conus tessulatus Born, 1778
Conus tiaratus Sowerby, 1833, ex Broderip, MS
Conus xanthicus Dall, 1910
Conus ximenes Gray, 1839
Terebra armillata Hinds, 1844
Terebra frigata Hinds, 1844
Terebra glauca Hinds, 1844
Terebra guayaquilensis E. A. Smith, 1880
Terebra hertleini Bratcher and Burch, 1970
Terebra jacquelinae Bratcher and Burch, 1970
Terebra ornata Gray, 1834
Terebra plicata Gray, 1834
Terebra purdyae Bratcher and Burch, 1970
Terebra robusta Hinds, 1844
Terebra stohleri Bratcher and Burch, 1970
Terebra strigata Sowerby, 1825
Agathotoma camarina (Dall, 1919)
Agladrillia badia McLean and Poorman, 1971
Borsonella abrupta McLean and Poorman, 1971
Borsonella agassizii (Dall, 1908)
Borsonella galapagana McLean and Poorman, 1971
Carinodrillia adonis Pilsbry and Lowe, 1932
Cerodrillia asymmetrica McLean and Poorman, 1971
Cerodrillia cybele (Pilsbry and Lowe, 1932)
Clathurella maryae McLean and Poorman, 1971
Clathurella rigida (Hinds, 1843)
Cleospira ochsneri (Hertlein and Strong, 1949)
Cochlespira cedonulli (Reeve, 1843)
Compsodrillia gracilis McLean and Poorman, 1971
Compsodrillia undatichorda McLean and Poorman, 1971
Cryptogemma benthima (Dall, 1908)
Cryptogemma eldorana (Dall, 1908)
Cryptogemma polystephanus (Dall, 1908)
Daphnella allemani (Bartsch, 1931)
Daphnella bartschi Dall, 1919
Daphnella gemmulifera McLean and Poorman, 1971
Daphnella mazatlanica Pilsbry and Lowe, 1932
Drillia albicostata (Sowerby, 1834)
Drillia clavata (Sowerby, 1834)
Drillia sinuosa McLean and Poorman, 1971
Drilliola zeuxippe (Dall, 1919)
Elaeocyma amplinucis McLean and Poorman, 1971
Elaeocyma melichroa McLean and Poorman, 1971
Elaeocyma splendidula (Sowerby, 1834)
Euclathurella acclivicallis McLean and Poorman, 1971
Fusiturricula andrei McLean and Poorman, 1971
Glyphostoma pustulosa McLean and Poorman, 1971
Glyphostoma scobina McLean and Poorman, 1971
Gymnobela brachis (Dall, 1919)
Gymnobela xylona (Dall, 1908)
Hindsiclava hertleini Emerson and Radwin, 1969
Ithycythara penelope (Dall, 1919)
Kermia informa McLean and Poorman, 1971
Kurtzia humboldti McLean and Poorman, 1971
Kylix rugifera (Sowerby, 1834)
Leptadrillia firmichorda McLean and Poorman, 1971
Leucosyrinx exulans (Dall, 1890)
Lioglyphostoma ericea (Hinds, 1843)
Microdaphne trichodes (Dall, 1919)
Mitrolumna keenae Emerson and Radwin, 1969
Mitrolumna mitriformis (Shasky, 1961)
Mitromorpha carpenteri Gilbert, 1954
Phymorhynchus castaneus (Dall, 1895)
Phymorhynchus cingulatus (Dall, 1890)
Pleurotomella dinora Dall, 1908
Pleurotomella hermione (Dall, 1919)
Rimosodaphnella deroyae McLean and Poorman, 1971

Splendrillia academica McLean and Poorman, 1971
Truncadaphne stonei (Hertlein and Strong, 1939)
Veprecula tornipila McLean and Poorman, 1971
Xanthodaphne agonia (Dall, 1890)
Xanthodaphne argeta (Dall, 1890)
Xanthodaphne suffusa (Dall, 1890)

Superfamily: Pyramidelloidea

Chrysallida rinella (Dall and Bartsch, 1909)
Cingulina evermanni (Baker, Hanna and Strong, 1928)
Herviera glirella (Melvill and Standen, 1896)
Menestho grijalvae (Baker, Hanna and Strong, 1854)
Miralda galapagensis (Dall and Bartsch, 1909)
Odostomia navisa Dall and Bartsch, 1907
Odostomia parella Dall and Bartsch, 1909
Pyramidella hastata (A. Adams in Sowerby, 1854)
Triptychus incantatus (Hertlein and Strong, 1939)
Turbonilla galapagensis Dall and Bartsch, 1909
Turbonilla houseri Dall and Bartsch, 1909
Turbonilla shimeki Dall and Bartsch, 1909
Turbonilla sinaloana Strong, 1949
Phasianema saxicola (C. B. Adams, 1852)

Subclass: Opisthobranchia

Microglyphis perconicus (Dall, 1890)
Bulla punctulata A. Adams, in Sowerby, 1850
Bulla rufolabris A. Adams, in Sowerby, 1850
Sulcoretusa galapagana (Dall, 1919)
Volvulella catharia Dall, 1919
Volvulella cylindrica (Carpenter, 1864)
Cylichna luticola (C. B. Adams, 1852)
Scaphander interruptus Dall, 1889
Navanax aenigmaticus (Bergh, 1894)
Cavolinia globulosa (Gray, 1850)
Cavolinia inflexa (Lesueur, 1813)
Cavolinia tridentata (Niebuhr, 1775)
Clio chaptalli Gray, 1850
Clio pyramidata Linnaeus, 1767
Diacria quadridentata (Blainville, 1821, ex Lesueur, MS)
Aplysia californica Cooper, 1863
Aplysia parvula Mörch, 1863
Dolabrifera dolabrifera (Rang, 1828)
Berthelinia chloris (Dall, 1918)
Julia thecaphora (Carpenter, 1857)
Lobiger souverbiei Fischer, 1856
Tylodina fungina Gabb, 1865
Umbraculum umbraculum (Lightfoot, 1786)
Berthellina engeli Gardiner, 1936
Pleurobranchus areolatus (Mörch, 1863)
Conualevia alba Collier and Farmer, 1964
Chromodoris baumanni Bartsch, 1970
Chromodoris sphoni (Marcus, 1971)
Glossodoris dalli (Bergh, 1894)
Glossodoris sedna (Marcus and Marcus, 1967)
Hypselodoris agassizi (Bergh, 1894)
Hypselodoris lapislazuli (Bertsch and Ferreira, 1974)
Doris peruviana d'Orbigny, 1837
Platydoris carolynae Mulliner and Sphon, 1974
Dendrodoris krebsii (Mörch, 1863)
Doriopsilla janaina Marcus and Marcus, 1967
Tambja mullineri Farmer, 1978
Flabellina iodinea (Cooper, 1863)
Phidiana lynceus Bergh, 1867
Aeolidiella alba Risbec, 1928

Subclass: Gymnomorpha

Hoffmannola lesliei (Stearns, 1892)
Onchidella steindachneri (Semper, 1882)

Subclass: Pulmonata

Melampus carolianus (Lesson, 1842)
Melampus tabogensis C. B. Adams, 1852
Pedipes angulatus C. B. Adams, 1852
Tralia vanderbilti Schwengel, 1938
Williamia peltoïdes (Carpenter, 1864)
Siphonaria gigas Sowerby, 1825

Class: SCAPHOPODA

Compressidens brevicornu (Pilsbry and Sharp, 1897)
Dentalium oerstedii Mörch, 1860
Episiphon innumerabilis (Pilsbry and Sharp, 1897)
Fissidentalium megathyris (Dall, 1890)
Siphonodentalium peruvianum (Dall, 1908)

Class: BIVALVIA

Nuculana elenensis (Sowerby, 1833)
Nuculana pontonia (Dall, 1890)
Anadara multicostata (Sowerby, 1833)
Arca mutabilis (Sowerby, 1833)
Arca pacifica (Sowerby, 1833)
Arca truncata (Sowerby, 1833)
Arcopsis solida (Sowerby, 1833)
Barbatia bailyi (Bartsch, 1931)
Barbatia gradata (Broderip and Sowerby, 1829)
Barbatia reeveana (d'Orbigny, 1846)
Barbatia rostae Berry, 1954
Fugleria illota (Sowerby, 1833)
Glycymeris lintea Olsson, 1961
Glycymeris maculata (Broderip, 1832)
Brachidontes adamsianus (Dunker, 1857)
Brachidontes puntarenensis (Pilsbry and Lowe, 1932)
Brachidontes semilaevis (Menke, 1849)
Modiolus capax (Conrad, 1837)
Myoforceps aristata (Dillwyn, 1817)
Mytella strigata (Hanley, 1843)
Septifer zeteki Hertlein and Strong, 1946
Atrina texta Hertlein, Hanna and Strong, 1943
Pinna rugosa Sowerby, 1835
Pinctada mazatlanica (Hanley, 1856)
Pteria sterna (Gould, 1851)
Isognomon recognitus (Mabille, 1895)
Malleus rufipunctatus (Reeve, 1858)
Ostrea fisheri Dall, 1914
Ostrea palmula Carpenter, 1857
Argopecten circularis (Sowerby, 1835)
Chlamys incantata Hertlein, 1972
Chlamys lowei (Hertlein, 1935)
Cyclopecten exquisitus Grau, 1959
Cyclopecten liriope (Dall, 1908)
Cyclopecten pernomus (Hertlein, 1935)
Delectopecten polyleptus (Dall, 1908)
Delectopecten zacae (Hertlein, 1935)
Hyalopecten neoceanicus (Dall, 1908)
Lyropecten magnificus (Sowerby, 1835)
Lyropecten subnodosus (Sowerby, 1835)
Pecten galapagensis Grau, 1959
Pecten hancocki Grau, 1959
Pecten sericeus Hinds, 1845
Acesta diomedae (Dall, 1908)
Lima tetrica Gould, 1851
Promantellum pacificum (d'Orbigny, 1846)
Pododesmus foliatus (Broderip, 1834)
Codakia distinguenda (Tryon, 1872)
Ctena galapagana (Dall, 1901)
Ctena mexicana (Dall, 1901)
Divalinga eburnea (Reeve, 1850)
Diplodonta subquadrata (Carpenter, 1856)
Phlyctiderma caelatum (Reeve, 1850)
Neaeromya stearnsii (Dall, 1899)
Cymatioa electilis (Berry, 1963)
Kellia suborbicularis (Montagu, 1803)
Halodakra subtrigona (Carpenter, 1857)
Cardita affinis Sowerby, 1833
Strophocardia megastropha (Gray, 1825)
Basterotia peninsularis (Jordan, 1936)
Chama buddiana C. B. Adams, 1852
Chama frondosa Broderip, 1835
Chama squamuligera Pilsbry and Lowe, 1932
Pseudochama clarionensis Willett, 1938
Pseudochama janus Reeve, 1847
Crassinella varians (Carpenter, 1857)
Laevicardium elenense (Sowerby, 1840)
Trachycardium consors (Sowerby, 1833)
Trigoniocardia biangulata (Broderip and Sowerby, 1829)
Chione compta (Broderip, 1835)
Chione subimbricata (Sowerby, 1835)
Chione undatella (Sowerby, 1835)
Dosinia dunkeri (Philippi, 1844)
Dosinia ponderosa (Gray, 1838)
Gouldia californica Dall, 1917
Megapitaria aurantiaca (Sowerby, 1831)
Periglypta multicostata (Sowerby, 1835)
Pitar consanguineus (C. B. Adams, 1852)
Pitar helenae Olsson, 1961
Pitar hoffstetteri Fischer-Piette, 1969
Protothaca grata (Say, 1831)
Protothaca pertincta (Dall, 1902)
Transennella galapagana Hertlein and Strong, 1939
Ventricolaria isocardia (Verrill, 1870)
Florimetis cognata (Pilsbry and Vanatta, 1902)
Florimetis dombei (Hanley, 1844)
Psammotreta viridotincta (Carpenter, 1856)
Tellina amianta Dall, 1900
Tellina chrysogona Dall, 1908
Tellina pacifica Dall, 1900
Cumingia lamellosa (Sowerby, 1833)
Semele flavescens (Gould, 1851)
Semele purpurascens (Gmelin, 1791)
Semele rupium (Sowerby, 1833)
Semele sowerbyi Tryon, 1869
Semele venusta (Reeve, 1853, ex A. Adams, MS)
Gari helenae Olsson, 1961
Cyathodonta galapagana Dall, 1915
Cardiomya californica (Dall, 1886)
Halicardia perplicata (Dall, 1890)

Class: CEPHALOPODA

Argonauta pacificus Dall, 1869
Octopus occidentalis Hoyle, 1896
Octopus oculifer (Hoyle, 1904)
Octopus roosevelti Stuart, 1941

References

Abbott, D. P., 1966, Factors influencing the zoogeographic affinities of Galápagos inshore marine fauna, in: *The Galápagos* (R. I. Bowman, ed.), pp. 108–122, Univerity of California Press, Berkeley.

Arp, A. J., and Childress, J. J., 1981, Blood function in the hydrothermal vent Vestimentiferan tube worm, *Science* **213**:342–344.

Baker, F., 1926, Expedition of the California Academy of Sciences to the Gulf of California in 1921: Mollusca of the family Triphoridae, *Proc. Calif. Acad. Sci.* (4th series) **15**:223–239.

Baker, F., Hanna, G. D., and Strong, A. M., 1928, Some Pyramidellidae from the Gulf of California, *Proc. Calif. Acad. Sci.* (4th series) **17**:205–246.

Baker, F., Hanna, G. D., and Strong, A. M., 1930a, Some Rissoid Mollusca from the Gulf of California, *Proc. Calif. Acad. Sci.* (4th series) **19**:23–40.

Baker, F., Hanna, G. D., and Strong, A. M., 1930b, Some mollusca of the family Epitoniidae from the Gulf of California, *Proc. Calif. Acad. Sci.* (4th series) **19**:41–56.

Baker, F., Hanna, G. D., and Strong, A. M., 1938a, Some mollusca of the families Cerithiopsidae, Cerithiidae and Cyclostrematidae from the Gulf of California and adjacent waters, *Proc. Calif. Acad. Sci.* (4th series) **23**:217–244.

Baker, F., Hanna, G. D., and Strong, A. M., 1938b, Columbellidae from Western Mexico, *Proc. Calif. Acad. Sci.* (4th series) **23**:245–254.

Bartsch, P., 1907, The West American mollusks of the genus *Triphoris*, *Proc. U.S. Natn. Mus.* **33**:249–262.

Bartsch, P., 1910, The West American mollusks of the genus *Alaba*, *Proc. U.S. Natn. Mus.* **39**:153–156.

Bartsch, P., 1911a, The recent and fossil mollusks of the genus *Cerithiopsis* from the west coast of America, *Proc. U.S. Natn. Mus.* **40**:327–367.

Bartsch, P., 1911b, The recent and fossil mollusks of the genus *Alvania* from the west coast of America, *Proc. U.S. Natn. Mus.* **41**:333–362.

Bartsch, P., 1915, The recent and fossil mollusks of the genus *Rissoina* from the west coast of America, *Proc. U.S. Natn. Mus.* **49**:33–62.

Bernard, F. R., 1989, Living Bivalvia of the Galápagos Islands, *Annu. Rep. west. Soc. Malac.* **21**:10.

Bertsch, H., Ferreira, A. J., Farmer, W. M., and Hayes, T. L., 1973, The *genera Chromodoris* and *Felimida* (*Nudibranchia*: Chromodorididae) in tropical West America: distributional data, description of a new species, and scanning electron microscopic studies of radulae, *Veliger* **15**:287–294.

Boss, K. J., and Turner, R. D., 1980, The giant white clam from the Galápagos Rift, *Calyptogena magnifica*, new species, *Malacologia* **20**:161–194.

Bouchet, P., 1976, Mise en évidence de stades larvaires planctoniques chez des Gastéropodes Prosobranches des étages bathyal et abyssal, *Bull. Mus. natn. Hist. nat. Paris*, 3e sér., n° 400, Zoologie **277**:947–972.

Bouchet, P., and Fontes, J.–C., 1981, Migrations verticales des larves de Gastéropodes abyssaux: arguments nouveaux dus à l'analyse isotopique de la coquille larvaire et postlarvaire, *C.R. Acad. Sc. Paris*, t. 292, sér. III: 1005–1008.

Bouchet, P., and Warén, A., 1979, Planktotrophic larval development in deep-water gastropods, *Sarsia* **64**(1–2):37–40.

Bratcher, T., 1979, Taxonomic changes in eastern Pacific Terebridae with the description of a new species (Mollusca: Gastropoda), *Veliger* **22**:61–64.

Bratcher, T., and Burch, R. D., 1970a, Five new species of *Terebra* from the eastern Pacific, *Veliger* **12**: 295–300.

Bratcher, T., and Burch, R. D., 1970b, Four new Terebrid Gastropods from eastern Pacific islands, *Contrib. Sci. Los Angeles*, **188** (May 1970):1–6.

Bratcher, T., and Burch, R. D., 1971, The Terebridae (Gastropoda) of Clarion, Socorro, Cocos, and Galápagos Islands, *Proc. Calif. Acad. Sci.* (4th series) **37**:537–566.

Brennan, E., 1973, Factors contributing to the unique environment of Galápagos marine mollusks, *Echo* **6**:45–57.

Breure, A. S. H., and Coppois, G., 1978, Notes on the genus *Naesiotus* Albers, 1850 (Mollusca, Gastropoda, Bulimulidae) *Neth. J. Zool.* **28**:161–192.

Briggs, J. C., 1966, Oceanic islands, endemism, and marine paleotemperatures, *Syst. Zool.* **15**:153–163.

Bullock, R. C., 1988, The genus *Chiton* in the New World (Polyplacophora: Chitonidae), *Veliger* **31**: 141–191.

Cate, C. N., 1969a, A revision of the eastern Pacific Ovulidae, *Veliger* **12**:95–102.

Cate, C. N., 1969b, The eastern Pacific Cowries, *Veliger* **12**:103–119.

Cate, C. N., 1973, A systematic revision of the recent Cypraeid family Ovulidae (Mollusca: Gastropoda), *Veliger* 15 (Suppl.), iv + 116 pp.

Cate, C. N., 1979, A review of the Triviidae (Mollusca: Gastropoda), *Mem. San Diego Soc. Nat. Hist.* **10**, 126 pp.

Cauquoin, M., 1969, Mollusques récoltés par M. Hoffstetter sur les côtes de l'Equateur et aux Iles Galápagos: Naticidae, *Bull. Mus. Nat. Hist. Nat. Paris* (2e série) **41**:565–570.

Cavanaugh, C. M., Gardiner, S. L., Jones, M. L., Jannasch, H. W., and Waterbury, J. B., 1981, Prokaryotic cells in the hydrothermal vent tube worm *Riftia pachyptila* Jones: possible chemoautotrophic symbionts, *Science* **213**:340–342.

Cernohorsky, W. O., 1975, The taxonomy of some West American and Atlantic Nassariidae based on their type-specimens, *Rec. Auckland Inst. Mus.* **12**:121–173.

Cinelli, F., and Colantoni, P., 1974, Alcune osservazioni sulla zonazione del benthos marino sulle coste rocciose delle isole Galápagos (Oceano Pacifico), in: *Galápagos, Studi e Ricerche*. Museo Zoologico dell'Università di Firenze, Spedizione "L. Mares G.R.S.T.S.", Gruppo Ricerche Scientifiche e Techniche Subacquee, 22 pp.

Coan, E. V., 1979, Recent eastern Pacific species of the Crassatellid bivalve genus *Crassinella*, *Veliger* **22**:1–11.

Coan, E. V., 1988, Recent eastern Pacific species of the bivalve genus *Semele*, *Veliger* **31**:1–42.

Coan, E. V., 1989, Some Bivalve records from the Galápagos Islands, *Annu. Rep. West. Soc. Malac.* **21**: 11–12.

Coppois, G., 1985, Etude de la spéciation chez les Bulimulidae endémiques de l'archipel des Galápagos (Mollusques, Gastéropodes, Pulmonés). Thèse de Doctorat, Université Libre de Bruxelles, 238 pp.

Coppois, G., and De Vos, L., 1986, Two different striation patterns of the protoconch in Galápagos Bulimulidae: An S.E.M. study, *J. molluscan Stud.* **52**:106–109.

Coppois, G., and Glowacki, C., 1982, Bulimulid land snails from the Galápagos: 1-factor analysis of Santa Cruz Island species, *Malacologia* **23**:209–219.

Dall, W. H., 1908, Reports on the dredging operations off the west coast of Central America to the Galápagos, to the west coast of Mexico, and in the Gulf of California . . . carried on by the U.S. Fish Commission Steamer "Albatross", during 1891: XIV—The Mollusca and Brachiopoda, *Bull. Mus. Comp. Zool. Harv.* **43**:205–487.

Dall, W. H., and Ochsner, W., 1928, Tertiary and Pleistocene mollusca from the Galápagos Islands, *Proc. Calif. Acad. Sci.* (4th series) **17**:89–140.

DuShane, H., 1974, The Panamic-Galápagan Epitoniidae, *Veliger* **16**(Supplement), 84 pp.

DuShane, H., 1979, The family Epitoniidae (Mollusca: Gastropoda) in the northeastern Pacific, *Veliger* **22**:91–134.

Emerson, W. K., 1967, Indo-Pacific faunal elements in the tropical eastern Pacific, with special reference to the mollusks, *Venus* **25**:85–92.

Emerson, W. K., 1978, Mollusks with Indo-Pacific faunal affinities in the eastern Pacific Ocean, *Nautilus* **92**:91–96.

Emerson, W. K., 1982, Zoogeographic implications of the occurrence of Indo-Pacific Gastropods on the West American Continental Borderland, *Annu. Rep. West. Soc. Malac.* **14**:13–14.

Emerson, W. K., 1983, New records of Prosobranch Gastropods from Pacific Panama, *Nautilus* **97**: 119–123.

Emerson, W. K., and D'Attilio, A., 1970, Three new species of Muricacean Gastropods from the eastern Pacific, *Veliger* **12**:270–274.

Emerson, W. K., and Old, W. E., Jr., 1965, New molluscan records for the Galápagos Islands, *Nautilus* **78**:116–120.

Emerson, W. K., Kennedy, G. L., Wehmiller, J. F., and Keenan, E., 1981, Age relations and zoogeographic implications of late Pleistocene marine invertebrate faunas from Turtle Bay, Baja California Sur, Mexico, *Nautilus* **95**:105–116.

Fechter, R., 1981, *Olivella dama*, new record: a Panamic element in the Indo-Pacific faunal Province?, *Spixiana* **4**:103–110.

Felbeck, H., 1981, Chemoautotrophic potential of the hydrothermal Vent Tube Worm, *Riftia pachyptila* Jones (Vestimentifera), *Science* **213**:336–338.

Ferreira, A. J., 1979, The family Lepidopleuridae (Mollusca: Polyplacophora) in the eastern Pacific, *Veliger* **22**:145–165.

Filatova, Z. A., 1980, Thermophilic deep sea benthic fauna communities of the Pacific rift zones, *Okeanologiya* **20**:520–524.

Finet, Y., 1985, Preliminary faunal list of the marine mollusks of the Galápagos Islands, *Docums Trav. Inst. r. Sci. Nat. Belg.*, Nº 20, 50 pp.

Finet, Y., 1987a, Living *Cypraea moneta* L. in the Galápagos Islands, *La Conchiglia* **19**(222–223):22–24.

Finet, Y., 1987b, An Indo-Pacific tonnid (Gastropoda:Tonnidae) recorded from the Galápagos Islands, *Revue suisse Zool.* **94**:819–827.

Finet, Y., 1988, Mollusques marins des Galápagos: Composition taxonomique et affinités biogéographiques, *Bull. Soc. int. Conch.* **10**(4):12–16.

Finet, Y., 1989, Mollusques marins des Galápagos (2): Facteurs influençant les affinités biogéographiques, *Bull. Soc. int. Conch.* **11**(1):6–14.

Fretter, V., Graham, A., and Mc Lean, J. H., 1981, The anatomy of the Galápagos Rift limpet, *Neomphalus fretterae*, *Malacologia* **21**:337–361.

Grau, G., 1959, *Pectinidae of the eastern Pacific*, Allan Hancock publications, Vol. 23, University of Southern California Press, Los Angeles, 308 pp., 57 pls.

Grove, J. S., 1989, El Niño 1982–1983 and new records of Indo-West Pacific fishes at the Galápagos, *Annu. Rep. West. Soc. Malac.* **21**:5.

Hanna, G. D., and Hertlein, L. G., 1961, Large species of *Terebra* (Mollusca) from the eastern Pacific, *Proc. Calif. Acad. Sci.* (4th series) **30**:67–80.

Hedgpeth, J. W., 1969, An intertidal reconnaissance of rocky shores of the Galápagos, *Wasmann J. Biol.* **27**:1–24.

Hertlein, L. G., 1937, A note on some species of marine mollusks occurring in both Polynesia and the western Americas, *Proc. Am. Phil. Soc.* **78**:303–312.

Hertlein, L. G., 1972, Pliocene fossils from Baltra (South Seymour) Island, Galápagos Islands, *Proc. Calif. Acad. Sci.* (4th series) **39**:25–46.

Hertlein, L. G., and Emerson, W. K., 1953, Mollusks from Clipperton Island (Eastern Pacific), with the description of a new species of Gastropod, *Trans. San Diego Soc. Nat. Hist.* **11**:345–364.

Hertlein, L. G., and Strong, A. M., 1939, Marine Pleistocene mollusks from the Galápagos Islands, *Proc. Calif. Acad. Sci.* (4th series) **23**:367–380.

Hertlein, L. G., and Strong, A. M., 1955, Marine mollusks collected at the Galápagos Islands during the voyage of the *Velero III*, 1931–1932, in: *Essays in the Natural Sciences in honor of Captain Allan Hancock on the occasion of his birthday, July 26, 1955*, pp. 111–145, 1 pl, University of Southern California Press, Los Angeles.

Hoagland, E., 1977, Systematic review of fossil and recent *Crepidula* and discussion of evolution of the Calyptraeidae, *Malacologia* **16**.:353–420.

Hoffstetter, R., 1952, Moluscos subfosiles de los estanques de sal de Salinas (Pen. de Santa Elena, Ecuador): Comparacion con la fauna actual del Ecuador, *Boln Inst. Cienc. nat. Univ. cent. Ecuad., Quito* **1**:5–79.

Houvenaghel, G. T., 1978, Oceanographic conditions in the Galápagos Archipelago, Ecuador, and their relationships with life on the islands, in: *Upwellings Ecosystems*, Third Symposium on Upwelling Ecosystems, Kiel, West Germany (R. Boje and T. Matthias, eds.), pp. 181–200, Springer-Verlag, New York.

Houvenaghel, G. T., 1984, Oceanographic setting of the Galápagos Islands, in: *Key Environments, Galápagos* (R. Perry, ed.), pp. 43–54, Pergamon Press, New York.

Houvenaghel, G. T., and Houvenaghel, N., 1974, Ecological aspects of the intertidal zonation of the rocky shores of the Galápagos Islands, *Mar. Biol. Berlin* **26**:135–152.

James, M. J., 1984, A new look at evolution in the Galápagos: Evidence from the late Cenozoïc marine molluscan fauna, *Biol. J. Linn. Soc.* **21**:77–95.

James, M. J., and Pitt, W. D., 1984, Biogographic affinities of Pleistocene and recent Galápagos marine molluscs, *Annu. Rep. west. Soc. Malac.* **20**:28.

Jones, M. L., 1981, *Riftia pachyptila* Jones: Observations on the Vestimentiferan worm from the Galápagos Rift, *Science* **213**:333–336.

Keen, A. M., 1971, *Sea Shells of Tropical West America* 2nd ed., Stanford University Press, Palo Alto, xiv + 1064 pp.

Keen, A. M., and Smith, A. G., 1961, West American species of the bivalved Gastropod genus *Berthelinia*, *Proc. Calif. Acad. Sci.* (4th series) **30**:47–66.

Kenk, V. C., and Wilson, B. R., 1985, A new mussel (Bivalvia, Mytilidae) from hydrothermal vents in the Galápagos Rift Zone, *Malacologia* **26**:253–271.

Kohn, A. J., 1971, Diversity, utilization of resources, and adaptive radiation in shallow-water marine invertebrates of tropical oceanic islands, *Limnol. Oceanogr.* **16**:332–348.

Laursen, D., 1981, Taxonomy and distribution of teleplanic Prosobranch larvae in the North Atlantic, *Dana Rep.* N° 89, 44 pp., 58 figs. and charts, 3 pls.

Lindberg, D. R., and Mc Lean, J. H., 1981, Tropical eastern Pacific limpets of the family Acmaeidae (Mollusca, Archeogastropoda): Generic criteria and descriptions of 6 new species from the mainland and the Galápagos Islands, *Proc. Calif. Acad. Sci.* **42**:323–340.

Lutz, R. A., and Jablonski, D., 1980, Molluscan larval shell morphology: Ecological and paleontological applications, in: *Skeletal growth of Aquatic Organisms: Biological Records of Environment Change* (D. C. Rhoads and R. A. Lutz, eds.), pp. 323–377, Plenum Press, New York.

Lutz, R. A., Jablonski, D., Rhoads, D. C., and Turner, R. D., 1980, Larval dispersal of a deep-sea hydrothermal vent bivalve from the Galápagos Rift, *Mar. Biol. Berlin* **57**:127–134.

Marincovich, L., Jr., 1977, Cenozoïc Naticidae (Mollusca: Gastropoda) of the Northeastern Pacific, *Bull. Am. Paleont.* **70**(294), 494 pp.

McFadien, M. S., 1973, Zoogeography and ecology of seven species of Panamic-Pacific Scaphopoda, *Veliger* **15**:340–347.

McLean, J. H., 1967, West American species of *Lucapinella*, *Veliger* **9**:349–352.

McLean, J. H., 1970a, New species of Panamic marine Gastropods, *Veliger* **12**:310–315.

McLean, J. H., 1970b, Description of a new genus and eight new species of eastern Pacific Fissurellidae, with notes on other species, *Veliger* **12**:362–367.

McLean, J. H., 1970c, Notes on the deep water Calliostomas of the Panamic Province with descriptions of six new species, *Veliger* **12**:421–426.

McLean, J. H., 1970d, New species of tropical eastern Pacific Gastropoda, *Malac. Rev.* **2**:115–130.

McLean, J. H., 1971, A revised classification of the family Turridae, with the proposal of new subfamilies, genera, and subgenera from the eastern Pacific, *Veliger* **14**:114–130.

McLean, J. H., 1981, The Galápagos Rift limpet *Neomphalus*: Relevance to understanding the evolution of a major paleozoïc–mesozoïc radiation, *Malacologia* **21**:291–336.

McLean, J. H., and Nybakken, J., 1979, Growth stages of *Conus fergusoni*, the reinstatement of *Conus xanthicus*, and *Conus kohni*, new species from the Galápagos Islands, Ecuador, *Veliger* **22**:135–144.

McLean, J. H., and Poorman, R., 1971, New species of tropical eastern Pacific Turridae, *Veliger* **14**:89–113.

Métivier, B., 1968, Mollusques marins récoltés par M. R. Hoffstetter sur les côtes de l'Equateur et aux îles Galápagos: Chitonidae et Fissurellidae, *Bull. Mus. Nat. Hist. Nat. Paris* (2e série) **40**:585–596.

Miller, R. C., 1966, Distribution of marine wood-boring organisms in the tropical eastern Pacific Ocean, in: *The Galápagos* (R. I. Bowman, ed.), pp. 145–148, University of California Press, Berkeley.

Morrison, J. P. E., 1963, Notes on American *Siphonaria*, *Annu. Rep. Am. Malac. Un.* (for 1963):7–9.

Nesbitt, E. C., and Pitt, W. D., 1986, *Nassarius* (Gastropoda: Neogastropoda) from the Galápagos Islands, *Veliger* **28**:294–301.

Nesis, K. N., 1973, Cephalopods of the East-equatorial and Southeastern Pacific, *Trudy Inst. Okeanol.* **94**:188–240.

Nybakken, J., 1970, Radular anatomy and systematics of the West American Conidae (Mollusca, Gastropoda), *Am. Mus. Novit.*, Nr. 2414 (1970):1–29.

Nybakken, J., 1978, Population characteristics and food resource utilization of *Conus* in the Galápagos Islands, *Pacif. Sci.* **32**:271–280.

Nybakken, J., 1979, Population characteristics and food resource utilization of *Conus* in the Sea of Cortez and West Mexico, *J. molluscan Stud.* **45**:82–97.

Odhner, N. H., 1950, Studies on Galápagos Bulimulids, *J. Conch. Paris* **90**:253–268.

Old, W. E., Jr., 1975, Living *Conus* from the New World, with special reference to "twin species", *Bull. Am. Malac. Un.* (1975):52.

Olsson, A. A., 1961, *Mollusks of the Tropical Eastern Pacific, particularly from the southern half of the Panamic-Pacific faunal Province (Panama to Peru). Panamic-Pacific Pelecypoda*, Paleontological Research Institution, Ithaca, N.Y., 574 pp.

Petit, R. E., 1970, Notes on Cancellariidae (Mollusca: Gastropoda)—II, *Tulane Stud. Geol. Paleont.* **8**: 83–88.

Pilsbry, H. A., and Lowe, H. N., 1932, West Mexican and Central American mollusks collected by H. N. Lowe, 1929–31, *Proc. Acad. Nat. Sci. Philad.* **84**:33–144, 17 pls.

Pitt, W. D., 1981, Two new Gastropod occurrences in the Ecuadorian Neogene, *Tulane Stud. Geol. Paleont.* **16**:155–156.

Pitt, W. D., James, M. J., Hickman, C. S., Lipps, J. H., and Pitt, L. J., 1986, Late Cenozoic marine mollusks from tuff cones in the Galápagos Islands, *Proc. Calif. Acad. Sci.* **44**:269–282.

Radwin, G. E., 1974, Columbellid genera of eastern Pacific, *Festivus* **5**:88–91.

Ramage, C., 1986, El Niño, *Pour la Science*, N° **106**:69–77.

Rau, G. H., 1981, Hydrothermal vent clam and tube worm carbon-13 to carbon-12 ratio further evidence of nonphotosynthetic food sources, *Science* **213**:338–340.

Rhoads, D. C., Lutz, R. A., Revelas, E. C., and Cerrato, R. M., 1981, Growth of bivalves at deep-sea hydrothermal vents along the Galápagos Rift, *Science* **214**:911–913.

Robertson, R., 1975, Faunal affinities of the Architectonicidae in the eastern Pacific, *Bull. Am. Malac. Un.* **(1975)**:51.

Robertson, R., 1976, *Heliacus trochoïdes*: an Indo-west-Pacific architectonicid newly found in the eastern Pacific (mainland Ecuador), *Veliger* **19**:13–18.

Robertson, R., 1979, *Philippia radiata*, new record: Another Indo-west-Pacific architectonicid newly found in the eastern Pacific (Colombia), *Veliger* **22**:191–193.

Rokop, F. J., 1972, Notes on abyssal Gastropods of the eastern Pacific, with descriptions of three new species, *Veliger* **15**:15–19.

Rost, H., 1955, A report on the family Arcidae (Pelecypoda), *Allan Hancock Pacif. Exped.* **20**:177–248.

Roth, B., and Coan, E. V., 1968, Further observations on the West American Marginellidae with the description of two new species, *Veliger* **11**:62–69.

Roth, B., and Coan, E. V., 1971, Marginellidae (Mollusca: Neogastropoda) from the Galápagos Islands and Cocos Island, *Proc. Calif. Acad. Sci.* (4th series) **37**:575–584.

Sabelli, B., and Tommasini, S., 1980, I molluschi marini raccolti dalla spedizione "L. Mares - G.R.S.T.S." alle Isole Galápagos: 2. Polyplacophora, in: *Galápagos, Studi e Ricerche*, Museo Zoologico dell'Università di Firenze: Spedizione "L. Mares G.R.S.T.S.", Gruppo Ricerche Scientifiche e Techniche Subacquee, 10 pp.

Schaeffer, M. B., 1962, Report on the investigations of the Inter-American Tropical Tuna Commission for the year 1961, *Annu. Rep. Inter-Am. Trop. Tuna Comm.* (for 1961), Appendix A:44–103.

Scheltema, R. S., 1966, Evidence for trans-Atlantic transport of Gastropod larvae belonging to the genus *Cymatium*, *Deep-Sea Res.* **13**:83–95.

Scheltema, R. S., 1968, Dispersal of larvae by equatorial ocean currents and its importance to the zoogeography of shoal-water tropical species, *Nature, Lond.* **217**:1159–1162.

Scheltema, R. S., 1971a, The dispersal of the larvae of shoal-water benthic invertebrate species over long distances by ocean currents, in: *Fourth European Marine Biol. Symposium, 1971, Cambridge* (D. J. Crisp, ed.), pp. 7–28, Cambridge University Press, Cambridge.

Scheltema, R. S., 1971b, Larval dispersal as a means of genetic exchange between geographically separated populations of shallow-water benthic marine gastropods, *Biol. Bull. Mar. Biol. Lab. Woods Hole* **140**:284–322.

Scheltema, R. S., 1972, Dispersal of larvae as a means of genetic exchange between widely separated populations of shoal-water benthic invertebrate species, in: *Fifth European Marine Biol. symposium, 1972* (B. Battaglia, ed.), pp. 101–114, Piccin Editore, Padua.

Scheltema, R. S., 1986, On dispersal and planktonic larvae of benthic invertebrates: An eclectic overview and summary of problems, *Bull. Mar. Sci.* **39**:290–322.

Scheltema, R. S., 1988, Initial evidence for the transport of teleplanic larvae of benthic invertebrates across the East Pacific Barrier, *Biol. Bull. Mar. Biol. Lab. Woods Hole* **174**:145–152.

Schwengel, J., 1938, Zoological results of the G. Vanderbilt South Pacific Expedition, 1937—Part I. Galápagos Mollusca, *Proc. Acad. Nat. Sci. Philad.* **90**:1–3.

Shasky, D. R., 1983, New records of Indo-Pacific mollusca from Cocos Island, Costa Rica. *Nautilus* **97(4)**:144–145.

Shasky, D. R., 1989, An update on the molluscs from the Galápagos Islands as listed in the preliminary faunal list of the marine mollusks of the Galápagos Islands by Yves Finet, *Annu. Rep. West. Soc. Malac.* **21**:8–10.

Smith, A. G., 1966, Land snails of the Galápagos, in: *The Galápagos* (R. I. Bowman, ed.), pp. 240–251, University of California Press, Berkeley.

Smith, A. G., and Ferreira, A. J., 1977, Chiton fauna of the Galápagos Islands, *Veliger* **20**:82–97.

Smith, E. A., 1877, Mollusca (in: A. Günther, Account of the zoological collections made during the visit of H.M.S. "Petrel" to the Galápagos Islands), *Proc. Zool. Soc. Lond.*, (1877):69–73, 91–93.

Soot-Ryen, T., 1932, The Norwegian Zoological Expedition to the Galápagos Islands 1925, conducted by Alf Wolleback - II. Pelecypods from Floreana (Santa Maria), Galápagos Islands, *Meddr zool. Mus., Oslo* N° 27 (*Nyt Mag. Naturvid.* 70):313–324.

Soot-Ryen, T., 1955, *A report on the family Mytilidae (Pelecypodal)*, Allan Hancock publications, Vol. 20, 174 pp., 10 pls., University of Southern California Press, Los Angeles.

Sphon, G. G., 1969, Notes on the Mitridae of the eastern Pacific—II. The genus *Thala*, with the description of a new species, *Veliger* **12**:84–88.

Sphon, G. G., 1976, The Mitridae of the Galápagos Islands, *Nautilus* **90**:63–64.

Sphon, G. G., and Mulliner, D. K., 1972, A preliminary list of known Opisthobranchs from the Galápagos Islands collected by the Ameripagos Expedition, *Veliger* **15**:147–152.

Stanley, S. M., 1978, Chronospecies' longevities, the origin of genera, and the punctuational model of evolution, *Paleobiology* **4**:26–40.

Stanley, S. M., 1979, *Macroevolution, Pattern and Process*, 332 pp., W. H. Freeman and Co., San Francisco.

Stearns, R. E. C., 1893, Scientific results of explorations by the U.S. Fish Commission steamer "Albatross". N° XXV— Report on the mollusk-fauna of the Galápagos Islands, with descriptions of new species, *Proc. U.S. Nat. Mus.* **16**:353–450.

Strong, A. M., and Hertlein, L. G., 1937, The Templeton Crocker Expedition of the California Academy of Sciences, 1932—N° 32: New species of recent mollusks from the coast of Western North America, *Proc. Calif. Acad. Sci.* (4th series) **22**:159–178.

Strong, A. M., and Hertlein, L. G., 1939, Marine mollusks from Panama collected by the Allan Hancock Expedition to the Galápagos Islands, 1931–1932, *Allan Hancock Pacif. Exped.* **2**:177–245.

Taviani, M., 1979, I molluschi marini raccolti dalla spedizione "L. Mares—G.R.S.T.S." alla Isole Galápagos: 1. Gastropoda e Bivalvia. Museo Zoologico dell'Università di Firenze: Galápagos, Studi e richerche. Spedizione "L. Mares G.R.S.T.S.", Gruppo Ricerche Scientifiche e Techniche Subacquee, 61 pp.

Thorson, G., 1950, Reproductive and larval ecology of marine bottom invertebrates, *Biol. Rev.* **25**:1–45.

Thorson, G., 1961, Length of pelagic life in marine invertebrates as related to larval transport by ocean currents, in: *Oceanography* (M. Sears, ed.), Publication n° 67, pp. 455–474, American Association for the Advancement of Science, Washington, D.C.

Tomlin, J. R. Le B., 1927, The Mollusca of the "St. George" Expedition—I. The Pacific Coast of South America, *J. Conch. Lond.* **18**:153–198.

Turekian, K. K., and Cochran, J. K., 1981, Growth rate of a vesicomyid clam from the Galápagos spreading center, *Science* 214:909–911.

Vagvolgyi, J., 1977, Six new species and subspecies of *Naesiotus* from the Galápagos Islands (Pulmonata: Bulimulidae), *Proc. Biol. Soc. Wash.* **90**:764–777.

Vermeij, G. J., 1987, The dispersal barrier in the tropical Pacific: Implications for molluscan speciation and extinction, *Evolution* **41**:1046–1058.

Wellington, G. M., 1984, Marine environment and protection, in: *Key Environments, Galápagos* (R. Perry, ed.), pp. 247–263, Pergamon Press, Oxford.

Wimmer, A., 1879, Zur Conchylien-Fauna der Galápagos-Inseln, *Sber. Akad. Wiss. Wien, Math.-Nat. Classe* (Abt. I) **80**:465–514.

Wittenberg, J. B., Morris, R. J., Gibson, Q. H., and Jones, M. L., 1981, Hemoglobin kinetics of the Galápagos Rift vent tube worm *Riftia pachyptila* Jones (Pogonophora; Vestimentifera), *Science* **213**:344–346.

Wooster, W. S., and Hedgpeth, J. W., 1966, The oceanographic setting of the Galápagos, in: *The Galápagos* (R. I. Bowman, ed.), pp. 100–107, University of California Press, Berkeley.

Wyrtki, K., 1965, Surface currents of the eastern tropical Pacific Ocean, *Bull. Inter-Am. Trop. Tuna Comm.* **9**:271–304.

Wyrtki, K., 1967, Circulation and water masses in the eastern equatorial Pacific Ocean, *Int. J. Oceanol. Limnol.* **1**:117–147.

Zinsmeister, W. J., and Emerson, W. K., 1979, The role of passive dispersal in the distribution of hemipelagic invertebrates, with examples from the tropical Pacific Ocean, *Veliger* **22**:32–40.

Chapter 13

The Opisthobranch Gastropod Fauna of the Galápagos Islands

TERRENCE M. GOSLINER

1. Introduction

The opisthobranch gastropod fauna of the Galápagos Islands has been relatively poorly studied. In her review of the marine molluscan fauna of the tropical eastern Pacific, Keen (1971) listed 11 species of shelled opisthobranchs from the Archipelago. Sphon and Mulliner (1972) recorded 10 more species of opisthobranchs and later (Mulliner and Sphon, 1974) described a new species of dorid nudibranch from the Galápagos. Bertsch and Ferreira (1974) and Farmer (1978) each recorded one additional species. Since then, little work has been undertaken on the opisthobranchs inhabiting the archipelago. Thirty-three opisthobranch species are presently known from the Galápagos, a number which will increase when undescribed material in hand is formally described and as additional field work is conducted in the islands.

In December 1971 I collected several opisthobranch species not previously known from the Galápagos. Additional material was collected by the Ameripagos Expedition by Sphon, Mulliner *et al.* and is housed at the Los Angeles County Museum. More material was collected by G. M. Wellington in 1974, and this material was recently donated to the California Academy of Sciences. Several other species were collected in 1988 by K. Kaiser and D. Shasky. From this material, an

TERRENCE M. GOSLINER • Department of Invertebrate Zoology and Geology, California Academy of Sciences, Golden Gate Park, San Francisco, California 94118.

additional 20–25 species have been recognized. Many of these taxa constitute undescribed species. Ten are identifiable with described species and are reported upon here. A discussion of the biogeographical implications of the opisthobranchs fauna of the Galápagos and related regions is provided.

Museum repositories:

- AHF—Allan Hancock Foundation, University of Southern California, Los Angeles, California (specimens housed at LACM).
- LACM—Los Angeles County Museum of Natural History, Los Angeles, California.
- CASIZ—Department of Invertebrate Zoology and Geology, California Academy of Sciences, San Francisco, California.

2. Species Descriptions

2.1. *Navanax aenigmaticus* (Bergh, 1893)

(see Gosliner, 1980 for complete synonymy)

- *Distribution*. *Navanax aenigmaticus* is known from the Pacific coast of North America from Bahía Vizcaino, Baja California (Behrens, 1983) to Peru (Gosliner, 1980) from Florida to Brazil in the western Atlantic (Marcus, 1955; Gosliner, 1980) and from Ghana in the eastern Atlantic (Edmunds, 1968).
- *Material*. One specimen, California Academy of Sciences, CASIZ 067093, from tidepool Bahía Academy, Isla Santa Cruz, Islas Galápagos, April 1974, G. M. Wellington. Two specimens, CASIZ 067094, 2m depth, Punta Espinosa, Isla Fernandina, Islas Galápagos, 21 June 1974, G. M. Wellington. 067095, from tidepools, Bahía James, Isla Santiago, Islas Galápagos, 21 July 1974, G. M. Wellington.

  Three specimens, Los Angeles County Museum, AHF 10, Bahía James, Isla Santiago, Islas Galápagos, 9–12 January 1932. One specimen, LACM 71-44, intertidal, Darwin Research Station, Bahía Academy, Isla Santa Cruz, Islas Galápagos, March 1971, Ameripagos Expedition. One specimen, LACM 71-69, 3–20 m depth, Isla Rábida, Islas Galápagos 24 March 1971. Five specimens, Los Angeles County Museum, AHF 796-38, intertidal, Bahía Sullivan, Isla Santiago, Islas Galápagos, 21 Jan, 1938. One specimen, AHF 76-33 on sand, n. shore Bahía Cartago, Isla Isabella, Islas Galápagos, 14 February 1933. Two specimens,
- *External morphology*. The preserved animals are up to 50 mm in length. The cephalic and posterior shields are approximately the same length. The posterior shield terminates in two subequal triangular lobes. A pair of eyes is situated at the anterodorsal surface of the cephalic shield. On other side of the mouth is a pair of bilobed sensory mounds.
- *Shell* (Fig. 1A). The shell is situated posteriorly, underneath the skin of the posterior shield. Basally (posteriorly) it is calcified and curved around the posterior end of the visceral mass. Extending anteriorly from the calcified portion is a broad conchiolin wing.
- *Reproductive system* (Fig. 1B). The hermaphroditic reproductive system is monaulic. From the ovotestis a thin, short preampullary duct expands into

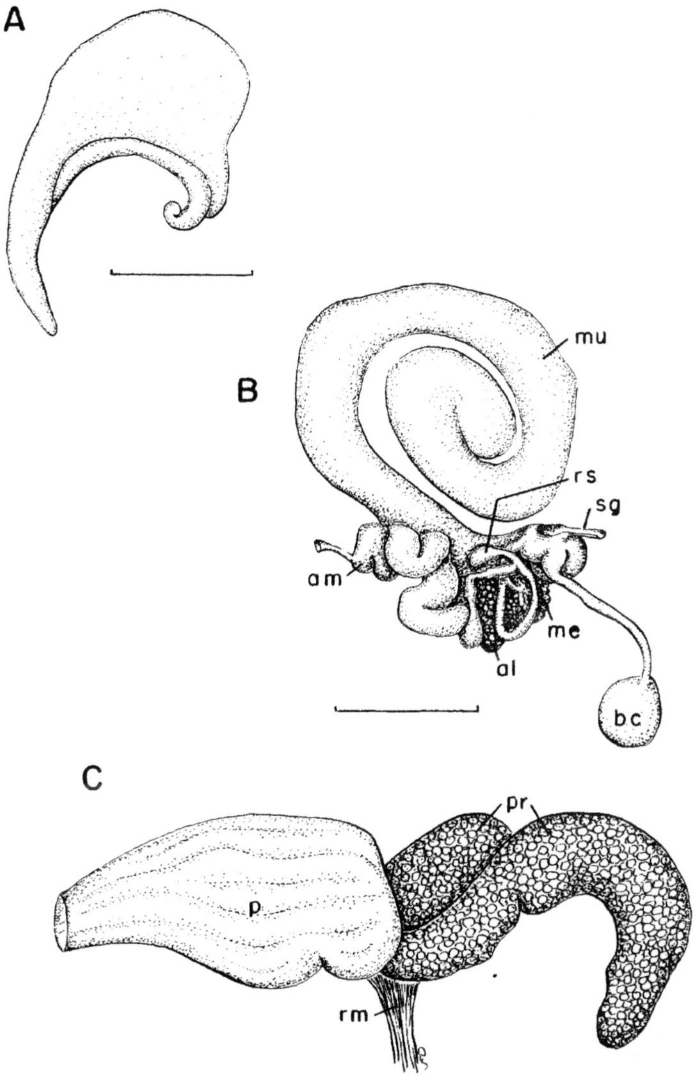

Figure 1. *Navanax aenigmaticus* (Bergh, 1893). (A). Shell, scale = 4 mm. (B). Reproductive system, al = albumen gland; am = ampulla; bc = bursa copulatrix; me = membrane gland; mu = mucous gland; rs = receptaculum seminis; sg = sperm groove, scale = 4 mm. (C). Penis, p = penis sac; pr = prostate; rm = retractor muscle, scale = 1 mm.

the thick, convoluted ampulla. It again narrow into the hermaphroditic duct and gives rise to a short, pyriform receptaculum seminis. The hermaphroditic duct curves over the surface of the albumen and membrane glands and branches to them before entering the genital atrium. The large coiled mucous gland also empties into the atrium as does the elongate duct of the spherical bursa copulatrix. The bursa is thin walled and partially embedded in the digestive gland. From the genital atrium, a ciliated sperm groove transports endogenous sperm to the cephalic penis. The penis (Fig. 1C) has a bilobed prostate, consisting of two unequal lobes. Near the entrance of the prostates into the penial sac is a thick retractor muscle. The penial sac contains numerous folds, but lacks a distinct penial papilla.

- *Discussion*. Gosliner (1980) discussed the systematics of this species and noted that material from the Galápagos was examined. He also listed the presumed synonyms of this species. This species is known from the Pacific coast of North America from Bahía Vizcaino, Baja California (Behrens, 1983), from Puerto Chileno near Cabo San Lucas in the Gulf of California (present study) and from Panamá and Peru (Gosliner, 1980). Gosliner noted that *Posterobranchaea maculata* D'Orbigny, 1837, from Valparaiso, Chile may in fact be a senior synonym of *Navanax aenigmaticus*. Should material from Chile prove to be conspecific with the present material application should be made to suppress *P. maculata* as this would severely disrupt the established nomenclature of the Aglajidae, as *Posterobranchaea* D'Orbigny, 1837, would have seniority over *Navanax* Pilsbry, 1895.

  Navanax aenigmaticus is also known from the western Atlantic from Florida to Brazil and from the eastern Atlantic coast of Ghana (Gosliner, 1980). I noted that the original description of *Chelidonura africana* Pruvot-Fol, 1953 actually consisted of specimens of two distinct species. I considered part of this material to be synonymous with *N. aenigmaticus*. However, subsequent study of material from the Mediterranean (Garcia and Garcia, 1984) has noted that some of Pruvot-Fol's material is properly placed in *Chelidonura* and none of it is likely to be synonymous with *N. aenigmaticus*. These authors have noted the difficulty in assigning the name *Chelidonura africana* to a single species and have noted the similarity of some material to *C. italica* Sordi, 1980. As a rèsult of this confusion, Cervera *et al.* (1988) considered *Chelidonura africana* as a *nomen dubium* and regarded the Spanish specimens of this species as *C. italica*. *Navanax berrieri* (Dieuzeide, 1935) has not been observed since its original description from Algeria. Certainly further study is required to clarify the systematics of the European aglajids. Nevertheless, *Navanax aenigmaticus* is represented on both sides of the Atlantic as well as from much of the tropical eastern Pacific.

2.2. *Aplysia parvula* Mörch, 1863

(for complete synonymy see Eales, 1960)

- *Distribution*. This species is circumtropical (Marcus and Marcus, 1967; Bebbington, 1977).

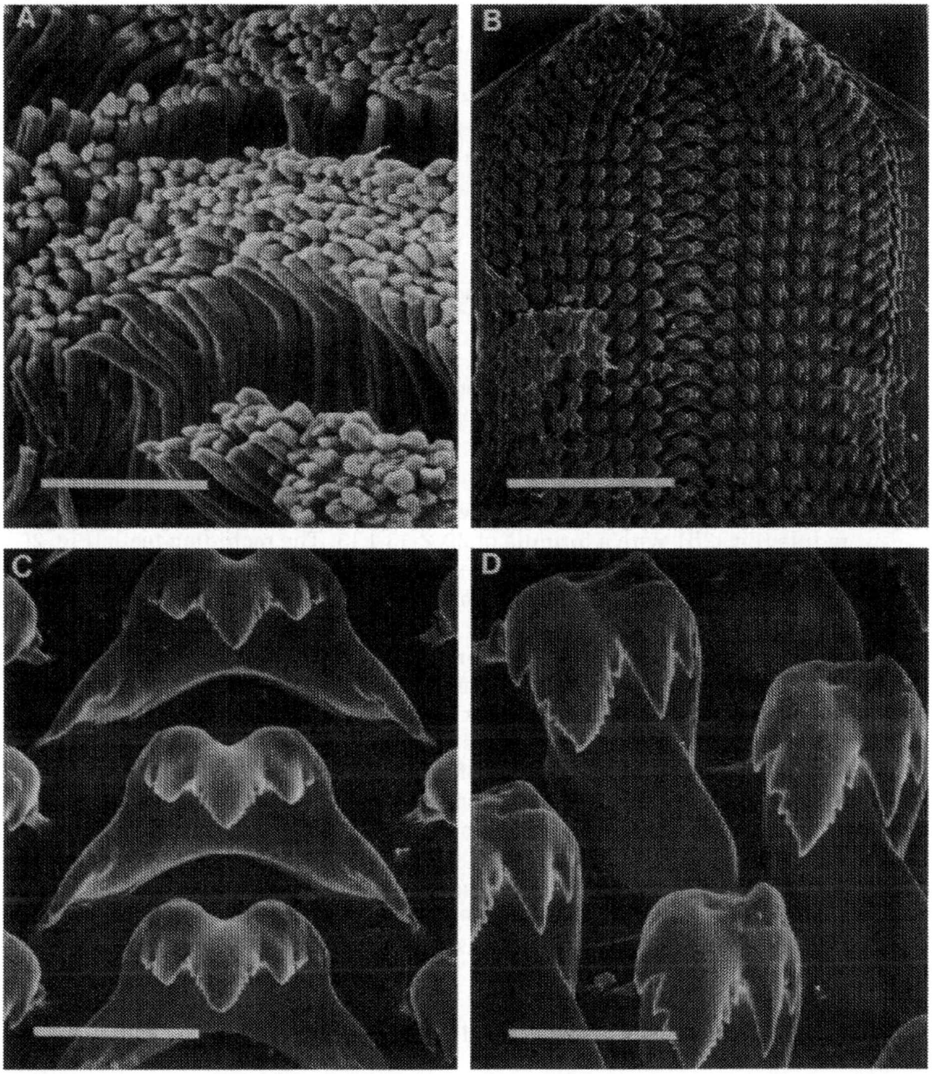

Figure 2. *Aplysia parvula* Mörch, 1863. Scanning electron micrographs. (A). Jaw rodlets, scale = 60 μm. (B). Entire radula, scale = 750 μm. (C). Rachidian teeth, scale = 100 μm. (D). Inner lateral teeth, scale = 75 μm.

- *Material*. Two specimens, California Academy of Sciences, CASIZ 067096, Punta Espinosa, Isla Fernandina, Islas Galápagos, 16 September 1974, G. M. Wellington. One specimen, Los Angeles County Museum, AHF 161-34, off Black Beach, Isla Floreana, 6 m depth, 17 January 1934.
- *External morphology*. The preserved specimens are contracted but reach a length of 20 mm. A black marginal line persists along the edge of the parapodia and around the margin of the large mantle opening.

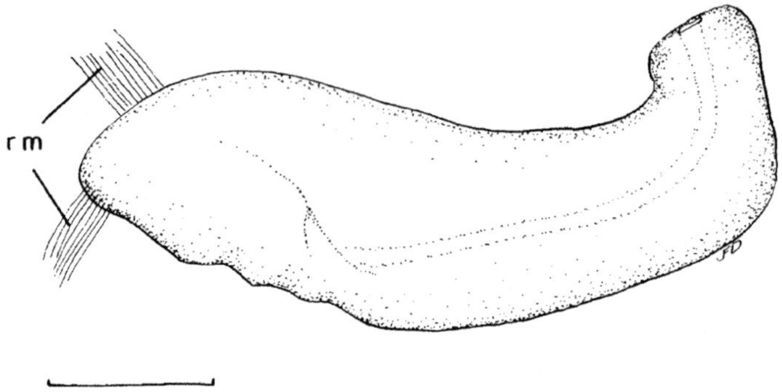

Figure 3. *Aplysia parvula* Mörch, 1863. Penis, rm = retractor muscles, scale = 0.5 mm.

- *Buccal mass.* Within the muscular buccal mass is an anterior pair of jaws, each of which is studded by minute rodlets (Fig. 2A). More posteriorly, is the radula (Fig. 2B), with a formula of 29 × 13.1.13. The rachidian teeth (Fig. 2C) consist of two cusps on either side of the large central cusp. The triangular central cusp bears 6–8 denticles on either side. The lateral cusps are smaller than the median one and possess only a trace of denticulation. The inner lateral teeth (Fig. 2D) bear three primary cusps. The innermost cusp is the largest and bears 6–7 triangular denticles on either side. The two outer cusps are smaller and narrower and lack any trace of denticles. The outer lateral teeth have poorly developed cusps. The outer three to four teeth are simply quadrangular, without any trace of denticles.
- *Reproductive system.* The posterior reproductive organs were not examined. The penis (Fig. 3) is unarmed and bears a simple papilla and a pair of retractor muscles.
- *Discussion.* This circumtropical species has been well studied over much of its range (Eales, 1960; Bebbington, 1974, 1977). The present material agrees with previous descriptions, in all aspects of its external and internal anatomy.

2.3. *Chromodoris sphoni* (Marcus, 1971)

Felimida sphoni Marcus, 1971:355, figs. 1–3
Chromodoris sphoni (Marcus, 1971), Bertsch, 1978a:321, fig. 9.

- *Distribution.* This species was previously known from Mazatlan, Mexico to Panama Bay (Bertsch, 1978a).
- *Material.* Two specimens, California Academy of Sciences, CASIZ 067097, Bahía Academy, Isla Santa Cruz, Islas Galápagos, intertidal zone, under lava rocks, 17 February 1988, Kirstie Kaiser.
- *External morphology.* From color transparencies provided by Kirstie Kaiser and deposited in the color slide collection of the California Academy of Sciences, it is readily apparent that the present specimens are conspecific with *C. sphoni*. The ground color is reddish purple. Numerous yellow spots

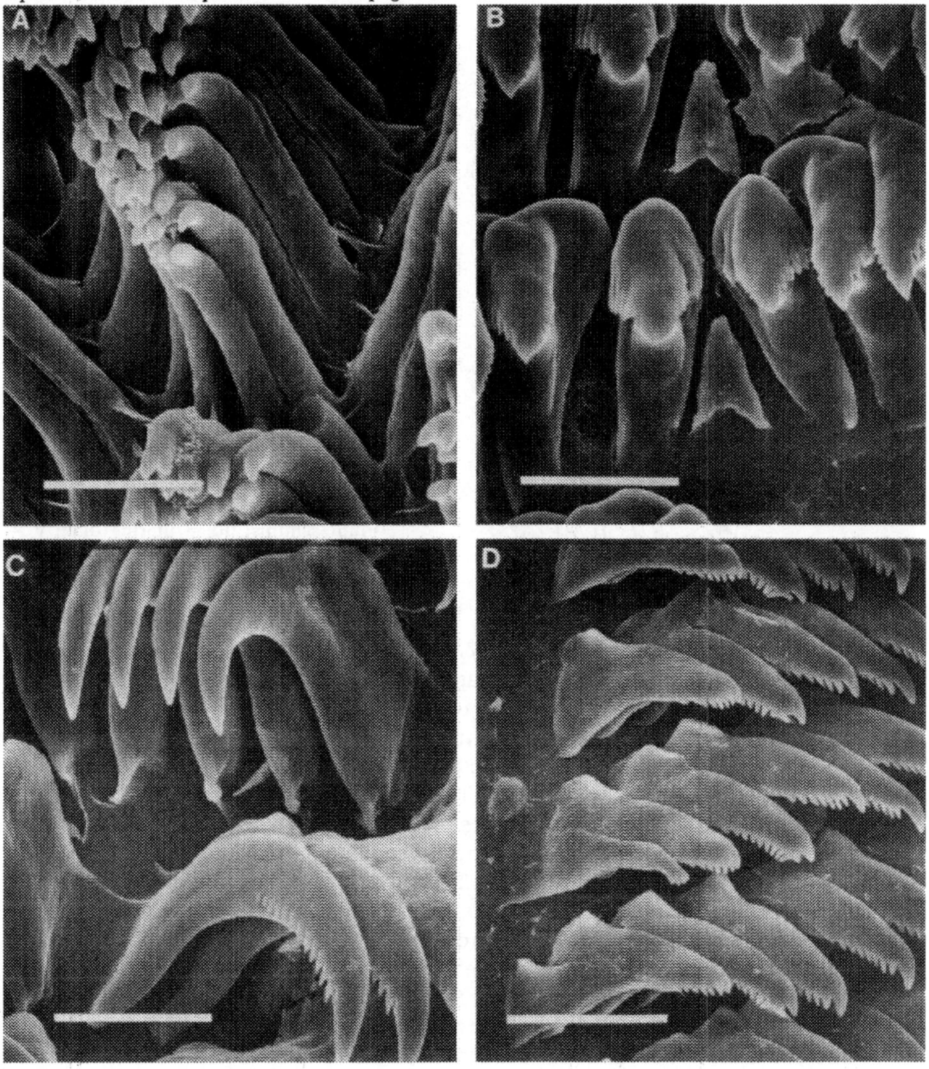

Figure 4. *Chromodoris sphoni* (Marcus, 1971). Scanning electron micrographs. (A). Jaw platelets, scale = 15 μm. (B). Rachis, scale = 25 μm. (C). Middle of half row, scale = 30 μm. (D). Outer lateral teeth, scale = 43 μm.

are present over the surface of the notum. A series of blue and white lines form marginal rings, which are interrupted on either side, midway between the rhinophores and gills. The rhinophores and ten unipinnate gills are whitish with charcoal gray apices.

- *Buccal apparatus*. The buccal mass is thick and muscular. Near its anterior limit is a ring of chitinous labial elements. Each element (Figure 4A) is distinctly bifid. More posteriorly, is the radula. Its formula in the one

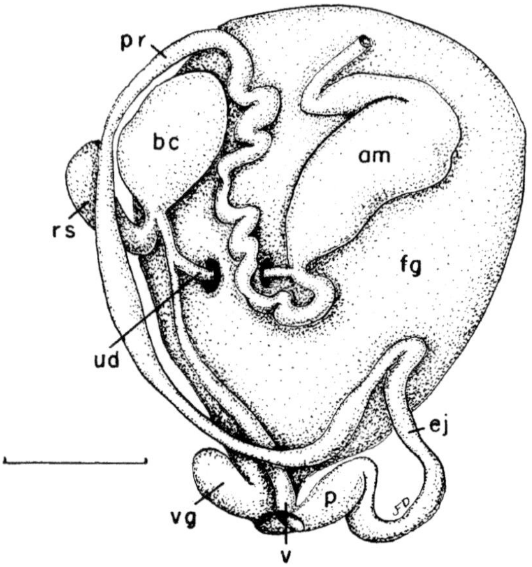

Figure 5. *Chromodoris sphoni* (Marcus, 1971). Reproductive system, am = ampulla; bc = bursa copulatrix; ej = ejaculatory portion of vas deferens; fg = female gland mass; p = penis; pr = prostatic portion of vas deferens; rs = receptaculum seminis; ud = uterine duct; v = vagina; vg = vestibular gland, scale = 0.5 mm.

specimen examined is 57 × 37.1.37. The central rachis of the radula (Fig. 4B) bears a row of distinct, triangular rachidian teeth. On other side of the rachidian tooth is an inner lateral tooth. The inner lateral are simply cuspidate with 4–7 denticles on the inner side of the central cusp and 5–6 denticles on the outer side. The next three lateral teeth also possess denticles on both sides of the central cusp. The lateral teeth from the middle of each half row (Fig. 4C) are simply hook shaped, bear 10–12 denticles on the outer surface of each tooth and lack denticles on their inner side. The outermost teeth are more elongate and have 3–10 denticles below the central cusp.

- *Reproductive system* (Fig. 5). The narrow preampullary duct widens into the saccate ampulla. The ampulla divides into a short oviduct, which enters the female gland mass, and an elongate vas deferens. The proximal portion of the vas deferens is prostatic and highly convoluted. It becomes ejaculatory near the point where the vas deferens traverses the surface of the bursa copulatrix and continues as a muscular duct to the hind end of the penial sac. The thin-walled bursa copulatrix and the thick, curved receptaculum seminis are joined at a common base. From this juncture the vaginal duct emerges, and after a short distance, divides into the short uterine duct. The uterine duct enters the female gland mass. The vagina is elongate and empties adjacent to the penis. The large female gland mass empties ventral to the penis and vagina and has a small vestibular gland at its base.
- *Discussion. Felimida sphoni* was placed in *Chromodoris* by Bertsch (1978a). Rudman (1984) also considered *Felimida* to be a junior synonym of *Chromodoris* The present material agrees with that described by Marcus (1971) and Bertsch (1978a) in its external and radular morphology. The reproductive anatomy was not described previously, but is consistent with that known for other members of the genus (Rudman, 1984).

2.4. *Glossodoris dalli* (Bergh, 1894)

(for complete synonymy see Bertsch, 1978a, Rudman, 1984)

- *Distribution.* *Glossodoris dalli* is known from Bahía Magdalena, Baja California Sur, and throughout the Gulf of California to Costa Rica (Bertsch, 1978b).
- *Material.* One specimen, California Academy of Sciences, CASIZ 067098, Punta Espinosa, Isla Fernandina, Islas Galápagos, intertidal, under rocks, 16 September 1974, G. M. Wellington.

  One specimen, Log Angeles County Museum, AHF 5633, Flamingo Bay, off Cormorant Bay, Isla Charles, intertidal zone, 5 February 1933. One specimen, AHF 808-38, Bahía Academy, Isla Santa Cruz, Islas Galápagos, intertidal zone, 25 January 1938.
- *External morphology.* The preserved specimens are up to 20 mm long with an undulate mantle margin and a few residual dark punctations on the notum.
- *Buccal mass.* The buccal mass is thick and muscular. The radular ribbon extends posteriorly from the mass for a considerable distance. At the anterior end of the mass, the labial cuticle bears numerous curved, bifid rodlets (Fig. 6A). The elongate radula has a formula of 148 × 35.1.35. The central portion of the radula (Fig. 6B) contains a row of short, triangular rachidian teeth. On either side of the rachidian are the inner lateral teeth. Each inner lateral bears 3–5 denticles on its inner face and 4–5 denticles on the outer side. The remaining laterals lack denticles on their inner side. The second laterals have 4–5 denticles on their outer face. Near the middle of each half row of the radula (Fig. 6C) the lateral teeth have an elongate cusp, which bears 8–9 denticles on their outer edge. The outermost 6–7 laterals (Fig. 6D) are elongate and entirely devoid of denticles.
- *Reproductive system* (Fig. 7). The narrow preampullary duct expands into a wide, tubular ampulla. The ampulla again narrows and bifurcates into the oviduct and vas deferens. The oviduct is extremely short and enters the female gland mass. The vas deferens is prostatic proximally and consists of numerous convolutions. It then becomes muscular and ejaculatory and widens and again narrows before entering the penial sac. The thin walled, spherical bursa copulatrix and the thick, curved receptaculum seminis share a common base. Also exiting from this common junction are the short uterine duct and the vagina. The vagina is elongate and curved and terminates at the genital atrium, adjacent to the penial sac. The large female gland mass exits ventral to the penis and vagina and has a saccate vestibular gland at its base.
- *Discussion.* The present material agrees with that described previously (Marcus and Marcus, 1967, as *Chromodoris banksi sonora*; Bertsch, 1978b).

2.5. *Doriopsilla janaina* Marcus and Marcus, 1967

- *Distribution;* This species is known from Bahía de los Angeles, Baja California (present study), Punta Lobos, Sonora, Mexico (Marcus and Marcus, 1967),

Figure 6. *Glossodoris dalli* (Bergh, 1894). Scanning electron micrographs. (A). Jaw platelets, scale = 15 μm. (B). Rachis, scale = 43 μm. (C). Middle of half row, scale = 30 μm. (D). Outer lateral teeth, scale = 60 μm.

the type locality off the Pacific coast of Panamá (Marcus and Marcus, 1967) and from the Galápagos (present study).

- *Material.* Two specimens, California Academy of Sciences, CASIZ 067099, Punta Espinosa, Isla Fernandina, Islas Galápagos, intertidal zone, under stones, 16 September 1974, G. M. Wellington. Two specimens, CASIZ 067109, S. James Bay, Isla Santiago, Islas Galápagos, intertidal zone, 21 July 1974, G. M. Wellington.

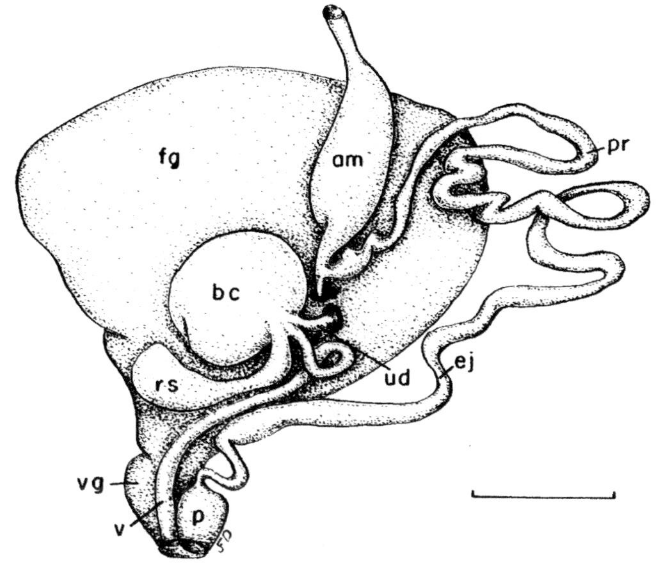

Figure 7. *Glossodoris dalli* (Bergh, 1893). Reproductive system, am = ampulla; bc = bursa copulatirx; ej = ejaculatory portion of vas deferens; fg = female gland mass; p = penis; pr = prostatic portion of vas deferens; rs = receptaculum seminis; ud = uterine duct; v = vagina; vg = vestibular gland, scale = 2 mm.

One specimen, Los Angeles County Museum, LACM 72-192, Bahía Urvina, Isla Isabela, Islas Galápagos, 1–3 m depth, 24 January 1972, Sylvia Earle. Three specimens AHF 11, Bahía Conway, Isla Santa Cruz, 12–14 January 1932. One specimen, LACM 34-53, Isla Bartholome, Islas Galápagos, intertidal zone, 23 January 1934. One specimen, AHF 803-38, Black Beach, Isla Floreana, 23 January, 1938. Five specimens, AHF 796-38, Bahía Sullivan, Isla Santiago, Islas Galápagos, intertidal zone on rock, 21 January 1934. Two specimens, AHF 800-38, n. shore Cartago Bay, Isla Isabela, 22 January 1938. Three specimens, LACM 25102, Darwin Research Station, Bahía Academy, Isla Santa Cruz, Islas Galápagos, September, 1973.

- *External morphology*. The living animals are bright orange with darker orange pigment on the rhinophores and gills. The dorsal surface of the body is covered with minute opaque white spots and slightly larger black ones. The white pigment may be concentrated in areas on the dorsal surface, such as on the apices of the tubercles. The entire notum is covered with simple rounded tubercles. The notum is densely spiculate, giving the animal a rigid leathery appearance. There are five tripinnate gills. The tentacles surrounding the mouth are poorly developed as in other porostomes.
- *Digestive tract* (Fig. 8A). The buccal mass is well developed and muscular. It is devoid of jaws or radula. Posteriorly, it narrows into an elongate muscular esophagus which forms one loop ventral to the central nervous system and continues posteriorly as a straight tube. There are no ptyaline glands present along the esophagus. The esophagus widens and forms a second loop just

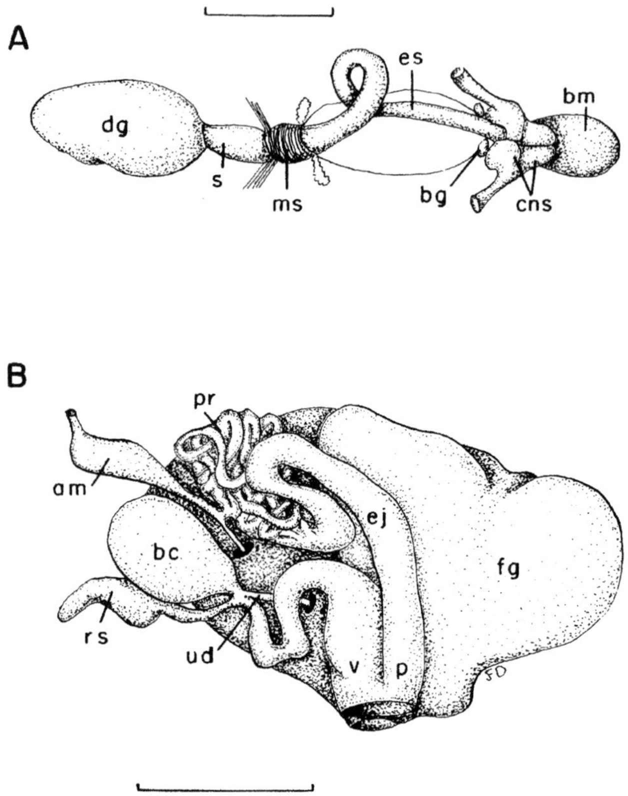

Figure 8. *Doriopsilla janaina* (Marcus and Marcus, 1967). (A). Digestive tract, bg = buccal ganglia; bm = buccal mass; cns = central nervous system; dg = digestive gland; es = esophagus; ms = muscular segement of esophagus; s = stomach, scale = 3 mm. (B). Reproductive system, am = ampulla; bc = bursa copulatrix; ej = ejaculatory portion of vas deferens; fg = female gland mass; p = penis; pr = prostate; rs = receptaculum seminis; ud = uterine duct; v = vagina, scale = 1 mm.

before it is differentiated into a short portion containing thick circular muscles. Posterior to this point the esophagus enters the stomach, which is surrounded by the digestive gland. At the junction of the highly muscular section of the esophagus with the stomach is a pair of retractor muscles. A large dorsal blood gland is found dorsal to the central nervous system. The ganglia of the central nervous system surround the esophagus. The buccal ganglia are present on the ventral surface of the anterior portion of the esophagus. From the posterior end of each buccal ganglion a buccal nerve extends posteriorly to the junction of the tubular esophagus with its highly muscular segment. At this point a gland is present along the posterior end of each nerve.

- *Reproductive system* (Fig. 8B). The ampulla is short and saccate. It divides into a short vas deferens and a more elongate oviduct. The vas deferens expands into a prostate gland consisting of many diffuse tubules. The prostate empties into a thick ejaculatory section of the vas deferens. The vas deferens terminates near the gonopore in a penis that it not well differentiated from the rest of the efferent duct. The penis contains several rows of distinct spines. The oviduct enters the female gland mass after a short distance. The large, thin walled bursa copulatrix joins the elongate duct of the thick walled receptaculum seminis with the vagina. Shortly after the junction of these two receptacles, the elongate uterine duct branches from the vagina. The uterine duct enters the female gland mass. The vagina widens considerable and curves to its gonopore adjacent to the penis. The large female gland mass exits from ventrally to the vagina and penis.

- *Discussion*. The external anatomy and reproductive system are identical to that described by Marcus and Marcus (1967). The digestive system was not described in the original description. The absence of a ptyaline gland and the presence of the buccal ganglia near the anterior end of the esophagus are consistent with its placement in the genus *Doriopsilla*.

2.6. *Tambja mullineri* Farmer, 1978

- *Distribution*. This species is known only from the Galápagos (Farmer, 1978).
- *Material*. Two specimens, California Academy of Sciences, CASIZ 067100, North Seymore Island, Islas Galápagos, 15 m depth, 13 February 1988, Kirstie Kaiser. One specimen, CASIZ 067101, Isla Floreana, Islas Galápagos, 14 m depth, 15 February 1988, Kirstie Kaiser.
- *External morphology*. The living specimens are 15–20 mm in length. Their ground color is uniformly black with a series of several longitudinal blue-green lines.
- *Buccal mass*. The buccal mass is thick and muscular. The chitinous labial cuticle is thin and weakly developed, devoid of any rodlets. The radular formula is 15 × 4.1.1.1.4. (Fig. 9A). The rachidian tooth is blunt and quadrangular, without denticles. The inner lateral tooth is elongate with a thickened central region and apical cusp. The outer lateral teeth (Fig. 9B) are roughly rectangular and decrease in size towards the outer margin.
- *Reproductive system* (Fig. 10). The ampulla is elongate and convoluted. It narrows and bifurcates into the thick, convoluted vas deferens and the short oviduct. The vas deferens enters the large, darkly pigmented penial sac. A simple penial papilla is present within the sac. The short oviduct enters the albumen portion of the female gland mass. The bulk of the female gland mass is composed of the many folds of the mucous gland. The thin walled bursa copulatrix and the thicker receptaculum seminis are of about equal size. The vaginal duct is straight and elongate. Near the distal third of its length, it branches into the uterine duct, which enters the albumen gland after a short distance. Near the place where the female glands empty into the nidamental aperture are two lobes of the vestibular gland.
- *Discussion*. *Tambja mullineri* is known only from the type locality, in the Galápagos. The specimens described here are identical with those described by Farmer (1978) in most details. The external morphology and radular formula are virtually identical in all respects. The reproductive system was

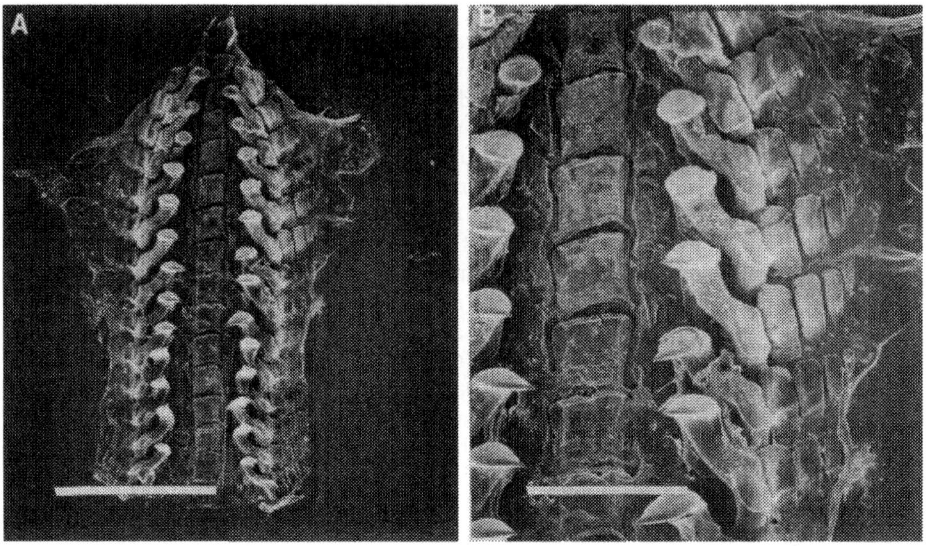

Figure 9. *Tambja mullineri* Farmer, 1978. Scanning electron microgaphs. (A). Entire radula, scale = 1.0 mm. (B). Half row, scale = 430 μm.

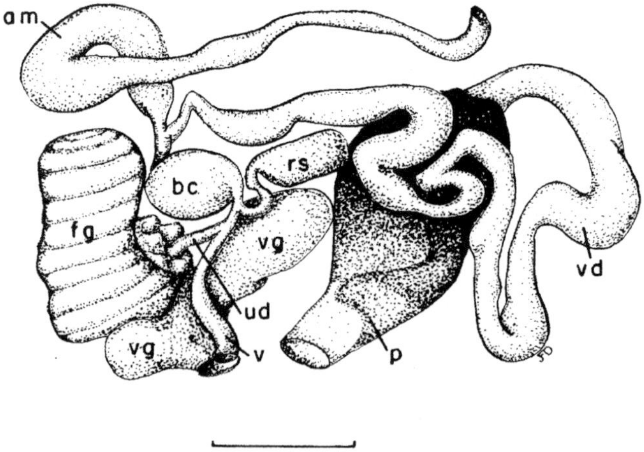

Figure 10. *Tambja mullineri* Farmer, 1978. Reproductive system, am = ampulla; bc = bursa copulatrix; fg = female gland mass; p = penis; rs = receptaculum seminis; ud = uterine duct; v = vagina; vd = vas deferens; vg = vestibular gland, scale = 1 mm.

also depicted by Farmer. Many of the organs are incorrectly labeled in his figure. Nevertheless, a bursa copulatrix and receptaculum seminis of equal size are present as are a large penial sac and a vestibular gland.

2.7. *Flabellina iodinea* (Cooper, 1863)

- *Distribution.* This species is known from the Pacific coast of North America from Vancouver Island, British Columbia to the Pacific and Gulf of California coasts of Baja California (Behrens, 1980). The occurrence of this species from the Galápagos represents the first record of it south of Mexico.
- *Material.* Five specimens, one dissected, California Academy of Sciences, CASIZ 067102, Punta Espinosa, Isla Fernandina, Galápagos Islands, 2–6 m depth, G. M. Wellington, 16 September 1974. Two specimens, CASIZ 067103, Punta Espinosa, Isla Fernandina, Galápagos Islands, 6 m depth, 21 June 1974, G. M. Wellington. One specimen, CASIZ 067104, Caleta Black, Isla Isabella, Galápagos, 2 m depth, 26 June 1974, G. M. Wellington.
- *External morphology.* The preserved specimens are 15–30 mm in length. The rhinophores are densely perfoliate. The cerata are arranged in elevated cushions. The anus is situated ventral to the first postcardiac ceratal cushion, in the pleuroproctic position.
- *Buccal mass.* The buccal mass is ovoid and highly muscular. Surrounded by muscular tissue are the paired, thick chitinous jaws. They bear 7–8 rows of denticles on their masticatory border (Fig. 11A). Between the jaws is the radular ribbon which has a formula of 22 × 1.1.1., in one specimen examined. The rachidian teeth (Fig. 11B) are roughly triangular with 11–12 denticles on either side of the narrow, elongate central cusp. The lateral teeth (Fig. 11C) are triangular with 10–13 elongate denticles on the inner face of the tooth. On the right side of the radula in the specimen examined an additional 9–10 denticles are present on the outer margin of the teeth. These denticles are smaller than the ones on the inner margin.
- *Reproductive system* (Fig. 12). The narrow preampullary duct emerges from the lobate ovotestis and expands into the convoluted ampulla. The postampullary duct narrows into the hermaphroditic duct and bifurcates into the thick, prostatic vas deferens and the narrow oviduct. The vas deferens continues proximally toward the gonopore and narrows immediately prior to its entrance into the penial sac. The penial papilla is short and thick. The oviduct runs distally until it enters the serial receptaculum seminis, which consists of approximately 10 small lobes. The oviduct enters the nidamental glands near the albumen gland. The membrane gland is situated ventrally, adjacent to the albumen gland. The bulk of the female gland mass is composed of the lobate mucous gland. Immediately, distal to the female gonopore is a saccate, sessile bursa copulatrix.
- *Discussion. Flabellina iodinea* is known along the Pacific coast of the North America from Vancouver Island to the Pacific Coast of Baja California and into the Gulf of California (Behrens, 1980). It has not been recorded previously from other portions of the Panamic Province.

  Flabellina iodinea has a very distinctive color pattern, with vivid purple

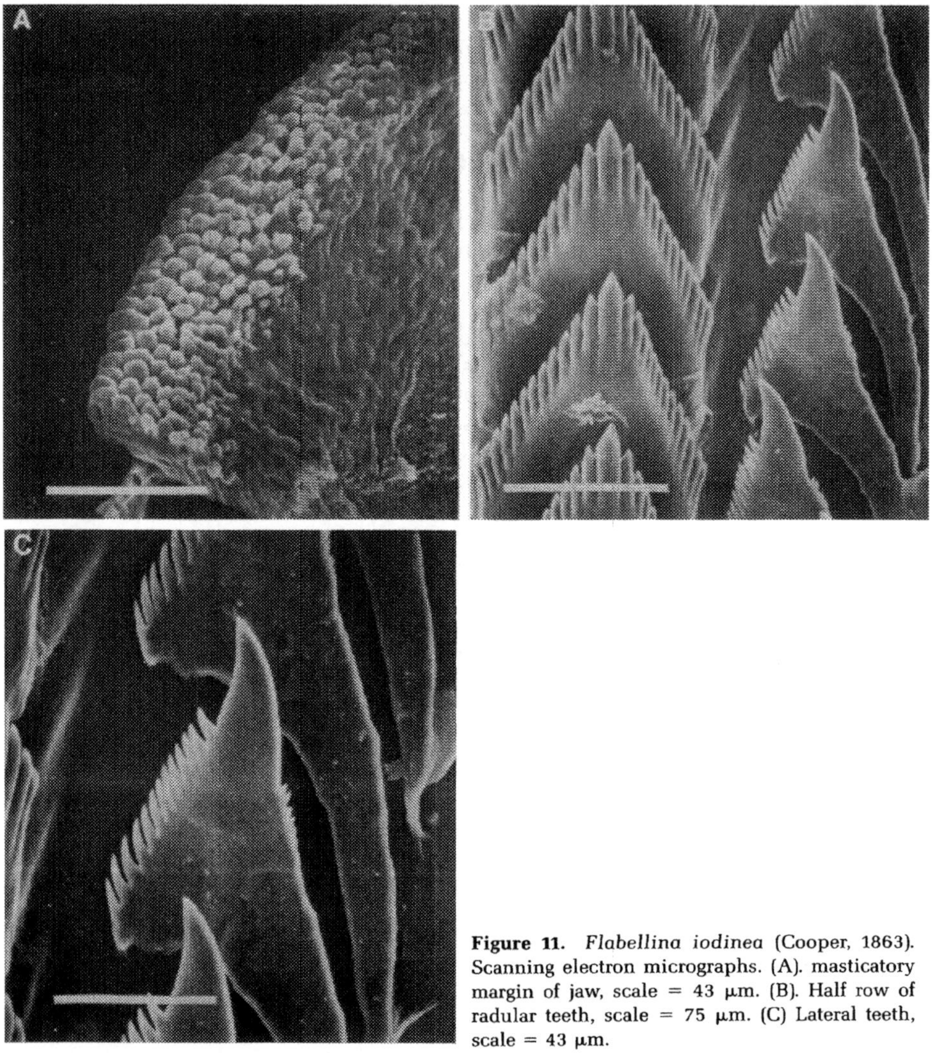

Figure 11. *Flabellina iodinea* (Cooper, 1863). Scanning electron micrographs. (A). masticatory margin of jaw, scale = 43 μm. (B). Half row of radular teeth, scale = 75 μm. (C) Lateral teeth, scale = 43 μm.

ground color and bright orange cerata (MacFarland, 1966; Behrens, 1980). Unfortunately, the color is not recorded for the present specimens. Nevertheless, there is little doubt as to the identification of the present material. Only about six species of *Flabellina* have densely perfoliate rhinophores. In most of these species the anus is situated in the interhepatic space. Only in *Flabellina iodinea* and *F. cynara* (Marcus and Marcus, 1967) is the anus situated more posteriorly. In *F. cynara* the anus is situated in the posterior half of the body, while in *F. iodinea* (including the present material) the anus is ventral to the first ceratal group of the first postcardiac ceratal cluster.

The radular morphology is virtually identical to that described by MacFarland (1966) and Marcus and Marcus (1967). Rachidian teeth with a large number of elongate denticles on either side of a narrow central cusp are

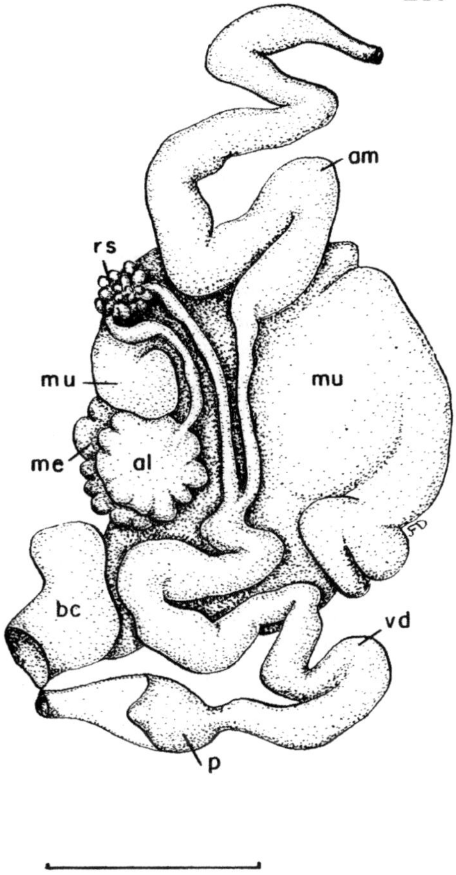

Figure 12. *Flabellina iodinea* (Cooper, 1863). Reproductive system, al = albumen gland; am = ampulla; bc = bursa copulatrix; me = membrane gland; mu = mucous gland; p = penis; rs = receptaculum seminis; vd = vas deferens, scale = 2 mm.

distinctive of *Flabellina iodinea* The lateral teeth are similar to those described previously, except that some of the laterals also have small denticles on their outer face.

The reproductive system differs from that described previously (MacFarland, 1966; Marcus and Marcus, 1967) in that a serial receptaculum seminis is present and consists of numerous lobes. Additional material from California was examined which revealed the presence of a multilobed receptaculum, which had been overlooked by MacFarland and Marcus and Marcus. There is little doubt that the material recorded from the Galápagos is conspecific with *F. iodinea*.

2.8. *Phidiana lynceus* Bergh, 1867

Phidiana lynceus Bergh, 1867:104, pl. 3B, pl 4A; Edmunds, 1964: 16, figs. 10B, 11A.

Phidiana selencae Bergh, 1879:560, pl. 6, figs. 10–18; Marcus and Marcus, 1962:479–480.

Phidiana brevicauda Engel, 1925:67, figs. 4–6.

- *Distribution. Phidiana lynceus* is known from the eastern Atlantic coast of Ghana (Edmunds, 1977) and is widely distributed in the western Atlantic (Edmunds and Just, 1983) from Florida to Brazil. It has also been recorded from the Pacific coast of Panamá (Marcus and Marcus, 1967). The present material constitutes the first record of this species from the Galápagos.
- *Material.* One specimen. dissected, California Academy of Sciences, CASIZ 067105, Punta Espinosa, Isla Fernandina, Islas Galápagos, intertidal zone, under stones, 16 September 1974, G. M. Wellington. Two specimens, CASIZ 067106, S. Bahía James, Isla Santiago, Islas Galápagos, intertidal zone, 21 July 1974, G. M. Wellington. One specimen, CASIZ 067107, Punta Espinosa, Isla Fernandina, Islas Galápagos, 16 September, 1974, G. M. Wellington.
- *External morphology.* The living animals reach 15–20 mm in length. The ground color is translucent white. Much of the head and notum are covered with rich orange pigment. Near the base of each oral tentacle an opaque white line is present. These are curved and converge between the rhinophores and continue posteriorly as a single medial longitudinal line. The perfoliate rhinophores are orange with opaque white apices. The oral tentacles are elongate, thickened at their bases. The cerata are arranged in oblique rows. They are translucent white with dark brown digestive gland showing through their surface. The cnidosacs are orange. Some of the cerata possess a subapical opaque white ring. The anterior digestive branch contains approximately 7–8 rows of cerata, with 3–6 cerata per row. The posterior digestive branch contains 8–9 rows of cerata. The gonopore is situated below the fifth row of cerata. The anus is situated in the posterior half of the body, posterior to the sixth row of the posterior digestive branch. The anterior end of the foot is simply rounded.
- *Buccal mass.* The muscular buccal mass contains a pair thick chitinous jaws. The jaws bear a single row of approximately 20 denticles (Fig. 13A) on the edge of the masticatory border. The radula formula is 15 × 0.1.0. The teeth (Fig. 13B,C) are broad along their masticatory edge and taper into a pair of posterior limbs. The masticatory border consists of 9–10 denticles on either side of the broad, triangular central cusp. Five or six denticles are present on either flank of the tooth and an additional four denticles are present on either edge of the central cusp.
- *Reproductive system* (Fig. 14). The lobed ovotestis narrows into a preampullary duct, which expands into a curved ampulla. The ampulla again narrows and divides into the thick, convoluted prostatic vas deferens and the short oviduct. The vas deferens continues towards the gonopore, where it enters the penial sac. Within the sac is the muscular penis, which terminates at a sharply recurved penial spine. Immediately adjacent to the bifurcation of the oviduct and vas deferens, is the narrow duct of the spherical receptaculum seminis. After uniting with the receptaculum duct the oviduct enters the female gland mass at the base of the albumen gland. Ventral to the albumen gland is a small membrane gland. Adjacent to both of these smaller glands is

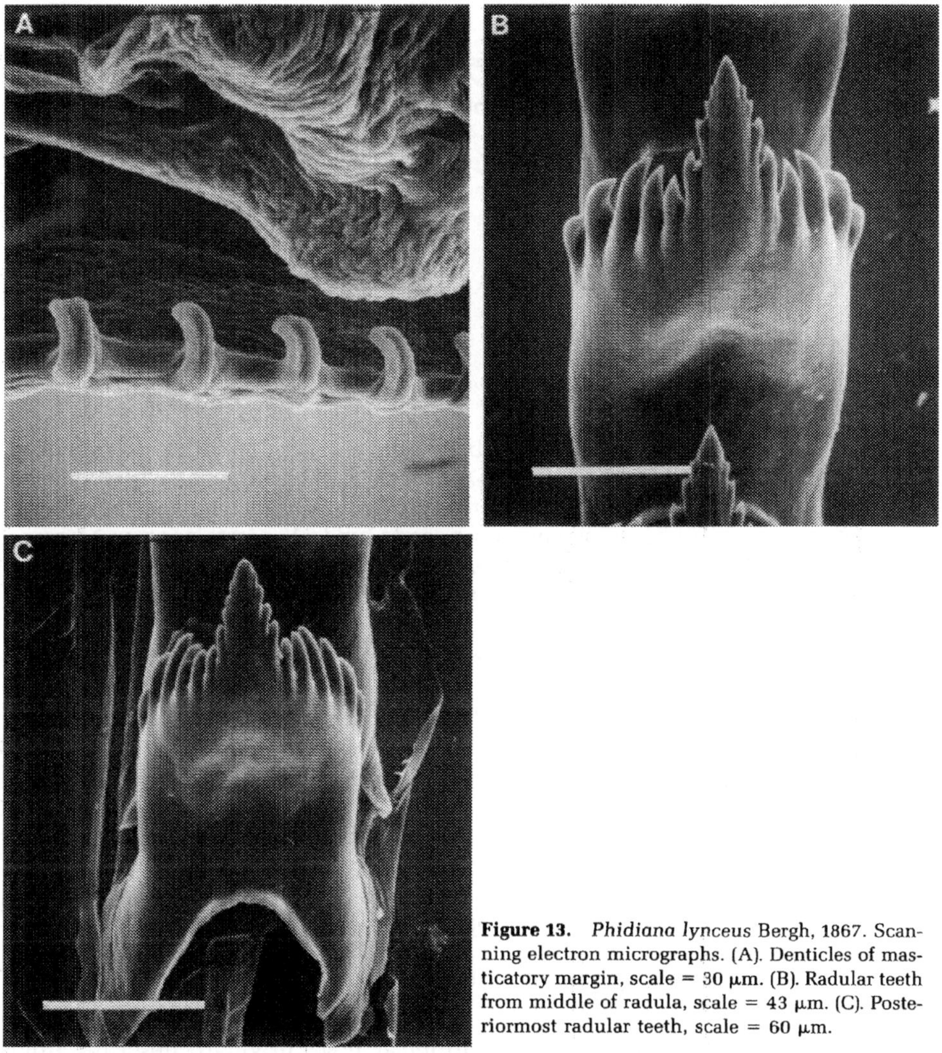

Figure 13. *Phidiana lynceus* Bergh, 1867. Scanning electron micrographs. (A). Denticles of masticatory margin, scale = 30 μm. (B). Radular teeth from middle of radula, scale = 43 μm. (C). Posteriormost radular teeth, scale = 60 μm.

the large, lobate mucous gland, which terminates at the female gonopore, adjacent to the penis.

- *Discussion.* Edmunds (1964) indicated that three species of *Phidiana* recorded from the western Atlantic (Engel, 1925) are conspecific. Marcus and Marcus (1967) first recorded this species from the Pacific coast of Panamá. Bertsch and Ferreira (1974) reviewed the Pacific species of *Phidiana* in their description of *P. lascrucensis*. *Phidiana hiltoni* (O'Donoghue, 1927), the senior synonym of *P. pugnax* Lance, 1962 and *P. nigra* MacFarland, 1966 (see McDonald, 1983), has a red line along each oral tentacle, which join anteriorly to the rhinophores. *Phidiana lascrucensis* lacks any lines on the head,

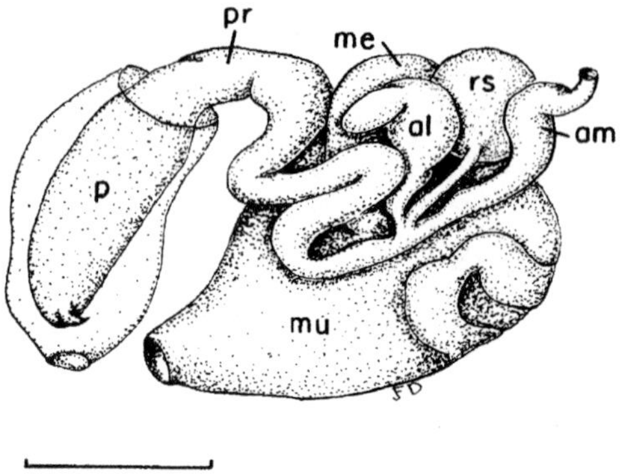

Figure 14. *Phidiana lynceus* Bergh, 1867. Reproductive system; al = albumen gland; am = ampulla; me = membrane gland; mu = mucous gland; p = penis; pr = prostate; rs = receptaculum seminis, scale = 1 mm.

though scattered white spots may be present and the tips of the oral tentacles are opaque white. *Phidiana inca* (D'Orbigny, 1837) has opaque white lines on the oral tentacles which join near the rhinophores. However, the animals are much larger and possess many more cerata (up to 13 cerata per row, Marcus, 1959). It also has fewer denticles on the radular teeth. Further examination of material from the mainland of Peru is necessary to verify that the two species are in fact distinct.

The present material agrees completely with that described by Marcus (1955, 1957, as *P. selenkai*) and Edmunds (1964), thus reaffirming the presence of this species in the tropical eastern Pacific.

2.9. *Aeolidiella alba* Risbec, 1928

Aeolidiella alba Risbec, 1928: 261, fig. 87, pl. 10 (fig. 9); Gosliner, 1985; 234, fig. 1A 2,3.

Spurilla alba (Risbec, 1928); Burn, 1966:98; Edmunds, 1969:465, fig. 9; Sphon, 1971:369; Edmunds and Just, 1983:200.

- *Distribution*. This species is circumtropical. It was originally described from New Caledonia (Risbec, 1928) and has been recorded from the Indo-Pacific tropics from southern Africa (Gosliner, 1985), Tanzania (Edmunds, 1969), Papua New Guinea (present study) and Australia (Willan and Coleman, 1984). It is also known from the Pacific coast of México (Sphon, 1971) and from the western Atlantic from the Barbados (Edmunds and Just, 1983).
- *Material*. One specimen, California Academy of Sciences, CASIZ 167108, near salt mine, Isla Santiago, Islas Galápagos, intertidal zone, under rocks, 19 February 1988, Kirstie Kaiser.
- *External morphology*. The living animal was approximately 5 mm in length. Its ground color, as recorded in color transparencies provided by Kirstie Kaiser and now in the color slide collection of the California Academy of Sciences is uniformly opaque white. Brown-orange markings are present on

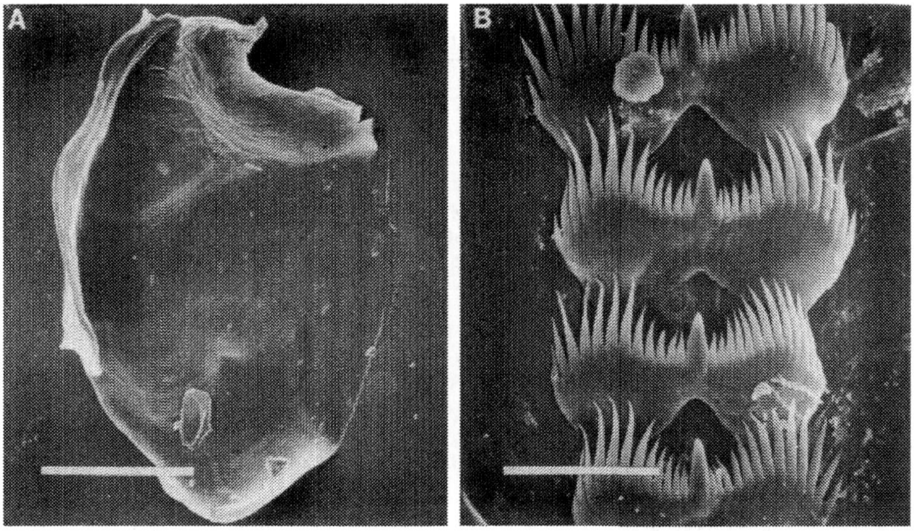

Figure 15. *Aeolidiella alba* Risbec, 1928. Scanning electron micrographs. (A). Jaw, scale = 150 μm. (B). Radular teeth, scale = 43 μm.

the rhinophores and on the head. Each rhinophore bears a pair of bulbous swellings along its length.

- *Buccal mass.* The buccal mass is thick and muscular with a pair of well developed chitinous jaws. Each jaw (Fig. 15A) is ovoid and lacks any trace of denticles along the edge of the elongate masticatory margin. The radular formula in the specimen studied here is 13 × 0.1.0. The teeth (Fig. 15B) are bilobed with a row of 11–14 denticles on either side of the triangular central cusp. The denticles vary considerable in length and are longest in the middle of each side.
- *Discussion.* The specimen was not adequately preserved to examine the reproductive organs. Nevertheless, the color of the living animal, the presence of rhinophoral swellings and the shape and number of radular teeth indicate that the Galápagos specimen is conspecific with *Aeolidiella alba*.

  This species has been placed in both *Aeolidiella* and *Spurilla*. Gosliner (1985) discussed the rationale for transferring this species back into *Aeolidiella*.

3. Discussion

The species of opisthobranchs recorded from the Galápagos are listed in Table 1. The biogeographic affinities of the Galápagos opisthobranch fauna were discussed briefly by Bertsch (1979). He stated that the few opisthobranchs known from there were also known from the mainland. Not enough is known about the distribution of most of the eastern Pacific cephalaspideans to include them in this discussion. The one exception is *Navanax aenigmaticus*, which has been well studied. Similarly, the pelagic thecosomes have different distributional patterns than their benthic counterparts and are generally excluded from discussion of

Table 1. Opisthobranch Gastropods Recorded fron the Galápagos Islands

Taxon[a]	Faunistic affinity[b]
Order CEPHALASPIDEA	
Family Acteonidae	
Microglyphis percònicus (Dall, 1890)	G
Family Bullidae	
Bulla punctulata A. Adams in Sowerby, 1850	EP
B. rufolabris A. Adams in Sowerby, 1850 (may be synonymous with *B. punctulata*)	EP
Family Retusidae	
Retusa galapagana Dall, 1919	G
Volvulella cylindrica (Carpenter, 1864)	EP
Family Scaphandridae	
Scaphander interruptus Dall, 1889	EP
Family Aglajidae	
Navanax aenigmaticus (Bergh, 1894),	A, EP
Order THECOSOMATA	
Family Cavoliniidae	
Cavolinia globulosa (Gray, 1850)	P
C. inflexa (Lesueur, 1813)	P
C. tridentata (Niebuhr, 1775)	P
Clio pyramidata Linnaeus, 1767	P
C. chaptalli Gray, 1850	P
Order ANASPIDEA	
Family Aplysiidae	
Aplysia parvula Mörch, 1863	C
Dolabrifera dolabrifera (Rang, 1828)	C
Order SACOGLOSSA	
Family Juliidae	
Berthelina chloris (Dall, 1918)	EP
Family Oxynoidae	
Lobiger souverbiei Fischer, 1856	C
Order NOTASPIDEA	
Family Umbraculidae	
Tylodina fungina Gabb, 1865	EP
Umbraculm umbraculum (Lightfoot, 1786)	C
Family Pleurobranchidae	
Berthellina engeli Gardner, 1936	A, EP
Pleurobranchus areolatus (Mörch, 1863)	WA, EP
Order NUDIBRANCHIA	
Suborder Doridacea	
Family Conualeviidae	
Conualevia alba Collier & Farmer, 1964	EP
Family Chromodorididae	
Chromodoris baumanni Bertsch, 1970	EP
C. sphoni (Marcus, 1971)	EP
Glossodoris sedna (Marcus and Marcus, 1967)	WA, EP
G. dalli (Bergh, 1894)	EP
Hypselodoris agassizii (Bergh, 1894)	EP
Hypselodoris lapislazuli (Bertsch & Ferreira, 1974)	G
Family Platydorididae	
Platydoris carolynae Mulliner and Sphon, 1974	G
Family Dendrodorididae	
Dendrodoris krebsii (Mörch, 1863)	WA, EP

continued

Table 1. *(continued)*

Taxon[a]	Faunistic affinity[b]
Doriopsilla janaina Marcus and Marcus, 1967	EP
Family Polyceridae	
Tambja mullineri Farmer, 1978	G
Suborder Aeolidacea	
Family Flabellinidae	
Flabellina iodinea (Cooper, 1863)	EP
Family Facelinidae	
Phidiana lynceus Bergh, 1867	A, EP
Family Aeolidiidae	
Aeolidiella alba Risbec, 1928	C

[a]Taxa in ***bold face*** = indicate species first recorded from the Galápagos in this chapter.
[b]Faunal affinities: A = Atlantic; C = circumtropical; EA = eastern Atlantic; EP = eastern Pacific; G = known only from the Galápagos; P = Planktonic; WA = western Atlantic.

oceanic island marine biota. This leaves 23 species of opisthobranchs for which ample distributional records are available. The largest proportion of these species (nine) are known from elsewhere in the eastern Pacific. Six species are known from both the tropical western Atlantic and the eastern Pacific. Five species are circumtropical in their distributions. Three species are known only from the Galápagos, though it is premature to consider them as endemics. It is interesting to note that none of the opisthobranchs recorded from the Galápagos have Indo-Pacific distributions, other than those species which are circumtropical and are also present in the tropical Atlantic. The biogeographical affinities of the Galápagos opisthobranch fauna are clearly with the tropical Americas. Certainly more information on other species present in the Galápagos is required to further elaborate upon the biogeography of the fauna. Similarly, information regarding the distribution of opisthobranchs along much of the Central and South American Pacific coast is virtually nonexistent. Additional fundamental distributional data must be collected from these areas to establish whether species presently known from the Galápagos are indeed endemic. Similarly, the opisthobranch mollusks from Cocos Island must be studied to ascertain their relationships with the Galápagos fauna.

ACKNOWLEDGMENTS. I thank several people for assisting in the collection of material. Virginia and Michael Gosliner accompanied me on my brief trip to the Galápagos in December, 1971, and provided considerable support, including financial assistance and assistance in the collection of specimens. Kirstie Kaiser provided material and color slides of specimens of material she and Don Shasky collected in the Galápagos. Gary McDonald provided specimens of material collected in the Galápagos by G. M. Wellington.

Gale Sphon of the Los Angeles County Museum generously assisted in the accumulation of information of Galápagos opisthobranchs and in sharing his knowledge of the fauna. Hans Bertsch enthusiastically assisted in the accumulation of data in the preparation of this manuscript. Jean De Mouthe kindly prepared the ink drawings and Mary Ann Tenorio prepared the final prints of the scanning electron micrographs. To these individuals I extend my thanks and appreciation.

References

Bebbington, A., 1974, Aplysiid species from East Africa with notes on the Indian Ocean Aplysiomorpha (Gastropoda: Opisthobranchia), *Zool. J. Linn. Soc.* **54**(1):63–99.

Bebbington, A., 1977 Aplysiid species from eastern Australia with notes on the Pacific Ocean Aplysiomorpha (Gastropoda, Opisthobranchia), *Trans. Zool. Soc. Lond.* **34**:87–147.

Behrens, D. W., 1980, *Pacific Coast Nudibranchs, a Guide to the Opisthobranchs of the Northeastern Pacific*, Sea Challengers, Los Osos.

Behrens, D. W., 1983, Report on the 1982 outer Baja California expedition, *Opisth. News* **15**(4):18–19.

Bergh, L. S. R., 1867, *Phidiana lynceus* og *Ismaila monstrosa*, *Vidensk. Medd. Danske Natur. For.* **7–9**:97–130.

Bergh, L. S. R., 1879, Bieträge zur Kenntniss der Aeolidiaden. VI. *Verh. der k. k.Zool. bot. Gesell. Wien* **28**:553–584.

Bergh, L. S. R., 1893, Die Gruppe der doridiiden, *Mitt. Zool. Stat. Neapel Zug. Rep. Mittelmeer.* **6**: 107–135.

Bergh, L. S. R., 1894, Die Opisthobranchien, Reports on the dredging operations off the west coast of Central America to the Galápagos, to the west coast of Mexico, and in the Gulf of California, in charge of Alexander Agassiz, carried on by the U.S. Fish Commission steamer "Albatross", during 1891, Lieut. Commander Z. L. Tanner, U. S. N. commanding, *Bull. Mus. Comp. Zool. Harv.* **25**(10):125–233.

Bertsch, H., 1978a, The Chromodoridinae nudibranchs from the Pacific coast of America. II. The genus *Chromodoris*, *Veliger* **20**(4):307–327.

Bertsch, H., 1978b, The Chromodoridinae nudibranchs from the Pacific coast of America. III. The genera *Chromolaichma* and *Mexichromis*, *Veliger* **21**(1):70–86.

Bertsch, H., 1979, Tropical faunal affinities of opisthobranchs from the Panamic Province (Eastern Pacific), *Nautilus* **93**(2–3):57–61.

Bertsch, H., and Ferreira, A. J., 1974, Four new species of nudibranchs from tropical West America, *Veliger* **16**(4):343–353.

Burn, R., 1966, Some opisthobranchs from southern Queensland, *J. Malac. Soc. Austr.* **9**:96–109.

Cervera, J., Templado, J., Garcia-Gomez, J., Ballesteros, M., Ortea, J., Garcia, F., Ros, J., and Luque, A., 1988, Catalogo actualizado y commentado do los opisthobranquios (Mollusca, Gastropoda) de la Peninsula Ibérica, Baleares y Canarías, con algunas referencias a Ceuta y la Isla de Alboran, *Iberus* **1**(suppl.):1–84.

Cooper, J. G., 1863, Some new genera and species of California Mollusca, *Proc. Calif. Acad. Nat. Sci.* **2**:202–207.

Dieuzeide, R. 1935, Sur une espece nouvelle de *Doridium* de la mediteranee, *Doridium berrieri* nov. sp., *Bull. Stat Aquic. Peche Cast.* **1**:105–117.

D'Orbigny, A., 1837, *Voyage dans l'Amérique Méridionale (Le Bresil, la republique orientale de l'Uruguay, la Republique argentine, la Patagonie, la republique du Chili, la republique de Bolivia, la republique de Perou) éxécuté pendant les annees 1826, 1827, 1828, 1829, 1830, 1831, 1832, et 1833, par Alcide d'Orbigny. 5. Mollusques*, pp. 377–408, Pitois-Leverault, Paris.

Eales, N. B., 1960, Revision of the world species of *Aplysia* (Gastropoda, Opisthobranchia), *Bull. Brit. Mus. Nat. Hist. Zool.* **5**(10):13–404.

Edmunds, M., 1964, Eolid Mollusca from Jamaica, with descriptions of two new genera and three new species, *Bull. Mar. Sci. Gulf. Carib.* **14**(1):1–32.

Edmunds, M., 1968, Opisthobranchiate Mollusca from Ghana, *Proc. Malac. Soc. Lond.* **38**:83–100.

Edmunds, M., 1969, Opisthobranchiate Mollusca from Tanzania, I. Eolidacea (Eubranchidae and Aeolidiidae), *Proc. Malac. Soc. Lond.* **38**:451–469.

Edmunds, M., 1975, Food currents and evolution of a West African nudibranch fauna, *J. Moll. Stud.* **42**(2):299.

Edmunds, M., 1977, Larval development, oceanic currents, and origins of the opisthobranch fauna of Ghana. *J. Moll. Stud.* **43**(3):301–308.

Edmunds, M., and Just, H., 1983, Eolid nudibranchiate Mollusca from Barbados, *J. Moll. Stud.* **49**(3):185–203.

Engel, H., 1925, Westindische opisthobranchiate Mollusken. Bijdragentot de kennis der fauna van Curaçao. Resultaten eener reis van Dr. C. J. van der Horst in 1920, *Bijd. Dierk. afl. Leiden* **24**:33–80.

Farmer, W. M., 1978, *Tambja* and *Roboastra* (Mollusca: Opisthobranchia) from the Gulf of California and the Galápagos Islands, *Veliger* **20**(4):375–385.

Garcia, J., and Garcia, F., 1984, Sobre la presencía de *Chelidonura africana* Pruvot-Fol (Mollusca: Opisthobranchia) en la litora íberico, *Boll. Malacologico Milano* **20**(1–4):77–82.

Gosliner, T. M., 1980, Systematics and phylogeny of the Aglajidae (Opisthobranchia: Mollusca), *Zool. J. Linn. Soc.* **68**:325–360.

Gosliner, T. M., 1985, The aeolid nudibranch family Aeolidiidae (Gastropoda: Opisthobranchia) from tropical southern Africa, *Ann. S. Afr. Mus.* **95**(6):233–267.

Keen, A. M., 1971, *Sea Shells of Tropical West America: marine mollusks from Baja California to Peru*, 2nd ed., Stanford Univ. Press, Stanford, California.

Lance, J. R., 1962, Two new opisthobranch mollusks from southern California, *Veliger* **4**(3):155–159.

MacFarland, F. M., 1966, Studies of opisthobranchiate mollusks of the Pacific coast of North America, *Mem. Calif. Acad. Sci.* **6**:1–546.

Marcus, Er., 1955, Opisthobranchia from Brazil, *Bol. Fac. Fil. Cien. Let. Univ. S. Paulo, Zool.* **20**:89–261.

Marcus, Er., 1957, On Opisthobranchia from Brazil, 2, *J. Linn. Soc. Lond., Zool.* **43**(292):390–486.

Marcus, Er., 1959, Reports from the Lund University Chile Expedition 1948–49. no. 36. Lamellariacea und Opisthobranchia, *Lunds Univ. Ars. (Ny Foljd), Avd. 2*, **55**(9):1–133.

Marcus, Er., and Marcus, Ev. D. B. R., 1967. American opisthobranch mollusks, *Stud. Trop. Oceanogr. Miami* **6**(1–2):1–256.

Marcus, Ev. D. B. R., 1971, On some euthyneuran gastropods from the Indian and Pacific Oceans, *Proc. Malac. Soc. Lond.* **39**:355–369.

Marcus, Ev. D. B. R., and Marcus, Er., 1962, Opisthobranchs from Florida and the Virgin Islands, *Bull. Mar. Sci. Gulf Carib.* **12**(3):450–488.

McDonald, G., 1983, A review of the nudibranchs of the California coast, *Malacologia* **24**(1–2):114–276.

Mörch, O., 1863, Contributions a la faune des Antilles Danoises, *J. Conchyl*, series 3, **11**:21–43.

Mulliner, D. K., and Sphon, G. G., 1974, A new *Platydoris* (Gastropoda: Nudibranchia) from the Galápagos Islands, *Trans. S. Diego Soc. Nat. Hist.* **17**(15):209–216.

O'Donoghue, C. H., 1927, Notes on a collection of nudibranchs from Laguna Beach, California, *J. Ent. Zool. Pomona Coll.* **19**(1–4):77–119.

Pilsbry, H. A., 1895, *Navanax* n.n. for *Navarchus*, *Nautilus* **8**:131.

Pruvot-Fol, A., 1953, Étude de quelques opisthobranches de la côte Atlantique du Maroc et du Sénégal, *Trav. Inst. Scien. Cherif. Zool.* **5**:1–105.

Risbec, J., 1928, Contribution a l'étude des nudibranches Néo-Calédoniens, *Faune des Col. Fran.* **2**(1): 1–328.

Rudman, W. B., 1984, The Chromodorididae (Opisthobranchia: Mollusca) of the Indo-West Pacific; a review of the genera. *Zool. J. Linn. Soc.* **81**(2–3):115–273.

Sordi, M., 1980, Una nuova specie di Aglajidae (Gastropoda: Opisthobranchia) vivente nel mare Tirreno: *Chelidonura italica* Sordi, *Att. Soc. Tosc. Sci. Nat. Mem., serie B*, **87**:285–197.

Sphon, G. G., 1971, New opisthobranch records from the eastern Pacific, *Veliger* **13**(4):368–369.

Sphon, G. G., and Mulliner, D. K., 1972, A preliminary list of known opisthobranchs from the Galápagos Islands collected by the Ameripagos Expedition, *Veliger* **15**(2):147–152.

Willan, R. C., and Coleman, N., 1984, *Nudibranchs of Australasia*, Australian Underwater Index, Sydney.

Chapter 14

Biogeography of Galápagos Land Snails

STEVEN M. CHAMBERS

1. Introduction

Land snails are among the earliest scientific collections recorded for the Galápagos. Although land snails have not always been the primary interest of collectors visiting the islands, material—mostly empty shells—made its way to private collections and museums in Europe and the United States, where it was described by specialists. Such prominent 19th century collectors as Hugh Cuming, Charles Darwin, and George Baur are well-known for their plant and/or animal collections, but only with respect to Cuming could it be said that mollusks exceeded or even rivaled his other interests in natural history. Despite this ancillary position among studies of the flora and fauna of the Galápagos, the land snails were the object of considerable biogeographical interest and speculation (Stearns, 1893; Dall, 1896). These speculations centered on how a remarkably diverse land snail fauna might develop on a small, arid, and extremely isolated archipelago.

The fauna considered in this chapter and referred to as "land snails" includes native terrestrial pulmonates of the order Stylommatophora, as well as two species of the terrestrial prosobranch family Helicinidae. Excluded from consideration are species of Stylommatophora that have been obviously introduced through human agencies during historical times (A. G. Smith, 1966) and marine or brackish water pulmonates.

The rationale for considering land snails in a work on marine invertebrates is based on their membership along with marine snails in the class Gastropoda and not on considerations of ecology or life history, which differ greatly between marine and terrestrial groups. The seas are barriers to land snail dispersal; unlike

STEVEN M. CHAMBERS • Division of Endangered Species, U.S. Fish and Wildlife Service, Albuquerque, New Mexico 87103. *Present address*: Southern California Field Station–Ventura Office, U.S. Fish and Wildlife Service, Ventura, California 93003.

their marine counterparts, land snails cannot survive in seawater, so they must be carried through the air on winds or by birds, or on the ocean's surface as passengers on floating materials in order to reach oceanic islands (Gulick, 1932; Carlquist, 1974).

The native land snail fauna of the Galápagos is summarized in Table 1. Most of the genera are represented by very small species, with shells that are a few millimeters or less in their greatest dimension. The exceptions are the bulimulids, which represent the most species-rich land snail family in the Galápagos, and the succineids. All of the native species, but none of the genera, are now considered to be endemic to the Galápagos Islands. The local radiation of the bulimulid genus *Naesiotus* is the most spectacular in the fauna of the islands, exceeding (at least at the species level) even that of the famous Darwin's finches. The extent of this radiation within such a small area is rivaled only by the radiation of similar bulimulids in Baja California (A. G. Smith, 1966). The magnitude of this radiation was initially recognized and described in the reviews of Reibisch (1892), Stearns (1893), and Dall (1896), and it has attracted the systematic and/or biogeographical interests of modern workers such as A. G. Smith (1966, 1971, 1972, 1974), whose collections have yet to be fully treated in publication, of Vagvolgyi (1977), and of Coppois (1984; Coppois and De Vos 1984, 1986; Coppois and Glowacki, 1983; Breure and Coppois, 1978).

This chapter describes the historical development of views of the origin of the Galápagos land snail fauna and continues with an assessment of attempts to interpret biogeographic relationships within the archipelago, including quantitative evaluations of equilibrium theories of island biogeography. Through this discussion, it should become apparent that the land snail fauna remains incompletely known, so that discussions of species diversity and inter-island similarities must remain largely speculative.

In an attempt to simplify a discussion that requires the use of a myriad of taxonomic names, modern names will be used throughout, except in cases where mention of obsolete names is required for the sake of discussion. Where obsolete names (with respect to Galápagos taxa) are used, they are enclosed within single quotation marks (').

The 80 species of land snails recognized are listed in Table 1. The bases for 68 species on this list are the treatments of Dall (1896) and Dall and Ochsner (1928). Those works can be consulted for synonyms of the recognized species. No attempt has been made here to critically review most of this taxonomy. It is very likely that some of the bulimulid species names recognized here are synonyms. The remaining 12 species recognized were described subsequent to Dall and Ochsner's 1928 compilation. A. G. Smith's (1974) recognition of synonyms of two of these species, *Naesiotus blombergi* and *Naesiotus eos*, is accepted here.

2. Origin of the Fauna

2.1. Initial Interpretations of Sources and Mechanisms

Many of the earliest published accounts (e.g., Pfeiffer, 1846, 1855; Sowerby, 1833b; E. A. Smith, 1877) of the Galápagos land snail fauna were simple descriptions of new species and did not include speculations on the origin or affinities of

Table 1. Recognized Species of the Native Land Snail Fauna of the Galápagos

Helicinidae (2 species)
Helicina nesiotca Dall, 1892
Helicina ochsneri Dall, 1917b
Achatinellidae (1 species)
Tornatellides chathamensis (Dall, 1892)
Pupillidae (5 species)
Gastrocopta munita (Reibisch, 1892)
Gastrocopta clausa (Reibisch, 1892)
Gastrocopta duncana Pilsbry, 1931
Pupisoma galapagorum Pilsbry, 1934
Nesopupa galapagensis Vagvolgyi, 1974
Strobilopsidae (1 species)
Strobilops helleri (Dall, 1900)
Succineidae (4 species)
Succinea bettii E. A. Smith, 1877
Succinea brevior E. A. Smith, 1877
Succinea producta Reibisch, 1892
Succinea corbis Dall, 1893
Bulimuidae (63 species)
Naesiotus nux (Broderip, 1832)
Naesiotus eschariferus (Sowerby, 1833a)
Naesiotus rugiferus (Sowerby, 1833b)
Naesiotus unifasciatus (Sowerby, 1833b)
Naesiotus calvus (Sowerby, 1833b)
Naesiotus ustulatus (Sowerby, 1833b)
Naesiotus jacobi (Sowerby, 1833b)
Naesiotus rugulosus (Sowerby, 1833b)
Naesiotus darwini (Pfeiffer, 1846)
Naesiotus sculpturatus (Pfeiffer, 1846)
Naesiotus perspectivus (Pfeiffer, 1846)
Naesiotus chemnitzioides (Forbes, 1850)
Naesiotus achatellinus (Forbes, 1850)
Naesiotus nucula (Pfeiffer, 1854)
Naesiotus galapaganus (Pfeiffer, 1855)
Naesiotus asperatus (Albers, 1857)
Naesiotus amastroides (Ancey, 1887)
Naesiotus planospira (Ancey, 1887)
Naesiotus invalidus (Reibisch, 1892)
Naesiotus pallidus (Reibisch, 1892)
Naesiotus nudus (Reibisch, 1892)
Naesiotus wolfi (Reibisch, 1892)
Naesiotus simrothi (Reibisch, 1892)
Naesiotus terebra (Reibisch, 1892)
Naesiotus canaliferus (Reibisch, 1892)
Naesiotus habeli (Dall, 1892)
Naesiotus duncanus (Dall, 1893)
Naesiotus olla (Dall, 1893)
Naesiotus tortuganus (Dall, 1893)
Naesiotus bauri (Dall, 1893)
Naesiotus reibischi (Dall, 1895)
Naesiotus tanneri (Dall, 1895)
Naesiotus nesioticus (Dall, 1896)
Naesiotus snodgrassi (Dall, 1900)
Naesiotus approximatus (Dall, 1900)
Naesiotus hoodensis (Dall, 1900)
Naesiotus indefatigabilis (Dall, 1900)
Naesiotus saeronius (Dall, 1917a)
Naesiotus rugatinus (Dall, 1917a)
Naesiotus cinerarius (Dall, 1917a)
Naesiotus trogonius (Dall, 1917a)
Naesiotus elaeodes (Dall, 1917b)
Naesiotus hemaerodes (Dall, 1917b)
Naesiotus perrus (Dall, 1917b)
Naesiotus cucullinus (Dall, 1917b)
Naesiotus albemarlensis (Dall, 1917b)
Naesiotus akamatus (Dall, 1917b)
Naesiotus adelphus (Dall, 1917b)
Naesiotus lycodus (Dall, 1917b)
Naesiotus alethorhytidus (Dall, 1917b)
Naesiotus cymatias (Dall, 1917b)
Naesiotus ochsneri (Dall, 1917b)
Naesiotus jervisensis (Dall, 1917b)
Naesiotus rabidensis (Dall, 1917b)
Naesiotus blombergi (Odhner, 1951)
Naesiotus eos (Odhner, 1951)
Naesiotus cavagnaroi A. G. Smith, 1972
Naesiotus hirsutus Vagvolgyi, 1977
Naesiotus pinzonensis Vagvolgyi, 1977
Naesiotus pinzonopsis Vagvolgyi, 1977
Naesiotus prepinguis Vagvolgyi, 1977
Naesiotus steadmani Chambers, 1986
Naesiotus kublerensis Chambers, 1986
Helicariondae (2 species)
Guppya bauri (Dall, 1892)
Habroconus? galapaganus (Dall, 1893)
Zonitidae (2 species)
Retinella? chathamensis (Dall, 1893)
Retinella? actinophora (Dall, 1900)
TOTAL NUMBER OF SPECIES: 80

the species. The history of early visits to the Galápagos by land snail collectors has been recounted by Dall (1896) and A. G. Smith (1966). The dates that the Galápagos species were described are included in Table 1. The sequence of collecting trips is summarized in Table 2 and discussed in the last section of this chapter.

Darwin (1839, 1860) was the first to discuss the biogeographical affinities of the land snail fauna. Based on his examination of the total flora and fauna of the

Table 2. Visits to the Galápagos that Resulted in Land Snail Records in the Scientific Literature[a]

Visit	San Cristóbal	Floreana	Santa Cruz	Santiago	Isabela
Cuming 1829	1	1		1	
Darwin 1835	2	2			
Kellett 1846	3				
Eugenie 1852		3			
Habel 1868	4				
Petrel 1875		4			
Wolf 1875	5	5	1	2	1
Albatross 1880	6	6	2	3	2
Baur 1890	7	7	3	4	3
Stanford 1898–1899	8	8	4	5	4
Ochsner 1905–1906	9	9	5	6	5
Pinchot 1929	10	10	6		6
Hancock 1931–1932	11				
Crocker 1932			7		
Blomberg 1934			8		
Blomberg 1948			9		
A. G. Smith 1964			10		
Van Mol 1970			11		
Vagvolgyi 1970			12	7	
Coppois 1973–1976			13	8	7
Steadman 1978–1980		11	14		

[a] Each visit is identified by name of a collector, expedition, or vessel, and the date of the visit. The numbers assigned to visits for each island are those employed in Fig. 1

archipelago, Darwin recognized that the Galápagos were volcanic, oceanic islands that are "zoologically part of America," and cited as evidence that the islands were situated only 500 to 600 miles (805 to 965 km) from the coast of South America, ocean currents flow through the archipelago in a westerly or WNW direction, and that driftwood and other buoyant plant materials could be found on the southeast shores of the islands. While accepting this generally American affinity, he also recognized that most of the land animals were endemic to the archipelago. He included the land snail fauna—of which he reported 16 "kinds"—in these generalizations, with the exception of one species of "*Helix*" that he believed was the same as one inhabiting Tahiti. This last observation has not been subsequently confirmed, as the specimen is apparently lost (Dall, 1896). With the exception of a few shells that became part of the Cuming Collection, now mainly at the British Museum (Natural History), the land snails that Darwin collected in the Galápagos have not been further described in print and may be lost (Dall, 1896; A. G. Smith, 1966).

In 1850 Forbes described two new Galápagos bulimulid species, one of which he acknowledged as having, along with the previously described bulimulid fauna (Broderip, 1832; Pfeiffer, 1846; Sowerby, 1833a–1839; 1833b), affinities with South American species. The other species, *Naesiotus achatellinus*, has relatively brightly colored banding and a glossy periostracum on the shell, which was the basis for his suggestion of a relationship to the beautiful "*Achatinellinae*" (italics used by Forbes) of the Hawaiian Islands.

Reibisch (1892) speculated on the origins of the Galápagos land snail fauna along with his formal descriptions of the collections made by Theodor Wolf in 1875 and 1878. Reibisch recognized both American and Polynesian affinities in different elements of the fauna. Although he recognized the bulimulid fauna as primarily American in origin, he believed that three of the bulimulid species (*Naesiotus canaliferus* [Reibisch, 1892]; '*Bulimulus cymatoferus*' Reibisch, 1892; and *Naesiotus achatellinus* [Forbes, 1850]) were Polynesian in origin. For the first two species he described the subgeneric "section" '*Pelecostoma*' (a synonym of *Naesiotus* Albers, 1850), stating that this group was very close to "den *Achatinellen*." The achatinnellines are today understood as a Hawaiian subfamily without a close relationship to the bulimulids. Reibisch lamented that soft parts were unavailable for study; had they been, he may have recognized that his '*B. cymatoferus*' was actually *Tornatellides chathamensis* (Dall, 1892), which is a member of an achatinellid subfamily that is widespread in the Pacific. Dall (1896) placed '*B. cymatoferus*' as a synonym of *T. chathamensis*; Reibisch's figure and stated measurements leave little doubt that Dall's assessment was correct. Reibisch was correct in his inference of a Polynesian origin for this species, but it is not a bulimulid, and there is no reason to believe that his other two "polynesian" bulimulids are not as American in origin as the rest of the bulimulid fauna of the Galápagos. In the non-bulimulid elements of the fauna, Reibisch believed that the two Galápagos pupillid species that he described were closely related to '*Pupa*' *wolfi* Miller, 1879, which he knew form Guayaquil Province, Ecuador. He stated that '*Helicina wolfi*' Reibisch, 1892 (placed in the synonymy of *Helicina nesiotica* Dall, 1892 by Dall [1896]) represented a bridge between Polynesian and American species, and that *Succinea bettii* Smith, 1877 was related to a species of *Succinea* from an island off the coast of Chile.

Reibisch was somewhat vague on the precise means of dispersal of land snails from the mainland, citing only oceanic currents ("Meeresströmungen"). He recognized that the present flow of the Humboldt Current northward along the coast of Chile and Peru, and then westerly towards the Galápagos explained the American elements of the fauna, but speculated that the currents may have flowed in a different direction at some earlier time to account for the Polynesian elements in the fauna.

In 1892 Dall published the first of two short papers describing new species from the collections of George Baur from the Galápagos. He noted "Panamic" affinities for two of three species described (Dall, 1892). One of these two species was *Tornatellides chathamensis*, which he originally placed in *Leptinaria* Albers, 1850 because all such mainland American forms were placed in *Leptinaria* rather than in the Pacific island genus '*Tornatellina*' (= *Tornatellides* Pilsbry, 1910). Dall (1900) later changed the generic placement of '*Leptinaria*' *chathamensis* to '*Tornatellina*' (= *Tornatellides*) based on radular characters. This reestablished the Polynesian affinity of this species. In the second paper, Dall (1893) described several new species and stated that Baur's collections had provided enough material to establish that the Galápagos bulimulids in particular and the land snail fauna in general constituted an American fauna.

Stearns (1893) provided an extensive discussion of the possible origin of the Galápagos land snail fauna in his report on the mollusks (terrestrial and marine) collected during the 1887–1888 voyage from Chesapeake Bay to San Francisco by the U.S. Fish Commission steamer *Albatross*. Stearns regarded the Galápagos land snail fauna as incompletely sampled, but complete enough to be recognized as distinctly American, as Ancey (1887) and others had previously indicated. He stated that the bulimulids were most similar to forms from Chile and Peru, dismissing both Polynesian origin and close relationship to Baja California bulimulids. He correctly recognized that *Naesiotus achatellinus* only superficially resembled brightly colored and smooth-surfaced Polynesian "achatinellas" and suggested that their similarities were a result of occupying similar upland, arboreal habitats. His inference on the lack of relationship was correct, although he was apparently unaware that *Achatinella* of Oahu in the Hawaiian Islands occupied lowland as well as highland habitats before agricultural development by native Hawaiians extirpated them from the lowlands (Welch, 1942). He viewed the superficial similarity between Galápagos and Baja California bulimulids to be the result of their occupying similar arid environments.

Stearns (1893) rejected Baur's (1891) contention that the Galápagos were continental islands [i.e., islands that originally were parts of continents (Wallace, 1892)] that had become fragmented through subsidence, favoring the arguments of Darwin, Wolf, and Agassiz that these were true oceanic islands. He found dispersal from continental America through both rafting on ocean currents and aerial distribution to be adequate to explain the composition of the land snail fauna. In favor of rafting, he recognized that from their directions of flow, the Humboldt, Mexican (= California), and Panama currents could contribute land snails to the Galápagos. He considered the archipelago to be a short trip from the American coast by these means. He enumerated sightings of rafting materials off the west American coast from flooding rivers as evidence that even parts of the west American coast that lacked large rivers could, in time of floods, provide materials to be carried upon ocean currents. Orr (1966) later described impressive mats of vegetation on the Guayas River of Ecuador. Stearns (1893) cited Agassiz's (1892) observations of sea bottom deposits of terrestrial origin between the Galápagos and Mexico as an indication of the amount of continental material that is constantly carried on ocean currents in the direction of the Galápagos. Stearns felt that land snails were well-suited to such ocean voyages, making passage as eggs or, in later stages, in dormancy. The ability to secrete an epiphragm to seal the aperture was seen as particularly important, and Stearns cited examples of land snails surviving in the dormant state in museum collections for over two and up to six years. Most curators of land snail collections can cite similar experiences with curated material that has seemingly come back to life. Stearns found aerial dispersal also a convincing mechanism, noting that some snail eggs are no larger than seeds and other wind-borne particles. He was impressed by the sand and dust carried aloft by North African winds across the Mediterranean and the Santa Ana winds of Southern California and recognized that occasional storm winds were especially likely to carry land snail colonists.

Dall's 1896 monograph on the Galápagos land snails included an extensive discussion of the morphological features and dispersal abilities of island land snails in general. He agreed with Stearns' major arguments about the oceanic nature of the Galápagos, the American origin of most of the fauna, and the likelihood of rafting on ocean currents and wind dispersal as mechanisms of colonization. He viewed rafts with snails attached by epiphragms of dried mucus to floating trees as especially convincing, and noted that some of the Galápagos bulimulids secrete two epiphragms in succession and that these structures conformed to irregularities of the shell to form a very tight seal. He extended Stearns' general observations of estivation in snails specifically to the Galápagos fauna, relating how specimens of *Naesiotus planospira* (Ancey, 1887) became reactivated after more than a year. Although he accepted both as plausible contributors to the Galápagos fauna, Dall favored the Panama Current over the Humboldt Current because he thought it required a shorter time for land snails to spend on the sea, and because the Gulf of Panama has a rich snail fauna closer to the sea and therefore would be a more likely source of snail colonists than the arid and relatively depauperate shores of Peru. He also saw great potential for minute species to be sheltered in the crevices of rafted material as well as being carried by strong winds, which he estimated could carry a small snail adhering to leaves from the American mainland to the Galápagos in eight to ten hours. Dall and Ochsner (1928) later reported the hibernation of Galápagos species of *Gastrocopta* Wollaston, 1898 attached to leaves, a situation that would aid dispersal by wind. Dall (1896) noted that the same families of small snails found in the Galápagos were frequent in other insular faunas, supporting the notion of successful long-distance dispersal.

Dall (1896) placed all Galápagos bulimulids by themselves in the subgenus *Naesiotus*, acknowledging that this was for the sake of taxonomic convenience and not a natural evolutionary grouping. He followed Stearns (1893) in rejecting contentions by Forbes (1850) and Reibisch (1892) of Polynesian affinities for any of the Galápagos bulimulids. He noted that the loose coiling of the apical whorls, to form an apical "dimple," was shared by bulimulids from the Galápagos, Peru, and Baja California. Despite his argument that many similar shell features of Baja California and Galápagos forms were the result of development in a similar, arid environment, he suggested that the Galápagos bulimulids were closer to the subgenus '*Orthotomium*' Fisher and Crosse, 1875 (today recognized as a synonym of *Rabdotus* Albers, 1850) of northern Mexico than to Peruvian forms, indicating a more northern origin than most other extant bulimulid groups.

Both American and Polynesian origins were recognized by Dall (1896) for the small snail species. He treated *Gastrocopta munita* as a synonym of the Ecuadorian species *Gastrocopta wolfi*, but stopped short of according the same treatment to *G. clausa*. He thought that *Helicina nesiotica* Dall, 1892 was very similar to Panamic forms of the genus. On the other hand, he speculated that *Guppya* (= '*Trochomorpha*') *bauri* (Dall, 1892) was similar to Polynesian species, and made the reasonable suggestion that it may have been the lost "*Helix*" that Darwin referred to as a Tahitian species.

In his treatment of the subgenera of *Bulimulus*, Pilsbry (1897–1898) held the subgenus *Naesiotus* to be a natural group, all descended from one invasion of the Galápagos through random dispersal. On the basis of similar apical sculpture, he placed *Naesiotus* in a "Division" that included groups from northern Mexico, Baja California, and eastern South America, and he stated that no anatomical feature distinguished the Galápagos species from these mainland forms. In his summary of bulimulid classification, Pilsbry (1902) recognized this Division as comprising three "Sections:" (1) Section '*Protoglyptus*' Pilsbry, 1897 (a synonym of *Naesiotus*), from eastern South America, the Caribbean, and Peru; (2) *Naesiotus*, restricted to the Galápagos; and (3) '*Orthotomium*' (a synonym of *Rabdotus*), from the southwestern United States, northern Mexico, and Baja California.

Dall (1900) reviewed the known Cocos Island, Costa Rica, land snail fauna of seven species, which included two *Leptinaria* species, but no *Tornatellides*, bulimulids, or any other taxa that might link the Cocos and Galápagos faunas. This conclusion has been supported by subsequent reviews of the Cocos land snail fauna (Hanna and Hertlein, 1938; Hertlein, 1963).

Dall (1917) restored the complete endemic status of the Galápagos land snail species by dropping his contention that any Galápagos *Gastrocopta* were conspecific with mainland South American species [Dall was confused on the precise nomenclature of this group (Pilsbry, 1931)]. He also described *Naesiotus trogonius*, noting that it was the smallest Galápagos bulimulid (height: 7 mm; maximum diameter: 3 mm) and speculated that it may represent the ancestor of the Galápagos bulimulids, presumable because its small size would aid its dispersal.

Dall (1920) divided the subgenus *Naesiotus* of the genus *Bulimulus* into 15 "groups." He considered the groups of *Naesiotus ustulatus* (Sowerby, 1833) (comprising 10 species) and *Naesiotus amastroides* (Ancey, 1887) (comprising four species, including *N. trogonius*) to be "near" or "close to" the ancestor of the Galápagos bulimulids, citing their small size and ground-loving habits. He further noted that this group was the most widely distributed among the islands and was particularly numerous on Isabela. His group of *B. (N.) hoodensis* consisted of the single species that, he remarked, resembled the "continental type" more than any other Galápagos bulimulid.

Gulick (1932) extensively discussed land snail faunas, including that of the Galápagos, in his general review of the biogeography of oceanic islands. His thesis was that, although land snails have low local rates of dispersal, they have large potential for passive dispersal over long distances. He stressed aerial dispersal, noting that small species can be dispersed by hurricane-like winds in the same manner as plant seeds. He noted that the islands of the Pacific are generally occupied by certain families of land snails. These include the cosmopolitan "micropulmonate" families Pupillidae, Zonitidae, Endodontidae, and Helicinidae, and the wide-ranging Pacific subfamily Tornatellininae (Family Achatinellidae), as well as a family or larger-size snails, the Succineidae. All of these families except the Endodontidae have representatives on the Galápagos. The absence of endodontids in the Galápagos should not be viewed as an anomaly, however, because those other micropulmonate families are themselves represented in the Galápagos by only a few species. The pupillids are the best represented of these families (five species), but all except *Nesopupa galapagensis* could have originated from the American mainland rather than from Pacific islands.

After reviewing the land snail faunas of the world's most remote islands, Gulick (1932) found little problem with deriving most of the Galápagos fauna from the American mainland. The exception he noted was *Tornatellides chathamensis*, the ancestors of which, he speculated, could have come 3300 miles (5300 km) from the Marquesas or 2150 miles (3460 km) from Easter Island. Gulick observed that, because of their low local dispersal rates, land snails are predisposed to becoming localized and to differentiate, accounting for such spectacular evolutionary radiations as in the achatinellids of the Hawaiian Islands, *Partula* Ferussac, 1821 of the Society Islands (Crampton, 1916, 1932), and the bulimulids of the Galápagos. He viewed the"harmonious" aspect of the Galápagos and certain other faunas as "newly harmonious," the result of radiation from relatively few original colonists. There is no reason, therefore, to hypothesize much larger and more ancient former land areas to account for the ecological diversity observed in these radiating groups. Lack (1947) similarly argued for an oceanic origin of the Galápagos and the *in situ* radiation of Darwin's finches from a "common stock." He cited similar radiations in other groups of Galápagos organisms, including *Naesiotus*, as evidence that adaptive radiation can mimic the patterns of highly diverse faunas that Baur (1891) and others regarded as harmonious.

2.2. Revival of Continental Island Theories

The concept of a continental origin of the Galápagos bulimulids through a direct land connection to South America was revived by Rehder (1940). He interpreted the great diversity of the fauna as evidence of its long isolation and held that a former presumed connection of the Galápagos to South America had subsided. He discounted the likelihood that the cold Humboldt Current could carry certain animal colonists to the Galápagos and rejected wind dispersal on the grounds, later rendered obsolete by the work of Weyrauch (1956), that the South American bulimulids that are most closely related to the Galápagos species were not found on the western slopes of the Andes, but in inter-Andean valleys to the east.

It is in his treatment of these inter-Andean populations that Rehder (1940) made his major contribution to the study of the Galápagos bulimulid fauna. He raised *Naesiotus* to full generic level and included within it the Galápagos bulimulids and *Naesiotus quitensis* (Pfeiffer, 1848) from the Interandean Plateau of Ecuador. This was the first inclusion of a mainland species in what had previously been considered an endemic Galápagos group. This taxonomic placement was based on the apical sculpture of the shells. Rehder noted the similar structure in '*Protoglyptus*' (a synonym of *Naesiotus*) from eastern South America and the Lesser Antilles and *Rabdotus* (= '*Orthotomium*') of Mexico and the southwestern United States and speculated that these groups had differentiated in the Tertiary from an ancient group of bulimulids that ranged from northern Mexico to Argentina.

The theory that the Galápagos were once a single continental island was revived yet again by Odhner (1950), who maintained that the land bridge between the archipelago and South America had subsided with ocean-bottom cooling during the "Glacial age." He believed that the nearest relatives of the Galápagos bulimulids were Chilean forms, which he did not specifically identify, and that even earlier ancestors were Pacific forms (how early was not indicated). Odhner noted two variations of apical sculpture were found in the Galápagos bulimulids, one much like the *Rabdotus* ('= *Orthotomium*') of Mexico. He rejected Dall's (1896) belief that the Galápagos fauna was derived from '*Orthotomium*,' finding it more likely, from his evaluation of the soft anatomy of several species, that the Mexican forms were derived from *Naesiotus*.

2.3. Reemergence of the Oceanic Island Theory

2.3.1. Bulimulids

Weyrauch (1956) studied several south American bulimulid species, including some that had previously been assigned to '*Protoglyptus*'. He considered two newly described species from Peru to be closer to the Galápagos bulimulids than to any other American continental species. For this and other reasons, he declared that '*Protoglyptus*' was a synonym of *Naesiotus*, which now had a range that included the Galápagos, Peru, Brazil, the Lesser Antilles, Bolivia, and Argentina. He speculated that *Naesiotus* arose in the Eocene, when conditions for dispersal from Peru were more favorable than in recent times; this undoubtedly preceded, however, the emergence of the Galápagos from the sea (Simkin, 1984). Of special significance to the origin of the Galápagos bulimulids were Weyrauch's reports of species of *Naesiotus* on the western slopes of the Andes, down to about 400 m in elevation; the previously observed absence of likely ancestral forms from near the western South American coast was thereby eliminated as an obstacle to the hypothesis that rafting materials carried one or more bulimulid colonists to the Galápagos. Furthermore, many of these Andean localities were in arid environments; snails that evolved in such habitats would have advantages for surviving the rigors of ocean travel and subsequent colonization of the low, arid zones of the Galápagos.

Steadman and Ray (1982) addressed a similar problem in the origin of the endemic Galápagos rodent, *Megaoryzomys curioi* (Niethammer, 1964), whose closest living relatives live in forested areas at higher elevations in South America. They hypothesized that the ancestor of the Galápagos species must have inhabited the lowlands of South America during or before the Pleistocene, when forested conditions may have existed in coastal areas.

On the proposed affinities between Baja California and Galápagos bulimulids, Weyrauch (1956) declined to place *Rabdotus* (= '*Orthotomium*') as a synonym of *Naesiotus* based on differences in microsculptural details of the apical whorls, and because he believed, partially based on its relatively confined distribution, that *Rabdotus* was a much younger group.

A. G. Smith (1966) presented views similar to Stearns, Dall, and Gulick on the agencies of rafting and winds in colonization of an oceanic Galápagos archipelago, followed by *in situ* radiation of the bulimulid fauna, and added in support of this general theory that estivating Galápagos species sometimes had two or three epiphragms, and large rafts of trees and other material have been observed floating towards the sea near Guayaquil, Ecuador (Orr, 1966).

A. G. Smith (1966) regarded the Galápagos bulimulids as members of the genus *Bulimulus*, subgenus *Naesiotus*, but doubted that *Naesiotus* represented a natural group and felt that the mainland fauna was not sufficiently known to interpret origins. Nevertheless, he restored the generic status of *Naesiotus* in his later publications (A. G. Smith, 1971, 1972, 1974). He suggested that certain bulimulids, which he did not specifically identify, from the arid zone near Guayaquil resembled the Galápagos species. Smith also noted that no bulimulids had been reported from Cocos Island, Costa Rica. Cocos has always been of interest to students of Galápagos biogeography because of the presence there of a finch thought to form a monophyletic group with the Galápagos finches until Steadman (1982) argued that Cocos and the Galápagos were derived from independent invasions by a mainland finch of the genus *Volatinia*.

A. G. Smith (1971) returned to an old question: the taxonomic status of *Naesiotus achatellinus*, which had at times been regarded as having affinities with Hawaiian achatinellids. From his observation of the apical sculpture of some newly discovered material of this rare species, he concluded that it was a *Naesiotus* and deserved no taxonomic distinction from other Galápagos *Naesiotus*.

Breure and Coppois (1978) again regarded *Naesiotus* as a genus, with '*Protoglyptus*' as a synonym, and distinct from *Bulimulus*. These conclusions were based on apical sculpture and the anatomy of the phallus complexes of 23 *Naesiotus* species, several of which were Galápagos species.

Breure's (1979) revision of the Bulimulinae, the bulimulid subfamily that includes the Galápagos species, retained the generic status of *Naesiotus*, with '*Protoglyptus*' but not *Rabdotus* (= '*Orthotomium*') as a synonym. The range of *Naesiotus* was reported as the Lesser Antilles, Venezuela, Brazil, Paraguay, Uruguay, Argentina, Bolivia, Peru, Ecuador (including the Galápagos), and Colombia. The apical sculpture was recognized as taxonomically useful, but not informative of phylogenetic relationships among the genera. Both *Bulimulus* and *Naesiotus* are regarded as existing in pre-Andean times, suggesting a lower Tertiary origin. Of interest to historical issues of Galápagos bulimulid origins and affinities are Breure's observations that the continental *Naesiotus* are generally ground-dwelling, whereas the Galápagos species have radiated to forms found on shrubs, rocks, and trees; and Andean *Naesiotus* are often found in low and arid portions of inter-Andean valleys. He also excluded from *Naesiotus* two species from Guadeloupe, Lesser Antilles, that Van Mol (1972) had considered anatomically similar to the Galápagos bulimulids, thus rejecting the implication of a particularly close relationship between Antillean and Galápagos bulimulids.

Hoffman (1988) has reassigned 21 former *Rabdotus* species to *Naesiotus* based on features of the apical whorls and reproductive anatomy. Three of these species occur in southern Arizona and Sonora, Mexico, while the remaining 18 species occur in Baja California Sur, Mexico. The presence of *Naesiotus* in western Mexico requires the reconsideration of Dall's (1896) conjecture that the Galápagos bulimulids were derived from Mexican forms. The extent of availability of continental South American *Naesiotus* for colonizing the Galápagos (Rehder, 1940; Weyrauch, 1956; A. G. Smith, 1966; Breure, 1979) was established after Dall's death. Although rafting from western Mexico remains a possibility, it seems more likely that *Naesiotus* colonized the Galápagos from the nearer shores of continental South America.

2.3.2. Non-Bulimulid Land Snails

A. G. Smith's (1966) overview of the entire land snail fauna led him to assert that all genera found on the Galápagos could also be found on mainland South America, except for *Tornatellides* [Vagvolgyi (1974) later added a second species with a Pacific origin]. He described the wide range of *Tornatellides* (including Hong Kong, Hawaii, New Zealand, Society Islands), and speculated that the Galápagos species was most similar to Hawaiian species, but that he had not compared members of that genus from the Revillagigedos Islands of Mexico, which are geographically closer to the Galápagos. Although Smith believed that the Galápagos had existed as long ago as the Miocene [they are probably much younger in geological age (Simkin, 1984)], he felt that colonization had occurred mainly within the Pleistocene, followed by inter-island colonizations and evolutionary radiation.

Vagvolgyi (1974) added a second species of non-American origin to the Galápagos land snail fauna when he described *Nesopupa galapaganus*. He regarded Hawaiian species as the most likely ancestors, but at the same time commented that

Nesopupa galapaganus was most similar to an undescribed species from Surinam, implying that the Surinam form may have been derived from the Galápagos species.

In his general view of Pacific land snail biogeography, Vagvolgyi (1975) promoted the importance of aerial dispersal in the distribution of Pacific island land snails. As part of his analysis, he compared faunas of young versus old islands. As examples of "young" islands, he presented the recolonization of Krakatoa after its 1883 explosion, and the Galápagos, on the basis that they had been habitable for no more than about one million years. His essential point was that, of 11 genera that colonized the Galápagos archipelago, nine comprised minute species, which he found similar to the size composition of faunas of older islands. Along with wind dispersal, Vagvolgyi regarded dispersal by birds as important colonizing agents. He reviewed the literature on bird dispersal of snails and noted that the Hawaiian Petrel (*Pterodroma phaeopygia*), which is restricted to the Hawaiian Islands and the Galápagos, was a good candidate for a dispersal agent between those two island groups.

The current, overall view of the origin of the Galápagos land snail fauna is remarkably close to the early interpretations of Stearns and Dall. With the exception of species of *Tornatellides* and *Nesopupa* with their Pacific Island origins, there is no difficulty in deriving the fauna from adjacent areas of the South American continent, although the precise method of dispersal, whether by rafting or by aerial transport (by winds or birds), for each colonizer is a matter of speculation.

3. Inter-Island Faunal Relationships

Darwin (1860) believed that each of the larger islands had an independent, volcanic origin, and the major islands of the archipelago were not the result of fragmentation or subsidence of a previously united land mass. This is the modern geological view and is accepted in this and the following section of this chapter.

Each of the larger islands of the Galápagos has its own endemic species (A. G. Smith, 1966). Dall (1896) listed species ranges known at the time and speculated that the larger number of endemics on the islands of San Cristóbal, Charles, and Española was a result of their southeastern position in the archipelago, which would make them the first to be reached by wind- or current-borne colonists from the east and southeast. It should be noted that these are probably also the oldest islands in the archipelago (Simkin, 1984). Dall's admonition that his interpretation should be treated with caution, because these were also the best-studied islands with the most completely known faunas, was borne out when A. G. Smith's (1966) tabulation of Dall's and subsequent records indicated that Santa Cruz, at the center of the Archipelago, had the highest rate of endemism. Smith did, however, note that non-endemic species on the southeastern group of islands were not found on the northwestern islands.

Dall's (1920) classification of Galápagos bulimulids include four "sections" that are confined to a single island. Ranges of the other sections include clusters of adjacent islands. Islands in these clusters are generally along east–west or northwest–southeast axes, as might be expected by prevailing wind and current directions, except the section of *Naesiotus nux* on Floreana and San Cristóbal, which lie on a southwest–northeast axis, and the section of *N. amastroides*, which occupies San Cristóbal, Floreana, and Isabela, bypassing Santa Cruz and Santiago.

Dall's sections may only roughly represent natural groups of bulimulids. This and other taxonomic uncertainties make inter-island faunal interpretations very difficult, as does the incomplete knowledge of the fauna, which is discussed at the end of the following section.

A. G. Smith (1966) recognized a general division between the southeastern group of islands and the northwestern group based on the observation that species occupying more than one island were as a rule confined to one or the other group.

Distributions of bulimulid species that occur on more than one island tend to include clusters of adjacent islands. R. P. Smith (1971) noted that most of these non-endemics were arid zone species; the habitat of these species is closer to the sea, and they probably are resistant to desiccation. These attributes make them more likely to be dispersed between islands.

4. Determinants of Inter-Island Species Diversity

The only quantitative attempt that I am aware of to analyze patterns of land snail diversity within the Galápagos is in R. P. Smith's (1971) unpublished study. Smith analyzed Galápagos bulimulid diversity in order to test aspects of MacArthur and Wilson's (1967) equilibrium theory of island biogeography concerning the effects of ecological diversity and inter-island distance. He contrasted the earlier findings of Hamilton and Rubinoff (1963, 1964), which concluded that species diversity in Galápagos finches was determined by insular isolation and maximum elevation, with studies on Galápagos plants, which either found (Hamilton *et al.*, 1963) or failed to find (Johnson *et al.*, 1968) an influence of island area on species diversity. R. P. Smith implied that influences on bulimulid species diversity should more closely follow those of plants than those of birds because of the passive nature of inter-island dispersal common to both plants and land snails. Smith regarded his analysis as a preliminary treatment of equilibrium theory in that the bulimulid fauna is incompletely known, both geographically and taxonomically, and because the islands varied in age.

R. P. Smith employed the stepwise multiple linear regression technique used by Hamilton and Rubinoff to test models of species diversity in the Galápagos bulimulids, using as dependent variables the number of species (S), number of endemic species, or the percent endemism for each island. As independent variables to predict species diversity, he used island area, maximum elevation, number of plant species, average distance from other islands in the archipelago, distance to the nearest neighboring island, and distance from the "central" island of Santa Cruz. He tested various linear models for each of the three measures of the three dependent variables, and various other models to predict S using log-transformed dependent and/or independent variables.

Of the models he tested, Smith found that the linear model best predicted S ($R^2 = 0.887$), with the number of plant species being the only variable having a significant regression coefficient ($p < 0.01$). When Isla Isabela was deleted from the model, because it was considered to be the island most likely to harbor a number of undiscovered species, number of plant species remained the most important predictor (partial regression coefficient significant, $p < 0.01$), but island area was also found to have a significant ($p < 0.05$) partial regression coefficient. Only number of plant species was found to account significantly for species diversity in the log-transformed models, which will not be further considered here.

On the basis of these analyses, Smith concluded that indices of ecological variation (plant diversity and area) were more important than inter-island distances in predicting species diversity, and that plant diversity should be a reasonable indicator of niche diversity for land snails. He correctly noted that island endemism at the level of species was higher in the bulimulids than in the finches, indicating that the bulimulid fauna was in a more advanced state of adaptive radiation than the finches as a result of *in situ* speciation on individual islands, while intra-island barriers are less likely to be sufficient to promote equivalent evolutionary differentiation in finches and other birds, which have much larger home ranges and routine dispersal capabilities than land snails. He linked this observation on speciation to MacArthur and Wilson's suggestion that ecological opportunity becomes a more important determinant of species diversity after the initial colonization opportunity has reached some equilibrium.

Connor and Simberloff (1978) critiqued the use of multiple regression studies to identify determinants of diversity in Galápagos finches and plants. They demonstrated the extreme sensitivity of stepwise multiple regression models to the completeness and quality of the data variables. With improved data, they reanalyzed earlier, contradictory studies and concluded that island area best predicted plant species diversity in the Galápagos. More importantly, they analyzed these models further by adding number of collecting trips as an additional independent variable, arguing that the collecting effort could bias species diversity indices if it were not proportional to the "real" number of species on each island. They concluded that the "number of botanical collecting trips to each of the Galápagos Islands is a better predictor of species number than are area, elevation, or isolation" (Connor and Simberloff, 1978, Abstract).

As with plants, differential collecting effort is an important potential bias in attempting to model species diversity of land snails, as R. P. Smith recognized. I have attempted to evaluate the potential effect of collecting bias on Smith's conclusions by using multiple linear regression on his original data, but without the stepwise procedure in order to avoid complications introduced by the order in which independent variables enter the regression. Smith's variables were first analyzed as he presented them, and then they were analyzed with number of trips as an additional independent variable. All analyses were carried out with untransformed variables. The numbers of land snail collecting trips to each island were inferred from literature accounts of the source of the material described (Table 2). Specific collectors on expeditions responsible for some very early collections, by whalers for example, were not identified in the descriptive literature and were not included in the counts of trips. The number of trips is therefore underestimated for some islands. As mentioned at the beginning of this chapter, relatively few of the identified trips were dedicated to land snail collecting, as most were by collectors of plants or vertebrates who incidentally picked up land snails during their primary efforts.

The reanalysis of Smith's variables resulted in a multiple regression with $R^2 =$

0.884, with only the number of plant species having a significant ($p < 0.05$) partial regression coefficient. Omitting Isabela, the island that is most likely to be undercollected, $R^2 = 0.957$, with only number of plant species ($p < 0.01$) and area ($p < 0.05$) having significant partial regression coefficients. These results are very similar to Smith's and bear out his finding on the importance of plant diversity and area.

When number of land snail collecting trips is added to Smith's data, R^2 is 0.974 (including Isabela) or 0.987 (excluding Isabela), both of which are higher than the corresponding values without trips included. Partial regression coefficients for number of trips are significant ($p < 0.01$ and $p < 0.05$, respectively) in both cases. Area is a significant predictor ($p < 0.05$) only when Isabela is excluded from the model. The best predictor of bulimulid species diversity in the Galápagos is therefore the number of collections made on each island, even though that effort is incompletely known and is not weighted for the highly variable degree of effort represented by individual trips.

An empirical approach to the problem of evaluating collection effort is to plot the increased numbers of species known over time, represented by number of visits. Such plots for five of the major Galápagos Islands are presented in Fig. 1. The collectors or expeditions responsible for each of the plotted trips to each island are given in Table 2. As the fauna of an island becomes known, these curves should level off as fewer new species records can be found during successive trips to an island. All curves but that for San Cristóbal have an upward inflection as a result of Ochsner's (1905–1906) several months of collecting land snails in the archipelago. Ochsner's addition of only a single species to the fauna of San Cristóbal may indicate that the recorded land snail fauna of this island was nearly complete. Ochsner, however, spent fewer days collecting land snails on San Cristóbal than any other major island, and San Cristóbal has been infrequently explored for land snails since Ochsner's visit. In addition, San Cristóbal was settled and farmed relatively early (Slevin, 1959), and the faunal diversity may have already been reduced by those activities.

Overall, the curves in Fig. 1 show some indications of leveling off owing to "collector saturation," although departures from a linear relationship between numbers of trips and species known are slight. That it is premature to conclude that any of these curves is approaching an asymptote representing the "actual"number of species for each island is illustrated by the case of Santa Cruz. The curve for Santa Cruz appeared to be about to level off, when the collections of Holocene fossil material by Steadman added new species records (Chambers, 1986; Chambers and Steadman, 1986). This is yet another example of paleontological studies (Pregill and Olson, 1981; Olson and Hilgartner, 1982; Olson and James, 1982a,b; Steadman *et al.*, 1984; Steadman, 1986) adding to the "historic" fauna of an island and questioning the species diversity data used in testing of equilibrium biogeographic theories.

It is clear that the Galápagos land snail fauna is too incompletely known to permit plausible, quantitative evaluation of biogeographic patterns with the techniques available at this time. It is unfortunate that loss and degradation of habitat through the action of humans and their commensals (Dall, 1896; Chambers and Steadman, 1986) may preclude the collection of the quality of species diversity data that might enable biogeographers to apply the methodology now at our disposal, or even to carry out simple empirical analysis of inter-island faunal relationships.

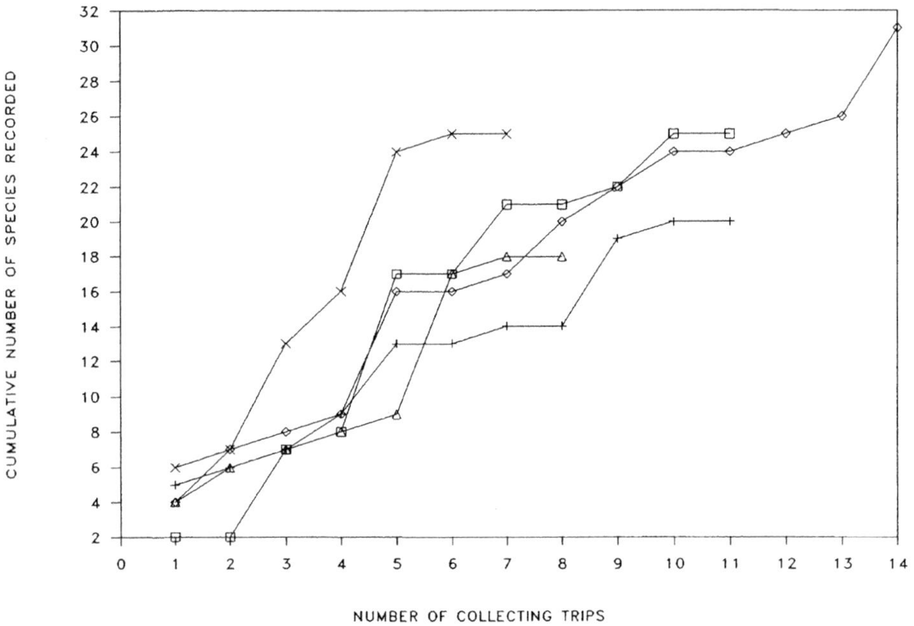

Figure 1. Cumulative numbers of land snail species recorded for five major islands as a function of successive collecting trips. □, San Cristóbal; Δ, Santiago; ◇, Santa Cruz; ×, Isabela; +, Floreana.

Acknowledgments. I thank M. J. James, D. W. Steadman, and an anonymous reviewer for comments on an earlier draft of the manuscript, and K. J. Boss for bringing R. P. Smith's unpublished work on Galápagos land snails to my attention.

References

Agassiz, A., 1892, Reports on the dredging operations off the west coast of Central America to the Galápagos, to the west coast of Mexico, and in the Gulf of California, in charge of Alexander Agassiz, carried on by the U.S. Fish Commission steamer "Albatross," Lieut. Commander Z. L. Tanner, U.S.N., commanding, *Bull. Mus. Comp. Zool. Harv. Univ.* **23**(1):1–90.

Albers, J. C., 1857, Diagnosen neuer Heliceen mit gelegentlichter Berichtigung einiger älteren Arten, *Malak. Bl.* **4**:89–100 [not seen].

Ancey, C. F., 1887, Nouvelles contributions malacologiques, vi: Etudes sur la faune malacologique des îles Galápagos, *Bull. Soc. Malac. de France* **4**:293–299 [not seen].

Baur, G., 1891, On the origin of the Galápagos Islands, *Amer. Natur.* **25**:217–229, 307–326.

Breure, A. S. H., 1979, Systematics, phylogeny and zoogeography of Bulimulinae (Mollusca), *Zool. Verhandel.* (Rijksmuseum van Natuurlikjke Historie, Leiden) **168**:1–215.

Breure, A. S. H., and Coppois, G., 1978, Notes on the genus *Naesiotus* Albers, 1850 (Mollusca, Gastropoda, Bulimulidae), *Neth. J. Zool.* **28**:161–192.

Broderip, W. J., 1832, New species of shells collected by Mr. Cuming on the western coast of South America and in the Islands of the South Pacific Ocean, *Proc. Zool. Soc. London* (1832), pp. 124–126 [not seen].

Carlquist, S., 1974, *Island Biology*, Columbia University Press, New York.

Chambers, S. M., 1986, Two new bulimulid land snails from Isla Santa Cruz, Galápagos, *Veliger* **28**: 287–293.

Chambers, S. M., and Steadman, D. W., 1986, Holocene terrestrial gastropod faunas from Isla Santa Cruz and Isla Floreana, Galápagos: Evidence for late Holocene declines. *Trans. San Diego Soc. Nat. Hist.* **21**:89–110.

Connor, E. F., and Simberloff, D., 1978, Species number and compositional similarity of the Galápagos flora and avifauna, *Ecol. Monogr.* **48**:219–248.

Coppois, G., 1984, Distribution of bulimulid land snails on the northern slope of Santa Cruz Island, Galápagos, *Biol. J. Linn. Soc.* **21**:217–227.

Coppois, G., and De Vos, L., 1984, Hairy periostracal ornamentations on Galápagos land snail shells: S.E.M. comparative study of five species (Pulmonata: Bulimulidae), *Arch. Moll.* **115**:143–149.

Coppois, G., and De Vos, L., 1986, Two different striation patterns of the protoconch in Galápagos Bulimulidae: An S.E.M. study, *J. Moll. Stud.* **52**:106–109.

Coppois, G., and Glowacki, C., 1983, Bulimulid land snails from the Galápagos: 1. Factor analysis of Santa Cruz Island species, *Malacologia* **23**:209–219.

Crampton, H. E., 1916, Studies on the variation, distribution, and evolution of the genus *Partula*; The species inhabiting Tahiti, Carnegie Inst. Washington, Publ. No. 228.

Crampton, H. E., 1932, Studies on the variation, distribution, and evolution of the genus *Partula*; The species inhabiting Moorea, Carnegie Inst. Washington, Publ. No. 410.

Dall, W. H., 1892, On some types new to the Fauna of the Galápagos Islands, *Nautilus* **5**:97–99.

Dall, W. H., 1893, Preliminary notice of new species of land-snails from the Galápagos Islands, collected by Dr. G. Baur. *Nautilus* **7**:52–56.

Dall, W. H., 1895, New species of land-shells from the Galápagos Islands, *Nautilus* **8**:126–127.

Dall, W. H., 1896, Insular landshell faunas, especially as illustrated by the data obtained by Dr. G. Baur in the Galápagos Islands, *Proc. Acad. Natur. Sci. Philad.* (1896), pp. 395–460.

Dall, W. H., 1900, Additions to the insular land-shell faunas of the Pacific coast, especially of the Galápagos and Cocos Islands, *Proc. Acad. Natur. Sci. Philad.* (1900), pp. 88–106.

Dall, W. H., 1917a, New Bulimulus from the Galápagos Islands and Peru, *Proc. Biol. Soc. Wash.* **30**:9–12.

Dall, W. H., 1917b, Preliminary descriptions of new species of Pulmonata of the Galápagos Islands, *Proc. Calif. Acad. Sci., 4th Series* **2**:375–382.

Dall, W. H., 1920, On the relations of the sectional groups of *Bulimulus* of the subgenus *Naesiotus* Albers, *J. Wash. Acad. Sci.* **10**:117–122.

Dall, W. H., and Ochsner, W. H., 1928, Landshells of the Galápagos Islands, *Proc. Calif. Acad. Sci., 4th Series* **17**:141–185.

Darwin, C., 1839, *Journal of Researches into the Geology and Natural History of the Various Countries Visited by H.M.S. Beagle, under the command of Captain Fitzroy, R.N. from 1832 to 1836*, Henry Colburn, London.

Darwin, C., 1860, *Voyage of the Beagle*, 1962 reprint by the Natural History Library, Doubleday, Garden City, New York.

Forbes, E., 1850, On the species of Mollusca collected during the surveying voyages of the Herald and Pandora, by Capt. Kellett, R. N., and Lieut. Wood, R.N. 1. On the land-shells collected during the Expedition, *Proc. Zool. Soc. Lond.* (1850), pp. 53–56.

Gulick, A., 1932, Biological peculiarities of oceanic islands, *Quart. Rev. Biol.* **7**:405–427.

Hamilton, T. H., and Rubinoff, I., 1963, Isolation, endemism, and multiplication of species in the Darwin Finches, *Evolution* **17**:388–403.

Hamilton, T. H., and Rubinoff, I., 1964, On models predicting abundance of species and endemics for the Darwin Finches in the Galápagos Archipelago, *Evolution* **18**:339–342.

Hamilton, T. H., Barth, R.H., and Bush, G. L., 1963, Species abundance: Natural regulation of insular variation, *Science* **142**:1575–1577.

Hanna, G. D., and Hertlein, L. G., 1938, Land and brackish water Mollusca of Cocos Island, *Allan Hancock Pac. Exped.* **2**(8):123–135.

Hertlein, L. G., 1963, Contribution to the biogeography of Cocos Island, including a bibliography, *Proc. Calif. Acad. Sci., 4th Series* **32**(8):219–289.

Hoffman, J. E., 1988, Assignment to genus *Naesiotus* (Albers, 1850) of several species formerly assigned to *Rabdotus* (Albers, 1850) (Gastropoda: Pulmonata: Bulimulidae), *Veliger* **30**:417–420.

Johnson, M. P., Mason, L. G., and Raven, P. H., 1968, Ecological parameters and plant species diversity, *Amer. Natur.* **102**:297–306.

Lack, D., 1947, *Darwin's Finches*, 1983 reissue by Cambridge Univ. Press, Cambridge.
MacArthur, R. H., and Wilson, E. O., 1967, *The Theory of Island Biogeography*, Princeton Univ. Press, Princeton, New Jersey.
Ochsner, W. H., 1905–1906, Land shell faunas, Unpublished transcript of Galápagos Islands field notes deposited in the library of the Division of Mollusks, National Museum of Natural History, Smithsonian Institution, Washington, D.C.
Odhner, N., 1951, Studies on Galápagos bulimulids, *J. Conchyliol.* **90**:253–268.
Olson, S. L., and Hilgartner, W. B., 1982, Fossil and subfossil birds from the Bahamas, in: Fossil vertebrates of the Bahamas (S. L. Olson, ed.), *Smithsonian Contrib. Paleobiol.*, No. 48.
Olson, S. L., and James, H. F., 1982a, Prodromus of the fossil avifauna of the Hawaiian Islands, *Smithson. Contrib. Zool.*, No. 365.
Olson, S. L., and James, H. F., 1982b, Fossil birds from the Hawaiian Islands: Evidence for wholesale extinction by man before Western contact, *Science* **217**:633–635.
Orr, R. T., 1966, Evolutionary aspects of the mammalian fauna of the Galápagos, in: *The Galápagos: Proceedings of the Symposia of the Galápagos International Scientific Project* (R. I. Bowman, ed.), pp. 276–281, Univ. California Press, Berkeley.
Pfeiffer, L., 1846, Descriptions of thirty new species of *Helicea*, belonging to the collection of H. Cuming, Esq., *Proc. Zool. Soc. Lond.*, **(1846)**:28–34.
Pfeiffer, L., 1854, Descriptions of sixty-six new land shells, from the collection of H. Cuming, Esq., *Proc. Zool. Soc. Lond.*, **(1852)**:56–70 [not seen].
Pfeiffer, L., 1855, Descriptions of four new species of Bulimulus from the collection of H. Cuming, Esq., *Proc. Zool. Soc. Lond.*, **(1854)**:57–58.
Pilsbry, H. A., 1897–1898, *Manual of Conchology*, 2nd Series: Pulmonata, Volume 11, Academy of Natural Sciences of Philadelphia.
Pilsbry, H. A., 1902, *Manual of Conchology*, 2nd Series: Pulmonata, Supplement to Volume 14, Academy of Natural Sciences of Philadelphia.
Pilsbry, H. A., 1927–1935, *Manual of Conchology*, 2nd Series: Pulmonata, Volume 28, Academy of Natural Sciences of Philadelphia.
Pregill, G. K., and Olson, S. L., 1981, Zoogeography of West Indian vertebrates in relation to Pleistocene climatic cycles, *Ann. Rev. Ecol. Syst.* (1981), pp. 75–98.
Rehder, H. A., 1940, New mollusks of the genus *Naesiotus* from Ecuador, *Nautilus* **53**:111–118.
Reibisch, P., 1892, Die conchyliogische Fauna der Galápagos-Inseln, *Sitzungsberichte und Abhandl. Naturwissenschaftlichen Gesellschaft Isis in Dresden* (1892) pp. 1–20.
Simkin, T., 1984, Geology of the Galápagos, in: *Galápagos* (R. Perry, ed.) pp. 15–41, Pergamon Press, Oxford.
Slevin, J. R., 1959, The Galápagos Islands: A history of their exploration, *Occ. Pap. Calif. Acad. Sci.*, No. XXV.
Smith, A. G., 1966, Land snails of the Galápagos, in: *The Galápagos; Proceedings of the Symposia of the Galápagos International Scientific Project* (R. I. Bowman, ed.) pp. 240–251, Univ. of California Press, Berkeley.
Smith, A. G., 1971, New record for a rare Galápagos land snail, *Nautilus* **85**:5–8.
Smith, A. G., 1972, Three new land snails from Isla Santa Cruz Indefatigable Island), Galápagos, *Proc. Calif. Acad. Sci.*, 4th Series **39**:7–24.
Smith, A. G., 1974, Galápagos bulimulids; A taxonomic correction, *Nautilus* **88**:67.
Smith, E. A., 1877, IV. Mollusca, in: Account of the zoological collection made during the visit of the H.M.S. 'Petrel' to the Galápagos Islands (communicated by A. Gunther), *Proc. Zool. Soc. Lond.*, **(1877)**:69–73.
Smith, R. P., Jr., 1971, Factors governing the dispersal, distribution, and microevolution of land snails of *Naesiotus* in the Galápagos archipelago, Unpublished B.A. Honors thesis, Harvard University, Cambridge, Mass.
Solem, A., 1976, *Endondontid Land Snails from Pacific Islands; Part I; Family Endodontidae*, Field Museum of Natural History, Chicago.
Sowerby, G. B., 1833a–1839, *Conchological Illustrations, Bulinus*, Sowerby, London [not seen].
Sowerby, G. B., 1833b, New species of shells collected by Mr. Cuming on the western coast of South America and among islands of the South Pacific Ocean, *Proc. Zool. Soc. Lond.* **(1833)**:34–38, 70–74.
Steadman, D. W., 1982, The origin of Darwin's finches (Fringillidae, Passeriformes), *Trans. San Diego Soc. Nat. Hist.* **19**:279–296.

Steadman, D. W., 1986, Holocene vertebrate fossils from Isla Floreana, Galápagos, *Smithson. Contrib. Zool.* **413**.

Steadman, D. W., and Ray, C. E., 1982, The relationships of *Megaoryzomys curioi*, an extinct cricetine rodent (Muroidea, Muridae) from the Galápagos Islands, Ecuador, *Smithson. Contrib. Paleobiol.* **51**.

Steadman, D. W., Pregill, G. K., and Olson, S. L., 1984, Fossil vertebrates from Antigua, Lesser Antilles: Evidence for late Holocene human-caused extinctions in the West Indies, *Proc. Nat. Acad. Sci. U.S.A.* **81**:4448–4451.

Stearns, R. E. C., 1893, Scientific results of explorations by the U.S. Fish Commission Steamer Albatross; XXV—Report on the Mollusk-Fauna of the Galápagos Islands, with descriptions of new species, *Proc. U.S. Nat. Mus.* **16**:353–450.

Vagvolgyi, J., 1974, *Nesopupa galapaganus*, a new Indo-Pacific element in the land snail fauna of the Galápagos Islands (Pulmonata: Vertiginidae), *Nautilus* **88**:86–89.

Vagvolgyi, J., 1975, body size, aerial dispersal, and origin of the Pacific land snail fauna, *Systematic Zool.* **24**:465–488.

Vagvolgyi, J., 1977, Six new species and subspecies of *Naesiotus* from the Galápagos Islands (Pulmonata: Bulimulidae), *Proc. Biol. Soc. Wash.* **90**:764–777.

Van Mol, J-J., 1972, Au sujet d'une nouvelle et remarquable espece de Bulimulidae des Iles Galápagos (Mollusca, Gastropoda, Pulmonata), *Bull. Inst. Roy. Sci. Nat. Belg.* **48**:1–7.

Wallace, A. R., 1892, *Island Life*, 2nd edition, Macmillan, London.

Weyrauch, W., 1956, The genus *Naesiotus*, with descriptions of new species and notes on other Peruvian Bulimulidae, *Proc. Acad. Natur. Sci. Philad.* **108**:1–17.

Welch, d'A., 1942, Distribution and variation of the Hawaiian tree snail *Achatinella apexfulva* Dixon in the Koolau Range, Oahu, *Smithson. Misc. Coll.* **103**:1–236.

Chapter 15

Charles Darwin's Contribution to the Molluscan Fauna of the Galápagos Islands

Historical Perspective on Endemicity and Biogeography

MATTHEW J. JAMES

1. Introduction

Since its appearance in 1859, many of the ideas and observations published by Darwin in the *Origin of Species* have served as the unifying program for biologists. I have endeavored to understand one small subset of the material presented by Darwin in three of his published works of 1839, 1844, and 1859. This material consists of observations about the taxonomic identity and biogeographic affinities of the terrestrial and marine molluscan fauna of the Galápagos Islands.

Charles Darwin's correspondence with the British conchologist Hugh Cuming resulted in an exaggerated estimate of endemicity for Galápagos marine gastropods based on a small collection of shells. References to high molluscan endemicity can be traced through Darwin's writings from 1839 to 1859. This small sample of mollusks made by Cuming in 1829, and the taxonomic information supplied by Cuming, eventually contributed to Darwin's evolutionary interpretation of the

MATTHEW J. JAMES • Department of Geology, Sonoma State University, Rohnert Park, California 94928.

Galápagos Islands. Although endemicity was overestimated, the biogeographic affinities of the non-endemic component of the Galápagos molluscan fauna as deduced by Darwin are remarkably similar to the affinities presently known for the fauna.

Oceanic islands played an important role in the development of Charles Darwin's ideas about evolution. It was on islands that Darwin observed the products of evolution, namely, unique species that had likely evolved in the islands themselves. The conceptual link between observed morphological difference and evolutionary divergence lies at the heart of Darwin's formulation of the mechanism of natural selection and of our perception of evolution today. I have examined the published observations made by Darwin in 1839 in the *Journal of Researches* concerning the biogeographic affinities, overall uniqueness, and degree of endemicity of the molluscan fauna of the Galápagos Islands, and have arrived at a better understanding, in part, of his predilection to attribute unique evolutionary status to the Galápagos as revealed 20 years later in the *Origin of Species* (1859). In the present chapter, I will concentrate on Darwin's observations about the terrestrial and marine mollusks of the Galápagos. In particular, where did Darwin receive specialized taxonomic information? How did he analyze and interpret that information? Subsequently, how did he incorporate that information into the *Journal of Researches* (Darwin, 1839), the *Essay of 1844* (see Darwin, 1909), and the *Origin of Species* (Darwin, 1859)?

2. Tracing a Thought from *Journal* to the *Origin*

In the *Journal of Researches* (popularly known as "The Voyage of the Beagle") first published in 1839, with a second, revised edition published in 1845 (see Freeman, 1977, for a chronology of the works of Darwin) Darwin described the geology and natural history of the Galápagos Islands. In the *Journal*, Darwin stated: "The natural history of these islands is eminently curious, and well deserves attention. Most of the organic productions are aboriginal creations, found nowhere else; there is even a difference between the inhabitants of the different islands; yet all show a marked relationship with those of America, though separated from that continent by an open space of ocean, between 500 and 600 miles in width. The archipelago is a little world within itself, or rather a satellite attached to America . . . . Hence, both in space and time, we seem to be brought somewhat near to that great fact—that mystery of mysteries—the first appearance of new beings on this earth." (Darwin, 1906, p. 363).

In the *Essay of 1844*, Darwin stated: "The Galápagos Archipelago is a remarkable instance of this latter fact [highly endemic insular faunas]; here almost every bird, its one mammifer, its reptiles, land and sea shells, and even fish, are almost all peculiar and distinct species, not found in any other quarter of the world: so are the majority of its plants." Comments on this passage were made previously by Williamson (1983) in reference to the avifauna of the islands.

In the 6th edition of the *Origin of Species*, Darwin stated: "[In the Galápagos Islands] almost every product of the land and of the water bears the unmistakable stamp of the American continent."

Thus, in Darwin's early publications about natural history and evolution, the Galápagos Islands are mentioned several times and are described in terms of the remarkable end products of natural selection. These islands, as well as others throughout the world, served as a paradigm for Darwin's evolutionary mechanism, although the role of isolation in Darwin's thinking was not as important then as we sometimes consider it today (Lesch, 1975).

3. Historical Background

Charles Darwin visited the Galápagos Islands from September 15 to October 20, 1835, during the *Beagle* voyage (Freeman, 1978; Keynes, 1979b; Burkhardt and Smith, 1985; Tasch, 1985). Particularly noteworthy of his natural history observations during these five weeks were those concerning the terrestrial biota of the archipelago, which brought the remarkable fauna of the islands to the attention of the scientific community and contributed to our present recognition of how the Galápagos exemplify the important evolutionary principles of dispersal and reproductive isolation leading to adaptive radiation and speciation. However, his comments about the marine invertebrate fauna of the islands have received much less attention in the biological literature. Part of the marine invertebrate fauna of the islands illustrates a separate evolutionary response to insular isolation from that displayed by the well-known terrestrial fauna (James, 1984), which usually is restricted to vertebrates (for example, finches, tortoises, and iguanas) and plants (in numerous genera, see Wiggins and Porter, 1971) when authors refer to the Galápagos biota. The evolutionary scenario so well exemplified by the Galápagos and other oceanic islands, such as Hawaii (Carlquist, 1970), consisting of dispersal and isolation (often involving a genetic bottleneck) followed by adaptive radiation and speciation, has been the dominant mechanism responsible for the origination of most of the terrestrial biota and some of the marine biota of the islands. However, as demonstrated by Ax (1977), not all Galápagos marine invertebrates exhibit the evolutionary pattern invoked by James (1984) to account for the pattern of molluscan evolution in the islands. Meiofaunal organisms (living interstitially in the littoral sand bottom) in the Galápagos exhibit a pattern of adaptive radiation and speciation similar to that of terrestrial organisms (Ax, 1977).

The result of Darwin's interaction with the conchologist Hugh Cuming left Darwin with the impression that Galápagos marine mollusks (particularly prosobranch gastropods) exhibited a pattern of evolutionary divergence comparable to the terrestrial biota, a pattern which is not consistent with present knowledge.

3.1. The Darwin–Cuming Correspondence

Following his return to England on October 2, 1836, Darwin received taxonomic information about various components of the marine fauna of the islands, including the molluscan fauna (Dance, 1986), from a variety of contemporary taxonomic specialists. Among those supplying information to Darwin about the Galápagos biota was the famous collector Hugh Cuming who visited the islands in 1829 (Howell, 1941; Dance, 1980; Burkhardt and Smith, 1985, p. 515), about five years prior to the *Beagle* voyage. Darwin wrote to Cuming in July 1845 concerning a collection of Recent mollusks made by Cuming in the Galápagos prior to the voyage of the *Beagle*. Although Darwin's letter to Cuming is apparently not preserved, Cuming's response is reproduced in *The Correspondence of Charles Darwin, Volume 3* (see Burkhardt and Smith, 1987, pp. 230–231). In the text of the letter reproduced below from Cuming to Darwin, I have italicized the names of genera for nomenclatural clarity, except for the first two instances where Oniscia and Stylifer are mentioned in the letter; apparently Cuming underlined these in the original for emphasis when responding to Darwin's inquiry.

80 Gower Street
Bedford Square, London
28th July 1845

My dear Sir

I was out on Saturday when your letter arrived, otherwise I would have answered it by that day's post—

The word *Oniscia* is correct—and *Stylifer*, is also correct—I did not find either of these Genera in the Pacific Ocean. They are both to be found in the West Indies—*Oniscia* is found also in China (two or three Species)—*Stylifer* is found in the Philippines and the Mauritius and according to Turton it is also British.

I have never seen a *Monoceras* any where but on the West coast of America and the Galapagus [sic] Islands. I have seen upwards of 50 parcels of Shells from the Sandwich Islands, but never one of the family amongst them. There is a constant Trade between those Islands and California, and no doubt they might have been carried there by Sailors. Pay no attention to the information you have gained from Books—

I have never found any of the following Genera, which I found both at the Galapagus and the West Coast of America in the Pacific Ocean, nor have I ever received them, from any Collector, viz: *Fissurella*, *Monoceras*, or *Cancellaria*.

Should you require any further information, I shall be most happy to give it, if in my power.

I remain my Dear Sir
Very truly yours
H Cuming

3.2. Darwin on Galápagos Mollusks

Darwin included only ten sentences in his *Journal of Researches* (1839) concerning the Galápagos molluscan fauna. His comments on land snails appear to be based on his own collections made in the islands and those on marine snails are clearly based on information supplied to him by the collector Cuming. However, this short passage in one of Darwin's earliest publications following the return of the *Beagle* contains considerable information about Darwin's perception of the biogeographic affinities of the fauna.

3.3. Sources of Information

While conducting this study of Darwin's contribution to the molluscan fauna of the Galápagos, I relied both on Darwin's published works and on the extensive correspondence published by Cambridge University Press (Burkhardt and Smith, 1985, 1986, 1987). My initial source of information was the section of ten sentences in Darwin (1839) and this was expanded using Darwin (1909) and Darwin (1859).

4. Analysis and Interpretation of Darwin's Comments

For each of the ten sentences reprinted here from Darwin (1839), I analyzed Darwin's remarks and interpreted them in the context of Darwin's contributions to knowledge of the Galápagos Islands. I also made additional comments about the present status of our knowledge of Galápagos biogeography and the molluscan fauna of the islands.

Darwin included only one sentence about the pulmonate fauna of the islands (see also Smith, 1966):

> Of land-shells I collected sixteen kinds (and two marked varieties), of which, with the exception of one *Helix* found at Tahiti, all are peculiar to this archipelago: a single fresh-water shell (*Paludina*) is common to Tahiti and Van Diemen's Land [= Tasmania].

Darwin considered fifteen of the sixteen species of land snails he collected to be endemic to the archipelago. It appears that Darwin based this observation on specimens he personally collected while in the islands in 1835. He comments on the wide geographic distribution of one species of the genus *Helix*, which was then also known from the island of Tahiti. Smith (1966) regarded the record of *Helix* in the Galápagos to be erroneous because no species of *Helix* has ever been reported from the islands since the report by Darwin and no species in that genus has ever been described from the Galápagos. Either a specimen of another genus was misidentified (by either Darwin or a taxonomist supplying information to Darwin) or the actual specimen, other than those specimens collected by Darwin and now in the Cuming Collection in the British Museum (Natural History), was subsequently lost (Smith, 1966). Historical changes in the Galápagos land snail fauna have been documented by Chambers and Steadman (1986), particularly in reference to habitat alteration resulting from the onset of human settlement in 1832 (Steadman, 1986), shortly after Cuming's Galápagos collecting trip in 1829 (Howell, 1941), but just before Darwin's visit in 1835. It is unclear from Darwin's comments what he considered the proper generic classification of the remaining fifteen land shells, although when he writes, "all are peculiar to this archipelago," this suggests that he considered them to be endemic because no described genus could be found in which they could be placed. The only aquatic pulmonate he collected was identified as belonging to the genus *Paludina*, which was known to be a widely distributed genus with populations also occurring in Tahiti and Van Diemen's Land (= Tasmania). As with *Helix*, specimens of *Paludina* have never been described or collected since Darwin's brief mention (Smith, 1966). Darwin's collection, identification, and interpretation of the terrestrial pulmonate fauna resulted in an endemicity value of 94% (15 of 16 "kinds"). This pattern is comparable to that found on other oceanic islands, such as the Hawaiian Islands and the Azores (Williamson, 1981, p. 52). Darwin recognized that the majority of Galápagos land snails were endemic to the archipelago, but he also realized the possibility of long distance dispersal bringing in taxa that would show affinities to distant faunas, such as the *Helix* and *Paludina*, he ascribed to Indo-Pacific insular faunas. A similar distributional pattern is known today for Galápagos land snails. All land snail genera occurring in the islands are Neotropical in origin, with the exception of *Tornatellides* and *Nesopupa* which are Indo-Pacific elements (Vagvolgyi, 1975). Fossil land snails of the genus *Gastrocopta* were reported by Hertlein (1972) and of the genus *Bulimulus* were reported from Isla Santa Fe by Pitt and Pitt (1989), although 'subfossil' land snail specimens were recovered from lava tubes in the islands (Steadman, 1986; Chambers and Steadman, 1986).

The remaining nine sentences in Darwin's treatment of the Galápagos molluscan fauna concern marine prosobranch gastropods:

> Mr Cuming, before our voyage, procured here ninety species of sea-shells, and this does not include several species not yet specifically examined, of *Trochus, Turbo, Monodonta,* and *Nassa*.

Darwin was aware of Cuming's collecting trip to the Galápagos and commented upon it in a letter (Burkhardt and Smith, 1985; p. 515) written to John Stephens Henslow in October 1836. Darwin wrote, "I have forgotten to mention one bad bit of news, namely that Cuming was at the Galápagos.- Did he collect plants, I doubt it, because the far greater part of the plants only live near the summit of mountains, some miles from the coast?" This "one bad bit of news" appears to refer to the anticipated meager size of Cuming's botanical collections which Darwin hoped would supplement the plants he himself had collected in the Galápagos and which would later be described by J. D. Hooker. The collecting activities of Hugh Cuming have been thoroughly analyzed by Dance (1980, 1986) who stated that "Among nineteenth-century men of action few contributed as much to the material advance of natural history as Hugh Cuming. . . " (Dance, 1980, p. 477). Regarding Cuming's Galápagos connection, Dance (1986, p. 123) wrote: "As a conchologist his name stood very high in the scientific world and his identifications were rarely questioned. Charles Darwin obtained from Cuming nearly all his published information on the distribution and frequency of shells and it was Cuming who named nearly all of the shells collected by him during the historic voyage of the *Beagle*." Cuming did not name the shells in the formal taxonomic sense of proposing binomial scientific names using the Linnean system for the Galápagos specimens, but rather he provided names or merely identified the shells for Darwin. Although as indicated by Dance (1986) and mentioned below, Cuming's list of names of the identified specimens was given to Darwin and exists today in the Darwin archives at Cambridge University.

The second sentence in Darwin (1839) states:

> He [Mr. Cuming] has been kind enough to give me the following interesting results: of ninety shells, no less than forty-seven are unknown elsewhere—a wonderful fact, considering how widely distributed sea-shells generally are.

This sentence, based on information provided to Darwin by Cuming, summarizes the view that Darwin came to hold concerning the degree of endemicity of the Galápagos malacofauna. Dance (1986, p. 135) wrote: "[Darwin] collected a number of shells, especially when he visited the Galápagos Islands, but shells were not of great interest to him; and even though he induced Cuming to name them for him they were dispersed without any attempt having been made to report them." Dance (1986, p. 135) further stated in a footnote that the identifications exist in manuscript form in the Darwin archive at Cambridge (see Burkhardt and Smith, 1987, pp. 230–231). Darwin's comments that "Mr. Cuming, before our voyage" and "He has been kind enough to give me the following interesting results," leaves unresolved the question of whether Darwin (1839) was reporting on shells collected during the

Beagle voyage as stated by Dance (1986, p. 135) or on material collected prior to the *Beagle* voyage and identified by Cuming.

According to the information supplied to Darwin by Cuming, over 52% (47 of 90 species) of the newly-collected marine mollusks were undescribed, endemic species. Darwin made a very important biogeographic and evolutionary observation by stating this to be a "wonderful fact." Darwin knew how widely distributed marine mollusks tend to be, although he was probably unaware of the details of the mechanism of this continuity, namely, planktonic veliger larvae that can be dispersed on ocean currents (Scheltema 1986a, 1986b, 1988; Zinsmeister and Emerson, 1979). However, Darwin (1909, p. 218) stated: "In all shell-fish (Gasteropods) [sic] the embryo passes through a state analogous to that of the Pteropodous Mollusca." Thus, prior to publication of the *Origin*, Darwin was clearly aware that some mollusks pass through a planktonic developmental stage similar to the mode of life of planktonic pteropods. Recent estimates of Galápagos marine molluscan endemicity range from a value close to the 52% cited Darwin [42% cited by James (1984) for the entire molluscan fauna or 47% for prosobranches] down closer to the more biologically reasonable value of 26% presented by Finet (1985) and modified slightly by Finet (1989) to 25.3%, by Shasky (1989) to 23.2%, and by Finet (Chapter 12, this volume) to 22.7% overall or 18.7% for shallow-water species. These latter values of endemicity are consistent with those known for other oceanic islands such as Hawaii (Kay, 1979, p. 15). It appears that Darwin did not carefully distinguish between the high rate of adaptive radiation and speciation exhibited by the terrestrial biota of the islands and the slower rates exhibited by some members of the marine invertebrate fauna. Williamson (1983) pointed out how Darwin used information about the terrestrial and marine biota to support his evolutionary mechanism in the essay of 1844: "The Galápagos Archipelago is a remarkable instance of this latter fact: here almost every bird, its one mammifer, its reptiles, land and sea shell, and even fish, are almost all peculiar and distinct species, not found in any other quarter of the world: so are the majority of its plants" (Darwin, 1909, p. 159). The importance of geographic isolation in the conceptual development of Darwin's mechanism of natural selection was analyzed by Lesch (1975) who concluded that isolation was *not* initially important to Darwin's conception of evolutionary divergence. Lesch (1975) traced the importance of isolation (such as in the Galápagos Islands) placed by Darwin in the 1st through 6th editions of the *Origin of Species* with reference to the ability of natural selection to produce new species. Only in the controversial years following publication of the *Origin* did Darwin begin to place greater emphasis on geographic isolation as an important component of the *mechanism* of natural selection. Although Darwin certainly incorporated observations on the *results* of natural selection seen on oceanic islands in some of his earliest notebooks on transmutation of species (Herbert, 1980 *passim*).

Darwin's fourth sentence reads:

> Of the forty-three shells found in other parts of the world, twenty-five inhabit the western coast of America, and of these eight are distinguishable as varieties; the remaining eighteen (including one new variety) were found by Mr. Cuming in the Low Archipelago, and some of them also at the Philippines.

Here Darwin addresses the non-endemic component of the Cuming collection. Comments made here by Darwin on biogeographic affinities are consistent with present knowledge of the Galápagos molluscan fauna. Twenty-five species are attributed to "the western coast of America", by which Darwin probably meant South America. Malacologists now consider the marine mollusks living in this geographic area to be part of the Panamic Molluscan Faunal Province which extends from Magdalena Bay in Baja California, Mexico south to Punta Aguja, Peru with a seaward extension including the Galápagos archipelago and other offshore islands (Keen, 1971, p. 4). Some of the non-endemic species mentioned by Darwin might have belonged to the Peruvian Molluscan Faunal Province which extends southward of the Panamic province. Darwin considered eight of the 25 Panamic species to be "varieties." The implication is that these varieties are morphologically distinct from Panamic populations, a notion consistent with Darwin's view of the uniqueness of the Galápagos fauna. These varieties could be recognized by modern taxonomists as valid species, as subspecies, or simply as geographic variants of mainland Panamic populations. Darwin listed 18 species collected in the Galápagos as also occurring in the Low Archipelago (= Tuamotu Archipelago) and in the Philippines. Both of these areas are located in the Indo-Pacific Molluscan Faunal Province. Emerson (1967, 1978) examined the factors that control the dispersal of Indo-Pacific species into the Panamic province.

Darwin's fifth sentence reads:

> This fact of shells from islands in the central part of the Pacific occurring here, deserves notice, for not one single sea-shell is known to be common to the islands of that ocean and to the west coast of America.

The phenomenon discussed here, namely, the disjunct distribution of Indo-Pacific mollusks occurring on islands in the eastern Pacific and not on the mainland of Central and/or South America puzzled biogeographers for many years. Reports of mollusks ranging from portions of the western Pacific and extending across vast distances to establish populations on islands in the eastern Pacific were difficult to explain in terms of long distance dispersal by planktonic larvae. The usual explanation was that although these taxa were able to disperse over long distances by way of teleplanic larvae, they were somehow unable to establish populations in the shallow waters of the mainland. The reason invoked, although without direct evidence, was that either the proper substratum, food resource, temperature regime or other vital resource was not available or that an unspecified predator or competitor (invoking the "ghost of competition-past") made establishment of a permanent population on the mainland impossible. Emerson (1967, 1978) and Emerson and Old (1965, 1968) discussed examples of Indo-Pacific taxa with disjunct distributions in the eastern Pacific. In recent years, however, extensive collecting along the western coasts of Central and South America has revealed a number of Indo-Pacific species that have successfully made the jump and established populations, although usually in very low numbers of individuals per species. Although for some taxa this distributional phenomenon appears to be an artifact of collecting, there are other examples of disjunct molluscan distributions between the Indo-Pacific and eastern Pacific that are not easily explained. There in fact does seem to be a distributional and biogeographic paradox consisting of taxa with teleplanic larvae that have been able to disperse across the East Pacific Barrier (Ekman, 1953; Abbott, 1966; Scheltema, 1988) and yet not establish mainland populations. Darwin's statement that "not one single sea-shell" occurring in the

Indo-Pacific was known to contradict this distributional paradox is founded in the collection bias discussed above.

Darwin's sixth sentence reads:

> The space of open sea running north and south off the west coast, separates two quite distinct conchological provinces; but at the Galápagos we have a halting place, where many new forms have been created, and whither these two great conchological provinces have each sent several colonists.

The conchological division here, once again, is between the Indo-Pacific molluscan faunal province and the Panamic molluscan faunal province. The Galápagos are considered by Darwin in this sentence to be a transition zone containing molluscan taxa from each of the two provinces. However, this view is somewhat influenced by the disproportionately large number of Indo-Pacific taxa attributed to the Galápagos by Cuming. Had a larger sample of mollusks been collected and a more accurate picture of their biogeographic affinities been known, then the Galápagos would have been viewed as principally part of the Panamic province with very few vagrant species from the Indo-Pacific, Californian, and Peruvian molluscan faunal provinces (Emerson, 1967, 1978; James, 1984). From an evolutionary perspective, Darwin viewed the Galápagos as the site of evolutionary divergence from source populations, as evidenced by his statement "where many new forms have been created." This picture of the Galápagos as the 'cradle of new species' is influenced by the large number of endemic species attributed to the fauna by Cuming in his written report to Darwin. This notion might have fit into Darwin's concept of the Galápagos of exemplifying his mechanism of natural selection leading to adaptive radiation and speciation. Blackwelder (1967, pp. 9–10) stated that many of the practical aspects of taxonomy, which was the basic activity indulged in by Cuming, were little changed by either Darwinism or Neo-Darwinism. So even if Cuming's identification of Galápagos mollusks had occurred after the publication of the *Origin*, it is possible that an estimate of high endemicity might still have been forthcoming.

Darwin's seventh sentence reads:

> The American province has also sent here representative species; for there is a Galapageian species of *Monoceras*, a genus only found on the west coast of America; and there are Galapageian species of *Fissurella* and *Cancellaria*, genera common on the west coast, but not found (as I am informed by Mr. Cuming) in the central islands of the Pacific.

Darwin discussed the "American province" in a manner that might lead the reader to conclude that it is different from the provinces discussed previously. Clearly, what modern malacologists call the Panamic province was referred to by Darwin in the quotes above. It is possible that Darwin was not discriminating between the Californian province and the Panamic province when he used the term "American province." According to Keen (1971, p. 4), the Panamic Province extends from "Magdalena Bay (Bahía Magdalena), Baja California, and Punta Aguja, Peru, with a seaward extension including the Galápagos archipelago and the other, nearer, offshore islands." Darwin's use of the phrase "sent here" in reference to the "American province" strongly suggests an element of long distance dispersal, which is consistent with the other observations made by Darwin in the seventh sentence of the passage above.

Darwin's eighth sentence reads:

> On the other hand, there are Galapageian species of *Oniscia* and *Stylifer*, genera common to the West Indies and to the Chinese and Indian seas, but not found either on the west coast of America or in the central Pacific.

Darwin noted here the apparent disjunct distribution (as discussed above) of species belonging to the genera *Oniscia* and *Stylifer* normally found in the Caribbean and more remote regions of the Indo-Pacific Province. The paradox Darwin drew attention to is the great distances between known populations of these genera and the difficulty of dispersal between distant geographic areas. Also in this sentence, Darwin extended the affinities of the Galápagos molluscan fauna to include the Caribbean fauna. Numerous cognate or sister species are known from either side of the Panamá land bridge (Vermeij, 1978, pp. 218–241 *et passim*; Smith, 1981). Closure of the Panama land bridge is considered to have taken place about 3.1 million years ago (Keigwin, 1978; Jones and Hasson, 1985), but several Pleistocene connections between the Pacific and Caribbean also occurred. These connections could have facilitated gene flow between the two areas and reduced morphological distinctions produced in isolation (see also Hamilton *et al.*, 1963).

Darwin's ninth sentence reads:

> I may here add, that after the comparison of Messrs. Cuming and Hinds of about 2000 shells from the eastern and western coasts of America, only one shell was found in common, namely, the *Purpura patula*, which inhabits the West Indies, the coast of Panama, and the Galápagos.

Darwin documents here both the apparent widespread distribution of a single species ["the *Purpura patula*" = *Purpura patula* (Linnaeus, 1758)] and a further similarity between the Caribbean molluscan fauna and the Panamic fauna in general and the Galápagos in particular. Keen (1971, p. 553) stated that "the West American form (*Purpura pansa* Gould, 1853) is easily separable from true *P. patula*, which is the Caribbean twin species, by the white in the aperture as well as by size." Jones and Hasson (1985) discussed the biological consequences of the closure of the Central American land bridge. Isthmian closure in late Pliocene time resulted in the evolutionary divergence of numerous sister or twin species as discussed by Keen (1971), Smith (1981) and Vermeij (1978).

The tenth and final sentence by Darwin on Galápagos mollusks reads:

> We have, therefore, in this quarter of the world, three great conchological sea-provinces, quite distinct, though surprisingly near each other, being separated by long north and south spaces either of land or of open sea.

In this final sentence, Darwin stressed the biogeographic superposition of species from three separate faunal provinces in the waters of the Galápagos Archipelago. Darwin also noted that although the three provinces are relatively close to each other, they are also separated by substantial physical barriers that necessitate a mechanism of long distance dispersal (such as planktonic or planktotrophic larvae) (Jablonski and Lutz, 1983) or a previous connection in past geologic time, as between the Caribbean Sea and the Pacific Ocean prior to about 3.1 million years ago (Jones and Hasson, 1985). The distinctiveness of the three conchological provinces is a reflection of the amount of evolutionary divergence that has occurred through time and of the absence of continuous direct genetic contact between provinces.

In this last sentence of the quotation from 1839, we see Darwin's insightful biogeographic conclusion that the molluscan fauna of the Galápagos is composed of faunal elements from several widespread faunal provinces. Although Darwin stated the Galápagos molluscan fauna is derived from "three great conchological sea-provinces", he actually mentions in sentences 6, 7, and 8 what modern malacologists would consider four faunal provinces. (1) The Indo West Pacific province ("Chinese and Indian seas" in sentence eight), (2) the Panamic Province ("west Coast of America" in sentences seven and eight; and "Panama" in sentence nine), (3) the Caribbean province ("West Indies" in sentence nine), and (4) perhaps the Californian province ("west coast of America" in sentences seven and eight).

5. Insular Evolution and Calculating Endemicity

Calculating the degree of endemicity for marine invertebrates in general and mollusks in particular on oceanic islands is fraught with accuracy problems, as discussed below. However, making such a calculation for island groups such as the Galápagos Islands or the Hawaiian Islands continues to be an important tool in evolutionary studies (Kay, 1980, 1984). Species-level endemicity on oceanic islands is often used as a direct measure of evolutionary divergence. Accurate estimates of endemicity are crucial to studies comparing the extent of evolution on different groups of oceanic islands that have different geological histories and which have varying degrees of geographic isolation. Simply summarizing species occurrences from published distributional lists and calculating the percentage of endemic species for an island group can result in either overestimation or underestimation for the following reasons:

1. Alpha taxonomic decisions will directly influence the number of endemic species described from a particular geographic area. Taxonomists who are "lumpers" or "splitters" will tend to describe either too few or too many new and endemic taxa, respectively. Crucial decisions are made by taxonomists to place new species in cosmopolitan genera or to describe a new genus. Likewise, decisions to accommodate populations from oceanic islands, such as the Galápagos, in widespread species will reduce apparent endemicity. Johnson and Raven (1973) arrived at a very different estimate of endemicity for Galápagos vascular plants than a previous study by Hamilton *et al.* (1963) by incorporating data from the important taxonomic monography by Wiggins and Porter (1971).
2. Geographically biased sampling will often overestimate endemicity. In the Galápagos, species have been described as new (and therefore endemic), only to be later found to occur elsewhere in the Panamic province and determined to be non-endemic. As discussed by James (1984), both field collection and new species descriptions have been disproportionately biased towards some Galápagos organisms.
3. Failure to find rare species or species of extremely small body size that are not readily collected using standard techniques will overestimate the degree of endemicity. Shasky (1989) modified the estimate of endemicity for Galápagos mollusks published by Finet (1985) by including species records for minute mollusks collected using a specialized sampling regime.

The pitfalls of endemicity estimation can be overcome by carefully analyzing and interpreting both the taxonomic descriptions of endemic taxa and by considering information from a wide geographic area when calculating the percentage of endemic species. This involves not only summarizing distributional information from the published literature but also by examining material in museum repositories and actual field collection of new material. The difference in endemicity values (42% vs. 26%) published by James (1984) and Finet (1985) for Galápagos marine mollusks is the result of these different approaches. In contrast, Gibson (1987) found no endemicity among the foraminifera of Enewetak.

6. Conclusions

Problems with accurately calculating the degree of endemicity notwithstanding, Darwin was left with the impression (provided by Cuming in the form of species identifications) that 52% (47 of 90 species) of the prosobranch fauna of the islands was endemic. This inflated estimate of endemicity is clearly the result of a small sample size, limited collecting area, and perhaps biased taxonomic treatment by Cuming. Darwin extended his observations from the mere taxonomic list provided by Cuming to broader concepts about biogeographic affinities (see Ghiselin, 1984, pp. 32–45) and mechanisms of long distance dispersal. Darwin commented on the biogeographic affinities of the Galápagos malacofauna by noting the known distribution of the non-endemic components of the fauna. He divided the remaining fauna into four of the principal biogeographic provinces recognized for Galápagos mollusks by Emerson (1967, 1978), James (1984), Finet (1985) and other authors: 1) Panamic Province (= American in Darwin's terminology), 2) Indo-Pacific (based on similarities with the Low or Tuamotu Archipelago and the Philippine Islands), 3) Caribbean Province (= West Indies in Darwin's terminology), and 4) Endemic component (based on the species identifications provided by Cuming.)

The present study was an exercise in historical analysis of a portion of Darwin's writing, and a means of further understanding the molluscan fauna of the Galápagos Islands. What is important here is that Darwin, at an early stage of his scientific career, accepted an estimate of high species-level endemism in one group of Galápagos marine invertebrates. Darwin's subsequent use of this estimate in pointing out the uniqueness of the Galápagos demonstrates the evolution of an idea that has been central to biologists for 130 years.

ACKNOWLEDGMENTS. Some of the ideas for this paper were developed while I was in residence at Worcester College, University of Oxford during the summer of 1982 while enrolled in a course on Charles Darwin and evolution. Grateful acknowledgment is extended to the Oxford/Berkeley Program of the University of California Extension for providing me with a scholarship to attend the course. Geoffrey Lewis, in his capacity as tutor for the Darwin course, contributed significantly to my knowledge, understanding, and appreciation of Darwin's life and evolutionary endeavors. Helpful discussions during the genesis of this paper were provided by D. R. Lindberg, W. Rainey, and J. W. Durham. M. A. Wilson and E. A. Kay commented on an early draft of the manuscript.

9. Appendix: Complete Passage from the *Journal of Researches*

The complete passage from the *Journal of Researches* is reprinted here in order to examine Darwin's statements as they originally appeared (generic names italicized here for clarity):

"Of land-shells I collected sixteen kinds (and two marked varieties), of which, with the exception of one *Helix* found at Tahiti, all are peculiar to this archipelago: a single fresh-water shell (*Paludina*) is common to Tahiti and Van Diemen's Land, Mr. Cuming, before our voyage, procured here ninety species of sea-shells, and this does not include several species not yet specifically examined, of *Trochus*, *Turbo*, *Monodonta*, and *Nassa*. He has been kind enough to give me the following interesting results: of ninety shells, no less than forty-seven are unknown elsewhere—a wonderful fact, considering how widely distributed sea-shells generally are. Of the forty-three shells found in other parts of the world, twenty-five inhabit the western coast of America, and of these eight are distinguishable as varieties; the remaining eighteen (including one new variety) were found by Mr. Cuming in the Low Archipelago, and some of them also at the Philippines. This fact of shells from islands in the central part of the Pacific occurring here, deserves notice, for not one single sea-shell is known to be common to the islands of that ocean and to the west coast of America. The space of open sea running north and south off the west coast, separates two quite distinct conchological provinces; but at the Galápagos we have a halting place, where many new forms have been created, and whither these two great conchological provinces have each sent several colonists. The American province has also sent here representative species; for there is a Galapageian species of *Monoceras*, a genus only found on the west coast of America; and there are Galapageian species of *Fissurella* and *Cancellaria*, genera common on the west coast, but not found (as I am informed by Mr. Cuming) in the central islands of the Pacific. On the other hand, there are Galapageian species of *Oniscia* and *Stylifer*, genera common to the West Indies and to the Chinese and Indian seas, but not found either on the west coast of America or in the central Pacific. I may here add, that after the comparison of Messrs. Cuming and Hinds of about 2000 shells from the eastern and western coasts of America, only one shell was found in common, namely, the *Purpura patula*, which inhabits the West Indies, the coast of Panama, and the Galápagos. We have, therefore, in this quarter of the world, three great conchological sea-provinces, quite distinct, though surprisingly near each other, being separated by long north and south spaces either ofland or of open sea".

References

Abbott, D. P., 1966, Factors influencing the zoogeographic affinities of Galápagos inshore fauna, in: *The Galápagos: Proceedings of the Symposia of the Galápagos International Scientific Project* (R. I. Bowman, ed.), pp. 108–122, University of California Press, Berkeley, California.

Ax, P., 1977, Problems of speciation in the interstitial fauna of the Galápagos, *Mikrofauna des Meeresboden* **61**:29–43.

Blackwelder, R. E., 1967, *Taxonomy*, John Wiley & Sons, Inc., New York.

Burkhardt, F., and Smith, S. (eds.), 1985, *The Correspondence of Charles Darwin, Volume 1, 1821–1836*, Cambridge University Press, Cambridge, England.

Burkhardt, F., and Smith, S. (eds.), 1986, *The Correspondence of Charles Darwin, Volume 2, 1837–1843*, Cambridge University Press, Cambridge, England.

Carlquist, S., 1970, *Hawaii, A Natural History*, Natural History Press, New York.

Chambers, S. M., and Steadman, D. W., 1986, Holocene terrestrial gastropod faunas from Isla Santa Cruz and Isla Floreana, Galápagos: Evidence for late Holocene declines, *Transactions of the San Diego Society of Natural History* **21**(6):89–110.

Dance, S. P., 1980, Hugh Cuming (1791–1865) Prince of Collectors, *J. Soc. Biblphy nat. Hist.* **9**(4):477–501.

Dance, S. P., 1986, *A History of Shell Collecting*, E. J. Brill, Leiden, The Netherlands.

Darwin, C., 1839, *Journal of Researches into the geology and natural history of the various countries visited by H. M. S. Beagle etc*, London, Henry Colburn.

Darwin, C., 1859, *On the origin of species by means of natural selection, or the preservation of favoured races in the struggle for life*, London, John Murray.

Darwin, C., 1906, *The Voyage of the Beagle*, J. M. Dent & Sons Ltd., Everyman's Library edition.

Darwin, C., 1909, *The Foundations of the Origin of Species*; Two Essays Written in 1842 and 1844, Edited by Francis Darwin, (Krauss Reprint of 1969), Cambridge Univ. Press, Cambridge, England.

Ekman, S., 1953, *Zoogeography of the Sea*, Sidgwick and Jackson, Ltd., London.

Emerson, W. K., 1967, Indo-Pacific faunal elements in the tropical eastern Pacific, with special reference to the mollusks, *Venus* **25**(3–4):87–93.

Emerson, W. K., 1978, Mollusks with Indo-Pacific faunal affinities in the eastern Pacific Ocean, *Nautilus* **92**(2):91–96.

Emerson, W. K., and Old, W. E., Jr., 1965, New molluscan records for the Galápagos Islands, *Nautilus* **78**(4):116–120.

Emerson, W. K., and Old, W. E., Jr., 1968, An additional record for *Cypraea teres* in the Galápagos Islands, *Veliger* **11**(2):98–99, 1 plate.

Finet, Y., 1985, Preliminary faunal list of the marine mollusks of the Galápagos Islands, *Documents de Travail, No. 20*, Institut royal des Sciences naturelles de Belgique, Bruxelles.

Freeman, R. B., 1977, *The Works of Charles Darwin*, Wm. Dawson & Sons–Archon Books, Folkestone, England.

Ghiselin, M. T., 1984, *The Triumph of the Darwinian Method*, University of Chicago Press, Chicago.

Gibson, T. G., 1987, Patterns of species origination in a Pacific atoll fauna: Late Cenozoic Foraminifera of Enewetak, in: *Fourth International Congress on Pacific Neogene Stratigraphy, Abstract Volume* (J. A. Barron and J. R. Blueford, eds.), p. 31, Math/Science Nucleus, Fremont, California.

Hamilton, T. H., Rubinoff, I., Barth, R. H., Jr., and Bush, G. L., 1963, Species abundance: Natural regulation of insular variation, *Science* **142**:1575–1577.

Herbert, S. (ed.), 1980, *The Red Notebook of Charles Darwin*, Cornell University Press, Ithaca, New York.

Hertlein, L. G., 1972, Pliocene fossils from Baltra (South Seymour) Island, Galápagos Islands, *Proc. Calif. Acad. Sci.* 4th Series, **39**(3):25–46.

Howell, J. T., 1941, Hugh Cuming's visit to the Galápagos Islands, *Lloydia* **4**(4):291–292.

Jablonski, D., and Lutz, R. A., 1983, Larval ecology of marine benthic invertebrates: Paleobiological implications, *Biological Reviews* **58**:21–89.

James, M. J., 1984, A new look at evolution in the Galápagos: evidence from the late Cenozoic marine molluscan fauna, *Biol Journ Linnean Soc* **21**(1–2):77–95.

Johnson, M. P., and Raven, P. H., 1973, Species number and endemism: The Galápagos Archipelago revisited, *Science* **179**:893–895.

Jones, D. S., and Hasson, P. F., 1985, History and development of the marine invertebrate faunas separated by the Central American isthmus, in: *The Great American Biotic Interchange* (F. G. Stehli and S. D. Webb, eds.), pp. 325–355, Plenum Press, New York.

Kay, E. A., 1979, *Hawaiian Marine Shells: Reef and Shore Fauna of Hawaii, Section 4: Mollusca*, Bernice P. Bishop Museum Special Publication 64 (4), Bishop Museum Press, Honolulu.

Kay, E. A., 1980, Little worlds of the Pacific, An essay on Pacific Basin biogeography, *University of Hawaii Lyon Arboretum, Lecture No. 9*, pp. 1–40.

Kay, E. A., 1984, Patterns of speciation in the Indo-West Pacific, in: *Biogeography of the Tropical Pacific* (F. J. Radovsky, P. H. Raven, and S. H. Sohmer, eds.) Bernice P. Bishop Museum Special Publication no. 72, pp. 15–31, Bernice P. Bishop Museum, Honolulu.

Keigwin, L. D., Jr., 1978, Pliocene closing of the Isthmus of Panama, based on biostratigraphic evidence from nearby Pacific Ocean and Caribbean Sea cores, *Geology* **6**:630–634.

Keynes, R. D., 1979a, Darwin and the Beagle, *Proc. Amer. Phil. Soc.* **123**(5):323–335.

Keynes, R. D. (ed.), 1979b, *The Beagle Record*, Cambridge University Press, Cambridge, England.

Lesch, J. E., 1975, The role of isolation in evolution: George J. Romanes and John T. Gulick. *Isis* **66**: 483–503.

Pitt, W. D., and Pitt, L. J., 1989, Notes on the marine molluscan fossil deposits of the Galápagos Islands, *Annual Report of the Western Society of Malacologists* **21**:6.

Porter, D. M., 1979, Endemism and evolution in Galápagos Islands vascular plants, in: *Plants and Islands* (D. Bramwell, ed.) Academic Press, London.

Shasky, D. R., 1989, Update on mollusks with Indo-Pacific faunal affinities in the tropical eastern Pacific—VIII, *Annual Report of the Western Society of Malacologists* **21**:8.

Smith, A. G., 1966, Land snails of the Galápagos, in: *The Galápagos: Proceedings of the Symposia of the Galápagos International Scientific Project* (R. I. Bowman, ed.), pp. 240–251, University of California Press, Berkeley, California.

Smith, J. T., 1981, Filtering through species list for significant distribution patterns in west Mexican and Caribbean cognates, *Annual Report of the Western Society of Malacologists* **13**:12.

Steadman, D. W., 1986, Holocene vertebrate fossils from Isla Floreana, Galápagos, *Smithsonian Contributions to Zoology*, No. 413.

Tasch, P., 1985, Darwin's *Beagle* voyage and Galápagos experience, *Journal of Geological Education* **33**(1):4–10.

Vagvolgyi, J., 1975, Body size, aerial dispersal, and origin of the Pacific land snail fauna, *Systematic Zoology* **24**:465–488.

Vermeij, G. J., 1978, *Biogeography and Adaptation*, Harvard University Press, Cambridge, Massachusetts.

Vinton, K. W., 1951, Origin of life on the Galápagos Islands, *American Journal of Science* **249**(5): 356–376.

Wellington, G. M., and Kuris, A. M., 1983, Growth and shell variation in the tropical eastern Pacific intertidal gastropod genus *Purpura*: Ecological and evolutionary implications, *Biological Bulletin* **164**:518–535.

Wiggins, I. L., and Porter, D. M., 1971, *Flora of the Galápagos Islands*, Stanford University Press, Stanford, California.

Williamson, M., 1981, *Island Populations*, Oxford University Press, Oxford, England.

Williamson, M., 1983, Darwin's finches worth a mention, *Nature* **302**:566.

Zinsmeister, W. J., and Emerson, W. K., 1979, The role of passive dispersal in the distribution of hemipelagic invertebrates, with examples from the tropical Pacific Ocean, *Veliger* **22**(1):32–40.

VII

Echinodermata

Chapter 16

Echinoderm Fauna of the Galápagos Islands

LINDA YVONNE MALUF

1. Introduction

There are nearly 200 species of echinoderms reported from the Galápagos Islands and Galápagos Ridge, including five crinoids, 44 asteroids, 74 ophiuroids, 37 echinoids, and 38 holothurians. This is nearly one-third of all echinoderm species known from the central eastern Pacific (CEP) between 34°30′N and 18°S (Maluf, 1988a). Seventy percent of all CEP families, and 45% of all CEP genera are recorded from the Galápagos. Holothurians are the most poorly represented class. In fact, two orders (Apodida and Molpadiida) containing five families and 14 CEP genera, as well as one large family in the Dendrochirotida (Cucumariidae) with ten CEP genera, are not reported from the Galápagos.

Roughly one-half of the Galápagos species are restricted to shallow water (< 200 m), and the other half are either restricted to water deeper than 200 m (36%) or range both above and below 200 m. Fewer than 20% of the shallow-water species are found exclusively at depths of < 20 m. The Galápagos fauna resembles that of the tropical Panamic region, but 17% of the species are endemic and the islands are recognized as a distinct subtropical province (Briggs, 1974; Maluf, 1988b).

Prior to this report information on the composition of this interesting fauna has been scattered among obscure and relatively inaccessible publications and museum records. Here I present a complete annotated species list with general

LINDA YVONNE MALUF • Department of Ecology and Evolutionary Biology, University of Arizona, Tucson, Arizona 85721. *Present address*: 4081 Minuteman, Boise, Idaho 83706.

distributional comments on Galápagos echinoderms in the context of the central eastern Pacific fauna.

2. Annotated Species List

The following species list is based on all known records in the literature and on published data from the catalogues of the Los Angeles County Museum's Allan Hancock Foundation Collections and on holdings at Scripps Institute of Oceanography (Luke, 1982). In the list, records based on these museum collections are indicated as either a 1 (LACM-AHF) or a 2 (SIO). These museum holdings may include erroneous identifications and the records must be considered as tentative. The ridges in this area of the eastern Pacific provide an extensive slope habitat (Fig. 1) and support a very high diversity of echinoderms, especially asteroids and ophiuroids.

This list is derived from the records for the 630 species of warm-water eastern Pacific echinoderms between Pt. Conception, California and southern Peru (Maluf, 1988a). Thus, detailed distributions within the Galápagos Islands are not given, and depth ranges refer to all known records, not to those recorded for the Galápagos Islands *per se*. Sources of Galápagos distribution records are cited within the list for each species. I also specify distributions in the Revillagigedos (Soccoro and Clarion), Clipperton, Cocos, and Malpelo Islands, and the Gulf of California. The interested reader is referred to Maluf (1988a) for detailed eastern Pacific distributions, a treatment of the classification (including synonyms), and more extensive literature citations.

2.1. Class Crinoidea

ORDER MILLERICRINIDA: FAMILY HYOCRINIDAE

Calamocrinus diomedae A. Agassiz, 1892—Off Mariato Point, Panama and the Galápagos (Agassiz, 1892). 717 to 1431 m.

ORDER BOURGUETICRINIDA: FAMILY BATHYCRINIDAE

Bathycrinus equatorialis A. H. Clark, 1908—East Pacific Rise west of the Galápagos at 4246 m (A. H. Clark, 1908).

ORDER COMATULIDA: FAMILY THALASSOMETRIDAE

Thalassometra agassizii (Hartlaub, 1895)—Cocos and Galápagos Ridges from 596 to 1429 m (Hartlaub, 1895).

FAMILY ANTEDONIDAE

Florometra serratissima (A. H. Clark, 1907)—Alaskan Peninsula to the Baja California Islands, Gulf of Panama and Galápagos (Ziesenhenne, 1937). 11 to 1252 m. 1,2

Psathyrometra bigradata (Hartlaub, 1895)—Off Mariato Point, Panama and the Galápagos from 705 to 1016 m (Hartlaub, 1895).

2.2. Class Asteroidea

ORDER PAXILLOSIDA: FAMILY LUIDIIDAE

Luidia (Alternaster) bellona Lütken, 1864—From Magdalena Bay, Baja California

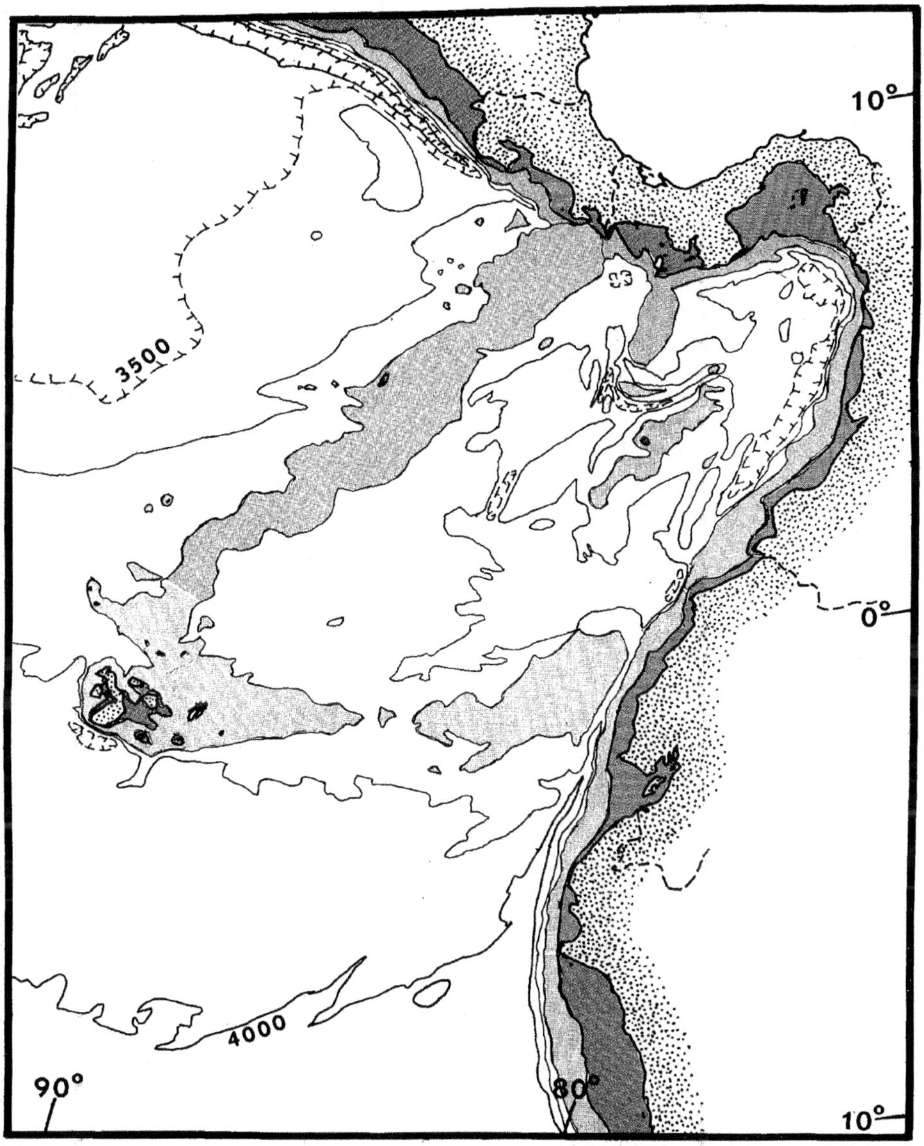

Figure 1. Bathymetry in the central eastern Pacific Ocean 10°N and 10°S of the Galápagos Islands. Dark shading is the extent of the shelf (shore to 200 m), and light shading indicates slope areas (200–2000 m). Redrawn from Geographic Map of the Circum-Pacific Region, 1977, Northeast Quadrant (map center point 35°N, 135°W), American Association of Petroleum Geologists.

and Puerto Refugio, Gulf of California to Sechura Bay, Peru, mostly on nearshore islands. Galápagos and Revillagigedos Islands records and most Gulf records are based on LACM-AHF holdings. Four to 55 m, extended to 201 m by museum holdings. 1

Luidia (Alternaster) superba A. H. Clark, 1917—Reported from Mazatlán, Manzanillo, Colombia, Ecuador, Galápagos (Downey and Wellington, 1978) and Chile. Three to 60 m.

Luidia (Petalaster) columbia (Gray, 1840)—From Cedros Island, Baja California and El Golfo de Santa Clara, Gulf of California to Sechura Bay, Peru, including the Revillagigedos and Galápagos Islands. Galápagos and most Gulf records are based on LACM-AHF holdings. Low tide to 146 m, extended to 220 m by museum holdings. 1,2

Luidia (Petalaster) foliolata Grube, 1865—From Kasaan Bay, Alaska to Taboga Island, Panama, including the Gulf of California south of Los Angeles Bay. Most central eastern Pacific records, including the Galápagos, are based on LACM-AHF holdings. Four to 348 m, extended to 476 m by museum records. 1,2

FAMILY ASTROPECTINIDAE

Astropecten fragilis Verrill, 1867—Reported at Isla San Pedro Nolasco in the Gulf of California and from Panama to Zorritos, Peru. Galápagos and Gulf records are based on LACM-AHF holdings. Low tide, extended to 203 m by museum records. 1

Dytaster gilberti Fisher, 1905—Patton Escarpment and Cortez Bank off Southern California, Galápagos (Ludwig, 1905), and off Central America to Trujillo, Peru. 1573 to 4335 m. 2

Leptychaster (Parastropecten) inermis (Ludwig, 1905)—Off Monterey and in the basins of the Southern California Borderland, Galápagos and Cocos Islands (Fisher, 1930), and off Panama. 732 to 1593 m. Tropical Pacific. 1,2

Psilaster sladeni Ludwig, 1905—The Galápagos Ridge at 1618 m (Ludwig, 1905).

FAMILY PORCELLANASTERIDAE

Eremicaster crassus gracilis (Sladen, 1883)—From Kamchatka, Alaska to Valparaiso, Chile in very deep waters (2420 to 5204 m), including those off Clipperton and Galápagos Islands (Ludwig, 1905). North Pacific. 2

Eremicaster pacificus (Ludwig, 1905)—Deep waters from the Pribilof Islands, Alaska to Valparaiso, Chile, including those off Cocos and Galápagos Islands (Fisher, 1911). 1463 m (museum record), 1571 to 5780 m. East Atlantic. 1,2

Porcellanaster ceruleus Thomson, 1877—Cocos and Carnegie Ridges (Ludwig, 1905; H. L. Clark, 1920a). 1158 to 6035 m. Indo-West Pacific and Atlantic.

Thoracaster cylindratus Sladen, 1883—Carnegie (Fisher, 1928) and Nazca Ridges from 2600 to 5303 m. Indo-Pacific and West Atlantic. 2

ORDER NOTOMYOTIDA: FAMILY BENTHOPECTINIDAE

Benthopecten pectinifer (Ludwig, 1905)—California, Guaymas Basin and the basins off Panama and the Galápagos (Ludwig, 1905). 1571 to 2323 m.

Benthopecten spinuliger (Ludwig, 1905)—Gulf of Panama, Galápagos and Cocos Ridges (Ludwig, 1905) from 1618 to 2323 m.

Pectinaster agassizii (Ludwig, 1905)—Southern California Borderland, Guaymas Basin, Cocos and Galápagos Ridges (Ludwig, 1905) and off Panama. 790 to 2323 m. Gulf of Aden. 1,2

ORDER VALVATIDA: FAMILY GONIASTERIDAE

Nymphaster diomedeae Ludwig, 1905—Off Rio Elata, Mexico (LACM-AHF), off Panama, Cocos and Galápagos Ridges (Ludwig, 1905). 702 to 1618 m. 1

Pseudarchaster pulcher Ludwig, 1905—Southern California Borderland (SIO tentative identification), off Acapulco and the Galápagos (Ludwig, 1905) from 702 to 1618 m. 2

FAMILY ASTERODISCIDIDAE

Paulia horrida Gray, 1840—Clarion Island, Cocos and Galápagos Ridges (Ludwig, 1905), Ecuador and Peru. Two to 123 m. *Pauliella aenigma* Ludwig, 1905, usually synonymized with *P. horrida*, may be a valid species (Rowe, 1977). It is found from the Tres Marías to Cocos Islands between 24 and 91 m. 1

FAMILY OREASTERIDAE

Nidorellia armata (Gray, 1840)—From Puerto Peñasco, Gulf of California and Bahía Margarita, Baja California to Zorritos, Peru and the Galápagos (H. L. Clark, 1902). Low tide to 73 m. 1,2

Pentaceraster cumingi (Gray, 1840)—From Isla Tiburón in the Gulf of California and Bahía Margarita, Baja California and the Revillagigedos Islands to Lobos Afuera Islands off Peru, and Cocos and Galápagos Islands (H. L. Clark, 1902). Especially common on islands and points. Low tide to 90 m, extended to 183 m by museum records. Hawaii. 1,2

FAMILY ASTEROPSEIDAE

Asteropsis carinifera (Lamarck, 1816)—Gulf of California islands, Clarion Island, Panama, Malpelo and Galápagos Islands (Downey, 1976). Intertidal to 36 m. Indo-Pacific. 1

FAMILY OPHIDIASTERIDAE

Linckia columbiae Gray, 1840—From San Pedro, California and the upper Gulf of California to northern Peru, including Clarion, Cocos and the Galápagos Islands (Fisher, 1911). Low tide to 100 m, extended to 156 m by museum holdings. 1,2

Narcissia gracilis gracilis A. H. Clark, 1916—Islands in the Gulf of California (LACM-AHF) and Cocos and Galápagos Islands (Downey, 1976). 57 to 91 m, shallow subtidal to 128 m based on museum holdings. 1

Pharia pyramidata (Gray, 1840)—From Catalina and Guadalupe Islands off southern California (LACM-AHF) and upper Gulf of California to Zorritos, Peru, including the Revillagigedos and Galápagos Islands (LACM-AHF). Low tide to 50 m, extended to 139 m by museum holdings. 1,2

Phataria unifascialis (Gray, 1840)—From Magdalena Bay, Baja California and throughout the Gulf of California to Zorritos, Peru including the Revillagigedos and Galápagos Islands (H. L. Clark, 1902). Low tide to 50 m. 1,2

Tamaria obstipa Ziesenhenne, 1942—Cocos and Galápagos Islands from 73 to 128 m (Ziesenhenne, 1942). 1

FAMILY MITHRODIIDAE

Mithrodia bradleyi Verrill, 1867—Islands off Baja California and in the Gulf of California south of the Midriff Islands, the Revillagigedos, Clipperton and Galápagos Islands (Steinbeck and Ricketts, 1941), Panama islands, and Arica, Chile (locality doubtful, Madsen, 1956). Low tide to 50 m. Central Pacific. This species has been provisionally referred to the synonymy of *M. clavigera* (Lamarck, 1816), a widespread Pacific species, by Pope and Rowe, 1977. 1,2

ORDER SPINULOSIDA: FAMILY SOLASTERIDAE

Lophaster furcilliger Fisher, 1905—From Alaska to Cortez Basin off southern California and in the Galápagos Islands (Fisher, 1911). 86 m (museum holdings), 351 to 2012 m. 1,2

Lophaster validus (Ludwig, 1905)—Islas Tres Marías, the Gulf of Panama and the Galápagos (Ludwig, 1905). 523 to 1244 m.

FAMILY PTERASTERIDAE

Hymenaster gracilis Ludwig, 1905—Off Colombia and the Galápagos Islands (Ludwig, 1905). 2418 to 3241 m.

Hymenaster platyacanthus Ludwig, 1905—Off Panama, Ecuador and the Galápagos Islands (Ludwig, 1905). 2487 to 2877 m.

FAMILY PORANIIDAE

Poraniopsis inflata (Fisher, 1906)—From Oregon to Guadalupe Island and in the Gulf of Panama to Mancora Bank and Galápagos Islands (Ludwig, 1905). 48 to 1094 m. 1,2

FAMILY ECHINASTERIDAE

Henricia gracilis (Ludwig, 1905)—Islas Tres Marías and Galápagos (Ludwig, 1905). 267 to 1244 m.

Henricia nana (Ludwig, 1905)—Galápagos Islands (Ludwig, 1905), 97 to 110 m (LACM-AHF). 1

ORDER FORCIPULATIDA: FAMILY ZOROASTERIDAE

Zoroaster longispinus Ludwig, 1905—Guaymas Basin, Gulf of California and off Panama and the Galápagos (Ludwig, 1905). 1427 to 2418 m.

FAMILY HELIASTERIDAE

Heliaster cumingii (Gray, 1840)—Galápagos Islands, low tide (H. L. Clark, 1920a). According to Madsen (1956), *Heliaster polybrachius* H. L. Clark, 1907 may be conspecific. This would extend the range of *H. cumingii* to Ecuador, Peru, and perhaps Acapulco. 1,2

Heliaster solaris A. H. Clark, 1920—Galápagos Islands (A. H. Clark, 1920), Puntarenas, Costa Rica (Boone, 1933). Intertidal. 1

FAMILY ASTERIIDAE

Astrometis sertulifera (Xantus, 1860)—From Vancouver Island to Cedros Island, Baja California, throughout the Gulf of California, Galápagos (Brusca, 1980 and LACM-AHF) and Lobos de Afuera Islands. Low tide to 65 m, extended to 156 m by museum holdings. 1

Coronaster marchenus Ziesenhenne, 1942—Cocos and Galápagos Islands (Ziesenhenne, 1942). Six to 91 m.

Distolasterias robusta (Ludwig, 1905)—Off the Galápagos at 704 m (Ludwig, 1905).

Hydrasterias improvisa (Ludwig, 1905)—Cocos Ridge, off the Galápagos (Ludwig, 1905). 1618 to 2418 m.

Tarsaster galapagensis (Ludwig, 1905)—Off Santa Catalina Island, California (LACM-AHF) and the Galápagos (Ludwig, 1905). 457 m (museum holdings) to 704 m. 1

ORDER BRISINGIDA: FAMILY BRISINGIDAE

Astrolirus panamensis (Ludwig, 1905)—Punta Santo Tomás, Baja California, off Sonora, Mexico and Panama, and from Cocos, Malpelo and Galápagos Islands (Ludwig, 1905). 48 to 2418 m.

Freyellaster scalaris (A. H. Clark, 1916)—Off the Galápagos at 1480 m (A. H. Clark, 1916).

2.3. Class Ophiuroidea

ORDER PHRYNOPHIURIDA: FAMILY OPHIOMYXIDAE

Ophiomyxa panamensis Lütken and Mortensen, 1899—Cedros Island, Baja California and Midriff islands, Gulf of California, off Panama to the Gulf of Guayaquil, Revillagigedos, Cocos and Galápagos Islands (H. L. Clark, 1940). 19 to 293 m (museum holdings), 46 to 155 m. 1

FAMILY ASTERONYCHIDAE

Asteronyx loveni Müller and Troschel, 1842—From Alaska to northern Baja California, Isla Tortuga and Altata, Gulf of California, off Panama and Colombia and at the Galápagos (Lütken and Mortensen, 1899). 152 to 2963 m. Indo-West Pacific and North Atlantic. 1,2

Astrodia plana Lütken and Mortensen, 1899—Gulf of Panama and off Malpelo and the Galápagos Islands (Lütken and Mortensen, 1899). 716 to 3200 m.

FAMILY ASTEROSCHEMATIDAE

Asteroschema sublaeve Lütken and Mortensen, 1899—Central and southern California, Gulf of California mouth, Gulf of Panama and Galápagos (Lütken and Mortensen, 1899). 605 to 1681 m. 2

FAMILY GORGONOCEPHALIDAE

Astrodendrum galapagense A. H. Clark, 1916—Galápagos Islands at 718 m (A. H. Clark, 1916).

Astrodictyum panamense (Verrill, 1867)—La Paz, Gulf of California and Panama to Zorritos, Peru, Malpelo and Galápagos Islands (Birkeland *et al.*, 1975). Records from La Paz and the Galápagos Islands are not well-substantiated (Gordon Hendler, pers. comm.). 14 to 64 m. Japan. 1

ORDER OPHIURIDA: FAMILY OPHIACANTHIDAE

Ophiacantha cyrena A. H. Clark, 1916—Galápagos Islands at 717 m (A. H. Clark, 1916).

Ophiacantha inconspicua Lütken and Mortensen, 1899—Gulf of Panama and Galápagos Islands (Lütken and Mortensen, 1899). 37 (museum holdings) to 1865 m. 1

Ophiacantha phragma Ziesenhenne, 1940—Southern and Baja California islands, Gulf of California, Panama, Cocos and Galápagos Islands (Ziesenhenne, 1940). 13 to 644 m. 1,2

Ophiacantha quadrispina H. L. Clark, 1917—Baja California islands, Gulf of California islands and Galápagos Islands (H. L. Clark, 1917a). 229 m, 183 to 549 m based on museum records. 1

Ophiacantha similis A. H. Clark, 1916—Galápagos Islands at 717 m (A. H. Clark, 1916).

Ophiacantha valenciennesi Lyman, 1879—Puerto Escondido, Gulf of California (questionable identification, LACM-AHF) and Galápagos Islands (H. L. Clark, 1917a) from 236 to 549 m. Caribbean. 1

Ophiolebes mortenseni A. H. Clark, 1916—Galápagos Islands at 717 m (A. H. Clark, 1916).
Ophiomyces multispinus Ziesenhenne, 1940—Galápagos Islands from 134 to 717 m (Ziesenhenne, 1940). 1
Ophiothamnus laevis Lütken and Mortensen, 1899—Galápagos Islands at 1008 m (Lütken and Mortensen, 1899).
FAMILY HEMIEURYALIDAE
Sigsbeia lineata Lütken and Mortensen, 1899—Hannibal Bank, Panama, and Cocos and Galápagos Islands (Lütken and Mortensen, 1899). 46 to 183 m. 1
FAMILY AMPHIURIDAE
Amphichondrius granulosus Nielsen, 1932—Morro Bay, California and central Gulf of California to La Plata Island, Ecuador and the Galápagos (LACM-AHF). Shore to 186 m. 1,2
Amphicontus minutus Hill, 1940—Galápagos Islands (Hill, 1940) and Independencia Bay, Peru, from nine to 22 m. 1
Amphilepis patens Lyman, 1879—Southern Alaska to Valparaiso, Chile and the Galápagos (H. L. Clark, 1917a). 385 to 4087 m. 2
Amphiodia tabogae Nielsen, 1932—Tangola-Tangola Bay, Mexico to Ecuador, Cocos and Galápagos Islands (LACM-AHF). Intertidal to 18 m, extended to 118 m by museum holdings. 1
Amphipholis platydisca Nielsen, 1932—Channel Islands, California and upper Gulf of California to Ecuador and the Galápagos Islands (LACM-AHF). Low tide to 112 m, extended to 137 m by museum holdings. 1,2
Amphipholis puntarenae (Lütken, 1856)—Dana Point, California and Isla Tiburón, Gulf of California to Ecuador and the Galápagos Islands (LACM-AHF). 18 to 183 m, shallow to 508 m (museum holdings). 1
Amphipholis squamata (Delle Chiaje, 1828)—Alaska to San Juan Bay, Peru, including the Gulf of California and the Galápagos Islands (LACM-AHF). Shore to 823 m. Cosmopolitan. 1,2
Amphiura arcystata H. L. Clark, 1911—Monterey, California and Gulf of California to Colombia, Cocos and Galápagos Islands (LACM-AHF). Six m (museum holdings) to 849 m. Japan. 1,2
Amphiura gymnogastra Lütken and Mortensen, 1899—Gulf of Panama and the Galápagos Islands (H. L. Clark, 1917a). 549 to 2323 m.
Amphiura polyacantha Lütken and Mortensen, 1899—Cabo Mala, Panama, and the Galápagos (H. L. Clark, 1902). 1271 m.
Amphiura serpentina Lütken and Mortensen, 1899—Monterey, California to Panama and the Galápagos (Lütken and Mortensen, 1899), central Gulf of California (SIO tentative identification). 770 to 1865 m. 2
Amphiura verticillata Ljungman, 1867—Galápagos Islands from 46 to 64 m (H. L. Clark, 1915). 1
Ophiophragmus disacanthus Ziesenhenne, 1940—Galápagos Islands from four to 37 m (Ziesenhenne, 1940).
Ophiophragmus marginatus (Lütken, 1859)—Puerto Peñasco, Gulf of California to Peru, Cocos and Galápagos Islands (LACM-AHF). Low tide, extended to 134 m by museum records. 1
Ophiophragmus paucispinus Nielsen, 1932—Lower Gulf of California to Panama,

Cocos and Galápagos Islands (LACM-AHF). Low tide, extended to 134 m by museum holdings. 1

Ophiophragmus tabogensis Nielsen, 1932—Midriff islands, Gulf of California to Colombia and the Galápagos Islands (LACM-AHF). Low tide, extended to 128 m by museum records. 1

FAMILY OPHIACTIDAE

Histampica duplicata (Lyman, 1875)—Cabo San Lucas, off Malpelo and Galápagos Islands (Lütken and Mortensen, 1899). 134 to 1966 m. West Atlantic. 1

Ophiactis plana Lyman, 1869—Off Malpelo and Galápagos Islands from 183 to 1918 m (Lütken and Mortensen, 1899). Japan, Indo-West Pacific, Atlantic.

Ophiactis savignyi (Müller and Troschel, 1842)—San Pedro, California and upper Gulf of California to Independencia Bay, Peru, including the Revillagigedos, Clipperton, Cocos, Malpelo and Galápagos Islands (A. H. Clark, 1939). Intertidal to 70 m, extended to 128 m by museum holdings. Tropicopolitan. 1,2

Ophiactis simplex (Le Conte, 1851)—Channel Islands, California and the Gulf of California to Independencia Bay, Peru, including the Revillagigedos, Cocos and Galápagos Islands (LACM-AHF). Intertidal to 229 m, extended to 302 m by museum records. 1,2

FAMILY OPHIOTRICHIDAE

Ophiothrix galapagensis Lütken and Mortensen, 1899—Isla Angel de la Guarda, Gulf of California to Islas Tres Marías, Revillagigedos, and Galápagos (Lütken and Mortensen, 1899). Intertidal to 549 m. 1

Ophiothrix magnifica Lyman, 1860—West Central America and the Galápagos Islands (H. L. Clark, 1902), Gulf of Guayaquil to Islas Lobos de Afuera, Peru. Intertidal to 11 m. 1

Ophiothrix spiculata Le Conte, 1851—Bering Sea to Lobos de Afuera Islands, Peru, including the Gulf of California and the Revillagigedos, Cocos and Galápagos Islands (H. L. Clark, 1902). Intertidal to at least 730 m, perhaps 2059 m. 1,2

FAMILY OPHIOCOMIDAE

Ophiocoma aethiops Lütken, 1859—Puerto Peñasco, Gulf of California and Malarrimo Point, Baja California to Lobos de Afuera Islands, Peru, including the Revillagigedos, Cocos, Malpelo, and Galápagos Islands (H. L. Clark, 1902). Intertidal to 30 m. 1,2

Ophiocoma alexandri Lyman, 1860—Catalina Island, California (LACM-AHF) and Magdalena Bay, Baja California, Puerto Peñasco, Gulf of California to Ecuador, including the Revillagigedos, Cocos, Malpelo and Galápagos Islands (Ziesenhenne, 1937). Intertidal to 30 m, extended to 70 m by museum records. 1,2

Ophiocomella schmitti A. H. Clark, 1939—Galápagos Islands, intertidal (A. H. Clark, 1939).

FAMILY OPHIOCHITONIDAE

"*Amphioplus*" *papillatus* (Lütken and Mortensen, 1899)—Galápagos Islands at 704 m (Lütken and Mortensen, 1899). This species, formerly in the Family Amphiuridae, was referred to Family Ophiochitonidae by A. M. Clark, 1970, without a genus designation. "*Amphioplus*" is used here pending designation of a genus within the Ophiochitonidae (A. M. Clark, pers. comm.).

FAMILY OPHIONEREIDAE

Ophionereis albomaculata E. A. Smith, 1877—Galápagos Islands, intertidal (H. L. Clark, 1902). 1

Ophionereis annulata (Le Conte, 1851)—San Pedro, California and the upper Gulf of California to Bahía Santa Elena, Ecuador, including the Revillagigedos, Cocos, and Galápagos Islands (H. L. Clark, 1902). Intertidal to 60 m, expanded to 229 m by museum holdings. 1,2

Ophionereis nuda Lütken and Mortensen, 1899—Islas Tres Marías, Gulf of Panama, Cocos and Galápagos Islands (Lütken and Mortensen, 1899). 73 to 155 m. 1

Ophionereis perplexa Ziesenhenne, 1940—Gulf of California islands, Central America, Manta, Ecuador, and the Galápagos Islands (Ziesenhenne, 1940). Shore to 73 m. 1

FAMILY OPHIODERMATIDAE

Ophioderma elaps Lütken, 1859—Galápagos Islands at 134 m (LACM-AHF), to 549 m. West Indies. 1

Ophioderma panamense Lütken, 1859—San Pedro, California and the upper Gulf of California to Paita, Peru, including the Revillagigedos, Cocos, and Galápagos Islands (Ziesenhenne, 1937). Intertidal to 20 m, perhaps to 73 m. 1,2

Ophioderma pentacantha H. L. Clark, 1917—Southern Gulf of California and the Galápagos Islands (H. L. Clark, 1917a). Shore to 183 m. 1,2

Ophioderma teres (Lyman, 1860)—Newport, California and the upper Gulf of California to La Plata Island, Ecuador, and the Galápagos Islands (H. L. Clark, 1902). Intertidal to 30 m, extended to 46 m by museum holdings. 1,2

Ophioderma variegatum Lütken, 1856—San Diego, California and the upper Gulf of California to Panama, Revillagigedos, Cocos and Galápagos Islands (H. L. Clark, 1940). Shore to 110 m. 1,2

Ophiuroconis bispinosa Ziesenhenne, 1937—Monterey, California and the upper Gulf of California to the Gulf of Fonesca and the Galápagos Islands (LACM-AHF). Four to 143 m (museum holdings). 1

FAMILY OPHIURIDAE

Amphiophiura irregularis Ziesenhenne, 1940—Galápagos Islands (Ziesenhenne, 1940), 106 to 146 m, extended to 238 m by museum records. 1

Amphiophiura obtecta (Lütken and Mortensen, 1899)—Off the Galápagos Islands, 2196 to 2487 m (Lütken and Mortensen, 1899).

Amphiophiura paucisquama Ziesenhenne, 1940—Galápagos Islands, 549 to 717 m (Ziesenhenne, 1940). 1

Gymnophiura mollis Lütken and Mortensen, 1899—Islas Tres Marías, Gulf of Panama, and east of the Galápagos (Lütken and Mortensen, 1899). 2487 to 2999 m. 2

Ophiocten hastatum Lyman, 1878—Washington to the Baja California islands, Costa Rica to Ecuador, Cocos, Malpelo, and Galápagos Ridges (Lütken and Mortensen, 1899). 825 to 4700 m. Japan, Southern Ocean, East Atlantic. 2

Ophiomisidium leurum Ziesenhenne, 1940—Gulf of Tehuantepec and the Galápagos Islands, 82 to 137 m (Ziesenhenne, 1940). 1

Ophiomusium diomedeae Lütken and Mortensen, 1899—Galápagos Islands at 704 m (Lütken and Mortensen, 1899).

Ophiomusium jolliensis McClendon, 1909—San Pedro, California to Cabo San Lucas, Gulf of California and the Galápagos Islands (H. L. Clark, 1917a). 17 m (museum holdings) to 1230 m. Japan. 1,2

Ophiomusium lymani Thomson, 1873—British Columbia to Chile, including the southern Gulf of California, east of Clipperton, and Cocos, Malpelo and Galápagos Ridges (Lütken and Mortensen, 1899). 51 to 2906 m. Indo-West Pacific, North Atlantic. 1,2

Ophiophyllum marginatum A. H. Clark, 1916—Galápagos Islands at 717 m (A. H. Clark, 1916).

Ophioplocus hancocki Ziesenhenne, 1935—Galápagos Islands, intertidal (Ziesenhenne, 1935). 1

Ophiosphalma glabrum (Lütken and Mortensen, 1899)—Southern California basins and Gulf of California basins to southern Chile, off Cocos, Malpelo, and Galápagos Islands (Lütken and Mortensen, 1899). 878 to 5203 m. 1,2

Ophiozonella alba (Lütken and Mortensen, 1899)—Off Cocos and Galápagos Islands (Lütken and Mortensen, 1899). 1408 to 2487 m.

Ophiozonella clypeata (Lyman, 1883)—South of the Galápagos Islands, 161 to 549 m (H. L. Clark, 1917a). Caribbean.

Ophiozonella contigua (Lütken and Mortensen, 1899)—Off the Galápagos Islands, 2418 to 2487 m (Lütken and Mortensen, 1899).

Ophiura flagellata (Lyman, 1878)—Aleutian Islands, southern Baja California, southern Gulf of California, and Galápagos Islands (H. L. Clark, 1917a). 128 to 2014 m. North Pacific, West Pacific. 2

Ophiura (Ophioglypha) irrorata (Lyman, 1860)—Bering Sea to southern Peru including Malpelo, Cocos, and Galápagos Ridges (Lütken and Mortensen, 1899). 405 to 5869 m. Indo-Pacific, Atlantic. 1,2

Ophiura plana (Lütken and Mortensen, 1899)—Guatemala Basin to the Gulf of Panama and southern Peru, off Malpelo and Galápagos Islands (Lütken and Mortensen, 1899). 1430 to 4082 m.

Ophiurolepis inornata (Lyman, 1878)—Galápagos Islands at 2174 to 2487 m (Lütken and Mortensen, 1899), 240 to 3385 m outside eastern Pacific. West Pacific.

FAMILY OPHIOLEUCIDAE

Ophiernus adspersus annectens Lütken and Mortensen, 1899—Southern Baja California to Panama and south of the Galápagos Islands (H. L. Clark, 1911). 770 to 1245 m, 20 to 3650 m outside the eastern Pacific. Japan, West Atlantic, Indian Ocean.

Ophiernus seminudus Lütken and Mortensen, 1899—Gulf of Tehuantepec to central Peru and the Galápagos (Lütken and Mortensen, 1899). 840 to 4082 m. 2

2.4. Class Echinoidea

ORDER CIDAROIDA: FAMILY CIDARIDAE

Aporocidaris milleri (A. Agassiz, 1898)—Upper Gulf of California and basins off southern Baja California to Ecuador, Malpelo and Galápagos Ridges (Mortensen, 1928). 851 to 3937 m. 2

Centrocidaris doederleini (A. Agassiz, 1898)—Off Cocos and Galápagos (Mortensen, 1928) and Mancora Bank, Peru. 91 to 550 m. 1,2

Eucidaris thourarsii (Valenciennes, 1846)—Catalina Island, California, and upper Gulf of California to Bahía Santa Elena, Ecuador, including the Revillagigedos, Cocos, Malpelo, and Galápagos Islands (Mortensen, 1928). Intertidal to 150 m. 1,2

Hesperocidaris houstoniana A. H. Clark, 1939—Galápagos Islands at 91 m (A. H. Clark, 1939).

Hesperocidaris panamensis (A. Agassiz, 1898)—Off Panama, Ecuador, Cocos and Galápagos Islands (A. Agassiz, 1898). 55 to 274 m. 1,2

Histocidaris cobosi (A. Agassiz, 1898)—Galápagos Islands, 615 to 704 m (A. Agassiz, 1898).

ORDER ECHINOTHURIOIDA: FAMILY ECHINOTHURIIDAE

Tromikosoma hispidum (A. Agassiz, 1898)—Guadalupe Island, Baja California, southern Gulf of California, Gulf of Panama, off Malpelo, Cocos and Galápagos Islands (A. Agassiz, 1904). 1820 to 2763 m.

ORDER DIADEMATOIDA: FAMILY DIADEMATIDAE

Astropyga pulvinata (Lamarck, 1816)—Upper Gulf of California and Magdalena Bay, Baja California, to Bahía Santa Elena, Ecuador, including the Revillagigedos and Galápagos Islands (Brusca, 1980). Low tide to 90 m. 1,2

Centrostephanus coronatus (Verrill, 1867)—Newport Bay, California and the upper Gulf of California to Peru, including Cocos and Galápagos Islands (H. L. Clark, 1948). Low tide to 125 m. This species may be synonymous with *Centrostephanus longispinus* (Philippi, 1845) from the western Atlantic (Pawson and Miller, 1983). 1,2

Diadema mexicanum A. Agassiz, 1863—Cedros Island and the upper Gulf of California to Colombia, and the Revillagigedos, Cocos, Malpelo and Galápagos Islands (H. L. Clark, 1902). Low tide to 113 m. 1,2

FAMILY ASPIDODIADEMATIDAE

Plesiodiadema horridum (A. Agassiz, 1898)—The Cocos, Malpelo and Galápagos Ridges from 1625 to 3241 m (A. Agassiz, 1898).

ORDER SALENIOIDA: FAMILY SALENIIDAE

Salenocidaris miliaris (A. Agassiz, 1898)—Gulf of Panama, and Cocos, Malpelo and Galápagos Ridges (A. Agassiz, 1898). 1200 to 3058 m.

ORDER ARBACIOIDA: FAMILY ARBACIIDAE

Arbacia incisa (A. Agassiz, 1863)—Newport Bay, California and upper Gulf of California to Islas Chinchas, Peru, and the Galápagos (Verrill, 1867). Intertidal to 90 m. 1,2

ORDER TEMNOPLEUROIDA: FAMILY TOXOPNEUSTIDAE

Lytechinus semituberculatus (L. Agassiz and Desor, 1846)—Clarion Island (record uncertain), Colombia to Peru, and the Galápagos (Verrill, 1867). Shore to 134 m.1

Toxopneustes roseus (A. Agassiz, 1863)—Lower Baja California and Guaymas, Gulf of California to Isla La Plata, Ecuador, Revillagigedos and Galápagos Islands (H. L. Clark, 1948). Intertidal to 55 m. 1,2

Tripneustes depressus A. Agassiz, 1863—Lower Baja California, northern Gulf of California to Panama, the Revillagigedos, Malpelo, and Galápagos Islands (H. L. Clark, 1902). Shore to 73 m. 1

ORDER ECHINOIDA: FAMILY ECHINOMETRIDAE

Caenocentrotus gibbosus (L. Agassiz and Desor, 1846)—Clarion Island (questionable locality), Panama to northern Chile, and the Galápagos Islands (H. L. Clark, 1902). Low tide to nine m. 1

Echinometra oblonga (Blainville, 1825)—Lower Gulf of California and the Revillagigedos and Galápagos Islands (H. L. Clark, 1948). Shore to 34 m. Indo-Pacific. 1

Echinometra vanbrunti A. Agassiz, 1863—Monterey, California and the upper Gulf of California to Lobos Afuera Islands, Peru, and the Revillagigedos, Cocos and Galápagos Islands (Grant and Hertlein, 1938). Intertidal to 53 m. 1,2

ORDER CLYPEASTEROIDA: FAMILY CLYPEASTERIDAE

Clypeaster elongatus H. L. Clark, 1948—Galápagos Islands at six m (H. L. Clark, 1948). 1

Clypeaster (Alexandria) europacificus H. L. Clark, 1914—Cabo Lobos, northern Gulf of California to Isla La Plata, Ecuador, and the Revillagigedos, Cocos, and Galápagos Islands (H. L. Clark, 1914). Low tide to 402 m. 1,2

Clypeaster (Stolonoclypus) ochrus H. L. Clark, 1914—Bahía Adair, Gulf of California to Bahía Santa Elena, Ecuador, and the Revillagigedos, Cocos, and Galápagos Islands (LACM-AHF). Shore to 91 m, extended to 162 m by museum records. 1

Clypeaster (Stolonoclypus) rotundus (A. Agassiz, 1863)—Bahía Adair, Gulf of California and Magdalena Bay, Baja California to Bahía Santa Elena, Ecuador, and the Revillagigedos, Cocos, and Galápagos Islands (H. L. Clark, 1902). Shore to 91 m. 1,2

Clypeaster (Stolonoclypus) speciosus Verrill, 1870—Puerto Peñasco, Gulf of California to Manzanillo, Mexico, Isla Cedros and Bahía Magdalena, Baja California, and at the Revillagigedos, Cocos, and Galápagos Islands (Ziesenhenne, 1937). Low tide to 100 m, extended to 128 m by museum holdings. 1,2

FAMILY MELLITIDAE

Encope galapagensis A. H. Clark, 1946—Galápagos Islands from four to 134 m (A. H. Clark, 1946).

Encope (Echinadesma) micropora L. Agassiz, 1841—San Diego, California and the upper Gulf of California to Callao, Peru, including the Revillagigedos, Cocos, and Galápagos Islands (H. L. Clark, 1902). Low tide to 82 m. This Galápagos record may be a misidentification for *E. galapagensis*. 1,2

Encope (Echinadesma) perspectiva L. Agassiz, 1841—Bahía Los Angeles, Gulf of California and Bahía Ballenas, Baja California to Costa Rica and the Galápagos Islands (Caso, 1980). Nine to 27 m. 1

Mellitella stokesii (L. Agassiz, 1841)—Guerrero, Mexico to the Gulf of Guayaquil and the Galápagos Islands (Verrill, 1867). Intertidal to 49 m.

ORDER CASSIDULOIDA: FAMILY CASSIDULIDAE

Cassidulus pacifica (A. Agassiz, 1863)—Southern Gulf of California to Panama, the Revillagigedos and Galápagos Islands (H. L. Clark, 1902). Seven to 134 m. 1

ORDER HOLASTEROIDA: FAMILY POURTALESIIDAE

Pourtalesia tanneri A. Agassiz, 1898—Southern Gulf of California basins, the Galápagos Islands (A. Agassiz, 1904), and Chiloe, Chile. 1450 to 2454 m. Greenland (Caso, 1983). 2

ORDER SPATANGOIDA: FAMILY SCHIZASTERIDAE

Agassizia scrobiculata Valenciennes, 1846—Bahía Adair, Gulf of California and Magdalena Bay, Baja California to Capon, Peru and the Galápagos Islands (Caso, 1983), Isla Juan Fernandez, Chile (fragment). Low tide to 62 m. 1

Brisaster latifrons (A. Agassiz, 1898)—Bering Sea to the Gulf of Panama and Galápagos Islands (H. L. Clark, 1948), and throughout the Gulf of California. 20 to 1900 m, possibly nine to 2817 m. 1,2

Paraster rotundatus (Döderlein, 1906)—Galápagos Islands at 20 m Chesher, 1972).

FAMILY BRISSIDAE

Brissopsis pacifica (A. Agassiz, 1898)—Southern Channel Islands, California and upper Gulf of California to Ecuador, including the Revillagigedos, Cocos, and Galápagos Islands (Caso, 1983). Nine to 1430 m. Hawaii, Zanzibar. 1,2

Meoma ventricosa grandis Gray, 1851—Santa Maria Bay, Baja California and from Angeles Channel, Gulf of California to Ecuador and the Revillagigedos, Cocos and Galápagos Islands (H. L. Clark, 1948). Intertidal to 200 m. 1

FAMILY SPATANGIDAE

Nacospatangus (Pseudomaretia) oblonga (Mortensen. 1950)—Galápagos Islands at 35 to 70 m (Mortensen, 1950).

FAMILY LOVENIIDAE

Lovenia (Lovenia) cordiformis A. Agassiz, 1872—Santa Barbara, California and the upper Gulf of California to northern Peru, the Revillagigedos, Cocos and Galápagos Islands (H. L. Clark, 1917b). Low tide to 201 m. 1,2

2.5. Class Holothuroidea

ORDER DENDROCHIROTIDA: FAMILY PSOLIDAE

Lissothuria veleronis (Deichmann, 1941)—Galápagos Islands at 55 m (Deichmann, 1941). 1

Psolus diomedeae Ludwig, 1894—Northern Baja California, central and southern Gulf of California, islands near the coasts of Colombia and Ecuador, and Cocos and Galápagos Islands (Ludwig, 1894). 13 to 302 m. 1,2

FAMILY PHYLLOPHORIDAE

Pentamera chierchia (Ludwig, 1887)—Puerto Peñasco, Gulf of California to Manta, Ecuador, including the Revillagigedos and Galápagos Islands (Deichmann, 1941). Intertidal to 78 m. 1

Pentamera chiloensis (Ludwig, 1887)—Central Gulf of California islands, Galápagos Islands (Deichmann, 1941), and La Plata Island, Ecuador to Cape Horn. 14 to 108 m. Southern Argentina. 1

Thyone neofusus Deichmann, 1941—Galápagos Islands, 37 to 110 m (Deichmann, 1941). 1

FAMILY SCLERODACTYLIDAE

Euthyonidiella zacae (Deichmann, 1938)—Lower Gulf of California, Tangola-Tangola Bay, Mexico, and the Galápagos Islands (Deichmann, 1941). 18 to 183 m. 1

Pachythyone lugubris (Deichmann, 1939)—Cedros Island, Baja California and central Gulf of California to Petatlán Bay, Mexico and the Galápagos Islands (Deichmann, 1941). 11 to 73 m. 1

ORDER DACTYLOCHIROTIDA: FAMILY YPSILOTHURIIDAE

Ypsilothuria bitentaculata (Ludwig, 1893)—Off Cristobal Bay, Baja California to

Malpelo and Cocos Ridge and off the Galápagos Islands (Ludwig, 1894). 225 to 4082 m. 2

ORDER ASPIDOCHIROTIDA: FAMILY HOLOTHURIIDAE

Holothuria (Cystipus) inhabilis Selenka, 1867—Cabo Tepopa, Gulf of California to Bahía Santa Elena, Ecuador, including the Revillagigedos, Cocos, and Galápagos Islands (Ludwig, 1894). One to 203 m. Indo-West Pacific. 1

Holothuria (Cystipus) rigida (Selenka, 1867)—Upper Gulf of California and southern Baja California to Bahía Honda, Panama and the Galápagos Islands (Deichmann, 1958). Low tide to 18 m. Circumtropical. 1

Holothuria (Halodeima) kefersteini (Selenka, 1867)—Puerto Escondido, Gulf of California to Peru, and the Revillagigedos, Cocos, and Galápagos Islands (Deichmann, 1938). Shore to 27 m, extended to 145 m by museum records. Records of *Holothuria atra* Jaeger, 1833 from Clipperton and Galápagos Islands are probably either *H. kefersteini* or *H. theeli* (D. L. Pawson, pers. comm.). 1

Holothuria (Lessonothuria) pardalis Selenka, 1867—Santa Cruz Island, California (locality questionable, LACM-AHF), Gulf of California, Tenacatita Bay, Mexico, islands off Panama and Colombia, and Cocos and Galápagos Islands (Deichmann, 1938). Intertidal to 127 m. Indo-West Pacific. This species may be conspecific with *Holothuria insignis* Ludwig, 1875, also of the Indo-Pacific (Rowe, 1969). 1

Holothuria (Mertensiothuria) leucospilota (Brandt, 1835)—Islas Tres Marías, Revillagigedos, Clipperton, Cocos, and Galápagos (Ludwig, 1894). Intertidal. Indo-Pacific. 1,2

Holothuria (Platyperona) difficilis Semper, 1868—Isla Espíritu Santo, Gulf of California to Islas Secas, Panama, and the Revillagigedos, Clipperton, Cocos, and Galápagos Islands (Deichmann, 1958). Intertidal to at least 44 m. Circumtropical. 1

Holothuria (Selenkothuria) lubrica Selenka, 1867—Puerto Peñasco, Gulf of California and Point Eugenia, Baja California, to Bahía Santa Elena, Ecuador, the Revillagigedos and Galápagos Islands (Deichmann, 1958). Intertidal to 55 m. 1

Holothuria (Selenkothuria) portovallartensis Caso, 1954—Guaymas, Gulf of California, to Zorritos, Peru, and the Revillagigedos and Galápagos Islands (Deichmann, 1958). Intertidal to 59 m. 1

Holothuria (Selenkothuria) theeli (Deichmann, 1938)—Central Gulf of California, Revillagigedos (record uncertain), Ecuador, and Cocos and Galápagos Islands (H. L. Clark, 1920b). Intertidal to 55 m, extended to 127 m by museum records. East Indies. Some records of *Holothuria atra* Jaeger, 1833 from Clipperton and Galápagos Islands may belong here (D. L. Pawson, pers. comm.). 1

Holothuria (Semperothuria) imitans Ludwig, 1875—San Carlos Bay, Gulf of California, to Port Utria, Colombia, and the Revillagigedos, Cocos, and Galápagos Islands (Deichmann, 1958). Intertidal and shallow. Indo-Pacific. 1

Holothuria (Semperothuria) languens Selenka, 1867—La Paz, Gulf of California, to Panama City, Panama and the Galápagos Islands (Deichmann, 1938). Intertidal to 91 m. Caribbean.

Holothuria (Thymiosycia) arenicola Semper, 1868—Cabo Colnett, Baja California and Puerto Peñasco, Gulf of California to Bahía Santa Elena, Ecuador, and the Revillagigedos, Cocos and Galápagos Islands (H. L. Clark, 1902). Intertidal to nine m. Circumtropical. 1

Holothuria (Thymiosycia) impatiens (Forskaal, 1775)—From Puerto Peñasco, Gulf of California, and Rosario Bay, Baja California to Bahía Santa Elena, Ecuador, and the Revillagigedos, Cocos, and Galápagos Islands (H. L. Clark, 1902). Intertidal to 11 m, extended to 36 m by museum holdings. Circumtropical. 1,2

Labidodemas americanum Deichmann, 1938—Southern Gulf of California, Gulf of Nicoya, Costa Rica, and the Cocos and Galápagos Islands (Deichmann, 1958). Intertidal. 1

FAMILY STICHOPODIDAE

Isostichopus fuscus (Ludwig, 1875)—Bahía Sebastian Vizcaino, Baja California, Puerto Peñasco, Gulf of California to the Gulf of Guayaquil, Ecuador, and the Revillagigedos, Cocos, and Galápagos Islands (Deichmann, 1938). Shore to 39 m. 1

FAMILY SYNALLACTIDAE

Capheira sulcata Ludwig, 1894—Off southern Colombia, northern Peru, and southwest of the Galápagos Islands (Ludwig, 1894). 2877 to 4334 m.

Mesothuria multipes (Ludwig, 1893)—Gulf of Panama, off Point Aguja, Peru, and the Cocos and Galápagos Ridges (Ludwig, 1894). 725 to 4064 m.

Pseudostichopus mollis Theel, 1886—Gulf of California (SIO), northern Channel Islands, California, to Chile, Malpelo, Cocos, Carnegie, and Galápagos Ridges (Ludwig, 1894). 100 to 5203 m. Antarctic, southwest Atlantic, Indian Ocean. 2

Synallactes aenigma Ludwig, 1894—Southwest of Malpelo, west of Peru, and east of the Galápagos Islands (Ludwig, 1894). 2404 to 4334 m.

Synallactes alexandri Ludwig, 1894—Gulf of Panama and the Galápagos Islands (Ludwig, 1894). 585 to 1018 m.

Synallactes triplax H. L. Clark, 1920—Galápagos Islands at 545 m (H. L. Clark, 1920b).

ORDER ELASIPODIDA: FAMILY DEIMATIDAE

Deima validum pacificum Ludwig, 1894—Cocos and Galápagos Ridges from 1618 to 2487 m (Ludwig, 1894).

FAMILY LAETMOGONIDAE

Laetmogone theeli Ludwig, 1894—West of the Galápagos Islands at 2417 m (Ludwig, 1894).

FAMILY PSYCHROPOTIDAE

Benthodytes incerta Ludwig, 1894—Off Puerto Angel, Mexico, the Carnegie Ridge, and west of the Galápagos (Ludwig, 1894). 2417 to 4087 m. Japan.

Benthodytes sanguinolenta Theel, 1882—Channel Islands, California to southern Peru, and Malpelo, Cocos and Galápagos Ridges (Ludwig, 1894). 1975 to 7250 m. Indo-West Pacific. Some of these records may be *Benthodytes typica* Theel, 1882 found on the Carnegie Ridge and in the Gulf of Panama, and in the North Atlantic and Indian Oceans (B. Hansen, pers. comm.). 2

Psychropotes verrucosa Ludwig, 1894—Off southern Colombia and west of the Galápagos (Ludwig, 1894). 2417 to 7250 m. Southeast Pacific and Indian Ocean.

FAMILY ELPIDIIDAE

Peniagone gracilis (Ludwig, 1894)—North of the Galápagos at 2475 m (Ludwig, 1894).

Peniagone intermedia Ludwig, 1894—Off Mexico, Carnegie Ridge and west of the Galápagos (Ludwig, 1894). 2418 to 5203 m.

Peniagone vitrea Theel, 1882—Southwest of Malpelo, Cocos Ridge, northern Peru to southern Chile, and the Galápagos (Ludwig, 1894). 1160 to 4507 m. Central Pacific.

FAMILY PELAGOTHURIIDAE

Pelagothuria natatrix Ludwig, 1894—Gulf of Panama, off Malpelo, Cocos and Galápagos Islands (Ludwig, 1894), northern Peru, and the East Pacific Rise. Pelagic, surface to 540 m (tow), and in dredges to 4504 m. Central Pacific.

3. Discussion

3.1. Biogeographical Affinities

Fifty-four percent of the non-endemic shallow-water Galápagos species are confined to the central eastern Pacific (CEP), and 17% are endemic (Table 1). Thus, 70% of all species are confined to the warm-water eastern Pacific. Cold-water East Pacific affiliations are predominantly northern. Of those species ranging outside the East Pacific, trans-Pacific distributions (Indo-Pacific, Central and West Pacific including Japan) are the most common. Ophiuroids and holothurians (genus *Holothuria*) are the most widespread groups. Sixty-four percent of all deeper-water species are confined to the CEP, and 18% are Galápagos endemics (Table 2). Crinoids show the greatest restriction. As with the shallow-water species, additional East Pacific distributions lie to the north, and ophiuroids and holothurians are the most wide-ranging groups outside the East Pacific. Species on other offshore islands of the CEP overlap broadly with the Galápagos Islands (Table 3). The Clipperton Island echinoderm fauna is very poorly known, and few conclusions can be drawn here. However, there are no cases of trans-Pacific species occurring on CEP offshore islands except for four species on Clipperton. This is a very high percentage of the known Clipperton echinoderms and is likely due to it being the only coral atoll in the East Pacific.

In a paper on the CEP shelf (< 200 m) echinoderms, Maluf (1988b) presents a dendrogram based on a cluster analysis using Simpson's index to assess the similarity of species. For all classes combined the dendrogram shows the greatest similarity between the Galápagos and Revillagigedos Islands, and an equal similarity to the East Pacific tropics and the subtropical Gulf of California (Fig. 2).

Table 1. Distribution of Shallow-Water (< 200 m) Galápagos Echinoderms

	Number of species					
Distribution	All	Crinoidea	Asteroidea	Ophiuroidea	Echinoidea	Holothuroidea
East Pacific						
Central	68	–	14	24	21	9
Northern	10	1	4	3	2	–
Southern	1	–	–	–	1	–
Galápagos	16	–	2	7	5	2
Trans-Pacific	14	–	3	4	2	5
Indian	1	–	–	–	–	1
Atlantic	5	–	–	3	–	2
Cosmopolitan	11	–	–	6	–	5
Totals	126	1	23	47	31	24

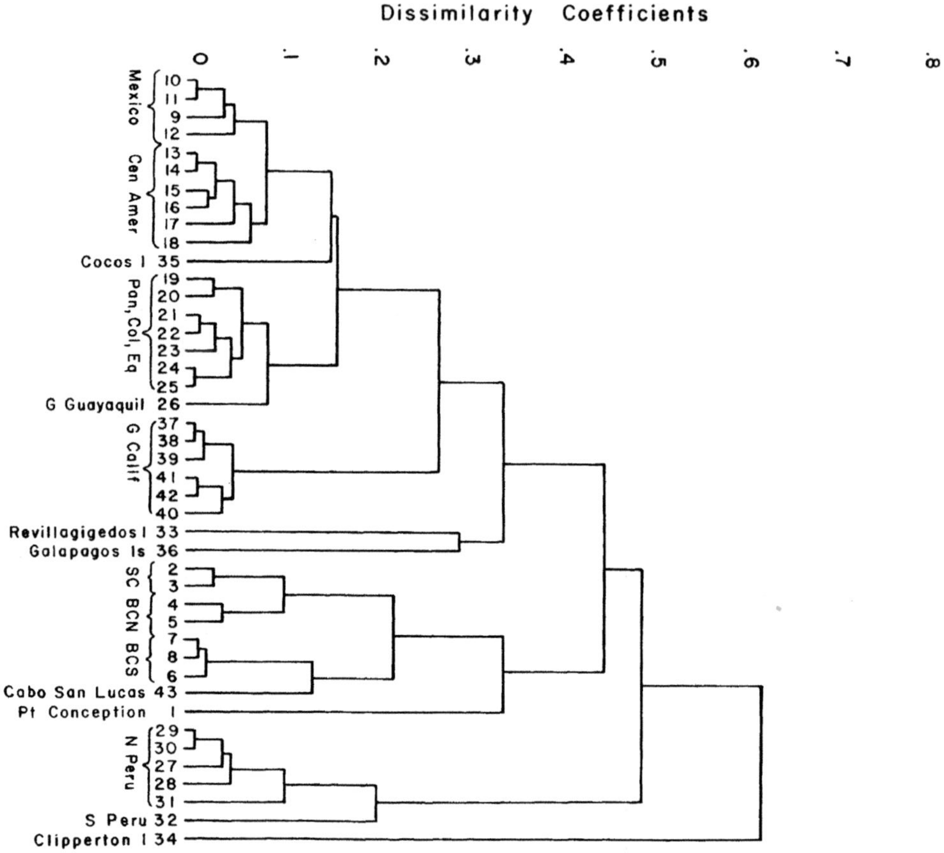

Figure 2. Dendrogram depicting clustering of CEP shelf echinoderms (N = 375). Dissimilarity coefficients are obtained by subtracting the calculated similarity coefficients from 1. For detailed explanation of sample sites and methods see Maluf, 1988b. (From Maluf, 1988b.)

Table 2. Distribution of Deep-Water (> 200 m) Galápagos Echinoderms

	Number of species					
Distribution	All	Crinoidea	Asteroidea	Ophiuroidea	Echinoidea	Holothu-roidea
East Pacific						
Central	47	4	12	16	7	8
Northern	9	1	4	3	1	–
Widespread	1	–	1	–	–	–
Galápagos	19	–	3	12	1	3
Trans-Pacific	10	–	1	4	1	4
Indian	2	–	1	–	–	1
Atlantic	6	–	1	4	1	–
Cosmopolitan	10	–	2	7	–	1
Totals	104	5	25	46	11	17

Table 3. Species Richness and Endemism of Central Eastern Pacific Offshore Islands[a]

Island	No. of species	% Endemic
Revillagigedos	73 (67)	3
Clipperton	14 (50)	0
Cocos	102 (70)	9
Malpelo	50 (62)	8
Galápagos	198	17

[a]Percentage of species shared with the Galápagos Islands in parentheses.

Similar analyses for each individual class fail to show such close similarity between the Galápagos and Revillagigedos Islands (unpublished data). Instead, the Revillagigedos show the greatest similarity to the Gulf of California (echinoids and asteroids), the tropics of western America (holothurians), and southern and Baja California (ophiuroids). The Galápagos and Cocos Islands share a similar echinoid and asteroid fauna. The Galápagos are, however, distinct in every class and show the least relationship to warm-temperate Peru, southern California, and Baja California.

3.2. Endemism

The Galápagos have a high percentage of endemics relative to other offshore islands of the central eastern Pacific (Table 3), but they are an archipelago and have increased environmental and spatial heterogeneity due to their size, number, and geographic location. Thirty-three species are restricted to the Galápagos, and their distribution within various depth ranges is given in Table 4. Eight endemic species occur at less than 20 m, and an additional seven species have a depth limit at less than 200 m (Table 5). Based on known nearest relations of the endemic shallow-water echinoids and holothurians, evolutionary relationships of these classes are primarily Atlantic whereas those of the asteroids and ophiuroids are East Pacific.

The deeper-water fauna (shallow limit greater than 200 m) has high endemism, with 19 species reported only from this area (Table 6). Thirty-seven species are confined to the Cocos, Galápagos, Carnegie, and Malpelo slope areas, and an additional 13 species are found only on the Cocos and Malpelo slopes. This general

Table 4. Degree of Endemism of the Galápagos Echinoderm Fauna[a]

Class	Number of species			
	< 20 m	< 200 m	> 200 m	Total
Crinoidea	–	1	5	5
Asteroidea	17 (2)	23 (5)	25 (7)	44
Ophiuroidea	28 (5)	47 (10)	46 (16)	74
Echinoidea	25 (8)	31 (14)	11 (3)	37
Holothuroidea	21	24 (5)	17 (8)	38
All	91 (4)	126 (8)	104 (10)	198

[a]Percent Endemic in Parentheses.

Table 5. Shallow-Water (< 200 m) Galápagos Endemic Species and Their Recorded Depths

Henricia nana	97–110
Heliaster cumingii	intertidal
Ophiomyces multispinus	134–717
Amphiura verticillata	46–64
Ophiophragmus disacanthus	4–37
Ophiocomella schmitti	intertidal
Ophionereis albomaculata	intertidal
Amphiophiura irregularis	106–238
Ophioplocus hancocki	intertidal
Hesperocidaris houstoniana	91
Clypeaster elongatus	6
Encope galapagensis	4–134
Paraster rotundatus	20
Nacospatangus oblonga	35–70
Lissothuria veleronis	55
Thyone neofusus	37–110

Table 6. Deep Water (> 200 m) Galápagos Endemic Species and Their Reported Depths

Psilaster sladeni	1618
Distolasterias robusta	704
Freyellaster scalaris	1480
Astrodendrum galapagense	718
Ophiacantha cyrena	717
Ophiacantha similis	717
Ophiolebes mortenseni	717
Ophiothamnus multispinus	134–717
Ophiothamnus laevis	1008
"Amphioplus" papillatus	704
Amphiophiura obtecta	2196–2487
Amphiophiura paucisquama	549–717
Ophiomusium diomedeae	704
Ophiophyllum marginatum	717
Ophiozonella contigua	2418–2487
Histocidaris cobosi	615–704
Synallactes triplax	545
Laetmogone theeli	2417
Peniagone gracilis	2475

geographical restriction of deep-water species is partly artifactual due to a paucity of samples at these depths over hard substrata. However, one half of all the deep-water Galápagos species have their distributions confined to the area when the slopes off Panama are included. The fractures and ridges in this region of the eastern Pacific provide an exceptionally large slope environment between 2°S and 10°N (Fig. 1). Geographical restriction of species in this area seems likely. Known evolutionary relationships of the deep-water endemics are mostly northern Pacific, but the holothurians also have a southern relationship and information is lacking for many ophiuroids. More intensive sampling and some much-needed systematics work will probably reduce the number of apparent endemics in both shallow and deep-water groups.

ACKNOWLEDGMENTS. I would like to express my thanks to Ailsa Clark (formerly at the British Museum of Natural History), Dr. David Pawson (National Museum of Natural History, Smithsonian Institution), and Dr. Bent Hansen (Zoologisk Museum, Copenhagen) for their assistance with systematics questions, and to Dr. Gordon Hendler (Natural History Museum of Los Angeles County) for both systematics assistance and initial review of this manuscript.

References

Agassiz, A., 1890, Notice of *Calamocrinus diomedeae*, a new stalked crinoid from the Galápagos, dredged by the U.S. Fish Commission steamer "Albatross," Lieut. Commander Z. L. Tanner, U.S.N., commanding, *Bull. Mus. Comp. Zool. Harv.* **20**(6):165–167.

Agassiz, A., 1892, Reports on an exploration off the west coasts of Mexico, Central and South America, and off the Galápagos Islands, in charge of Alexander Agassiz, by the U.S. Fish Commission steamer "Albatross," during 1891, Lieut. Commander Z. L. Tanner, U.S.N., commanding, I. *Calamocrinus diomedeae*, a new stalked crinoid, with notes on the apical system and the homologies of echinoderms, *Mem. Mus. Comp. Zool. Harv.* **17**(2):1–95.

Agassiz, A., 1898, Reports on the dredging operations off the west coast of Central America, Mexico and the Gulf of California by the U.S.S. "Albatross" during 1891. Preliminary report on the Echini, *Bull. Mus. Comp. Zool. Harv.* **32**(5):71–86.

Agassiz, A., 1904, The Panamic Deep Sea Echini. Report XXXII on an exploration off the west coasts of Mexico, Central and South America, and off the Galápagos Islands, *Am. Mus. Comp. Zool.* **31**:1–243.

Birkeland, C., Meyer, D. L., Stames, J. P., and Buford, C. L., 1975, Subtidal communities of Malpelo Island, in: The Biological Investigation of Malpelo Island, Colombia (J. B. Graham, ed.), *Smithson. Contr. Zool.* **176**:55–68.

Boone, L., 1933, Scientific results of cruises yachts "Eagle" and "Ara," 1921–1928. Coelenterata, Echinodermata, and Mollusca, *Bull. Vanderbilt Mar. Mus.* **4**:1–217.

Briggs, J. C., 1974, *Marine Zoogeography*, McGraw-Hill, New York.

Brusca, R. C., 1980, *Common Intertidal Invertebrates of the Gulf of California*, 2nd Ed., University of Arizona Press, Tucson.

Caso, M. E., 1980, Los Equinoideos del Pacífico de México. Parte 3. Órden Clypeasteroida, *An. Centro Cienc. Mar Limnol.*, Publ. Esp. **4**:1–252.

Caso, M. E., 1983, Los Equinoideos del Pacífico de México. Parte 4. Órdenes Cassiduloida y Spatangoida, *An. Centro Cienc. Mar Limnol.*, Publ. Esp. **6**:1–200.

Chesher, R. H., 1972, The status of knowledge of Panamanian echinoids, 1971, with comments on other echinoderms, *Bull. Biol. Soc. Wash.* **2**:139–158.

Clark, A. H., 1908, Notice of some crinoids in the collection of the Museum of Comparative Zoology, *Bull. Mus. Comp. Zool. Harv.* **51**(8):233–248.

Clark, A. H., 1916, One new starfish and five new brittle stars from the Galápagos Islands, *Ann. Mag. Nat. Hist.* **(8)17**:115–122.

Clark, A. H., 1920, A new name for *Heliaster multiradiatus* (Gray), *Proc. Biol. Soc. Wash.* **33**:183.
Clark, A. H., 1939, Echinoderms (other than holothurians) collected on the Presidential Cruise of 1938, *Smithson. Misc. Collect.* **98**(11):1–18.
Clark, A. H., 1946, Echinoderms from the Pearl Islands, Bay of Panama, with a revision of the Pacific species of the genus *Encope*, *Smithson. Misc. Collect.* **106**(5):1–11.
Clark, A. M., 1970, Notes on the family Amphiuridae (Ophiuroidea), *Bull. Br. Mus. Nat. Hist. (Zool.)* **19**:1–81.
Clark, H. L., 1902, Papers from the Hopkins Stanford Galápagos expedition, 1898–1899. XII. Echinodermata, *Proc. Wash. Acad. Sci.* **4**:521–531.
Clark, H. L., 1911, North Pacific Ophiurans in the collection of the United States National Museum, *Bull. U.S. Natl. Mus.* **75**:1–302.
Clark, H. L., 1914, Hawaiian and other Pacific Echini. The Clypeastridae, Arachnoididae, Laganidae, Fibulariidae and Scutellidae, *Mem. Mus. Comp. Zool. Harv.* **46**(1):1–78.
Clark, H. L., 1915, Catalogue of recent Ophiurans based on the collection of the Museum of Comparative Zoology, *Mem. Mus. Comp. Zool. Harv.* **25**(4):165–376.
Clark, H. L., 1917a, Ophiuroidea. Report XVIII and XXX on the scientific results of the expedition of the "Albatross" to the tropical Pacific, 1899–1900 and 1904–1905, *Bull. Mus. Comp. Zool. Harv.* **61**(12):429–453.
Clark, H. L., 1917b, Hawaiian and other Pacific Echini. The Echinoneidae, Nucleolitidae, Urechinidae, Echinocorythidae, Calymnidae, Pourtalesiidae, Palaeostomatidae, Aeropsidae, Palaeopneustidae, Hemiasteridae and Spatangidae, *Mem. Mus. Comp. Zool. Harv.* **46**(2):85–283.
Clark, H. L., 1920a, Asteroidea. Report XXXII on the scientific results of the expedition of the "Albatross" to the tropical Pacific, 1904–1905, *Mem. Mus. Comp. Zool. Harv.* **39**(3):69–114.
Clark, H. L., 1920b, Holothuroidea. Report XXXIII on the scientific results of the expedition of the "Albatross" to the tropical Pacific, 1899–1900 and 1904–1905, *Mem. Mus. Comp. Zool. Harv.* **39**(4):121–154.
Clark, H. L., 1940, Notes on Echinoderms from the West Coast of Central America. Eastern Pacific expeditions of the New York Zoological Society No. 21, *Zoologica, N.Y.* **25**:331–352.
Clark, H. L., 1948, A report on the Echini of the warmer Eastern Pacific, based on the collections of the "Velero" III, *Allan Hancock Pacif. Exped.* **8**(5):225–351.
Deichmann, E., 1938, Holothurians from the western coasts of Lower California and Central America, and from the Galápagos Islands; Eastern Pacific Expeditions of the New York Zoological Society, *Zoologica, N.Y* **23**(18):361–387.
Deichmann, E., 1941, The Holothuroidea collected by the "Velero" III during the years 1932 to 1938. Part 1. Dendrochirota, *Allan Hancock Pacif. Exped.* **8**(3):61–153.
Deichmann, E., 1958, The Holothuroidea collected by the "Velero" III and IV during the years 1932 to 1954. Part 2. Aspidochirota, *Allan Hancock Pacif. Exped.* **11**(2):249–331.
Downey, M. E., 1976, Asteroidea from Malpelo Island with a description of a new species of the genus *Tamaria*, *Smithson. Contrib. Zool.* **176**:86–90.
Downey, M. E., and Wellington, G. M., 1978, Rediscovery of the giant seastar *Luidia superba* A. H. Clark in the Galápagos Islands, *Bull. Mar. Sci.* **28**(2):375–376.
Fisher, W. K., 1911, Asteroidea of the North Pacific and adjacent waters, Part 1. Phanerozonia and Spinulosa, *Bull. U.S. Natl. Mus.* **76**:1–419.
Fisher, W. K., 1928, Sea stars from the "Arcturus" Oceanographic Expedition, *Zoologica, N.Y.* **8**(9):487–493.
Fisher, W. K., 1930, Asteroidea of the North Pacific and adjacent waters. Part 3. Forcipulata, *Bull. U.S. Natl. Mus.* **76**:1–356.
Grant, U. S., IV, and Hertlein, L. G., 1938, The West American Cenozoic Echinoidea, *Univ. Calif. Publ. Math. Phys. Sci.* **2**:1–225.
Hartlaub, C., 1895, Reports on the dredging operations off the west coast of Central America to the Galápagos, to the west coast of Mexico, and in the Gulf of California, in charge of Alexander Agassiz, carried on by the U.S. Fish Commission steamer "Albatross," during 1891, Lieut. Commander Z. L. Tanner, U.S.N., commanding. 13. Die Comatuliden, *Bull. Mus. Comp. Zool. Harv.* **27**(4):129–152.
Hill, A., 1940, A new genus of brittle stars, *Amphicontus*, *Rep. Allan Hancock Pacif. Exped.* (1932–38) **8**(1):1–4.

Los Angeles County Museum, Natural History—Allan Hancock Foundation catalog of the Echinoderm Collections, 1984, Unpublished card file of cataloged specimens, identified primarily by F. Ziesenhenne and E. Deichmann.

Ludwig, H., 1894, Reports on an exploration off the west coasts of Mexico, Central and South America, and off the Galápagos Islands, in charge of Alexander Agassiz, by the U.S. Fish Commission steamer "Albatross," during 1891, Lieut. Commander Z. L. Tanner, U.S.N., commanding. XII. The Holothurioidea, *Mem. Mus. Comp. Zool. Harv.* 17(3):1–183.

Ludwig, H., 1905, Asteroidea. Reports on an exploration off the west coasts of Mexico, Central and South America, and off the Galápagos Islands during 1891 by the "Albatross," A. Agassiz XXXV; Reports on the scientific results of the expedition to the Tropical Pacific in charge of A. Agassiz aboard the "Albatross," from Aug. 1899 to Mar. 1900. VII, *Mem. Mus. Comp. Zool. Harv.* **3**:21–292.

Luke, S. R., 1982, Catalog of the Benthic Invertebrate Collections, Echinodermata, SIO Reference Series No. 82-5, (66 pp.), University of California, Berkeley, California.

Lütken, C. F., and Mortensen, T., 1899, Reports on an exploration off the west coasts of Mexico, Central and South America, and off the Galápagos Islands, in charge of A. Agassiz, during 1891 on the "Albatross." XXV. The Ophiuridae, *Mem. Mus. Comp. Zool. Harv.* **23**(2):93–208.

Madsen, F. J., 1956, Reports of the Lund University Chile Expedition 1948–49, 24. Asteroidea, with a survey of the Asteroidea of the Chilean Shelf, *Acta Univ. lund Andra. Afd. N.S.* **52**(2):1–53.

Maluf, L. Y., 1988a, Composition and distribution of the Central Eastern Pacific echinoderms, Technical Report 2, pp. 1–242, Nat. Hist. Mus. L.A. County, Los Angeles, California.

Maluf, L. Y., 1988b, Biogeography of the central eastern Pacific shelf echinoderms, in: *Echinoderm Biology* (R. D. Burke, P. V. Mladenov, P. Lambert, and R. L. Parsley, eds.), pp 389–398, A. A. Balkema, Rotterdam/Brookfield.

Mortensen, T., 1928, *A. Monograph of the Echinoidea. I. Cidaroidea*, C. A. Reitzel, Copenhagen.

Mortensen, T., 1950, New Echinoidea (Spatangoidea), (Preliminary Notice), *Vidensk. Meddr dansk Naturh. Foren.* **112**:157–163.

Pawson, D. L., and Miller, J. E., 1983, Systematics and ecology of the sea-urchin genus *Centrostephanus* (Echinodermata: Echinoidea) from the Atlantic and eastern Pacific Oceans, *Smithson. Contrib. Mar. Sci.* **20**:1–15.

Pope, E. C., and Rowe, F. W. E., 1977, A new genus and two new species in the family Mithrodiidae (Echinodermata: Asteroidea) with comments on the status of species of *Mithrodia* Gray, 1840, *Aust. Zool.* **19**(2):201–216.

Rowe, F. W. E., 1969, A review of the family Holothuriidae (Holothuroidea: Aspidochirotida), *Bull. Br. Mus. Nat. Hist. (Zool.)* **18**(4):119–170.

Rowe, F. W. E., 1977, A new family of Asteroidea (Echinodermata), with the description of five new species and one new subspecies of *Asterodiscides*, *Rec. Aust. Mus.* **31**(4–6):187–234.

Steinbeck, J., and Ricketts, E. F., 1941, *Sea of Cortez. A leisurely journal of travel and research*, Viking Press, New York.

Verrill, A. E., 1867, Notes on the echinoderms of Panama and west coast of America, with descriptions of new genera and species, *Trans. Conn. Acad. Arts Sci.* **1**(2):251–322.

Verrill, A. E., 1871, On the echinoderm fauna of the Gulf of California and Cape San Lucas, *Trans. Conn. Acad. Arts Sci.* **1**(9):593–596.

Ziesenhenne, F. C., 1935, A new brittle star from the Galápagos Islands, *Allan Hancock Pacif. Exped.* **2**(1):1–4.

Ziesenhenne, F. C., 1937, Echinoderms from the west coast of Lower California, the Gulf of California and Clarion Island, *Zoologica, N.Y.* **22**(15):209–239.

Ziesenhenne, F. C., 1940, New ophiurans of the Allan Hancock Pacific Expeditions, *Allan Hancock Pacif. Exped.* **8**(2):9–42.

Ziesenhenne, F. C., 1942, New Eastern Pacific sea stars, *Allan Hancock Pacif. Exped.* **8**(4):197–224.

VIII

Bryozoa

Chapter 17

The Bryozoa of the Galápagos

WILLIAM C. BANTA

1. Introduction

One hundred eighty-four species of bryozoans are present in the Galápagos Islands (Banta and Redden, 1990), a number comparable to that of the most diverse bryozoan faunas known. This number is probably an underestimate, especially for ctenostomes. Variations in species lists among the Islands are small, so at least for the purposes of this report, the Galápagos can be treated as a single zoogeographical unit. Although the relative numbers of anascan and ascophoran cheilostome is near the average for the rest of the world, the proportion of stenolaemates (18%) is about twice as high as than in the rest of the Panamic. Comparable proportions of stenolaemates are found in the Magellanic and Chilean waters off South America, but otherwise there is zoogeographic affinity between these areas and the Galápagos. Relatively low water temperatures and elevated productivities associated

WILLIAM C. BANTA • Department of Biology, The American University, Washington, D.C. 20016.

with upwelling do not fully explain the high proportion of stenolaemates in Galápagos waters.

At the family level, few absences are conspicuous, except for the Steginoporellidae, characteristic of coral communities, rare in the Galápagos.

Zoogeographically, the Galápagos Islands belong to the North Hemisphere, squarely in the Panamic Province, with 46% of Galápagos species shared with the rest of the Panamic. The distribution of Panamic bryozoans seems to divide the Panamic into two subprovinces near 10°N. Within the Panamic, strong affinities also are seen among the Galápagos and Cocos Island and the Gulf of California. More distant affinities are seen with the Gulf of Mexico and Indo-Pacific. The fauna seems unrelated to that of the Chilean and Magellanic Provinces of South America. The few shared species are mostly circumglobal or cosmopolitan in distribution.

Thirty-four of 184 species (18%) of Galápagos bryozoans are endemic to the Islands. This proportion is comparable to that of other major groups of shallow-water animals. There is one endemic genus among 53 in the islands, about 2%. Proportionally, there are more endemic stenolaemates (14 of 33 species, 42%) than endemics in anascans or ascophorans. This difference does not seem to be a simple case of more lumping in the Stenolaemata than among cheilostomes; on the contrary, lumping probably has been more severe among stenolaemates.

Affinities with neighboring zoogeographical provinces are stronger for cheilostomes than for stenolaemate Bryozoa. This, taken together with the higher rate of endemism for stenolaemates, suggests that the dispersive abilities of stenolaemates are inferior to those of cheilostomes. The difference cannot be accounted for by the presence of planktotrophic larvae among some cheilostomes, because Panamic species with planktotrophic larvae are no better represented in the Islands than are those with lecithotrophic larvae.

In general, the Galápagos bryozoan fauna seems to be proportionately overrepresented by species which range into latitudes higher than about 30°. There also are proportionally more species with geographical ranges wider than those indigenous to waters nearer the Galápagos.

The bryozoan distribution data taken as a whole conveys a picture of the Galápagos as a biological isolate belonging securely in the Panamic Province, but represented disproportionally by colder water species of the Northern Hemisphere with weedier, r-selected tendencies. There is evidence of difficulty in arriving at the Islands and of genetic isolation and local adaptation, especially among stenolaemates. The rarity of some habitats, notably estuaries and coral reefs, has contributed to a unique species composition.

The stenolaemates of the Galápagos in particular are disproportionally abundant for the Panamic. There is evidence that the Galápagos stenolaemate fauna may be more isolated and/or more rapidly evolving than the cheilostome fauna.

2. General

Bryozoans are aquatic, suspension-feeding lophophorate animals with nearly microscopic zoids which multiply asexually to produce macroscopic colonies, sometimes of considerable size. Although apomictic populations are known, outbreeding occurs.

The phylum Bryozoa can be divided into two subphyla: the exclusively freshwater Phylactolaemata and the largely marine Gymnolaemata. Most living and fossil Bryozoa belong to the Gymnolaemata, subdivided into two classes, the Stenolaemata and the Eurystomata. Eurystomes include the Ctenostomata, which form non-calcareous colonies, and the Cheilostomata, whose colonies are impregnated with varying amounts of calcium carbonate. The cheilostomes are further subdivided into two orders, the Anasca, whose delicate frontal surfaces remain uncalcified, and the Ascophora, with calcified frontal walls and generally more heavily calcified colonies. Stenolaemates have tubular zoids which generally produce strongly calcified colonies.

3. Collections

There are 5 published works describing collections from the Galápagos Islands. Hastings (1930) had 37 taxa from 11 shallow-water sites collected by C. Crossland (1927). Canu and Bassler (1930) studied 50 species from four stations collected in 1888 and 1891 from the *Albatross* collections of the U.S. Fisheries Commission. One of these stations was from relatively deep water (684 fathoms) Osburn (1950, 1952, 1953) and Osburn and Soule (1953) had extensive collections, from at least 25 stations, some of great depth, made in the 1930s by the *Velero III* expeditions of the Allan Hancock Foundation (see Fraser 1943). In all, they recorded 177 species from the Galápagos. Soule (1963b) adds 3 additional records from the Hancock Galápagos collections in his monograph on Bryozoa of the Gulf of California. Banta and Redden (1990) reported seven new records from a single site in the Galápagos. They also listed all known species for the islands and details of their distribution and taxonomy. Their data comprise the basic data for this chapter.

For purposes of this chapter, zoogeographical provinces are designated as (1) cosmopolitan: recorded worldwide from the equator into Arctic and Antarctic waters; (2) circumglobal: worldwide, but absent from high latitudes, above about 60°N and S; (3) Pan American: recorded in both the eastern Pacific and the Caribbean or Gulf of Mexico; present or absent elsewhere in the Atlantic and the Mediterranean; (5) Pacific: known from both the eastern and western Pacific; (6) Eastern Pacific: found exclusively along the western coasts of the Americas, including areas outside the Panamic; (7) Panamic: coastal areas of the Americas from about the U.S. Mexican border to Guayaquil in Ecuador; (8) Endemic: not recorded outside the Galápagos.

4. Number of Species

In all, 184 species and subspecies of Bryozoa have been recorded from the Islands. This number is probably an underestimate because it is based on less than 35 sites among all the Islands. The proportion of new species described based on new collections is one indication of how thoroughly an area has been collected. Of the 37 species recorded by Hastings from the Galápagos, 3 were of new species, about 12%. Similar figures for Canu and Bassler (1930) are 21 of 50 species (42%); Osburn (1950, 1952, 1953): 38 of 143 species (27%). Banta and Redden (1990) were unable to identify 3 of 30 species (10%).

A species number of 184 appears to be a substantial number for a bryozoan fauna in so small a geographical area. Comparative data, however, are difficult to evaluate. Gordon (1964), for example, counts 672 species for New Zealand and conservatively estimates the fauna at well over 800. New Zealand has been collected with less intensity than the Galápagos, but is spread over 30 degrees of latitude; the Galápagos span only about 3°. Soule (1963, p. 28) counts 200 species in the Gulf of California, 10 degrees of latitude. Kluge (1964) records 375 species for the northern seas of the U.S.S.R., which encompasses some 20 degrees of latitude and half the globe in longitude. Winston (1982, p. 102) found 84 bryozoan species near Indian River, Florida. The area sampled was smaller, but she collected more thoroughly than the Galápagos has been collected. Examples could be multiplied but it seems evident that in terms of species number, the bryozoan fauna of the Galápagos is at least comparable with the most diverse faunas known.

5. Disharmony

Disharmony is the situation whereby, compared to the mainland biota, an isolated community possesses a greater (or lesser) proportion of species belonging to some given taxon. For example, mangroves are well represented in the Galápagos because of their water-borne seeds, but pines are absent from the indigenous Galápagos flora because their seeds are poor travelers.

I looked for evidence of disharmony among major taxa in the Galápagos bryozoan fauna, compared to a list of 19 studies of the recent bryozoan fauna world wide (Table 1). The list encompasses only studies where all bryozoans were treated, including stenolaemates, and excludes studies from incomparable habitats, such as exclusively estuarine environments, deep-water, and the like. I regarded cribrilinids as ascophorans.

5.1. Phylactolaemates

No phylactolaemates are known from the Galápagos, probably because of the scarcity of fresh water on the Islands. However, phylactolaemates can survive in temporary streams and ponds, partly because of their ability to form statoblasts. Statoblasts are asexually produced propagules similar to gemmules in sponges. Wading birds could be expected to carry the statoblasts, so phylactolaemate bryozoans may be present in the Galápagos, but unreported.

5.2. Ctenostomes

Ctenostomes are hardly represented at all in the Galápagos: only two species (*Amathia vidovici* and *Buskia seriata*). Ctenostomes are soft-bodied, noncalcified animals and are therefore less likely to be collected or preserved than cheilostomes or stenolaemates. Many ctenostomes are epizoic on other invertebrates and normally go to specialists in the host's taxon. If specimens are dry, ctenostomes are almost unrecognizable. Hastings (1930) reported on cheilostomes only; Canu and

Table 1. Proportions of Bryozoan Taxa from Various Parts of the World[a,b,c]

Author	Date	Locality	Steno	Ana	Ascoph	Cten	TOT	%Sten	%Ana
Galápagos	–	Galápagos	33	50	99	2	184	18	33
Canu and Bassler	1928	Gulf of Mexico	18	45	87	0	150	12	34
Canu and Bassler	1929	Philippines	58	96	217	0	371	16	31
Barbosa (1964)	1930	Brazil	15	90	130	34	269	6	41
Marcus	1940	Denmark	28	44	48	21	141	20	48
Osburn	1950	W. Pacific	86	175	239	32	532	16	42
Soule	1963a	Cocos Id.	1	8	11	0	20	5	42
Soule	1963b	Gulf Calif.	15	3	82	13	173	9	43
Soule and Soule	1964	Baja Calif.	3	6	28	0	37	8	18
Cook	1968	West Africa	18	69	87	9	183	10	44
Banta	1969	California	9	19	37	1	66	14	34
Schopf	1974	Caribbean	3	9	21	0	33	9	30
d'Hondt	1979	W. Africa	4	25	18	3	50	8	58
Hayward and Cook	1979	So. Africa	2	18	30	1	51	4	38
Moyano	1982	Magellanic	46	48	86	15	195	24	36
Winston	1982	Florida	3	31	35	15	84	4	47
Hayward and Cook	1979	So. Africa	16	13	76	0	105	15	15
Mayano	1983	Chile	46	70	132	19	267	17	35
Winston and Heimberg	1986	Indonesia	0	8	25	0	33	0	24
Sum			371	837	1389	163	2760		
Average			20.6	46.5	77.2	9.1	153.3	10.8	36.6
SD			23.1	41.9	64.3	11.1	132.6	6.1	10.5
Max			86	175	239	34	532	23.6	58
Min			0	6	11	0	20	0.0	15

[a]Abbreviations: Steno, Stenolaemata; Ana, Anasca; Ascoph, Ascophora; Cten, Ctenostomata; % Steno, percentage of Stenolaemata relative to all Bryozoa; % Ana, percentage of anascans relative to all Bryozoa.

[b]Data from Banta and Redden, 1990; not included in the statistics.

[c]A compilation of Brazilian records dating mostly from around 1930.

Bassler (1930) apparently worked largely on dried material, as I did. Only Osburn and Soule (1953) had opportunity to record ctenostomes. It is likely, therefore, that ctenostomes have been under-reported in the Galápagos, relative to other higher bryozoan taxa. On the other hand, there are some areas in the world where ctenostomes would be expected, but are rare or altogether lacking (Table 1).

5.3. Stenolaemates (cyclostomes)

Thirty-three of 184 (18%) of the Galápagos bryozoan fauna consists of stenolaemates (Cyclostomata) (Table 1). This percentage is close to twice the proportion seen elsewhere in the Panamic; stenolaemates comprise 5% of the bryozoan fauna of Cocos Island, 9% in the Gulf of California and 8% for Baja California. The proportion of stenolaemates increases southward from the Galápagos. In the Magellanic Province, stenolaemates make up over a quarter of the fauna (26%). Within the Antarctic and Subantarctic regions, stenolaemates comprise between 29 and 53% (Moyano 1982, p. 83). Although the proportion of stenolaemates seems to link the Galápagos with the southern hemisphere, there is little in the data to suggest strong zoogeographical affinities. For example, only two of 79 stenolaemate species listed by Moyano (1983) are shared between the Chilean region and the Galápagos.

Climatic peculiarities of the Islands offer a possible partial explanation of the apparent surplus of stenolaemates. Waters in the Galápagos are notably cooler and more productive than elsewhere in the Panamic, and therefore more comparable to the cool, productive waters of more southern latitudes (Colinvaux 1984). Climate alone, however, may not be enough to explain the high stenolaemate:eurystome ratios. Moyano (1982) points out that the proportion of stenolaemates is much higher around 60°W (South America) than it is elsewhere in the Antarctic and Subantarctic.

5.4. Cheilostomes

Among cheilostome species in this data set, the ratio of anascans to total cheilostomes averages about 37%, close to the 34% (53:98) for Galápagos cheilostomes. The ratio varies from the 11% for Cocos Island (Soule 1963a) to an excess of anascans over ascophorans in West Africa (58%; d'Hondt, 1979). The 34% ratio in the Galápagos is somewhat smaller than that for Eastern Pacific bryozoans, about 73% for the fauna as a whole (Osburn 1950, 1952) and 43% for the Gulf of California (Soule, 1963b, 1964), but larger than the ratios obtained in smaller collections in the subtropical Eastern Pacific (Banta 1969; Soule and Soule, 1964). There seems to be no evidence for disharmony in the anascan:ascophoran ratio among Galápagos bryozoans.

5.5. Widely Distributed Species

Thirty-nine of the 184 Galápagos species (21%) are of circumglobal or cosmopolitan species. A few Galápagos species (*Aetea anguina*, *Bugula neritina*, *Celle-*

porella hyalina) are common in harbors and on ship bottoms world wide, at least in warmer waters (Ryland 1971). These 'feral' species may have been introduced within the last few centuries by shipping and the genetic mixing may be a continuing process. If so, some of these fouling species may represent single world-wide gene pools. On the other hand, my collections from Academy Bay, Santa Cruz, are the only ones I know of made in a Galápagos fouling environment. The only fouling species there was *Bugula neritina*. If *Aetea anguina* and *Celleporella hyalina* are introduced harbor species, then it appears that they have escaped from the harbor environment into more typical Galápagos waters. An alternative explanation, at least as likely, is that these widely distributed species represent a number of morphologically similar, but distinct species.

The ctenostomes, *Bowerbankia gracilis* and *Victorella pavida*, expected in warm harbors, are absent from the Galápagos. Although they occur in full salt water, these species are more characteristic of brackish water, which is all but absent in the Islands. Furthermore, since these are ctenostomes and therefore soft-bodied, they may have been missed by collectors. The absence of *Cryptosula pallasiana* and species of *Electra* and *Conopeum* May reflect the lack of typical harbor environments in the Galápagos, or of collections from those environments. Like the fouling ctenostomes, these animals prefer brackish water, but are not limited to it (Winston, 1977).

5.6. Representation at the Family Level

Of 76 bryozoan families found in the Panamic Province, 19 are not represented in the Galápagos: Catenicellidae, Clavoporidae, Epistomiidae, Eurystomellidae, Exochonellidae, Faciminariidae, Flustridae, Frondiporidae, Hiantoporidae, Horneridae, Labioporellidae, Metrarabdotosidae, Nolellidae, Onchosoeciidae, Penetrantiidae, Pherusellidae, Reptadeonellidae, Steginoporellidae, Stomochaetosellidae, and Walkeriidae. None of these families are represented in the Panamic by more than two species. Five are ctenostome families, probably underrepresented because of a collecting artifact (see above). Most of the species in the list are represented by uncommon or rare species. A conspicuous exception is the family Steginoporellidae, which contains the very common circumtropical coral-loving genus, *Steginoporella*. Its absence from the bryozoan species list may reflect the rarity of coral reefs in the Galápagos.

6. Endemism

6.1. Endemic Species

Thirty-four of 184 bryozoan species (18%) are endemic to the Galápagos (Table 2). These numbers compare with 27% of the Galápagos shore fishes (Walker and Rosenblatt, 1961), 15% of the brachyuran crabs (Garth, 1946) and 23% for molluscs (Finet, 1989, p. 7; Shasky, 1989, p. 8; Kay, this volume, ca. 18%; Finet, this volume, 18.7% for shallow-water species). The proportion of endemism, therefore, appears to be roughly the same for bryozoans as it is for most other marine animals.

Table 2. Endemism in the Galápagos Bryozoan Fauna

Bryozoa	Number of species	% of fauna	Endemic	% Endemic
Stenolaemata	33	18	14	42
Anasca	50	27	5	10
Ascophora	99	54	15	15
Ctenostomata	2	1	0	0
Total	184	100	34	18

Among taxa, by far the highest degree of endemism is among stenolaemates, 42%, compared 10% for anascans and 15% for ascophorans. The cause of this difference among higher taxa does not seem to be a simple matter of splitters at work in one taxon, with lumpers in another. In my experience, stenolaemates are not more finely split than eurystomes (anascans and ascophorans); if anything, the opposite is true because of the relative lack of commonly used taxonomic character states among stenolaemates. If this experience is correct, it suggests that geographical isolation and/or differences, in evolutionary rates are more important to endemism than taxonomic splitting or lumping. If the relative degree of lumping or splitting were important, one would expect the degree of endemism to be more nearly the same among higher taxa.

A difference in ability of larvae to disperse over long distances might account for the differences in endemism between stenolaemates and eurystomes. Both groups each possess a number of different larval types (Nielsen, 1970). Planktotrophic larvae are unknown among stenolaemates, but present in a small number of eurystomes. Planktotrophic larvae have long larval lives spent feeding in the plankton, in contrast to the usual lecithotrophic or brooded larvae, which settle within minutes or hours soon after release. Presumably, planktotrophic larvae would have a better change of reaching the Islands than other larval types and one would expect a disproportionally large representation of species with planktotrophic larvae.

To test this idea one can look at the proportion of the Panamic fauna represented in the Galápagos which belong to the Electrinidae and Membraniporidae, the only families represented in the Panamic known to possess planktotrophic larvae. Of the 11 Panamic species, 6 are represented in the Galápagos—representing 55%. This percentage is close to that for the Galápagos Bryozoa as a whole (54%). Available evidence, therefore, provides no support for the idea that species with planktotrophic larvae are better represented in the Galápagos than those with other larval types.

6.2. Endemic Genera

Only one of the 53 genera represented in the islands, *Pachycleithonia* is endemic, about 2%. Even this instance is questionable, because some authors, including Osburn (1952, p. 472), synonomize *Pachycleithonia* with *Watersipora*.

7. Zoogeographic Affinities of the Galápagos

7.1. Indices of Similarity

Tables 3 and 5 employ indexes of zoogeographical affinities. For each region is given the number of species shared by the two regions (Tc), the total number of species in the Galápagos (T1), and the number in the region being compared (T2). K2 is the Kulczinsky-2 Index (Moyano, 1983):

$$K2 = \frac{Tc(T1 + T2)}{2(T1)(T2)} \times 100$$

Values of this index vary from 0 (no shared species) to 100 (all species shared.) Also given is the Similarity Coefficient (SC) (Greig-Smith, 1964):

$$SC = \frac{2(Tc)}{T1 + T2}$$

Values vary from 0 (no shared species) to 1 (all species shared). Although both K2 and SC give similar results, K2 is less sensitive than SC to large differences between faunal sizes in the two populations.

7.2. Distribution within the Galápagos

I compared Kulczinsky-2 indexes and Similarity Coefficients for the bryozoan fauna of the Islands for islands in which there are at least 25 species records. No clear patterns were evident. Even distant Wenman Island has a bryozoan fauna as closely related to the fauna of Isabela as have the neighboring islands of Santiago and Santa Cruz. Within the resolution of the sparse data available, therefore, the Galápagos can be treated as a single biogeographical unit.

7.3. Affinities with Other Zoogeographical Provinces

Table 3 indicates that the Galápagos bryozoan fauna belongs securely in the Panamic Province, with a K2 index of 63. Next is Cocos Island, about 400 miles northwest of the Galápagos on the Cocos Ridge. The Cocos Ridge is of interest because it may have served as a zoogeographical connection for some shallow-water invertebrates during periods of lowered sea level. If this is true, one would expect conspicuous similarity between the faunas of the Galápagos and Cocos Island. All 20 species known from Cocos Island are represented in the Galápagos, but the sample size is small and the K2 index is less than that for the Panamic as a whole, about 55.

Panamic bryozoans comprise 46% of the Galápagos bryozoan fauna (Table 4). This is a substantially greater proportion than the 25% for molluscs (2803/688; data from James, 1984, p. 81, and Finet 1989, p. 7). The percentage is lowest for the Ctenostomata (15%), but the Ctenostomata have been poorly collected in the Galápagos, (see Section 5.2). The Stenolaemata are slightly less well represented in

Table 3. Zoogeographic Affinities of Galápagos Bryozoa[a]

Locality or Province	TC	T1	T2	K2	SC	Reference
Panamic	148	184	323	63	0.58	Osburn, 1950–1953
Cocos Islands	20	184	20	55	0.20	Soule, 1963a
Gulf of California	85	184	166	50	0.49	Soule, 1963b
Western Atlantic	79	151	278	40	0.37	Schopf, 1973
Gulf of Mexico	35	184	136	22	0.22	Canu and Bassler, 1928
Indo-Pacific	51	184	437	20	0.16	Harmer, 1915–1957
Chile	24	184	267	11	0.11	Moyano, 1982
Magellanic area	10	184	195	5	0.05	Moyano, 1983

[a]Abbreviations: TC number of species in common; T1, T2, number of species recorded from tha locality; K2 Kulczinsky-2 coefficient (see text); SC Similarity coefficient (see text).

Table 4. The Galápagos Bryozoan Fauna Compared to the Bryozoan Fauna of the Panamic Province as a Whole[a,b]

	Steno	Ana	Ascoph	Ctenost	Total
Panamic	52	104	158	13	327
Galápagos	19	45	84	2	150
% Galápagan	37	43	53	15	46

[a]Abbreviations: Ana Anasca, Ascoph, Ascophora; Ctenost, Ctenostomata; Steno, Stenolaemata; Galap, Galápagos.
[b]Species endemic to the Galápagos are excluded.

Table 5. Zoogeographic Affinities of Galápagos Bryozoa Stenolaemata only[a]

Locality or Province	TC	T1	T2	K2	SC	Reference
Panamic	19	33	50	48	0.46	Osburn, 1950–1953
Gulf of California	5	33	15	24	0.21	Soule, 1963b
Gulf of Mexico	1	33	12	6	0.04	Canu and Bassler, 1928
Magellanic	1	33	45	3	0.03	Moyano, 1983
Cocos Island	1	33	20	4	0.04	Soule, 1963a

[a]Abbreviations: TC, Number of species in common; T1, T2, number of species recorded from tha locality; K2, Kulczinsky-2 coefficient (see text); SC, Similarity Coefficient (see text).

the Galápagos than anascans or ascophorans (37% compared to 43% and 53%, respectively).

Within the Panamic, the well-known similarity between the Galápagos and the shallow-water fauna of the Gulf of California is indicated by a K2 index of 50. I do not have good comparative data for the bryozoan fauna of the western Atlantic; the best available compilation is that of Schopf (1973), who listed cheilostomes only. The K2 index for this comparison is 40. However, omission of stenolaemates probably leads to an exaggerated estimate of similarity. When stenolaemates are included and the fauna of the Gulf of Mexico alone is compared, K2 drops to 22, not much better than the similarity to the distant Indo-Pacific, with a K2 of 20. The similarity of the Galápagos bryozoan fauna with that of the Indo-Pacific is probably also exaggerated because of the difficulty in estimating T2 (total number of bryozoan species present) for the Indo-Pacific. I made a conservative estimate of T2 based on perusal of the works of Canu and Bassler (1929), Harmer (1915, 1926, 1934, 1957), and others.

Perhaps the most striking figures are those for the Panamic and Magellanic Provinces of South America; the K2 similarity index is only 11 and 5, respectively. Moyano (1982) lists 195 species of bryozoans known from the Magellanic fauna (southern Chile, Argentina, the Juan Fernandez Islands, the Antarctic Peninsula and surrounding areas). Only 10 species are in common with the Galápagos fauna: *Lichenopora fimbriata*, Aetea anguina, *A. ligulata*, *Bugula neritina*, *Celleporella hyalina*, *Fenestrulina malusi*, *Hippopodinella adpressa*, *Hippothoa divaricata*, *Microporella ciliate*, and *Scruparia ambigua*. All 10 are circumglobal or cosmopolitan in distribution. Only one species (*L. fimbriata*) is a stenolaemate, despite the high proportion of stenolaemates in both the Galápagos (33 of 184 species, 18%) and the magellanic (46 of 180 species, 26%).

The bryozoan fauna of central and northern Chile is more similar to the Galápagos (Moyano, 1983), but still only 24 of the 184 Galápagos species (18%) are shared. Most of these shared species (15) have circumglobal or cosmopolitan distributions. There are five Pan American species, and four restricted to the eastern Pacific. Only two stenolaemates, *Lichenopora octoradiata* and *Diaperoecia major*, are shared between the Galápagos and northern Chile, even though a high proportion of the Chilean fauna (46 of 248 species, 19%) consists of stenolaemates.

Unfortunately, the bryozoan fauna of Peru and southern Ecuador still is too poorly known to permit meaningful comparison with the Galápagos.

7.4. Distribution of Galápagos Stenolaemates

Table 5 resembles Table 3, except that it contains only data for stenolaemate bryozoa. In each case proportionally fewer stenolaemate species are shared between areas than are bryozoans as a whole. For stenolaemates the K2 index among the sites varies from 48 to 3. In both tables the rank order is almost the same; that is, the Panamic stenolaemates display the most affinity with the Galápagos, and the same is true of Panamic bryozoans as a whole. Next in order is the Gulf of California, and so on. An exception is Cocos Island, which shares only one known stenolaemate with the Galápagos, but which shows considerable affinity among the other bryozoan species. On the other hand, the sample size for Cocos Island is very small (20 species total).

These data indicate that stenolaemate bryozoans are less well dispersed, at least among nearby zoogeographical areas than are bryozoans as a whole.

7.5. Ranges of Galápagos Species

Figure 1 is a frequency distribution of the maximum latitude of the range of species found in the Galápagos overlain by that for the bryozoans of the Panamic as a whole. Consider first the Panamic bryozoan species. There are peaks at 0–10

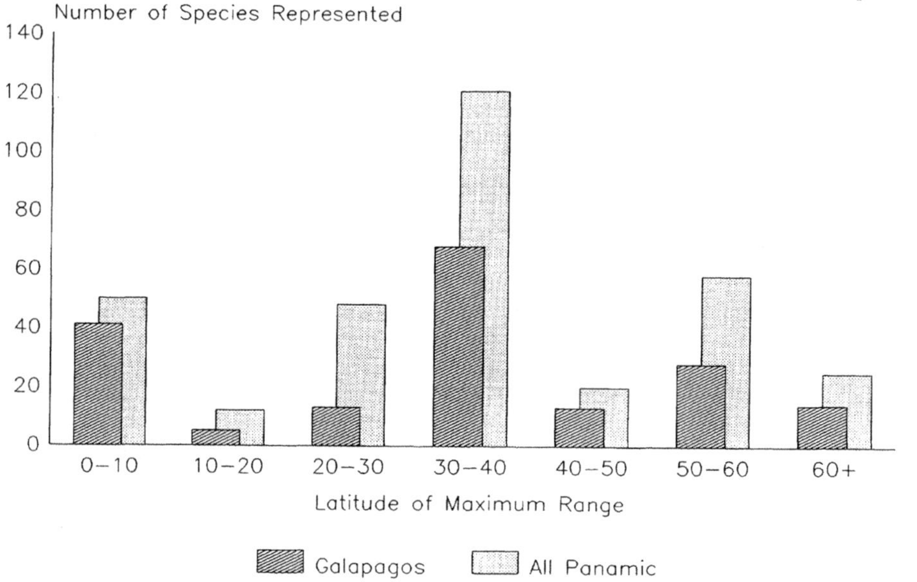

Figure 1. Histogram of numbers of bryozoan species arranged by latitude of maximum range. 'Galápagos': species found in the Galápagos; 'All Panamic': species found anywhere in the Panamic Province, including Galápagos species.

degrees and at 30–40 degrees; there are minima at 10–20 and 40–50 degrees. The peak at 0–10 days is due at least in part to the high degree of endemism in the Galápagos. Endemic bryozoans show a maximum latitude of distribution of 0 because the Islands all lie near the Equator. The drop in species number between 30–40 and 40–50 degrees corresponds to the approximate latitude of Cape Mendocino, California, where there is a major change in the biota (Hedgepeth, 1957). The break at about 10 degrees may represent a natural faunal subdivision of the Panamic Province. 8°N is the approximate northern border of the Gulf of Panama; between it and the Equator lie the Galápagos and Cocos Island.

The darker bars in Fig. 1 represent maximum latitude distribution ranges for species represented in the Galápagos. With two exceptions the proportion of Galápagos species relative to the rest of the Panamic Bryozoan fauna is near about 0.5 within each range. One exception is among species with latitudes of maximum distribution between 0 and 10 degrees; this is an effect of the 34 endemic species, all of whose latitude of maximum distribution is near 0. The second exception is the bar at 20–30 degrees; only about a third of the species with maximum latitudes of distribution between 20 and 30 degrees are found in the Galápagos.

This latter observation is paralleled by data represented in Fig. 2. This histogram represents the percentage of Panamic species represented in the Galápagos versus the geographic range category for that species. The first bar, for the 34 endemic species, is 100% because all species restricted to the Galápagos obviously occur there. The other categories represent progressively larger ranges. About 40% of exclusively Panamic and Eastern Pacific species are found in the Galápagos.

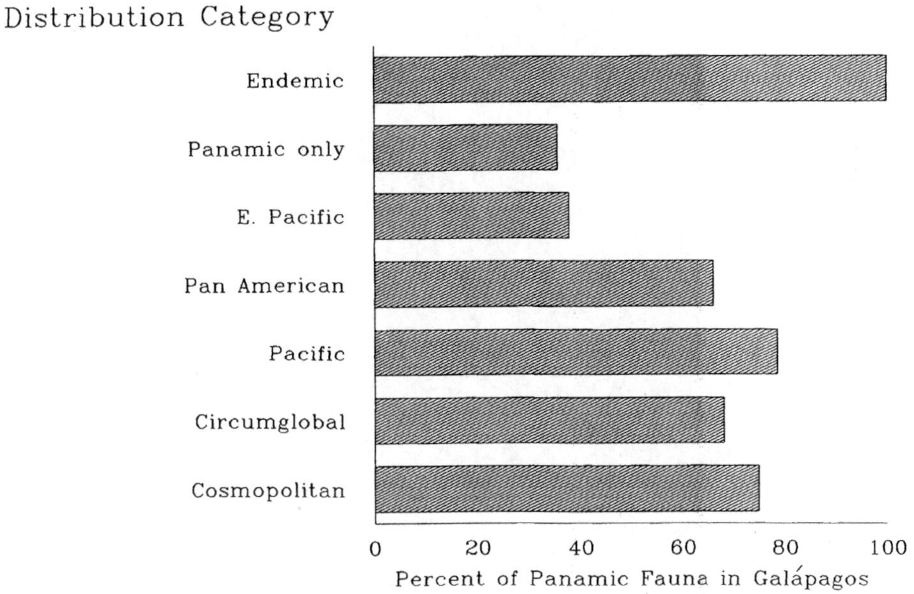

Figure 2. Histogram of numbers of bryozoan species found in the Galápagos arranged by distribution category. Section 2.3 in the text gives an explanation of ranges of species in each distribution category.

Higher proportions, above about 60% are found for species with wider ranges. The low proportion for exclusively Panamic species corresponds to the low representation of species whose maximum latitude of distribution of is between 20 and 30 degrees; this is because I took the 30 degrees North as the limit between the Panamic and Californian Province for species found on the Pacific side of Baja California (extended north to 35 degrees for species in the northern Gulf of California; see Soule, 1963b). Species with ranges extending farther north I considered 'Eastern Pacific' species.

These data seem to suggest that bryozoan species with ranges limited to the Eastern Pacific, including exclusively Panamic species, are proportionally less represented in the Galápagos Islands than species with wider ranges, including Pan American, Pacific, Circumglobal, and Cosmopolitan species. Presumably, this is an effect of (1) greater powers of distribution of more widely distributed species, (2) the greater antiquity of widely distributed species, or (3) a combination of both phenomena.

7.6. Influence of Coloniality on Bryozoan Distributions

Bryozoans are sessile and feed on plankton—so they can attach to floating objects, such as rafts of vegetation. They therefore can be transported to the Islands

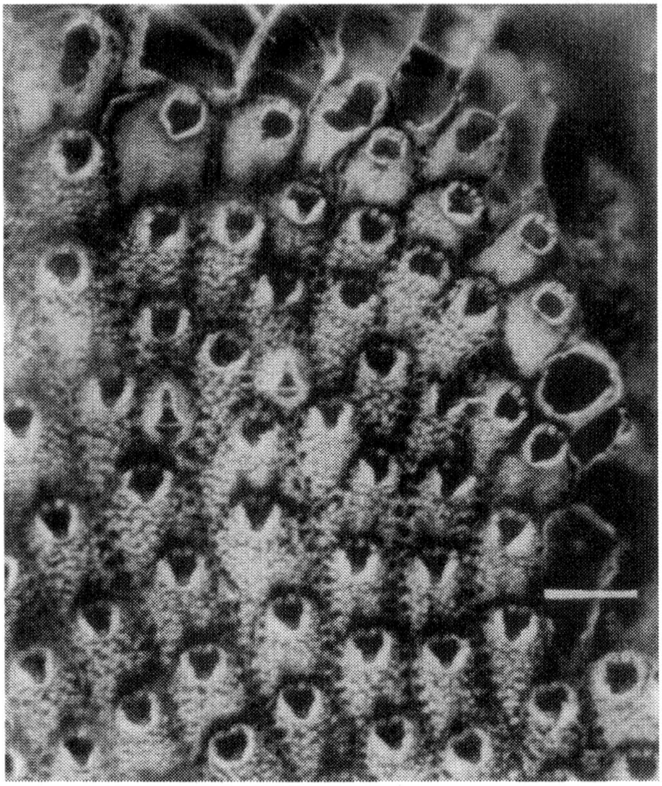

Figure 3–6. Four representative ascophoran bryozoans from the Galápagos. Light photomicrographs of KOCl-cleaned specimens stained with congo red and coated with sublimed ammonium chloride. Collected by the author from Academy Bay, Santa Cruz Island, June 2, 1980. Scale bars, 0.5 mm.

Figure 3. *Parasmittina* sp. aff. *hastingsae* Osburn.

in the adult stage, as well as larvae. Their planktotrophic diet of the adults presumably would make survival during transport more likely and increase the likelihood of establishment of a permanent population once transport had been accomplished. Most invertebrates would be transported only as larvae.

7.7. The Implications of Lumping

Most quantitative zoogeographical conclusions depend utterly on the validity of the species identifications. This is especially the case when we attempt comparisons by counting lower taxa. Recent work on bryozoan taxonomy above the alpha level (karyotypes, biochemical taxonomy, and maternal genetics) is not reassuring (Backus 1979; Thorpe and Ryland, 1979). So far the picture emerging is one of a large number of sibling species which can be recognized morphologically only on

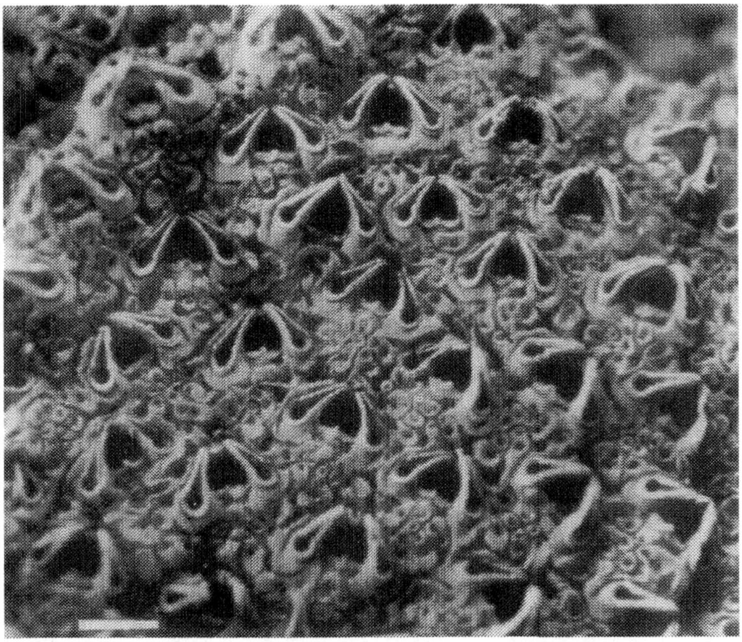

Figure 4. *Enantiosula manica* Canu and Bassler.

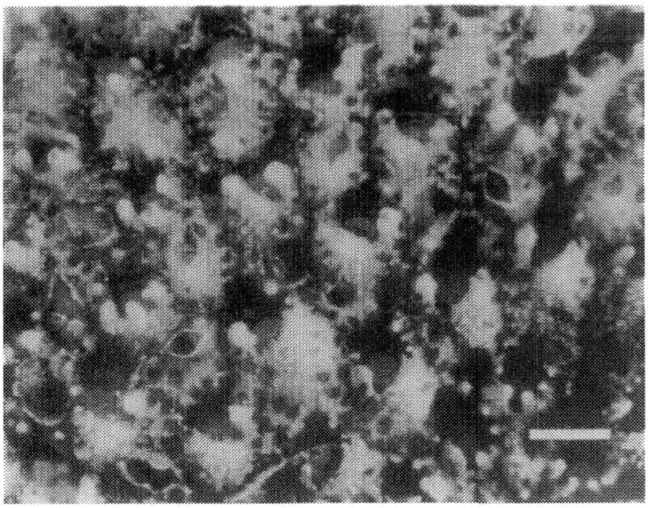

Figure 5. *Rhynchozoon rostratum* (Busk).

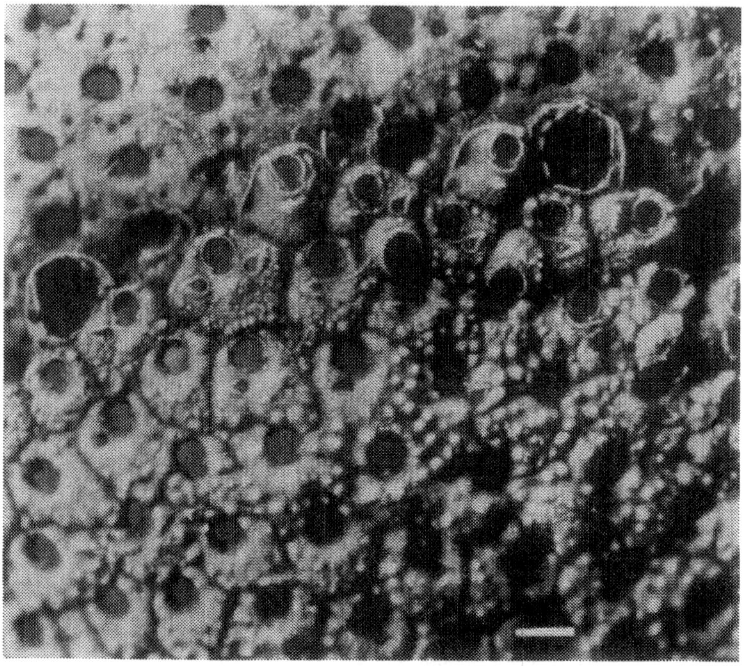

Figure 6. *Cleidochasma porcellanum* (Busk).

the basis of anatomical details not usually described adequately, at least in the older literature. In the tropics, preliminary work of J. B. Jackson and A. H. Cheetham in Panama provides electrophoretic evidence and maternal inheritance data on cultured bryozoan larvae to indicate that a number of hitherto unrecognized bryozoan species exist within presently recognized species which differ in relatively subtle morphological features (Jackson and Cheetham, 1989; Jackson and Cheetham, 1990; A. H. Cheetham, personal communication, 1989). The implication is that any conclusions we draw about Galápagos bryozoans should be muted by the caution that many 'species' may represent complexes of more than one biological species. There are at least three important implications of cryptic lumping.

1. Diversity and endemism may be much higher among Galápagos bryozoans than the data now suggests.
2. Unless lumping has been significantly more or less severe in the Galápagos than elsewhere (and there is no reason to suppose it has), conclusions based on proportional representation of taxa may not be affected.
3. Many of the widely distributed species may be species groups comparable to genera. However, unless the species in these groups differ substantially in their biologies, conclusions we may draw about these 'species' may well turn out to be as applicable to those genera.

ACKNOWLEDGMENTS. Mike Temkin, Donald Bonyai, and Rebecca Beerman helped collect specimens in the Galápagos. JoAnn Sanner at the U.S. National Museum of Natural History helped with the collections and literature and identified some *Parasmittina* specimens. Alan H. Cheetham, also of the Museum, discussed the subject with me and was a great help with both details and broad ideas included in this chapter. Lisa Turni and Laura Gelpey helped with photos. Thanks also are due to the Government of Ecuador, the Charles Darwin Foundation, and the Smithsonian Institution for assistance in collecting and studying specimens. Supported in part by an American University Research Award. Contribution 18-89, The American University Biology Department.

References

Backus, B. T., 1979, Karyotypic variation in the bryozoan, *Paludicella articulata*, *Cytologia* **46**:561–566.

Banta, W. C., 1969, Systematic list of Bryozoa identified from Farnsworth Bank, Santa Catalina Island, in: Hartman, O., Quantitative survey of the benthos of San Pedro Basic, southern California. Part II, Final results and conclusions. *Allan Hancock Pac. Exped.* **19**(2):419–422.

Banta, W. C., and Carson, R. J. M., 1977, Bryozoa from Costa Rica, *Pac. Sci.* **31**(4):381–424.

Banta, W. C., and Redden, J., 1990, A checklist of the Bryozoa of the Galápagos, Proc. *Biol. Soc. Wash.* **103**(4):789–802.

Barbossa, M. M., 1964, Catalog das especies atuais de Bryozoa do Brasil com indicacoes bibliograficas, *Mus. Nac. Rio de Janeiro*, v + 46 pp.

Canu, F., and Bassler, R. S., 1928, Fossil and Recent Bryozoa of the Gulf of Mexico region, *Proc. U.S. Nat. Mus.* **72**(14):1–199.

Canu, F., and Bassler, R. S., 1929, Contributions to the biology of the Philippine Archipelago and adjacent regions, *Bull. U.S. Nat. Mus.* **100**(9):1–683.

Canu, F., and Bassler, R. S., 1930, The bryozoan fauna of the Galápagos Islands, *Proc. U.S. Nat. Mus.* **76**(13):1–78.

Colinvaux, P. A., 1984, The Galápagos climate: present and past, in: *Key Environments: Galápagos*, pp. 55–69, (R. Perry, ed.), Pergamon Press, New York.

Cook, P. L., 1968, Bryozoa (Polyzoa) from the coasts of tropical west Africa, *Atlantidae Rep.* (Copenhagen) **10**:115–262.

Crossland, C., 1927, The expedition to the south Pacific of the S.Y. 'St. George,' *Trans. Roy. Soc. Edinb.* **55**(2):531–554.

Finet, Y., 1985, Preliminary faunal list of the marine molluscs of the Galápagos Islands, *Docum Trav. Inst. Roy. Soc. Nat. Belg.* **29**:(50 pp.).

Finet, Y., 1989, Recent molluscan fauna in the Galápagos: Taxonomic composition of the fauna (abstract), *Ann. Rep. West. Soc. Malacol.* **21**:7.

Fraser, C. M., 1943, General account of the scientific work of the *Velero III* in the eastern Pacific, 1931–41. Part III. A ten-year list of the *Velero III* collecting stations, *Allan Hancock Pac. Exped.* **1**(3):259–431.

Garth, J. S., 1946, Distribution studies of Galápagos Brachyura, *Allan Hancock Pac. Exped.* **5**:603–648.

Gordon, D. P., 1984, The marine fauna of New Zealand: Bryozoa: Gymnolaemata from the Kermadec Ridge, *Mem. N. Z. Oceanogr. Inst.* **91**:1–52.

Greig-Smith, R., 1964, *Quantitative Plant Ecology*, Butterworth, London, 245 pp.

Harmer, S. F., 1915, Polyzoa of the Siboga Expedition. Part I. Entoprocta, Ctenostomata and Cyclostomata, *Siboga-Exped. Monogr.* 37a: pp. vi + 180.

Harmer, S. F., 1926, The Polyzoa of the Siboga Expedition. Part II. Cheilostomata Anasca. *Siboga-Exped. Monogr.* 28b: pp. 181–501.

Harmer, S. F., 1934, Polyzoa of the Siboga Expedition. Part III. Cheilostomata Ascophora. I. Family Reteporidae, *Siboga-Exped. Monogr.* 38c: vii + pp. 503–640.

Harmer, S. F., 1957, Polyzoa of the Siboga Expedition. Part IV. Cheilostomata Ascophora. II. (Ascophora, except Reteporidae, with additions to part II, Anasca), *Siboga-Expedite Monogr.* 28d (Leiden), xv + pp. 641–1147.

Hastings, A. B., 1930, Cheilostomatous Bryozoa from the vicinity of the Panama Canal collected by Dr. C. Crossland on the cruise of the S.Y. "St. George.", *Proc. Zool. Soc. London* **1929**(4):1–17.

Hayward, P. J., and Cook, P. L., 1979, The South African Museum's *Meiring Naude* cruises. Part 9. Bryozoa, *Ann. So. African Mus.* **79**(4):43–130.

Hayward, P. L., and Cook, P. L., 1983, The South African Museum's *Meiring Naude* cruises. Part 13. Bryozoa II, *Ann. So. African Mus.* **91**(1):1–161.

Hedgepeth, J. W., 1957, Marine Biogeography, in: *Treatise on Marine Ecology and Paleoecology* (J. W. Hedgepeth, ed.), *Geol. Soc. Am.* **67**(1), pp. 359–382.

d'Hondt, J.-L., 1979, Catalogue des collections de bryozoaires de l'Ifan, *Bull. de l'Efan* **41**(A2):330–348.

Jackson, J. B. C., and Cheetham, A. H., 1989, Are skeletons enough to tell species apart? A test using cheilostome Bryozoa (abstract), *Geo. Soc. Am., Abstract with Programs* **21**(6):A289.

Jackson, J. B. C., and Cheetham, A. H., 1990, Evolutionary significance of morphospecies: A test with cheilostome Bryozoa, *Science* **248**:579–583.

James, M. J., 1984, A new look at evolution the Galápagos: Evidence from the late Cenozoic marine molluscan fauna, *Biol. J. Linn. Soc.* **21**(1–2):77–95.

Kluge, G. A., 1962, Bryozoa of the northern seas of the USSR., Opred Fauna SSSR 76:1–584. [In Russian. English translation (1975) available from N.T.I.S., U.S. Dept. Commerce, Springfield, VA 22161.]

Marcus, E., 1940, Mosdyr (Bryozoa eller Polyzoa), *Danmarks Fauna* (Copenhagen) **46**:1–401.

Moyano, H. I., 1982, Magellanic Bryozoa: Some ecological and zoogeographical aspects, *Mar. Biol.* **67**:81–96.

Moyano, H. I., 1983, Southern Pacific Bryozoa: A general view with emphasis on Chilean species, *Gayana* **46**:3–45.

Nielsen, C., 1970, On metamorphosis and ancestrula formation in Cyclostomatous Bryozoa, *Ophelia* **7**(2):217–256.

Osburn, R. C., 1950, Bryozoa of the Pacific coast of America. Part I, Cheilostomata-Anasca, *Allan Hancock Pac. Exped.* **14**(1):1–269.

Osburn, R. C., 1952, Bryozoa of the Pacific coast of America. Part 2, Cheilostomata-Ascophora, *Allan Hancock Pac. Exped.* **14**(2):271–611.

Osburn, R. C., 1953, Bryozoa of the Pacific coast of America. Part III. Cyclostomata, Ctenostomata, Entoprocta and Addenda, *Allan Hancock Pac. Exped.* **14**(3):613–841.

Osburn, R. C., and Soule, J. D., 1953, Order Ectoprocta, Suborder Ctenostomata, pp. 726–758, in: R. C. Osburn: Bryozoa of the Pacific coast of America. Part III. Cyclostomata, Ctenostomata, Entoprocta and Addenda. *Allan Hancock Pac. Exped.* **14**(3), 228 pp.

Ryland, J. S., 1971, Bryozoa (Polyzoa) and marine fouling, in: *Marine, Borers, Fungi and Fouling Organisms of Wood* (E. B. Jones, S. K. Gareth, and S. K. Eltringham, eds.), pp. 137–154, Org. Econ. Coop. Devel., Paris.

Shasky, D. R., 1989, An update on the molluscs from the Galápagos as listed in the preliminary faunal list of the marine molluscs of the Galápagos by Yves Finet (abstract), *Annu. Rep. West. Soc. Malacol.* 21:7.

Schopf, T. J. M., 1973, Ergonomics of polymorphism: Its relation to the colony as the unit of natural selection in species of the phylum Ectoprocta, in: *Animal Colonies: Their Development and Function through Time* (R. S. Boardman, A. H. Cheetham, and W. A. Oliver, eds.), pp. 247–294, Dowden, Hutchinson and Ross, Shroudsburg, Pennsylvania.

Soule, D. F., and Soule, J. D., 1964, The Ectoprocta (Bryozoa) of Scammon's Lagoon, Baja California, *Am. Mus. Novit.* **2199**:1–56.

Soule, J. D., 1959, Results of the Puritan-American Museum of Natural History expedition to western Mexico. 6. Anascan Cheilostomata (Bryozoa) of the Gulf of California, *Am. Mus. Novit.* **1969**:1–54.

Soule, J. D., 1961, Results of the Puritan-American Museum of Natural History expedition to western Mexico. Ascophoran Cheilostomata (Bryozoa) of the Gulf of California, *American Mus. Novit.* **2053**:1–66.

Soule, J. D., 1963a, Phylum Bryozoa, in: *Contribution to the biogeography of Cocos Island, including a bibliography* (L. G. Hertline, ed.), *Proc. Calif. Acad. Sci.* **32**(8):232–234.

Soule, J. D., 1963b, Results of the Puritan-American Museum of Natural History expedition to western

Mexico. 18. Cyclostomata, Ctenostomata (Ectoprocta) and Entoprocta of the Gulf of California, *Am. Mus. Novit.* **2144**:1–34.

Thorpe, J. P., and Ryland, J. S., 1979, Cryptic speciation detected by biochemical genetics in three ecologically important intertidal bryozoans, *Estuar. Coast. Mar. Sci.* **8**:395–398.

Walker, B. W., and Rosenblatt, R. H., 1961, The Marine fishes of the Galápagos islands (abstract), pp. 470–471, *Tenth Pacific Science Congress*, Honolulu, Hawaii.

Winston, J. E., 1977, Distribution and ecology of estuarine ectoprocts: A critical review, *Chesapeake Science* **18**(1):34–57.

Winston, J. E., and Heimberg, B. F., 1986, Bryozoans from Bali, Lombok, and Komodo, *Am. Mus. Novit*, **2641**:1–49.

Winston, J. E., 1982, Marine bryozoans (Ectoprocta) of the Indian River area (Florida), *Bull. Am. Mus. Nat. Hist.* **173**(2):100–176.

Winston, J. E., 1982, Marine bryozoans of the Indian River area (Florida), *Bull. Am. Mus. Nat. Hist.* **173**(2):100–176.

IX

Taphonomy and Paleoecology

Chapter 18

The Past as a Key to the Present

Taphonomy and Paleoecology of the Urvina Bay Uplift

DAVID L. MALMQUIST

1. Overview

Taphonomy is the study of the processes of preservation and how they affect information in the fossil record (Behrensmeyer and Kidwell, 1985). A common tactic in taphonomic research is comparison of living communities and fossil assemblages in order to measure the information that taphonomic processes remove from a living community as it passes into an assemblage of fossil remains (Johnson, 1960, 1962, 1965; Lawrence, 1968; Warme, 1969; Warme *et al.*, 1976;

DAVID L. MALMQUIST • Earth Sciences Board of Studies, University of California–Santa Cruz, Santa Cruz, California 95064.

Lasker, 1976; Peterson, 1976, 1977; Schopf, 1978; Koch and Sohl, 1983; Staff *et al.*, 1986; Fürsich and Flessa, 1987). Understanding this information loss is necessary in order to "strip away the taphonomic overprint" when using fossil remains to reconstruct a once-living community (Lawrence, 1968).

Inherent in many living/fossil comparisons is the assumption that biases arise only from taphonomic processes acting on the fossil assemblage, and that the living community is bias-free, in essence a "control group" that is, after death, subjected to the vagaries of preservation and sampling. Taphonomists have paid less attention to the potential biases affecting characterization of the living community than to those affecting the dead assemblage.

In this chapter, I use an uplifted, exquisitely preserved assemblage of invertebrate skeletal remains at Urvina Bay to quantify the degree and direction of the biases affecting characterization of the living "control" group. Comparison of the uplifted and living communities at Urvina Bay demonstrates that 1) in cases of exceptional fossil preservation and accessibility, the fossil assemblage gives a more accurate picture of the original living community than does a living control group and 2) paleoecologists must be aware of potential bias in living analogs as well as in fossil assemblages.

2. Introduction

Early in 1954, rising magma beneath Isabela Island in the Galápagos Archipelago of Ecuador uplifted and subaerially exposed a shallowly submerged shelf in the island's Urvina Bay (Fig. 1; Couffer, 1956; Richards, 1957; Glynn *et al.*, 1983; Simkin, 1984; Colgan and Malmquist, 1987; see Chapter 4, this volume). The maximum vertical elevation of roughly 7 m stranded and killed the organisms inhabiting the bay's waters (Malmquist *et al.*, 1986).

Present-day exquisite preservation of skeletonized organisms from the uplifted area demonstrates minimal post-uplift destruction, justifying a study of differences between the assemblage on the uplifted area and the present living community in the nearby intertidal and shallow subtidal areas. Despite excellent preservation, several taphonomic processes have left distinctive traces on the uplifted organisms. These can be used to infer possible periods of subaerial exposure in the post-mortem history of marine fossil assemblages in tectonically active areas elsewhere.

Because access to the uplifted area is so much easier than in the adjacent submarine environment, detailed sampling on the uplifted area gives a more complete and accurate picture of the skeletonized fraction of a shallow, benthic Galápagos marine community than does underwater sampling. Moreover, redundant information permits inferences regarding the presence of many of the soft-bodied or poorly skeletonized organisms that were uplifted but not subsequently preserved.

This Galápagan "invertebrate Pompeii" inverts Lyell's uniformitarian dictum that the present is the key to the past, because at Urvina Bay the past of 37 years ago provides a key to the present. We can use the uplifted area at Urvina Bay as a baseline from which to judge the inherent bias of underwater sampling on the subsequent calculation of such community parameters as species richness, diversity, and taxonomic composition and to clarify between-habitat variations in species composition of a living Galápagos marine community.

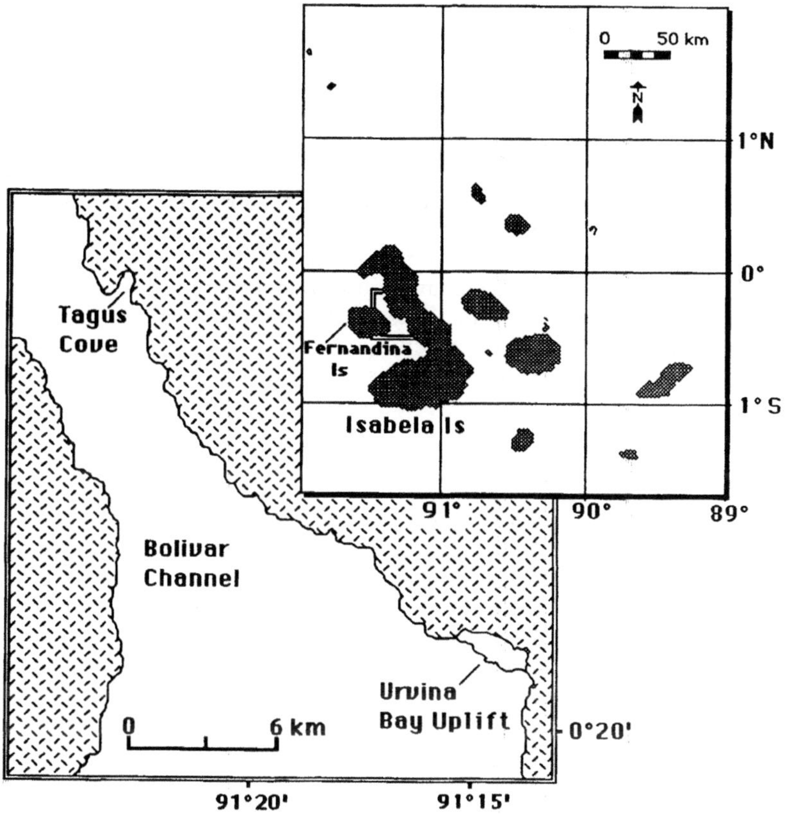

Figure 1. Location map of the Urvina Bay uplift, west-central coast of Isabela Island.

2.1. Necessary Conditions

In order to use Urvina Bay's uplifted assemblage as a proxy for the present living community, we must first be sure that it records the instantaneous, catastrophic death of the once-living community from which it derives, and has not been time-averaged in its accumulation. Because sampling of a living community effectively gives a "snapshot" of its composition and structure at a single moment in time, we must, if we are to make valid comparisons between the uplifted and living communities, show that the assemblage on the uplifted area gives the same instantaneous view of the once-living community from whence it came. The presence of attritionally derived organisms on the uplifted area would blur its otherwise clear signal of catastrophic death (Malmquist *et al.*, 1986).

Second, we must also be able to demonstrate that the uplifted assemblage has suffered only minimally from its 37-year exposure to taphonomic processes. Otherwise, we would not be able to distinguish whether the differences between the uplifted and living communities arose from taphonomic alterations to the uplifted assemblage or from other causes.

Finally, we must be certain that post-uplift changes in the composition of the living community have been minimal. This is apparently the case, at least in terms of gastropods. Of the 83 gastropod species at Urvina Bay, only 9 occur solely in the living marine community and not on the uplifted area (see Table 2), and of these, 5 are indeterminate species that may be synonymous with known uplifted species.

Having ruled out other possible reasons for the observed differences between the uplifted and living communities, we can justifiably argue that their differences stem from biases in sampling. In order to demonstrate that the fossil assemblage on the uplifted area meets these necessary conditions, I first present evidence for the young age and catastrophic nature of the uplift at Urvina Bay, and then show how the taphonomy of uplifted regular echinoids argues for minimal post-uplift destruction of Urvina Bay's skeletal remains.

2.2. Date of the Uplift Event

Sedimentologic and paleontologic evidence, along with accounts of visitors, indicate that the uplift of Urvina Bay was a single, discrete event that took place within a matter of hours during February, 1954.

Pre-uplift visits to Urvina Bay by local residents in December, 1953 (Karl Angermeyer, pers. comm., 1987), coupled with the post-uplift visit by members of a nature film crew* in March 1954 (Couffer, 1956), bracket the interval during which the uplift event occurred. The film crew used three lines of evidence to estimate that the uplift event had occurred two to three months before their arrival (Couffer, 1956). When they found the site in March of 1954 they noted that 1) the smell of decay was still present, 2) in shaded locations, the soft parts of shelled organisms were still moist, and 3) a land plant had taken root in the marine sediments.

Several facts, however, suggest that their estimate may have been too great, and that the uplift event probably took place in February, 1954. First, taphonomic studies have shown that the odor of decaying invertebrate remains disappears within four weeks (Plotnick, 1986). It is likely that the odor of the organisms at Urvina Bay disappeared even sooner, because their decaying remains would have been desiccated by the equatorial sun. Second, although Couffer did not specify the type of plant he found growing in the recently uplifted sediments, it was most likely a halophyte, given the saline nature of those sediments. The salt bush *Cryptocarpus pyriformis* HBK, 1817, thrives in salt-rich substrates (Wiggins and Porter, 1971; Jackson, 1985), including supratidal areas on present-day Galápagos beaches, and could have colonized the uplifted area almost immediately.

Examination by Richards (1957) of photographs taken by Couffer in March showed that most of the club-spined sea urchins (*Eucidaris thouarsii* Valenciennes) stranded by the uplift event retained their spines. Schäfer (1972) reports that the spine musculature of dead echinoids (*Echinus*) in the waters of the North Sea decayed in only seven days. Mummification of echinoid tests on the uplifted area, coupled with their removal from disturbance by waves and currents, may

*The crew was there to film footage for Walt Disney's "True-Life Adventure Feature, Islands of the Sea."

have delayed spine loss. However, when we placed tests of recently killed specimens of *Eucidaris* on the uplifted area, they shed their spines within two weeks. All evidence thus indicates that the uplift event took place in February, 1954.

2.3. Rapidity of the Uplift Event

There is abundant sedimentologic and biologic evidence for the rapidity of the uplift event. First, the now land-locked beaches bordering the inner margins of the uplifted area retain their original angles of repose. This is true not only for the high-energy beaches of basaltic cobbles that rim most of the uplifted area, but for the coralline-sand beach that lies along the uplift area's northern margin (Fig. 2). If the uplift event at Urvina Bay had occurred gradually, as did the 1974 uplift of Punta Espinosa on Fernandina Island (Glynn *et al.*, 1983), the beaches would be terraced, with each bench marking the level at which the uplifted area temporarily paused. Second, we found large-scale, well-preserved ripple marks on the uplifted area. These ripples would have been destroyed or modified if they had been slowly lifted through the swash zone. Third, Couffer (1956) discovered the remains of uplifted fish that were trapped on narrow ledges above what had been the local sea floor. Had the uplift event occurred slowly, these fish would have been able to either escape the uplifted area or at least swim away from the ledges into adjacent pools. The possibility remains, however, that the fish were exposed to volcanic gases and thus killed before they could escape (Couffer, 1956). Finally, Richards (1957) reports that E. P. Chace of the San Diego Museum of Natural History examined a collection of uplifted gastropod shells from Urvina Bay and concluded, based on their "appearance," that they had died rapidly. Although Richards does not define Chace's criteria, it may well have been that the shells were unworn and found with their opercula *in situ*. We found many such undisturbed shells on the uplifted area.

The presence of many sea turtle skeletons (*Chelonia mydas agassizii* Bocourt) on the uplifted area suggests that the uplift event lowered the water level more rapidly than does the outgoing tide, because otherwise a trapped turtle could have swum into deeper water. Large numbers of *C. agassizii* feed today in the shallow intertidal pools of the modern Urvina Bay, but we found a turtle temporarily trapped by the outgoing tide on only two occasions. Thus, we can refine our estimate of the rapidity of the uplift event to a period of less than or equal to six hours, the time between high and low tides in the diurnal tidal cycle of the Galápagos (Hedgpeth, 1969).

3. Taphonomy of the Urvina Bay Uplifted Assemblage

3.1. Synopsis

We examined the preservation of uplifted organisms to determine whether taphonomic processes had significantly biased the uplifted assemblage *vis-á-vis* the original living community. The results of this study show 1) uplifted skeletal remains were almost perfectly preserved and 2) the soft-bodied or poorly skeletonized fraction of the original living community was almost totally destroyed,

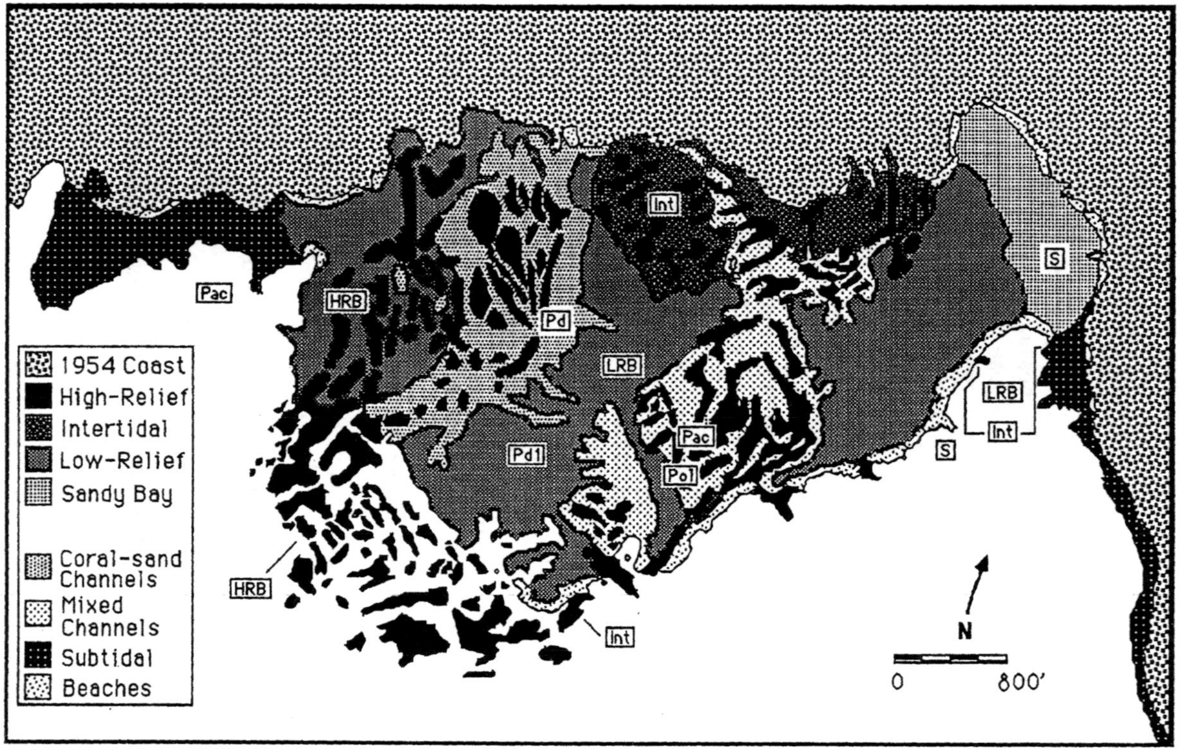

Figure 2. Detailed field map of the Urvina Bay uplift, showing the configuration of the coast prior to 1954, and the present-day distribution of uplifted habitats. Water is in white. The location of uplifted and submarine sampling sites is shown by the boxed letters: HRB = high-relief basalt, Int = intertidal, LRB = low-relief basalt, S = sand, Pac = *Pavona clavus*, Pd1 = *Pocillopora damicornis*, Pd = pocilloporid flats, Pol = *Porites lobata*. Coral-sand channels, mixed channels, subtidal areas, and beaches are not discussed in this chapter.

although soft tissues were rarely preserved and the presence of many nonpreserved organisms can be inferred from redundant information. Exquisite preservation of skeletal remains was crucial to the subsequent comparison of uplifted and living communities because it largely removed taphonomic bias as a possible cause of the differences found to exist between the two.

3.2. Background

Taphonomic and actualistic studies have shown that numerous processes act to confound information in the fossil record (e.g., Johnson, 1960, 1965; VanValen, 1964; Lawrence, 1968; Peterson, 1976, 1977; Bosence, 1979; Koch and Sohl, 1983). Post-mortem transport and mixing produce thanatocoenoses, death assemblages of fossil organisms that did not naturally occur together while alive. Time-averaging, the incorporation of organisms that lived at different times into the same fossil assemblage, may invalidate abundance, diversity, and time-range data, as may differing lengths of generation times, which can produce an overabundance of short-lived taxa. The instantaneous death and undisturbed post-mortem history of Urvina Bay's marine community, however, preclude these biases.

3.3 Differential Nonpreservation

One taphonomic process clearly biased the Urvina Bay fossil assemblage. Differential nonpreservation, which leads to over-representation of hard-shelled as opposed to soft-bodied taxa (Lawrence, 1968; Lasker, 1976; Schopf, 1978) is very apparent on the uplifted area.

When Couffer discovered the uplifted area in March of 1954, he reported both hard-shelled and soft-bodied taxa. He found calcareous algae, crabs, lobsters, asteroids and ophiuroids, regular and irregular echinoids, sea turtles, gastropods, and corals; and, in addition, fish, eels, and marine worms (Couffer, 1956). In addition to the organisms reported by Couffer, we frequently encountered the remains of bivalves, chitons, and barnacles; and, much more rarely, tunicates, sponges, and fleshy algae.

Thirty-seven years after the uplift event, the calcareous algae, crabs, regular and irregular echinoids, gastropods, and stony corals (Chapter 4, this volume) remain exquisitely preserved. Lobsters, asteroids and ophiuroids, fish, eels, sponges, and marine worms, on the other hand, are extremely rare on the surface, although they do occur as disarticulated fragments within the uplift area's surficial sediments. Sea turtles are locally abundant, but poorly preserved.

Differential nonpreservation is largely explained by differences in skeletal architecture and mineralogy between preserved and nonpreserved organisms. Nonpreservation of the marine worms on the uplifted area clearly resulted from their lack of a skeleton, whereas poor preservation of asteroids and ophiuroids was fostered by the easy disarticulation of their numerous ossicles after decay of the connecting collagenous material (Kier, 1977; Smith, 1984).

Poor preservation of fish and eels is explicable in light of their physically fragile and chemically unstable skeletons of calcium phosphate. Oxidation or decomposition by microorganisms is the normal fate of fish remains in most reef sediments (Laporte, 1979); subaerial exposure has apparently had the same effect.

The absence of lobsters despite the remains of numerous xanthid and porcellanid crabs is more difficult to explain. Skeletal mineralogy of the two groups is quite similar. Both possess exoskeletons of moderately calcified chitin. Furthermore, exquisite preservation of the crabs in normal, or nontraumatic, life positions (Bishop, 1986) indicates that they, and presumably the lobsters also, suffered only minimal post-uplift disturbance. Lobsters are perhaps absent because they were more attractive to terrestrial scavengers.

3.4. Scavenging and Trampling

Scavenging is responsible for the fragmentation and transport of skeletal remains in many terrestrial settings (Behrensmeyer, 1975; Behrensmeyer *et al.*, 1979; Hill, 1980; Behrensmeyer and Dechant Boaz, 1980; Laporte and Behrensmeyer, 1980). The rarity of large terrestrial scavengers in the Galápagos has thus undoubtedly aided the *in situ* preservation of the uplifted assemblage. The only scavengers that Couffer encountered on his first visit to the uplifted area were arthropods (flies, spiders, and carrion-feeding beetles) and birds (fly catchers, warblers, and mocking thrushes). Although these animals may have removed soft tissues from the uplifted remains, their relatively small size apparently kept them from significantly altering the position or location of those remains.

In addition to the paucity of terrestrial scavengers in the Galápagos, another factor that aided the preservation of the uplifted assemblage was the superabundance of dead remains produced by the uplift event. The sudden appearance of so many dead organisms very likely exceeded the ability of the few local scavengers to fully use the temporarily abundant resource (Behrensmeyer, 1975; Coe, 1978). Moreover, desiccation by the equatorial sun probably made the uplifted remains unattractive to scavengers rather quickly. When Couffer returned to the uplifted area in June, four months after his first visit, the flies and spiders on the decaying remains were gone. The only living animals that he encountered on the uplifted area were doves, hawks, other land birds, and lava lizards (Couffer, 1956).

Trampling by large terrestrial vertebrates is another important process in the breakdown of skeletal remains in many terrestrial environments (Laporte and Behrensmeyer, 1980). At Urvina Bay, however, trampling is minimal. The small amount that does occur is concentrated in sandy areas on the uplifted area and stems from activity by a few feral donkeys. Burrowing by land iguanas, and wallowing by Galápagos tortoises also plays a small role.

3.5. Redundant Information

Even though the physical remains of many soft-bodied or poorly skeletonized taxa were altered or destroyed, their original presence is sometimes recorded in the form of redundant information (Lawrence, 1968), including tracks, borings, and drill holes. For instance, even though intact uplifted sponges are rare (although sponge spicules are fairly abundant within sediments beneath basaltic boulders), the presence of clionid sponge borings in molluscan shells indicates that these sponges were originally present. Likewise, the presence of distinctive wear patterns (Boekschoten, 1967; Müller, 1979) or particular bionts on gastropod shells (Walker, 1988) suggests appropriation of the shells by marine hermit crabs. Drill holes in gastropod and bivalve shells, as well as in echinoid tests (Smith, 1984) indicates that naticid, muricid, or thaidid gastropods were present (Lawrence, 1968; Vermeij, 1987).

On the uplifted area, other, less usual sorts of redundant information exist. For instance, the presence of fleshy algae is suggested by large areas of barren basalt that lie between the remaining patches of calcareous red algae (Mitchell Colgan, pers. comm., 1989).

The presence of damselfish is recorded in the morphology of coral colonies that the fish help to protect from the encroachment of grazing urchins. These coral colonies are highly bioeroded on their sides and undersurfaces by urchins, but less so on their tops, where the damselfish exclude urchins in the process of guarding territories (Glynn *et al.*, 1983; Colgan and Malmquist, 1988; Malmquist, under review). On the other hand, the presence of echinoids, particularly *Eucidaris thouarsii*, is clearly marked by their excavations in the uplifted corals (Fig. 3E,F).

A common coral morphology on the uplifted area—colonies of *Pocillopora* with cropped branches—suggests the presence of fish corallivores. Three species of parrotfish, *Scarus ghoban* Forskål, *S. perrico* Jordan and Gilbert, and *S. rubroviolaceous* Bleeker, as well as the puffers *Arothron meleagris* (Lacépede) and *A. hispidus* (Linnaeus), are common croppers of *Pocillopora*, particularly along the coast of Panama, but also in the Galápagos (Glynn *et al.*, 1983; see Plate 2 in that volume).

Fish left additional evidence of their presence in the original living community through predation on urchins. Although the rate of fish predation on *Eucidaris* in the Galápagos is low compared with predation rates in Panama (Glynn *et al.*, 1983), we found several uplifted *Eucidaris* tests with either broken spines or breakage in the vicinity of the peristome—both signs of fish predation (Smith, 1984; Glynn *et al.*, 1983). Of course, broken spines could also result from post-uplift trampling, although this appears to have been minimal.

Another type of redundant information on the uplifted area comes from the differential weathering caused by vermetid mollusks. Where they attach to the basaltic substratum, these gastropods act as protective, erosion-resistant caps, so that the surrounding basalt weathers away more readily, leaving vermetid colonies standing on basaltic pedestals. Even though weathering may eventually loosen the vermetid colony from its perch, the pedestal survives, at least temporarily, leaving indirect evidence of its maker.

The taxonomic resolution provided by redundant information is poor—seldom beneath the family level. Nevertheless, redundancy is important because it provides information about the original living community that skeletal remains do not.

3.6. Taphonomic Signature of the Uplifted Assemblage

The subaerial taphonomic processes operating on the uplifted marine assemblage at Urvina Bay have left a distinctive signature. This signature, if present in other Galápagos fossil assemblages, would present strong evidence that they

Figure 3. The five uplifted habitats (A–E) used in the uplift/living comparison; the *Pocillopora damicornis* colony is also figured in (F). Abbreviations as in Fig. 2. (A) High-relief basalt shows relief of up to 7 m, paleo-water depths of 0–9 m, and a substrate of linear basaltic pressure ridges. Visible is the white "bathtub ring" of bleached calcareous algae. (B) The intertidal habitat shows relief of less than 1 m, paleo-water depths of 0–3 m, and a substrate of basaltic cobbles and bedrock. Shown are uplifted barnacles. (C) Low-relief basalt has relief of less than 1 m, paleo-water depths from 3–9 m, and substrates of basaltic bedrock, boulders, and cobbles; bioclastic and basaltic sand; and scattered colonies of *P. damicornis*. It is the most extensive and heterogenous of the sampled habitats. (D) The sand habitat has no appreciable relief, paleo-water depths of 2–8 m, and a substrate of basaltic and bioclastic sand. In the background is a newly formed, post-uplift beach berm. (E) The 3 m-high *Pavona clavus* colony stood in 4 m of water, and was exposed at low tide. The numerous excavations on this colony were produced by the echinoid *E. thouarsii*. (F) Colony of *P. damicornis*. Note the high-degree of echinoid bioerosion.

formed in the same manner as the uplifted assemblage at Urvina Bay, or at least experienced a period of subaerial exposure at some time in their early histories, and would suggest that they may be similarly well preserved.

Given the active tectonic setting of the Galápagos, episodes of catastrophic uplift are not unlikely. In fact, an uplifted area similar to the one at Urvina Bay was discovered by Gerry Wellington in 1974 at Punta Espinosa on Fernandina Island, six weeks after an earthquake shook the area (Glynn *et al.*, 1983; Simkin, 1984). Measurement of this uplifted area suggests that it rose in a series of small steps to a maximum height of 80–90 cm above sea level (Simkin, 1984). Uplift and consequent subaerial exposure of skeletal remains is also evident in the late-Pleistocene white shell-sand beach near Villamil on the southern coast of Isabela Island (Dall and Ochsner, 1928; Hickman and Lipps, 1985).

One of the most characteristic signs of subaerial exposure among Urvina Bay's uplifted gastropods is removal of the shells' columellas by the terrestrial hermit crab *Coenobita compressus* H. Milne Edwards. The presence of large numbers of such shells in an assemblage of marine fossils provides clear evidence for a period of terrestrial exposure. At Urvina Bay, the most common hermited shells on the uplifted area are from the thaids *Neorapana grandis* (Sowerby, 1835) (34%), *Purpura pansa* Gould, 1853 (30%) and *Thais melones* (Duclos, 1832) (13%).

The terrestrial hermit crabs and their hollowed shells are not restricted to supratidal areas. Shell-exchange experiments on the uplifted area showed that *Coenobita* regularly ventures as far as 150 m inland, and will travel much farther inland if enticed there by food. Members of our field party saw large numbers of hermit crabs feeding on drought-killed donkeys as far as 500 m inland during the 1986 field season.

Vermetid pedestals, in addition to indicating the presence of vermetids, also suggest subaerial exposure. The type of weathering that produces vermetid pedestals on the uplifted area results from the nucleation of salt crystals within pores in scoriaceous basalt (Wentworth and MacDonald, 1953; McBirney and Williams, 1969; Simkin, 1984). This type of weathering follows evaporation of sea spray and is common in coastal zones (Ollier, 1984), but does not act under the water surface.

Crushing of skeletal remains, without subsequent transport of the fragments by moving water, is a sign of trampling by terrestrial organisms under arid conditions such as those that prevail in the Galápagos. Although not common at Urvina Bay, because of the rarity of large terrestrial organisms, such breakage provides good evidence of subaerial exposure under dry conditions.

Association of marine and terrestrial fossils within basaltic lava flows, in conjunction with the other evidence presented here, indicates subaerial exposure. For instance, if the uplifted area at Urvina Bay were suddenly covered with lava or ash from nearby volcanoes Alcedo or Darwin, the resulting deposit might contain impressions of trees along with uplifted marine invertebrates. Burial of the uplift at Urvina Bay is not an unlikely event. Recent volcanic activity in the Galápagos includes the 1979 eruption of Cerro Azul on Isabela Island (Moore, 1980b; Simkin, 1984; Jackson, 1985), the eruption of Volcano La Cumbre on Fernandina Island in 1984 (David Day, pers. comm., 1987), and the collapse of the caldera on Fernandina Island in 1968 (Simkin, 1984; Jackson, 1985). Volcanic activity in the Galápagos has occurred in 7 of the 15 years between 1969 and 1984 (Simkin, 1984), and 53 eruptions from eight volcanoes have been recorded during historic times (Richards, 1962; Simkin, 1984). Because of the islands' remote location and small human population, many eruptions have probably gone unrecorded, and thus these numbers are most likely conservative estimates.

4. Taphonomy of Regular Echinoids

Qualitative examination of Urvina Bay's uplifted assemblage, as just reviewed, strongly suggests that its skeletonized portion experienced minimal post-uplift taphonomic bias. To test this assumption further, I chose to quantitatively study the taphonomy of uplifted regular echinoids.

I focused on regular echinoids for several reasons. First, echinoids are, after gastropods, the most abundant uplifted organisms. The three most abundant uplifted echinoid species are *Eucidaris thouarsii*, *Lytechinus semituberculatus* (Valenciennes), and *Echinometra vanbrunti* A. Agassiz, 1863. On uplifted pocilloporid patch reefs, density values for *E. thouarsii* reach 36/m^2 (median = 19/m^2, n = 20); densities of *L. semituberculatus* in areas of low-relief basalt reach 172/m^2 (median = 5/m^2, n = 25; D. L. Malmquist, unpubl. field data, 1987). The *Eucidaris* density data corroborate those of Glynn (1979) and Glynn *et al.* (1983), who reported median densities of *Eucidaris* in the Galápagos from 2–8 individuals/m^2 on pocilloporid reef flats to 34 individuals/m^2 on the reef edge. Second, the high-Mg calcite of the echinoid test is more soluble than the aragonite or low-Mg calcite shell of gastropods, and their test architecture of porous, sutured ossicles is physically weaker (Chave, 1964; Currey, 1975; Kier, 1977; Smith, 1984).

Echinoids are thus more fragile than mollusks and represent a potentially weaker link in Urvina Bay's taphonomic chain, making them more sensitive indicators of post-mortem processes than gastropods.

4.1. Methods

I divided the uplifted area into six microhabitats (based on the composition and topography of the substratum) and studied the taphonomy of echinoids within each. The microhabitats were high-relief basalt, low-relief basalt, pocilloporid flats, and discrete colonies of *Pavona clavus* Dana, 1846, *Pocillopora damicornis* (Linnaeus) 1758, and *Porites lobata* Dana, 1846 (Figs. 2 and 3). I excluded sand and intertidal microhabitats from this analysis because they contain no regular echinoids.

In the high-relief basalt, low-relief basalt, and pocilloporid-flat microhabitats I counted echinoids within blindly thrown 0.25 m^2 quadrats. On the discrete colonies of *P. clavus*, *P. damicornis*, and *P. lobata* I counted all echinoids present. To determine mode of death and taphonomic history of the individuals within each echinoid species, I recorded the following states of echinoid preservation:

1. Test condition (intact or broken, and, if broken, whether along sutures or across plates).
2. Test orientation (periproct down, peristome down, or on edge).
3. Test exposure (cryptic or exposed).

I also noted whether spines were present or absent. I explain the significance of these four parameters in detail below.

4.2. Results and Discussion

4.2.1. Test Condition

The high percentage of intact specimens of *E. vanbrunti* and *L. semituberculatus* at Urvina Bay indicates that these uplifted echinoids died catastrophically from subaerial exposure and have suffered only minimal post-uplift transport or destruction (see Smith, 1984). In four of the six microhabitats studied, more than 90% of the individuals of these two species are intact. The lowest percentage of intact tests recorded for either of these two species within a microhabitat was 75% (Fig. 4A). The percentage of intact specimens of *E. thouarsii* was lower, but this is to be expected, as interdigitation between ossicles in this species is less than in *L. semituberculatus* and *E. vanbrunti* (Smith, 1984). Tests of uplifted *Eucidaris* break along sutures at the slightest touch, whereas the other echinoids are quite durable.

That 98% of the uplifted *E. thouarsii* with broken tests are broken along sutures (Fig. 4A) again indicates that the uplifted echinoids died due to suffocation and desiccation after the uplift event, and that they suffered minimal post-mortem transport. Because the sutures between ossicles in an echinoid's test are strengthened by stereom interdigitation and collagenous material, they are, in most living echinoids, stronger than the ossicles themselves. After death, however, when the collagenous material decays, the sutures become lines of weakness. Therefore, a test that is broken along the sutures suggests that the echinoid died with the test intact, with breakage occurring after death (Kier, 1974; Smith, 1984).

As noted above, because ossicles in living echinoids are commonly weaker than sutures, breakage while the echinoid is alive occurs preferentially across the ossicles. Therefore, a test that is broken across the ossicles suggests that it was broken prior to, or coincident with, death. This form of breakage most commonly occurs through predation or storm damage. The low percentage of uplifted echinoids that were broken across ossicles (Fig. 4A) is the last piece of evidence drawn from the condition of the test to indicate that few of the uplifted echinoids died due to pre-uplift predation or storm damage.

4.2.2. Test Orientation

The orientation of uplifted echinoids at Urvina Bay indicates that they suffered only minor post-uplift transport. The evidence, however, is less compelling than that provided by the test-condition data.

Living, exposed individuals of *E. thouarsii* in the modern Urvina Bay generally occur with their peristome (mouth) facing downward against the substratum. Less commonly, *E. thouarsii* lives or feeds "upside down" with its peristome facing up beneath overhangs or on the ceilings of caves or recesses (D. L. Malmquist, unpubl. field data, 1987). Thus, uplifted echinoids found with their peristome facing down most likely died *in situ*, particularly if the tests retain their spines,

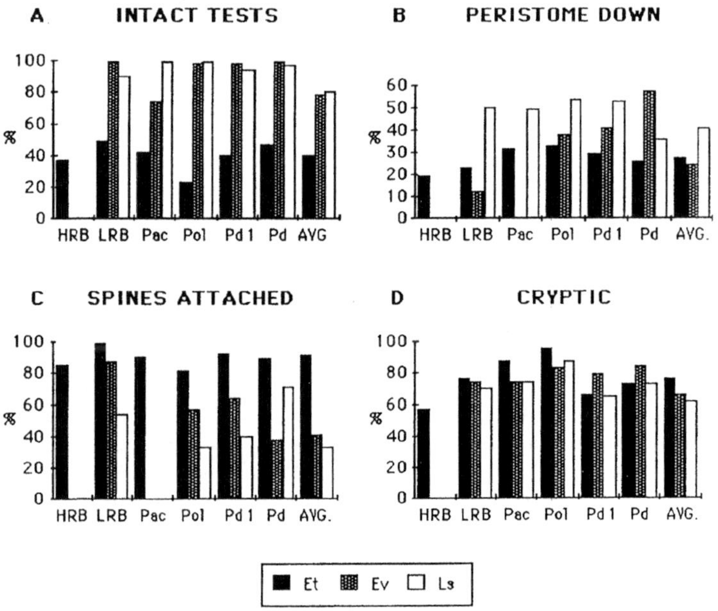

Figure 4. The quality of preservation of the three most abundant uplifted echinoid species at Urvina Bay. Et = *Eucidaris thouarsii*, Ev = *Echinometra vanbrunti*, Ls = *Lytechinus semituberculatus*. (A) Intact tests; (B) peristome down; (C) spines attached; (D) cryptic. Abbreviations as in Fig. 2. See text for detailed explanation.

whereas "upside down" echinoids were more likely due to post-mortem transport, particularly if the test has also lost its spines (Aslin, 1968). If inverted (i.e., peristome-up) urchins retain their spines, the distance of transport and degree of exposure was certainly slight.

Averaged over the five microhabitats in which they occur, 48% of uplifted *L. semituberculatus* and 38% of uplifted *E. vanbrunti* are peristome-down (Fig. 4B). These are high percentages given that most of these uplifted urchins (75% *L. semituberculatus* and 80% *E. vanbrunti*) occur cryptically (Fig. 4D), where many were probably inverted (peristome-up) while alive. Uplifted *E. thouarsii* are also commonly inverted. This may be a result of their lower diameter-to-height ratio (i.e., more spherical), which would make them more easily overturned, or from their cryptic habit (Fig. 4D).

"Upside down" echinoids at Urvina Bay commonly retain their spines (Fig. 4C), indicating that they suffered little post-uplift transport. *E. thouarsii*, the echinoid least commonly found with the peristome facing down, is also the echinoid most likely to be found with spines present. This may, however, simply reflect the better developed spine musculature in this massive-spined echinoid.

The orientation data, like the data on test condition, provide good evidence for minimal post-uplift transport or destruction.

4.2.3. Spines

The high percentage of uplifted echinoids that retained their spines corroborates the hypothesis that they suffered only minor post-uplift transport or exposure. In all microhabitats studied (Fig. 4C), more than 80% of the *E. thouarsii* tests had at least some spines attached or in contact with the test as a skirt-like fringe.

Absence of spines, on the other hand, indicates some post-mortem disturbance (Aslin, 1968). While uplifted *L. semituberculatus* and *E. vanbrunti* are less likely to retain their spines than is *E. thouarsii* (Fig. 4C), this is not unexpected given these species' smaller, less-massive spines and their more poorly developed spine musculature. The spines of these two echinoids are quicker to disarticulate, easier to transport, and more likely to be missed by the observer (Smith, 1984; D. L. Malmquist, unpubl. field data, 1987).

4.2.4. Test Exposure

The main importance to this study of finding cryptic subfossil echinoids is that it argues against their being brought to the uplifted area through the feeding activities of sea birds. Although living echinoids can be either cryptic or exposed, it is very unlikely that sea birds would deposit the tests of consumed echinoids in cryptic locations. Accordingly, the high percentage (Fig. 4D) of cryptic uplifted echinoids is strong evidence for their *in situ* death and lack of post-uplift transport.

4.3. Summary

The preservation of fossil echinoids on the uplifted area at Urvina Bay indicates that they died catastrophically through exposure and suffocation and suffered minimal post-uplift destruction. The quantitative examination of the uplifted echinoids thus corroborates the qualitative evidence that the skeletonized fraction of the subfossil marine community at Urvina Bay is extremely well preserved. This is especially true given that echinoids are more fragile than mollusks and thus more sensitive indicators of post-mortem processes.

5. Comparison of Uplifted and Living Gastropods

5.1. Synopsis

Having established instantaneous death and minimal taphonomic overprinting, we can now examine how the uplifted assemblage compares with the community presently inhabiting the shallow waters of Urvina Bay. The closely drawn analogy between the two communities, spatially, ecologically, and temporally (Lawrence, 1968), permits a detailed, quantitative comparison that raises several important questions:

- Can we, given such a perfectly preserved skeletal community, expect a "perfect" paleoecologic reconstruction?
- If not, how and why will the uplifted and living communities differ?

- What might these differences say about the relative importance of taphonomic versus "living" biases, and their effects on paleoecologic reconstructions?

These are the crucial questions behind the molluscan study presented here.

Analysis of the uplift/living comparison indicates that significant differences do exist between the two communities, and that they arise mainly from the greater ease of sampling on the uplifted portion of Urvina Bay as opposed to in the water of the present bay. Because sampling on the uplifted area is free from the difficulties inherent in underwater sampling, the uplifted assemblage gives an unusually accurate picture of a Galápagos marine community. This picture demonstrates that underwater sampling tends to overestimate the abundance of common living species, and to underestimate the abundance of rare ones. The apparent dominance of Urvina Bay's living microhabitats by a few abundant species has two notable affects on the living community. First, it decreases its actual species richness and diversity values. Second, it exaggerates between-habitat variations in its taxonomic composition. These variations diminish on the uplifted area, where numerically rare but widely dispersed species blur between-habitat distinctions.

5.2. Methods

I focused the uplift/living comparison on gastropods because they are 1) the most abundant uplifted organisms and thus amenable to statistical analysis, 2) one of the best-studied groups in the eastern Pacific and Galápagos (Keen, 1971; Finet, 1985; Chapter 11, this volume), which minimizes taxonomic inconsistencies, and 3) the most common fossil organisms in the Galápagos (Dall, 1924; Dall and Oschner, 1928; Hertlein and Strong, 1939; Hertlein, 1972; Lipps and Hickman, 1982; Pitt and James, 1983; Pitt, 1984; Hickman and Lipps, 1985; Pitt *et al.*, 1986; James, 1984).

We counted only those mollusks that died catastrophically during the uplift event, and not beforehand. Organisms that died due to predation or senescence before the uplift event occurred (and whose remains were then uplifted) are rare compared to the organisms actually killed by the uplift event, but counting them would nonetheless cloud the uplifted assemblage's otherwise clear signal of catastrophic death (Malmquist *et al.*, 1986). Our criteria for catastrophic death were 1) *in situ* opercula, 2) absence of epibionts or borings within the aperture of shells, and 3) intact shells that retain their color, periostracum, or sculpture. Evidence for attritional death came from signs of pre-uplift exposure on the sea floor, and included within-aperture epibionts or borings, broken shells, and highly worn shells, even if intact.

In order to generate baseline data on the number, types, and gross distribution of gastropod species, we ran a total of 20 transects: 10 across the uplifted area, 6 in the modern shallow subtidal, and 4 in the modern intertidal. We identified and counted all gastropods within 0.25 m^2 quadrats thrown every 50 ft along the transects, and within a 20-cm swath along each transect. Identifications were based on Keen (1971), as well as on the Galápagos molluscan collections of the Charles Darwin Research Station (CDRS) and the California Academy of Sciences (CAS). Voucher specimens of these gastropods will be placed at CAS upon the completion of this study.

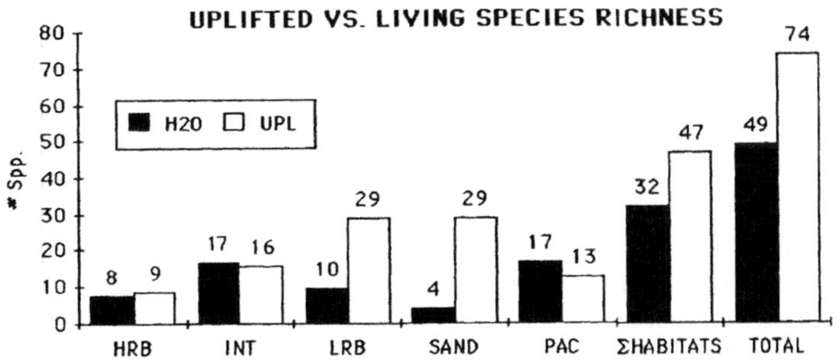

Figure 5. Comparison of species richness in the five uplifted and living habitat pairs (HRB, INT, LRB, SAND, PAC—see Fig. 2 for abbreviations), the summed habitats, and the total uplifted and living gastropod communities, including species found during transect sampling only. H2O = living species, UPL = uplifted species. Note that in five of the seven comparisons, the uplift has the larger number of species. The exceptions are the intertidal (INT) and *Pavona clavus* (PAC) habitats.

In order to compare specific uplifted and living environments, we used topographic relief, water depth, and substrate measured every 10 ft along each transect to define five microhabitats common to the uplifted area and present-day Urvina Bay. We inferred paleo-water depths on the uplifted area from the white "bathtub ring" of calcareous algae that marks the level of the pre-uplift mean low tide (Fig. 3A). The five delineated microhabitat pairs (one on the uplifted area, one underwater) are high-relief basalt, intertidal, low-relief basalt, sand channels and bays, and colonies of *Pavona clavus* (Figs. 2 and 3). Three of the uplifted coral microhabitats that we used in our analysis of echinoid taphonomy (pocilloporid flats and discrete colonies of *Pocillopora damicornus* and *Porites lobata*) had no living analogs, and were thus excluded from the uplift/living comparison.

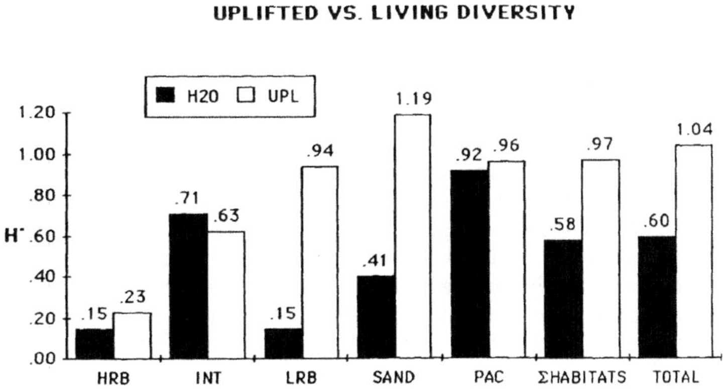

Figure 6. Comparison of Shannon diversity values (H') for the five uplifted and living habitat pairs (HRB, INT, LRB, SAND, PAC—see Fig. 2 for abbreviations), the summed habitats, and the total uplifted and living gastropod communities. Note the consistently higher values for the uplifted communities, and the intertidal exception.

5.3. Results

Statistical comparison of the uplifted and living gastropods at Urvina Bay demonstrates that, overall, species richness values are higher for the uplifted assemblage than for the living community (Fig. 5). Comparisons between microhabitats show, however, that in two of the five microhabitats studied (intertidal and *Pavona clavus*) living species outnumber uplifted species. Uplifted species are more numerous in the high-relief basalt, low-relief basalt, and sand microhabitats.

Figure 6 shows values of the Shannon Diversity Index for five uplifted and living microhabitat pairs, for the sum of the microhabitats, and for the uplift and living communities *in toto*, including transect data. In four of the five microhabitats sampled, significant differences exist between the uplifted and living groups ($p = 0.05$; see Brower and Zar, 1984, p. 159, for explanation of the t-test used to compare the two indices). The summed-microhabitat and *in toto* comparison of the fossil and living communities likewise indicates significant differences. There is no significant diversity difference in only one microhabitat—colonies of the massive coral *Pavona clavus*. In the cases where significant differences exist, the greatest diversity values in all but one instance occur in the uplifted microhabitats. The exception is the uplifted rocky intertidal, which is less diverse than its living analog.

5.4. Discussion

The results of the uplift/living comparison at Urvina Bay show that species richness and diversity are generally greater on the uplifted area than in the water. These results corroborate the findings of previous live/dead studies (Johnson, 1965; Warme, 1969; Peterson, 1976; Warme *et al.*, 1976).

Urvina Bay, however, differs significantly from previous studies in that it subaerially preserves the catastrophic death of a high-energy, rocky bottom subtropical community dominated by epifaunal gastropods, whereas previous live/dead studies focused on attritionally derived marine assemblages of infaunal bivalves in low-to-moderate energy, soft-bottom, temperate lagoons (Johnson, 1960, 1965; Warme, 1969; Peterson, 1976). Moreover, several of the individual microhabitat comparisons at Urvina Bay deviate both from the general findings of the Urvina Bay study and the findings of other live/dead studies. Alternate explanations are thus needed.

Several lines of evidence indicate that the greater species richness and diversity values in the uplifted assemblage stem simply from the greater ease of sampling on the uplifted area. The first line of evidence is the nature of the microhabitats in which the calculated values are greater for the living as opposed to the uplifted communities (Figs. 5 and 6). The higher values in the living intertidal microhabitat are most likely due to the greater ease of sampling in intertidal as opposed to subtidal microhabitats. In other words, the efficiency of sampling in the living intertidal, unlike that underwater, approaches that of sampling on the uplifted area (Connell, 1972).

The similar number of species in the uplifted and living *Pavona clavus* microhabitats most likely stems from their small areal extent and uniform substrate. Our sampling of the uplifted and submerged *P. clavus* colonies was probably equally thorough, whereas sampling of areally extensive and heterogenous submarine microhabitats such as low-relief basalt is inherently uneven. The uplifted and submerged colonies are much smaller than the other microhabitats, and remarkably similar in size. The submerged colony is 12.5 m long, 9.6 m wide, and 3 m high. The uplifted colony is 12 m in diameter, and likewise 3 m high (Fig. 3E). Both corals grew in approximately 4 m of water (Mitchell Colgan, pers. comm., 1989).

The equality of sampling in uplifted and living intertidal microhabitats is also apparent in the comparisons of rank correlation among species (Table 1). Here, the living intertidal is the only microhabitat that is more similar to its paired analog than to any other microhabitat. In the nine other uplift/living comparisons (columns and rows in the similarity matrix), either 1) the uplifted microhabitat is more similar to another uplifted or living microhabitat than it is to its living counterpart, or 2) the living microhabitat is more similar to another living or uplifted microhabitat than it is to its uplifted counterpart.

Presence/absence data tell a similar story (Table 1): Jaccard's coefficient shows greatest similarity between uplifted and living microhabitat pairs in two microhabitats, the living intertidal and *P. clavus*, again suggesting that the efficiency of sampling in these two living microhabitats approaches that possible on the uplifted area. The lack of closest similarity in these two microhabitats in terms of Simpson's coefficient is to be expected: this similarity coefficient is expressly designed to minimize species richness differences like those produced by underwater as opposed to uplift sampling at Urvina Bay.

These comparisons confute our expectation, given the minimal destruction of uplifted skeletal remains, that the members of each uplift/living microhabitat pair would be more similar to each other than to any other uplifted or living microhabitat. On the other hand, this lack of similarity is in itself good evidence for the biasing effects of underwater versus uplift sampling.

The second indication that sampling bias is the cause of the observed differences is that the uplifted area preserves an assemblage of organisms killed instantaneously: the higher species richness and diversity values in the uplifted high-relief basalt, low-relief basalt, sand, and summed microhabitats are inexplicable by previous live/dead studies that attribute similar results to time-dependent biases such as time-averaging, shifting microhabitats, and short-term temporal variations in the taxonomic composition of the living community (Johnson, 1965; Warme, 1969; Peterson, 1976; Warme *et al.*, 1976). Because of the instantaneous nature of the uplift event, such biases cannot have operated at Urvina Bay. Thus, these differences must be due to the more complete sampling possible on the uplifted area.

5.4.1. Sources of Sampling Bias

Sampling on the uplifted area could be more effective than underwater sampling for a number of reasons. One possibility is that uplift sampling simply allows total greater numbers of specimens to be counted, with this increased sample size producing higher species richness and diversity values (Koch, 1987). Such, however, is not the case, because no statistically significant relationship exists between number of species and sample size among the five uplifted and five living microhabitats ($R = 0.261$, $df = 8$, $p = 0.01$).

A more likely source of sampling bias is the greater probability of finding rare species on the uplifted area due to increased search time and easier access. The rarity of the species that occur solely on the uplifted area supports this view (Fig. 7).

Table 1. Coefficients of Community Similarity for the 5 Uplifted Habitat Pairs, the Summed Habitats, and the *in toto* Living Gastropod Communites[a]

CCj											ΣH2O	
	LHRB	LINT	LLRB	S	LPAC	FHRB	FINT	FLRB	FS	FPAC	HAB.	TOTAL
LHRB		*0.18*	*0.39*	*0.20*	*0.39*	***0.42***	*0.33*	*0.23*	*0.16*	*0.15*		
LINT	0.18		*0.17*	*0.05*	*0.25*	*0.17*	***0.26***	*0.15*	*0.15*	*0.15*		
LLRB	0.39	0.17		*0.27*	*0.29*	*0.27*	*0.18*	***0.26***	*0.11*	*0.18*		
LS	0.20	0.05	0.27		*0.11*	*0.18*	*0.11*	*0.10*	***0.06***	*0.06*		
LPAC	0.39	0.25	0.29	0.11		*0.30*	*0.32*	*0.24*	*0.12*	***0.43***		
FHRB	**0.42**	0.17	0.27	0.18	0.30		*0.25*	*0.23*	*0.15*	*0.47*		
FINT	0.33	**0.26**	0.18	0.11	0.32	0.25		*0.32*	*0.25*	*0.38*		
FLRB	0.23	0.15	**0.26**	0.10	0.24	0.23	0.32		*0.45*	*0.36*		
FS	0.16	0.15	0.11	**0.06**	0.12	0.15	0.25	0.45		*0.20*		
FPAC	0.50	0.15	0.28	0.06	**0.43**	0.47	0.38	0.36	0.20			
ΣUPL											0.36	0.48
CCs												
	LHRB	LINT	LLRB	S	LPAC	FHRB	FINT	FLRB	FS	FPAC		
LHRB		*0.50*	*0.63*	*0.50*	*0.88*	***0.63***	*0.75*	*0.88*	*0.63*	*0.88*		
LINT	0.50		*0.40*	*0.25*	*0.41*	*0.44*	***0.44***	*0.33*	*0.33*	*0.31*		
LLRB	0.63	0.40		*0.75*	*0.60*	*0.44*	*0.40*	***0.80***	*0.40*	*0.50*		
LS	0.50	0.25	0.75		*0.50*	*0.50*	*0.50*	*0.75*	***0.50***	*0.25*		
LPAC	0.88	0.41	0.60	0.50		*0.67*	*0.50*	*0.53*	*0.29*	***0.69***		
FHRB	**0.63**	0.44	0.44	0.50	0.67		*0.56*	*0.78*	*0.56*	*0.78*		
FINT	0.75	**0.44**	0.40	0.50	0.50	0.56		*0.69*	*0.56*	*0.62*		
FLRB	0.88	0.33	**0.80**	0.75	0.53	0.78	0.69		*0.62*	*0.85*		
FS	0.63	0.33	0.40	**0.50**	0.29	0.56	0.56	0.62		*0.54*		
FPAC	0.88	0.31	0.50	0.25	**0.69**	0.78	0.62	0.85	0.54			
ΣUPL											0.66	0.82
RHO												
	LHRB	LINT	LLRB	S	LPAC	FHRB	FINT	FLRB	FS	FPAC		
LHRB		*0.17*	*0.51*	*0.29*	*0.60*	***0.57***	*0.53*	*0.45*	*0.14*	*0.65*		
LINT	0.17		*0.14*	*0.00*	*0.15*	*0.11*	***0.25***	*-0.16*	*-0.23*	*0.05*		
LLRB	0.51	0.14		*0.46*	*0.37*	*0.35*	*0.23*	***0.36***	*-0.39*	*0.34*		
LS	0.29	0.00	0.46		*0.18*	*0.26*	*0.18*	*0.16*	***-0.47***	*0.05*		
LPAC	0.60	0.15	0.37	0.18		*0.41*	*0.42*	*0.26*	*-0.22*	***0.55***		
FHRB	**0.57**	0.11	0.35	0.26	0.41		*0.38*	*0.40*	*0.06*	*0.61*		
FINT	0.53	**0.25**	0.23	0.18	0.42	0.38		*0.49*	*0.17*	*0.55*		
FLRB	0.45	-0.16	**0.36**	0.16	0.26	0.40	0.49		*0.32*	*0.55*		
FS	0.14	-0.23	-0.39	**-0.47**	-0.22	0.06	0.17	0.32		*0.08*		
FPAC	0.65	0.05	0.34	0.05	**0.55**	0.61	0.55	0.55	0.08			

[a]See text for explanation. CCj = Jaccard's coefficient, CCs = Simpson's coefficient, RHO = Spearman's coefficient of rank correlation. Abbreviations as in Fig. 2, with prefixes added: L = living, F = fossil (uplifted). Comparisons between uplift and living habitat pairs shown in boldface. Values are repeated in italics. For each coefficient, higher values indicate closer similarity, with perfect similarity at 1. Spearman's Rho compares communities in terms of their rank order of species abundance (Sokal and Rohlf, 1969; Neave and Worthington, 1988). The Jaccard and Simpson coefficents measure similarity in terms of presence and absence of species, with Simpson's emphasizing faunal similarities and Jaccard's faunal differences (Simpson, 1947; Cheetham and Hazel, 1969; Hazel, 1970; Fallaw, 1979). These two indices represent the extremes in a range of possible indices; the values of other indices would lie between the values calculated here (Campbell and Valentine, 1977).

Note that of the 59 species in the 5 abundance ranges to the left of the ordinate (i.e., ≤ 32 specimens), 28 were found only on the uplifted area (47%), whereas only 3 of the 15 species in the 8 highest abundance ranges (20%), are uplift-only species (Fig. 7A). The microhabitat comparison gives similar results (Fig. 7B). Uplift sampling is thus roughly twice as likely as underwater sampling to produce rare, as opposed to abundant, species (Koch, 1978).

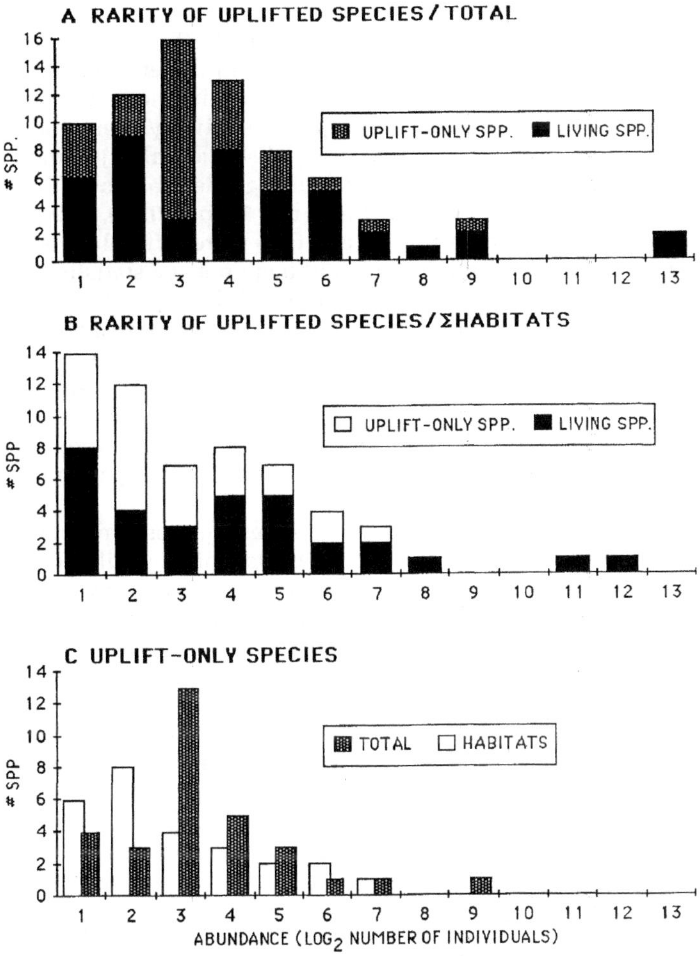

Figure 7. Frequency of living versus uplift-only gastropod species. (A) Total uplifted and living gastropod species, including transect data; (B) summed habitats (HRB, INT, LRB, S, and PAC—see Fig. 2 for abbreviations). The blackened portions of the columns in (A) and (B) show the frequency distribution of living species at Urvina Bay. This is the distribution that would result from underwater sampling alone. The grid (A), and white (B), portions show the frequency and abundance of species added through sampling on the uplift. Note that the uplift-only species tend to be rare (i.e., $\leq$ 32 individuals). (C) Frequency distribution of the uplift-only species shown in (A) and (B), with living species removed. Again note the rarity of species found only on the uplift.

5.4.2. Effects of Underwater Sampling Biases

Given that uplift sampling is more complete than underwater sampling, how might this affect our depiction of the two communities? First, as noted above, the difficulties inherent in working underwater decrease the chance of sampling rare gastropod species. This increases the apparent proportional representation of abundant species in the underwater community, which in turn acts to decrease the community's apparent species richness and diversity (Figs. 5 and 6).

Another outcome of the difficulties inherent in underwater sampling is that they generate artifactual distinctions among living microhabitats. Statistical comparison of presence/absence values for equivalent living/living and uplift/uplift microhabitat pairs (i.e., LINT/LHRB vs. FINT/FHRB; Table 1) demonstrates that uplift/uplift comparisons generate consistently greater values (Wilcoxon Signed Rank Test for Matched Pairs, one-tailed test, $n = 10$, $p = 0.05$), indicating that the uplifted microhabitats are more similar, and thus more difficult to tell apart. Rank correlation values follow a similar trend, although not at a statistically significant level. However, in three of the four cases in which living rank correlation values exceed uplift values, thus contradicting the trend, sample-size effects are probably responsible: the three cases involve comparison with the living sand microhabitat, which contains only four species, each of which occurs in at least one, and in as many as eight other microhabitats (Table 2). Thus comparisons between the living sand microhabitat and other living microhabitats are likely to produce high similarity values. The uplifted sand microhabitat, by comparison, contains 29 species, 8 of which occur only there.

By excluding the background "noise" of rare species, underwater sampling generates artificial differences among microhabitats. This is similar to what happens when one views the nighttime sky from a large city—contrary to what one might expect, the light pollution there actually makes constellations more visible, because it blocks out the stars of lesser magnitude that blur the constellations' distinctive patterns. Likewise, sampling underwater, where less abundant species are less likely to be found, allows us to define distinctive constellations of species for each microhabitat. Sampling on the uplifted area, however, where rare (and abundant) species are more likely to be found in a greater variety of microhabitats, demonstrates that the apparent underwater constellations are only artifactual.

6. Conclusions

1. The uplift of Urvina Bay was a rapid, single-pulse event that occurred in February, 1954. The sudden subaerial exposure of the previously submerged shelf catastrophically killed the organisms living there.
2. In the 37 years since the uplift event took place, the skeletal remains of the uplifted organisms have suffered minimal taphonomic overprinting. Soft-bodied or poorly skeletonized organisms have been nearly totally destroyed, although they occur rarely as desiccated remains.
3. Because of its ease of access, sampling on the uplifted area is much more complete than sampling underwater. The uplifted area thus gives an unusually complete and accurate picture of the skeletal portion of a Galápagos marine gastropod community. The biases inherent in underwater sampling, on the other hand, tend to misrepresent the living gastropod community, producing artificially low species richness and diversity values by underestimating the abundance of rare species and thus overestimating the relative proportion of abundant species. These biases decrease in more easily sampled marine microhabitats such as the intertidal.

Table 2. Abundance of Gastropod Species in the 5 Uplifted and Living Habitat Pairs, the Summed Habitats, and the *in toto* Uplifted and Living Gastropod Communites[a]

		HRB%		INT%		LRB%		SAND%		Pac%		HABITAT #		TOTAL #	
SPECIES	AUTHOR, YEAR	H2O	UPL	H2O	UPL	H2O	UPL	H2O	UPL	H2O	UPL	ΣH2O	ΣUPL	ΣH2O	ΣUPL
*Bulla rufolabris**	A. Adams, 1825						5.4		19.7				76		359
Cancellaria darwini*	Petit, 1970														5
*Cancellaria haemostoma**	Sowerby, 1832						0.3		0.6				3		22
*Cantharus gemmatus**	(Reeve, 1846)						0.6						5		5
Cantharus janelli	Kiener, 1835-36					0.1	0.3					1	2	1	31
Cantharus sanguinolentis	(Duclos, 1833)			1.9		0.2	0.5			0.4		11	4	18	142
Cantharus spp.*															28
Cassis coarctata*	Sowerby, 1825														47
*Cassis tenuis**	Wood, 1828				0.3								1		5
Cerithium adustum	Kiener, 1841	6.8	1.2		0.8	93.5	43.2	70.0	1.2	36.5	14.3	1438	359	4465	3906
Cerithium uncinatum	(Gmelin, 1791)								0.6				1	270	2
Cheilea cepacea	(Broderip, 1834)						0.1		0.6				2	P	50
Colubraria lucasensis*	Strong & Hertlein, 1937														3
Columbella castanea*	Sowerby, 1832														7
Columbella fuscata	Sowerby, 1832				0.8		3.5		2.9				35	3	241
Columbella haemostoma	Sowerby, 1832	0.03	0.2			2.9	8.0		7.5	1.1	1.4	39	77	85	1455
Columbella paytensis	Lesson, 1830				0.3		1.0			0.4	2.9	1	11	1	79
Columbellids														14	4
*Columbella sp. 3***														60	
Conus brunneus	Wood, 1828		0.2			0.2	0.4	10.0				3	4	14	223
Conus dalli*	Stearns, 1873														1
*Conus ?emersoni**	Hanna, 1963								4.0				7		14
Conus nux	Broderip, 1833						0.5	10.0	0.6			1	5	8	177
Conus patricius*	Hinds, 1843														1
Conus perplexus*	Sowerby, 1857														1
*Conus spp.***														2	
Conus purpurascens	Sowerby, 1833		0.5	0.2								1	3	2	49
Conus tiaratus	Sowerby, 1833													1	153
Coralliophila nux	(Reeve, 1846)			0.2	42.8					1.4		5	154	5	340
*Coralliophila parva***	(E. A. Smith, 1877)									7.0		20		20	
Crepidula aculeata*	(Gmelin, 1791)						2.1		1.7				19		144
*Crucibulum spp.**							0.3						2		2

[a] Habitat values are percentages of the total within-habitat abundance; summed-habitat and total values show actual numerical frequency.

Table 2. *(continued)*[a]

		HRB%		INT%		LRB%		SAND%		Pac%		HABITAT #		TOTAL #	
SPECIES	AUTHOR, YEAR	H2O	UPL	H2O	UPL	H2O	UPL	H2O	UPL	H2O	UPL	ΣH2O	ΣUPL	ΣH2O	ΣUPL
Cymatium p. keenae*	(Beu, 1970)								0.6				1		17
Cymatium lineatum*	(Broderip, 1833)														P
Cypraea cervinetta	Kiener, 1843					0.1				0.4		2		2	32
Cypraea (Pseudozonaria) spp.									1.2				2	10	18
Diodora inaequalis	(Sowerby, 1835)				0.3		0.5		2.9				10	P	87
Drillia spp.*															2
Engina maura	(Sowerby, 1832)			0.2					1.2			1	2	11	138
Engina pulchra	(Reeve, 1846)				0.6		0.8			7.7	4.3	22	11	22	12
Engina pyrostoma	(Sowerby, 1832)	0.2					0.1			8.4	1.4	30	2	30	2
Fasciolaria princeps*	Sowerby, 1825														2
*Fissurella obscura**	Sowerby, 1835						0.1		0.6				2		6
*Heliacus bicanaliculatus***	(Valenciennes, 1832)													P	
Hipponix grayanus	Menke, 1853	91.6	84.6		35.0		10.5		2.9	2.1	8.6	2633	772	4603	4319
Latirus sanguineus	(Wood, 1828)					0.1	0.8		0.6		4.3	1	10	3	100
Leucozonia tuberculata	(Broderip, 1833)	0.2	0.2	0.5	0.6		1.4		0.6	6.7	12.9	27	24	47	335
*Littorina spp.***				0.9								4		4	
Malea ringens*	(Swainson, 1822)														1
*Mitra effusa**	Broderip, 1836		0.2								1.4		2		8
Mitra gausapata*	Reeve, 1845														4
*Mitra spp.***														P	
*Mitra sp.1***														2	
Mitra tristis	Broderip, 1836		0.2	2.1	5.3		3.0		2.3	6.3	20.0	27	61	50	739
Mitrella guttata	(Sowerby, 1832)			0.7	0.3				9.2			3	17	3	92
Modulus cerodes	(A. Adams, 1851)				0.3		2.3		0.6				20	P	75
*Morum tuberculosum**	(Reeve, 1842)						0.3						2		77
*Murexiella vittata**	(Broderip, 1833)										1.4		1		8
Muricanthus princeps	(Broderip, 1833)								0.6				1	2	101
Muricopsis zeteki	Hertlein & Strong, 1951									2.1		6		6	15
*Nassarius versicolor**	(C. B. Adams, 1852)								5.2				9		19
Neorapana grandis	(Sowerby, 1835)													P	21
*Nerita scabricosta***	Lamarck, 1822			1.9								8		8	
Nodilittorina galapagiensis	(Stearns, 1892)			52.8								224		225	P

[a] Habitat values are percentages of the total within-habitat abundance; summed-habitat and total values show actual numerical frequency.

Table 2. (*continued*)[a]

SPECIES		HRB%		INT%		LRB%		SAND%		Pac%		HABITAT #		TOTAL #	
		H2O	UPL	H2O	UPL	H2O	UPL	H2O	UPL	H2O	UPL	ΣH2O	ΣUPL	ΣH2O	ΣUPL
Phyllocoma scalariformis*	(Broderip, 1833)														P
Polinices uber	(Valenciennes, 1832)			0.2			0.3		1.2			1	4	1	14
Purpura columellaris	(Lamarck, 1822)													45	1
Purpura pansa	Gould, 1853			0.5								2		2	28
Rissoina inca	Orbigny, 1840						0.4		16.8				32	1	133
Serpulorbis eruciformis*	(Mörch, 1862)														5
Serpulorbis margaritaceus	(Chenu, 1844)													22	49
Strombus granulatus*	Swainson, 1822														1
Tegula snodgrassi	(Pilsbry & Vanatta, 1902)	0.03		0.7	0.8	0.5	4.2		6.4		8.6	10	53	15	473
*Terebra ornata**	Gray, 1834								2.3				4		4
*Terebra plicata**	Gray, 1834								5.2				9		9
Thais melones	(Duclos, 1832)	0.03		15.8	0.8	2.0		10.0		6.3		111	3	337	17
Thais planospira	(Lamarck, 1822)			1.7						0.4		8		8	6
*Thais speciosa***	(Valenciennes, 1832)									0.4		1		1	
Thais spp.				13.9	0.3							59	1	60	1
Trivia maugeriae*	(Sowerby, 1832)														4
*Turrid spp.**									0.6				1		5
Vermicularia p. eburnea	(Reeve, 1842)	1.1	13.0	5.7	10.8	0.6	8.9			12.6	18.6	98	206	103	258
*Volvarina t. taeniolata**	Mörch, 1860						0.4						3		4
TOTAL %		100	100	100	100	100	100	100	100	100	100				
TOTAL n		2868	654	424	360	1211	778	10	173	285	70	4798	2035	10592	14738
TOTAL s		8	9	17	16	10	29	4	29	17	13	32	47	49	74
*uplift-only species															
**living-only species															
transect-only species															

[a] Habitat values are percentages of the total within-habitat abundance; summed-habitat and total values show actual numerical frequency.

4. Because underwater sampling underestimates the species richness and diversity of a living community, using a sampling-biased living community as the present-day analog in a paleoecologic reconstruction will produce results that overestimate how well the fossil community is actually preserved.

ACKNOWLEDGMENTS. This research was partially funded by NSF grant EAR-850896 to Mitchell Colgan, and by a Chancellor's Special Research Fund from the University of California–Santa Cruz to Léo F. Laporte. Permission to conduct this study was granted by the Ecuadorian Ministerio de Agricultura y Granaderia and the Departmento de Parques Nacionales y Vida Silvestre, as well as the Charles Darwin Foundation. I thank Mitchell Colgan for his help in the field and in the lab, and for his generosity in making this project possible; and Léo Laporte, for his conceptual and editorial help. I also thank Sain Colgan, Thomas Smalley, Melissa Keppel, Allison Kay, Margaret Liniecki Laporte, and the staff of the Charles Darwin Research Station for their help in the field; and Nan Rosenbloom for her help in typing references and for her patience. This work is in partial fulfillment of the requirements of a Ph.D. dissertation at the University of California–Santa Cruz.

References

Aslin, C. J., 1968, Echinoid preservation in the Upper Estuarine Limestone of Bilsworth, Northamptonshire, *Geol. Mag.* **105**:506–518.

Behrensmeyer, A. K., 1975, The taphonomy and paleoecology of Plio-Pleistocene vertebrate assemblages of Lake Rudolf, Kenya, *Bull. Mus. Comp. Zool.* **146**(10):473–578.

Behrensmeyer, A. K., Western, D., and Dechant Boaz, D. E., 1979, New perspectives in vertebrate paleoecology from a recent bone assemblage, *Paleobiology* **5**:12–21.

Behrensmeyer, A. K., and Dechant Boaz, D. E., 1980, The recent bones of Amboseli National Park, Kenya, in relation to East African paleoecology, in: *Fossils in the Making* (A. K. Behrensmeyer and A. P. Hill, eds.), pp. 72–94, University of Chicago Press, Chicago.

Behrensmeyer, A. K., and Kidwell, S. M., 1985, Taphonomy's contribution to paleobiology, *Paleobiology* **11**(1):105–119.

Bishop, G. A., 1986, Taphonomy of the North American decapods, *J. Crust. Biol.* **6**:326–355.

Boekschoten, G. J., 1967, Palaeoecology of some mollusca from the Tielrode Sands (Pliocene, Begium), *Palaeogeogr. Palaeoclimatol. Palaeoecol.* **3**:311–362.

Bosence, D. W., 1979, Live and dead faunas from coralline algal gravels, County Galway, *Palaeontology* **22**:449–478.

Brower, J. E., and Zar, J. H., 1984, *Field and Laboratory Methods for General Ecology*, 2nd ed., William C. Brown Publishers, Dubuque, Iowa.

Campbell, C. A., and Valentine, J. W., 1977, Comparability of modern and ancient marine faunal provinces, *Paleobiology* **3**:49–57.

Chave, K. E., 1964, Skeletal durability and preservation, in: *Approaches to Paleoecology*, (J. Imbrie and N. D. Newell, eds.), pp. 377–387, Wiley, New York.

Cheetham, A. E., and Hazel, J. E., 1969, Binary (Presence-Absence) Similarity Coefficients, *J. Paleontol.* **43**(5):1130–1136.

Coe, M. J., 1978, The decomposition of elephant carcasses in the Tsavo (East) National Park, Kenya, *J. Arid Environ.* **1**:71–86.

Colgan, M. W., and Malmquist, D. L., 1987, The Urvina Bay Uplift: A dry trek through a Galápagos coral reef, *Oceanus* **30**(2):61–66.

Colgan, M. W., and Malmquist, D. L., 1988, The growth and destruction of a Galápagos coral community: A study of the Urvina Bay Uplift, Isabela Island, Galápagos, Ecuador, *Geological Society of America Abstracts with Program, Cordilleran Section* **20**:151.

Connell, J. H., 1972, Community interactions on marine rocky intertidal shores, *Annu. Rev. Ecol. Syst.* **3**:169–192.

Couffer, J., 1956, The disappearance of Urvina Bay, *Nat. Hist.* **65**:378–383.

Currey, J. D., 1975, A comparison of the strength of echinoid spines and mollusc shells, *J. Mar. Biol. Assoc. U.K.* **55**:419–424.

Dall, W. H., 1924, Note on fossiliferous strata on the Galápagos Islands explored by W. H. Ochsner of the Expedition of the California Academy of Sciences in 1905–1906, *Geol. Mag.* **61**:428–429.

Dall, W. H., and Ochsner, W. H., 1928, Tertiary and Pleistocene Mollusca from the Galápagos Islands, *Proceedings of the California Academy of Sciences*; 4th series, **17**(4):89–185, plus 2–7, 5 text figures (pp. 91–92, 96–97).

Fallaw, W. C., 1979, A test of the Simpson coefficient and other binary coefficients of faunal similarity, *J. Paleontol.* **53**(4):1029–1034.

Finet, Y., 1985, Preliminary faunal list of the marine mollusks of the Galápagos Islands, *Documents de Travail, No. 20, Patrimoine de l'Institut royal des Sciences naturelles de Belgique, Bruxelles.*

Fürsich, F. T., and Flessa, K. W., 1987, Taphonomy of tidal flat molluscs in the northern Gulf of California: paleoenvironmental analysis despite the perils of preservation, *PALAIOS* **2**:543–559.

Glynn, P. W., Wellington, G. M., and Birkeland, C., 1979, Coral reef growth in the Galápagos: Limitation by sea urchins, *Science* **203**:47–49.

Glynn, P. W., Wellington, G. M., and Wells, J. W., 1983, *Corals and Coral Reefs of the Galápagos Islands*, University of California Press, Berkeley, California, 330 pp.

Hazel, J. E., 1970, Binary coefficients and clustering in biostratigraphy, *Geol. Soc. Am. Bull.* **81**:3237–3252.

Hedgpeth, J. W., 1969, An intertidal reconnaissance of rocky shores of the Galápagos, *Wasmann J. Biol.* **27**:1–24.

Hertlein, L. G., 1972, Pliocene fossils from Baltra (South Seymour) Island, Galápagos Islands, *Proc. Calif. Acad. Sci.* **39**:25–46.

Hertlein, L. G., and Strong, A. M., 1939, Marine Pleistocene mollusks from the Galápagos Islands, *Proc. Calif. Acad. Sci.*, 4th series, **23**(24):367–380.

Hickman, C. S., and Lipps, J. H., 1985, Geologic youth of Galápagos Islands confirmed by marine stratigraphy and paleontology, *Science* **227**(4694):1578–1580.

Hill, A. P., 1980, Early postmortem damage to the remains of some contemporary East African mammals, in: *Fossils in the Making* (A. K. Behrensmeyer and A. P. Hill, eds.), pp. 131–152, University of Chicago Press, Chicago.

Jackson, M. H., 1985, *Galápagos, a Natural History Guide*, University of Calgary Press, Calgary, Alberta, 284 pp.

James, M. J., 1984, A new look at evolution in the Galápagos: Evidence from the late Cenozoic marine molluscan fauna, *Biol. J. Linn. Soc.* **21**:77–95.

Johnson, R. G., 1960, Models and methods for the analysis of the mode of formation of fossil assemblages, *Geol. Soc. Am. Bull.* **71**:1075–1086.

Johnson, R. G., 1962, Mode of formation of marine fossil assemblages of the Pleistocene Millerton Formation of California, *Geol. Soc. Am. Bull.* **73**:113–130.

Johnson, R. G., 1965, Pelecypod death assemblages in Tomales Bay, California, *J. Paleontol.* **39**(1):80–85.

Keen, A. M., 1971, *Sea Shells of Tropical West America*, 2nd ed., Stanford University Press, Stanford, California, i–xvi + 1064 pp.

Kier, P. M., 1974, Evolutionary trends and their functional significance in the post Paleozoic echinoids, *J. Paleontol.* **48**(suppl. No. 3):1–95.

Kier, P. M., 1977, The poor fossil record of the regular echinoid, *Paleobiology* **3**:(2):168–175.

Koch, C. F., 1978, Bias in the published fossil record, *Paleobiology* **4**(3):367–372.

Koch, C. F., 1987, Prediction of sample size effects on the measured temporal and geographic distribution patterns of species, *Paleobiology* **13**(1):100–108.

Koch, C. F., and Sohl, N. F., 1983, Preservational effects in paleoecological studies: Cretaceous mollusk examples, *Paleobiology* **9**(1):26–34.

Laporte, L. F., 1979, *Ancient Environments*, 2nd ed., Prentice Hall, Englewood Cliffs, New Jersey, 163 pp.

Laporte, L. F., and Behrensmeyer, A. K., 1980, Tracks and substrate reworking by terrestrial vertebrates in Quaternary sediments of Kenya, *J. Sediment. Petrol.* **50**(4):1337–1346.

Lasker, H., 1976, Effects of differential preservation on the measurement of taxonomic diversity, *Paleobiology* **2**:84–93.

Lawrence, D. R., 1968, Taphonomy and information losses in fossil communities, *Geol. Soc. Am. Bull.* **79**:1315–1330.

Lipps, J. H., and Hickman, C. S., 1982, Paleontology and geologic history of the Galápagos Islands, *Geol. Soc. Am. Abstracts with Program* **14**(7):548.

Malmquist, D. L., 1991, The role of fragmentation in the bioerosion of Galápagos corals, *Coral Reefs* (under review).

Malmquist, D. L., Colgan, M. W., and Espinoza, R. E., 1986, Evidence for catastrophic versus attritional mortality in the fossil record: Urvina Bay, Isabela Island, Galápagos Archipelago, Ecuador, *Geol. Soc. Am. Abstracts with Program* **18**:680.

McBirney, A. R., and Williams, H., 1969, Geology and petrology of the Galápagos Islands, *Geol. Soc. Am. Mem. 118*, 197 pp.

Moore, T. De Roy, 1980b, The awakening volcano, *Pac. Discovery* **33**(4):25–31.

Müller, M. H., 1979, Fossilization (Taphonomy), in: *Treatise on Invertebrate Paleontology, Part A (Introduction)* (R. C. Moore and C. Teichert, eds.), A2-A78, Geological Society of America and University of Kansas Press, Lawrence, Kansas.

Neave, H. R., and Worthington, P. L., 1988, *Distribution-Free Tests*, Unwin Hyman, London, 430 pp.

Ollier, C., 1984, *Geomorphology Texts: Weathering*, 2nd ed., Longman Group Ltd., London, 270 pp.

Peterson, C. H., 1976, Relative abundances of living and dead molluscs in two California lagoons, *Lethaia* **9**:137–148.

Peterson, C. H., 1977, The paleoecological significance of undetected short-term temporal variability, *J. Paleontol.* **51**(5):976–981.

Pitt, W. D., 1984, Late Cenozoic invertebrate paleontology of the Galápagos Islands, *Annual Report of the Charles Darwin Research Station* (for 1982) (Informe Anual i.982), pp. 258–268.

Pitt, W. D., and James, M. J., 1983, Late Cenozoic marine invertebrate paleontology of the Galápagos Islands, *Annu. Rep. West. Soc. Malacologists* **15**(for 1982):14–15.

Pitt, W. D., James, M. J., Hickman, C. S., Lipps, J. H., and Pitt, L. J., 1986, Late Cenozoic marine mollusks from tuff cones in the Galápagos Islands, *Proc. Calif. Acad. Sci.* **44**(12):269–282, 18 figs.

Plotnick, R. E., 1986, Taphonomy of a modern shrimp, *PALAIOS* **1**:286–293.

Richards, A. F., 1957, Volcanism in eastern Pacific Ocean Basin: 1945–1955, *Int. Geol. Congr.*, 20th session, section 1, Vulcanología del Cenozóico, **1**:19–31.

Richards, A. F., 1962, Active volcanoes of the Archipelago de Colon (Galápagos), Part XIV, *Catalogue of the Active Volcanoes of the World*, International Association of Volcanology, Naples.

Schäfer, W., 1972, *Ecology and Paleoecology of Marine Environments*, University of Chicago Press, Chicago, 568 pp.

Schopf, T. J. M., 1978, Fossilization potential of an intertidal fauna: Friday Harbor, Washington, *Paleobiology* **4**:261–270.

Simkin, T., 1984, Geology of the Galápagos Islands, in: *Key Environments Series: Galápagos Islands*, (J. E. Treherne and R. Perry, eds.), pp. 15–41, Pergammon Press, Oxford.

Simpson, G. G., 1947, Holarctic mammalian faunas and continental relationships during the Cenozoic, *Geol. Soc. Am. Bull.* **58**:613–688.

Smith, A., 1984, *Echinoid Palaeobiology: Special Topics in Palaeontology*, (C. T. Scrutton and C. P. Hughes, eds.), George, Allen and Unwin, London, 190 pp.

Sokal, R. R., and Rohlf, F. J., 1969, *Biometry: The Principles and Practice of Statistics in Biological Research*, W. H. Freeman and Company, San Francisco, 776 pp.

Staff, G. M., Stanton, R. J. Jr., Powell, E. N., and Cummins, H., 1986, Time-averaging, taphonomy, and their impact on paleocommunity reconstruction: death assemblages, in: Texas Bays, *Geol. Soc. Am. Bull.* **97**:428–443.

Van Valen, L., 1964, Relative abundances of species in some fossil mammal faunas, *Am. Naturalist* **98**:109–116.

Vermeij, G. J., 1987, *Evolution and Escalation: an Ecological History of Life*, Princeton University Press, Princeton, New Jersey, 527 pp.

Walker, S. E., 1988, Taphonomic significance of hermit crabs (Anomura: Paguridae): epifaunal hermit crab—infaunal gastropod example, *Palaeogeogr. Palaeoclimatol. Palaeoecol.* **63**:45–71.

Warme, J. E., 1969, Live and dead molluscs in a coastal lagoon, *Journal of Paleontology* **43**(1):141–150.
Warme, J. E., Ekdale, A. A., Ekdale, S. F., and Peterson, C. H., 1976, Raw material of the fossil record, in: *Structure and Classification of Paleocommunities* (R. W. Scott and R. R. West, eds.), Dowden, Hutchinson & Ross, Stroudsburg, Pa., 291 pp.
Wentworth, C. K., and MacDonald, G. A., 1953, Structures and forms of basaltic rocks in Hawaii, *U.S. Geol. Surv. Bull. 994*, 98 pp.
Wiggins, I. L., and Porter, D. M., 1971, *Flora of the Galápagos Islands*, Stanford University Press, Stanford, California, 998 pp.

Chapter 19

Taphonomy and Paleoecology of Villamil Fossil Megagastropods of Isla Isabela

SALLY E. WALKER

1. Introduction

The study of Galápagos molluscan fossils has chiefly been one of documenting the systematics, biogeographic affinities, and distribution of the fauna. In conjunction with radiometric age dating, these fossils have also provided evidence for the geological youth of the islands, thereby setting the evolutionary time scale for this insular fauna (Dall and Ochsner, 1928; Hertlein, 1972; Emerson, 1978; James, 1984; Hickman and Lipps, 1985; Taviani, 1979).

In contrast, a paleoecological and taphonomic (i.e., preservational) study of Galápagos fossils has not been done. Such a study would elucidate community structure (the components and ecological roles of these past organisms) and the postmortem events that led to fossilization of the fauna. These factors are intrinsic to understanding insular evolution of community structure and how they may differ from mainland communities. For instance, a major corallivore in the Galá-

SALLY E. WALKER • Department of Paleontology, University of California–Berkeley, Berkeley, California 94720. *Present address:* Department of Zoology, Oregon State University, Corvallis, Oregon 97331-2914.

pagos is the club-spined sea urchin, *Eucidaris thouarsii* (Glynn *et al.*, 1979). This urchin produces dramatic community effects on Galápagos corals (e.g., bioerosion with sediment production, morphological change of the living coral, and subsequent death of the coral from induced algal growth) (Glynn *et al.*, 1979; Glynn and Wellington, 1983, pp. 130–144). In contrast, mainland (Panama) populations of *E. thousarii* rarely graze on live coral and do not have such dramatic community effects.

The *Eucidaris* example clearly shows how geographically different populations, with associated differences in diet and behavior, may significantly affect the ecological structure of communities, and potentially the evolution of the fauna and how it may be preserved in the fossil record. For example, *Eucidaris* taphonomic artifacts such as abraded coral fragments or morphologically induced changes of coral terminal branches may be preserved in the fossil record. For similar examples, see Kaufman, 1981, and Jackson, 1983.

This chapter provides an account of the paleoecology and taphonomy of megagastropods from the Villamil fossil locality located on Isla Isabela (Albemarle). Megagastropods were chosen for three reasons: (1) to determine if fossil gastropod taxonomic composition may indicate past community structure; (2) to examine the extent of shell-boring or encrusting organisms and predation "events" recorded on the shell, indicating bioerosive activity or heavy predation; and (3) to determine if the boring and encrusting organisms on the shells were also indicative of occupation by hermit crabs. Despite their ubiquity in modern communities, hermit crabs have a poor fossil record. However, the symbionts encrusting a hermit crab's shell can fossilize, and their presence on gastropod shells indicates a hermit crab once occupied the shell (Walker, 1988; 1989).

1.1. What Is Taphonomy?

> "We who cannot grow lichens athwart us should consider the case of the tortoise who though scratched front and side still has lichens that ride on his rump, in a crescent, of sort-is!" Weber and Gradstein, 1984

The above bit of doggerel was coined to explain the peculiar distribution of lichens on Galápagos land tortoises. Lichen occur in a limited area on the upper rear portion of the tortoises' carapace, which Weber and Gradstein (1984) attribute to the "tortoises habitually . . . crashing through dense underbrush, so lichens cannot easily survive on the front, top, and sides of the carapace where abrasion is most likely to occur . . ." In a like manner, the distribution and type of organisms that live on or in marine gastropod shells leave telltale clues about the shell maker's life style or the postmortem history of the shell (e.g., if the shell was worn either by a secondary occupant such as a hermit crab or left unoccupied). The study of these clues, a paleontological foray into the field of forensics, is part of the field called taphonomy (the "hows" and "whys" of fossil preservation) and is complementary to paleoecology (the study of ancient communities and their ecological associations) (see Behrensmeyer, 1984; Behrensmeyer and Kidwell, 1985; Kidwell and Behrensmeyer, 1988).

1.2. Villamil Locality Description and History of Discovery

The Villamil deposit is located approximately 1.25 miles (2 km) NE of the fishing village of Villamil (Fig. 1; Villamil is sometimes written as "Vilamil" although General Jose Villamil, a notable orchilla grower and purveyor for one of the first settlements in the Galápagos, spelled his name with two 'l's; Hickman, 1985). The locality consists of white to tan loose silty sand with interspersed, indurated coquina surrounded on all sides by a lava flow.

Villamil fossils were first discovered by Ochsner during a California Academy of Sciences (CAS) expedition (1905–1906) to the Galápagos and were subsequently described by Dall (1924) and Dall and Ochsner (1928). At that time (before 1928), the volcanic islands were considered destitute of fossil-bearing sedimentary rocks. The discovery, therefore, provided exciting new evidence for the minimum age of the islands and also for the derivation of the fauna, controversies that continue to this day (Emerson, 1978; James, 1984; Hickman and Lipps, 1985).

Dall and Ochsner (1928) described the Villamil locality as a "shell-sand" from an old "weathered beach" about 40 feet (12–15 m) above sea level. They reported 34 gastropod species, with 11 new species, and 14 bivalves from this locality (CAS loc. 802 and loc. 803). A recent (1982) paleontological expedition to the Villamil fossil locality produced 38 megagastropod species and 11 species of bivalves (Hickman and Lipps, 1985; Pitt *et al.*, 1986). Specimens referred to in this paper are from this 1982 expedition.

1.3. Age of Villamil Deposit

Dall and Ochsner (1928) considered Villamil fossils to be "young" and suggested a Pleistocene age. Hertlein and Strong (1939; pp. 369, 372–373) mentioned the discovery of Dall and Ochsner, but described another locality from Isabela Island, not to be confused with the Villamil locality (Tagus Cove, CAS loc. 1306; Fig. 1) composed of beach rock and considered to be late Pleistocene or "possibly subfossil in age." Although the lava flow around the Villamil fossil locality has not been dated, Hickman and Lipps (1985; their figure 2) infer, presumably using radiometric age dates from the Cape Berkeley volcano (dated at 0.09 ± 0.04 million years after Cox and Dalrymple, 1966; Fig. 1) that the Villamil deposit is approximately .72 million years old. However, the deposit may be even younger, based on the dating of the Sierra Negra volcano adjacent to the Villamil site (0.503 ± 0.084 million years: R. Duncan, pers. commun., 1989).

It is quite possible that the Villamil fossil deposit was uplifted during volcanic activity in much the same manner as the recent coral communities of Urvina Bay were uplifted in 1954 from the west side of Isabela Island (Glynn and Wellington, 1983; Colgan and Malmquist, 1987; Chapter 4, this volume; Chapter 18, this volume) and at the Punta Espinosa uplift, Isla Fernandina (Glynn and Wellington, 1983; p. 121). In seismically active areas such as the Galápagos, uplift may be a major mode of preservation of *in situ* marine communities and beach deposits (Darwin examined this mode of preservation in great detail at Valparaiso; Darwin, 1897).

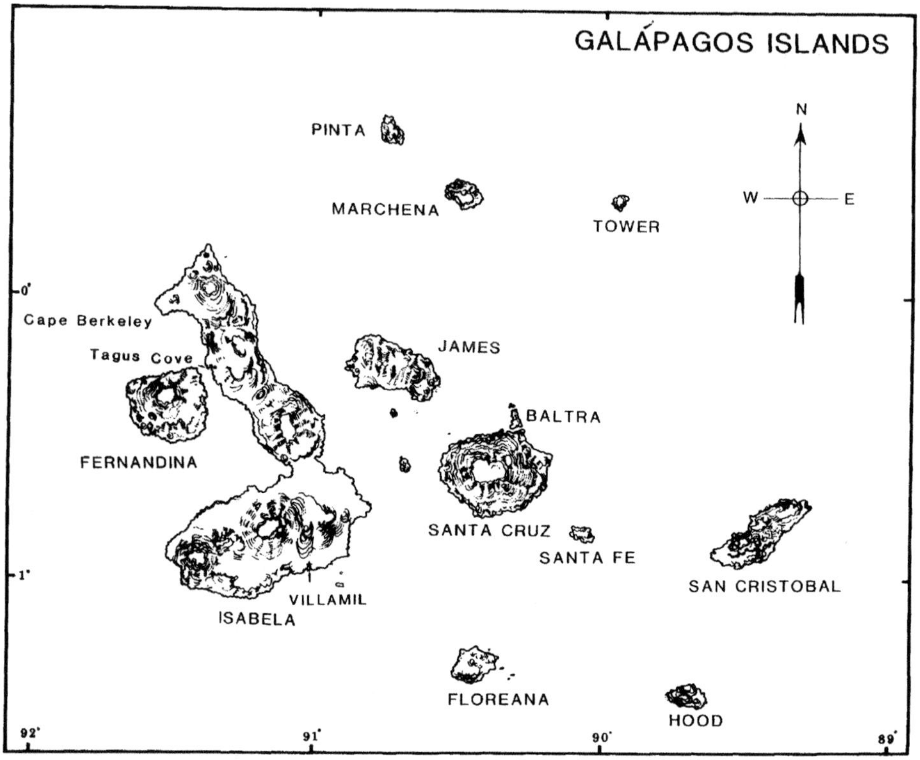

Figure 1. Map of part of the Galápagos archipelago. The Villamil fossil deposit is located on the southeast sector of Isla Isabela.

2. Methods

Specimens for this study were derived from CAS locality 61229 (collected in 1982, by C. S. Hickman, J. Lipps, W. and L. Pitt, and M. James). Thirty-eight gastropod species were recorded from the 1982 collection (W. D. Pitt, pers. commun.). I surveyed 22 of these taxa that were available at the time of this work to determine the type of community structure represented at Villamil. Microgastropods from a similar locality at Villamil representing 15 additional species and 1000 specimens were also abundant but not a part of this CAS loc. 61229 collection (Walker, unpubl. data).

The gastropod shells were examined for the following taphonomic artifacts: (1) the presence or absence of fossil epi- and endobionts (bionts) which would indicate the presence of hermit crabs, such as the exoskeletons of serpulids (including spirorbids), bryozoans, crustose coralline algae, and trace fossils from boring bryozoans and spionid polychaetes in particular locations around the aperture of the shell (see Walker, 1988; 1989); (2) the presence of predatory drill holes from naticid gastropods and octopus as well as shell repair marks from other predators (e.g., presumably crabs); and (3) evidence of subaerial exposure (e.g., lichens) and overall physical appearance of the shell, whether it was shiny or chalky in appearance.

Recent Galápagos intertidal hermit crabs (*Calcinus explorator*) from Academy Bay, Santa Cruz Island, have potentially fossilizable bionts within and surrounding the shell aperture: spirorbids (*Spirorbis* cf. *pagenstecheri*, *S.* cf. *translucens*, and *S.* cf. *regalis*) (after Bailey and Harris, 1968) and a cheilostome bryozoan (*Stomatochetosella* sp.) (Walker, unpubl. data, 1989). Subtidal hermit crabs have fewer spirorbids but have cyclostome bryozoans present in the aperature (*Disporella stellata* var. *pacifica*) (Walker and Moore, unpubl. data, 1989). Therefore, it is reasonable to suggest that some of the Villamil fossil shells might also have some of these encrusting species represented as fossils.

3. Results

The sediment fraction contained sea urchin spines (?*Eucidaris*), small gastropods (< .5mm) composed chiefly of nassariids, olivids, *Hipponix* spp., crepidulids, bivalves, coral fragments (abraded ?*Pocillopora*), and numerous barnacle plates. A large whale barnacle, *Coronula diadema*, was also described from this deposit (*C. diadema* is an epibiont on humpback, fin, blue, and sperm whales; Zullo, 1986, p. 58).

3.1. Taxonomic Composition of the Fauna

The taxonomic composition, habitat (if known), and biogeographic affinities of Villamil fossils are reported in Table 1. The species represent an offshore (very low intertidal-to-subtidal deposit to 40 m) assemblage of gastropods. Only one Villamil Fossil, *Terebra litorea* Dall and Ochsner 1928, is thought to be extinct and is limited to the Villamil locality (Dall and Ochsner, 1928). Many of the Villamil fossil gastropods are characteristic of coral reefs (e.g., *Latiaxis hindsii* (Carpenter, 1857) is a predator on *Pocillopora* corals, Glynn and Wellington, 1983; *Epitonium* species) or coral-derived sediments (e.g., *Nassarius caelolineatus* Nesbitt and Pitt 1986, and terebrids).

3.2. Biogeographic Affinities

Villamil fossils have living conspecifics chiefly from the California and Panamic provinces (Table 1). Only four of the Villamil fossils are considered endemic to the Galápagos Islands (*Calliostoma santacruzanum*, *Nassarius caelolineatus*, *Terebra frigata*, and the extinct *T. litorea*).

3.3. Taphonomic Results

Most of the Villamil fossils represented at CAS loc. 61229 were nearly free of boring and encrusting bionts. Shells had a white, or in some cases, slightly chalky appearance, which I consider to be essentially pristine (Table 2). Despite careful examination, I rarely found any evidence of encrusting organisms. In one instance,

Table 1. Taxonomic Composition, Habitat (if Known) and Biogeographic Distribution of Pleistocene Gastropods from Villamil Terrace Deposit, Isa Isabela[a]

Gastropod species	Habitat/Range
Calliostoma santacruzanum Mclean, 1970	45 m, rocky areas; Santa Cruz Island, Galápagos; endemic
Cancellaria cf. *gemmulata* (Sowerby, 1832)	Offshore, to 45m; Gulf of California to Panama Bay and Galápagos
Cassis centiquadrata (Valenceinnes, 1832)	Very low water, in sand; Gulf of California to Peru and Galápagos
Colubraria lucasensis Strong and Hertlein, 1937	In 37 to 45 m; Baja California, Panama to Galápagos
Conus tornatus (Sowerby, 1832)	Offshore, 37 m; Cedros Island, Baja to south of Ecuador
Crepidula aculeata (Gmelin, 1791)	Intertidal, under rocks; California through Gulf of California and south to Valpraiso, Chile
Epitonium cf. *hindsii* (Carpenter, 1856)	Panama
Epitonium minuticostatum (DeBoury, 1912)	California to Galápagos
Epitonium politum (Sowerby, 1844)	San Pedro, California to Ecuador and Galápagos
Fusinus dupetitthouarsi (Kiener, 1840)	Extreme low tide to 50 m, mudflats, sandbars; outer coast of Baja through Gulf and Ecuador
Hipponix spp.	—
Latiaxis cf. *hindsii* (Carpenter, 1857)	Mostly offshore; coral associate (Glynn and Wellington, 1983); Puertecitos, Baja California to Panama
Nassarius caelolineatus Nesbitt and Pitt, 1986	To depths of 70 m in coral sand and mud; widespread as a fossil on Galápagos Islands; endemic
Natica cf. *grayi* Philippi, 1852	Mostly offshore to 37 m; Magdalena Bay, Baja California, south to Manta, Ecuador
Oliva cf. *slendidula* Sowerby, 1825	Rare, Tres Marias Islands, West Mexico to Panama
Polinices uber (Valenciennes, 1832)	Intertidal to offshore from 4 to 90 m; Scammon's Lagoon, Baja California, south to Peru
Sinum cf. *cymba* (Menke, 1828)	Ecuador to Chile
Subcancilla erythrogramma (Tomlin, 1931)	Offshore, 18 to 37 m; Baja California to Columbia
Terebra frigata Hinds, 1844	Offshore, 4 to 82 m, in sand; endemic
Terebra litorea Dall and Ochsner, 1928	Pleistocene only; endemic
Thais cf. *callaoensis* (Gray, 1828)	Panama to Peru; Galápagos
Trivia cf. *fusca* (Sowerby, 1832)	Galápagos; maybe Panama and Ecuador

[a]Habitat and biogeographic distribution based on recent specimens from Keen (1971) unless otherwise specified.

an encrusting bryozoan fell out of the aperture intact (retaining with its colonial growth the curvature of the shell). However, no remnant or etching on the shell interior remained. Only one specimen, *Latiaxis* cf. *hindsii*, contained encrusting bryozoans on the spines and callus. The pristine quality of the shells may be due to the taphonomic nature of these tropical encrusting species: they do not preserve well. This pristine quality is a taxa-wide characteristic shared, as well, by the micro/juvenile mollusk component taken from a bulk sample (Walker, in prep.).

The three most common gastropods collected were *Nassarius caelolineatus*, *Terebra frigata*, and *Terebra litorea*. Of these, the small size class represented by *N. caelolineatus* were all pristine (from twenty-five randomly selected specimens;

Table 2. Taphonomic Characters Present on Villamil Fossils

Gastropod species	Number[a]	Size and mean (mm)[b]	Drilled[c]	Hcrabs[d]	Taphonomic remarks[e]
Coiled gastropods:					
Calliostoma cf. *santacruzanum*	2	6.3, 4.0	0	0	Worn, no bionts
Cancellaria cf. *gemmulata*	4	12.0–20.0 (14.3)	1	1	Hcrab: spionid bore hole in columella; pitted (1); chalky (3) to pristine (1)
Cassis centiquadrata	1	—	0	0	Fragmented
Colubraria lucasensis	1	22.6	0	0	Last whorl and penultimate whorl damage
Conus tornatus	3	19.7–23.6 (21.9)	0	0	Pigment pattern present
Epitonium cf. *hindsii*	2	12.6, 11.7	0	0	Pristine
Epitonium minuticostatum	3	8.0–11.0 (9.3)	0	0	Worn; damaged
Epitonium politum	2	14.0, 16.0	0	0	Very worn, chalky
Fusinus dupetitthouarsi	3	49.4, 122.5, 143.0	1	?1	Hcrab: bivalves attached to aperture; spionids in drag mark; drill hole from octopus (122.5)
Laxtiaxis cf. *hindsii*	2	12.0, 19.6	0	0	12.0; encrusting bryozoan near callus; 19.6: encrusting bryozoan on spines
Nassarius caelolineatus	25	7.8–19.4 (12.54)	3	0	No bionts, worn to shiny; naticid drill holes
Natica cf. *grayi*	8	6.2–10.5 (7.7)	1	0	All pristine; naticid drill hole
Oliva cf. *splendidula*	2	40.1, 41.2	0	0	One pitted, other pristine
Polinices uber	2	18.7, 20.0	0	0	Pristine
Sinum cf. *cymba*	1	18.0	0	0	Pristine
Subcancilla erthrogramma	3	16.0–18.3 (17.3)	0	0	Black ?lichens present on one specimen, others pristine
Terebra frigata	25	22.0–36.5 (29.0)	4	0	Robust shells; pitted; shell repair on 4 specimens; naticid drillholes
Terebra litorea	14	28.0–90.4 (53.7)	1	2	Hcrab: spionid in columella; boring bryozoan present; shell repair on 4 specimens; naticid drill hole
Thais cf. *callaoensis*	1	8.9	0	0	Black ?lichens present on shell exterior
Trivia cf. *fusca*	4	(9.0)	0	0	Worn, no bionts
Limpet-like gastropods:					
Crepidula aculeata	1	—	0	0	No bionts
Hipponix spp.	2	—	0	0	Worn, no bionts
Total	111 (108 coiled gastropods)		11	4 (3% of coiled gastropods)	

[a]Number = number of individuals.
[b]Size: mean is given in parentheses.
[c]Drilled = number of predatory gastropod drill holes present.
[d]Hcrabs = number of shells with biont characters that indicate a former hermit crab occupant.
[e]Number in parentheses represents number of shells which have specified character.

mean shell size: 12.54 mm). In contrast, the terebrids, a larger size class (> 30 mm), had evidence of hermit crab occupation (Table 2) and were often pitted by boring clionid sponges (smaller pittings were probably from either minute sponges or fungus) (Fig. 2A,B; Tables 3 and 4). These pittings appear to have a pattern: dorsal surfaces of *Terebra frigata* were more pitted than aperture areas (36% top pitting, 9/25 specimens; 4%, 1/25 specimens, bottom pitting). *Terebra litorea* had both sides equally pitted (36% top pitting, 5/14 specimens; 43% bottom pitting, 6/14).

Few of the shells were drilled by predatory mollusks (e.g., naticids, muricids, or octopods). Few repair marks were observed on shells and all evidence of shell repair was on terebrids (Table 2). These shell repair marks are similar to those produced by callapid crabs from Guam (Vermeij *et al.*, 1980). Few brachyuran crab chelipeds were present in this deposit and have not been identified.

An unusual finding was the occurrence of a polychaete worm trace fossil *Helicotaphrichnus commensalis* Kern, Grimmer, and Lister, 1974. Three gastropod species had this trace fossil in the columella of the shell, *Cancellaria* cf. *gemmulata* (Fig. 3), *Terebra litorea*, and *T. frigata*.

Other possible hermit crab-shell associates are the encrusting bivalves (*Ostrea* sp.) on *Fusinus dupetitthouarsi* (Fig. 4A,B). These ostreids practically enclose the

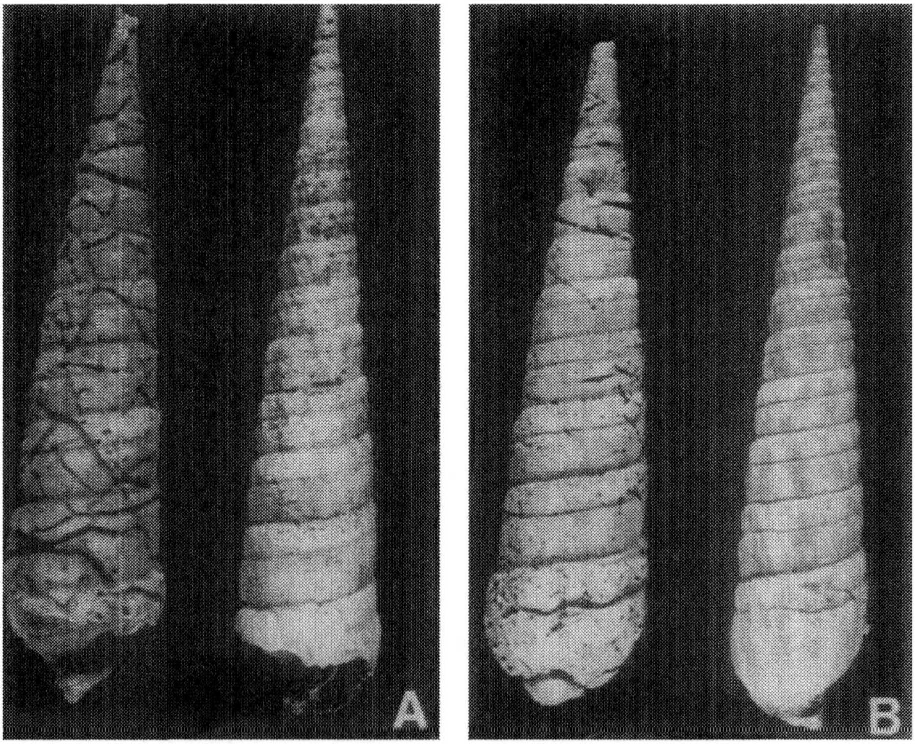

Figure 2. Shell pitting on *Terebra litorea* Dall and Ochsner, 1928. (A) Clionid borings on aperture side of specimen on left, fungal-like pits on right specimen. (B) Same specimens as in (A), dorsal view. Specimen on left has less clionid bore holes on top than on aperture side shown in (A). Specimen on right of (B) has pristine top with few fungal-like pits. Both specimens CAS loc. 61229, left specimen shell height, 85 mm; right specimen, 90 mm.

Table 3. Taphonomic Patterns on *Terebra litorea* by Size Class[a]

Size class	n	Clean	Drilled	Borgbry	Spioncol	Shellrep	Hcrab	Pits
1	3	0	1	0	0	0	0	2
2	8	2	0	0	0	2	0	5
3	1	0	0	1	1	1	1	1
4	2	0	0	0	1	1	1	1
Totals	14	2	1	1*	2*	4	2	9

[a] Size classes represented are: (1) 20.0–39.9 mm; (2) 40.0–59.9 mm; (3) 60.0–79.9 mm; (4) 80.0–99.9 mm. n = Number of individuals; clean = no bionts; drilled = predatory bore holes; borgbry = boring bryozoan trace fossil; shellrep = shell repair; hcrab = minimum number of shells with hermit crab-associated bionts; pits = pitting from clionid sponges and fungi. Kruskal-Wallis one-way analysis of variance indicates none of these varibles are significantly different between size clases except for *borgbry (p =.005, 3d.f.) and *spioncol (p =.027, 3d.f.; these spionid bore holes indicate an hcrab).

aperture of the gastropod, except for a small opening, where presumably a hermit crab lived. Boring spionid trace fossils and boring bryozoans were also present on *Fusinus*.

The possible presence of lichen or other unidentified black fungal organisms on some of the shells (Table 2) suggests subaerial exposure. These organisms may also produce pits on shells.

4. Discussion

4.1. Taxonomic Composition Indicates Paleoecological Setting

The taxonomic composition of this deposit indicates a subtidal assemblage (to approximately 40 m) of coral reef-associated mollusks, either on the reef itself or in sediments surrounding reef habitats. This molluscan community was diverse, composed of predatory snails (naticids), scavengers (nassariids) and coral predators (*Latiaxis* and epitoniids). In addition, the sediments also contained spines from the sea urchin, ?*Eucidaris* spp. (a major corallivore in the Galápagos today), and coral fragments. Coral reefs occur on the eastern side of Isla Isabela today (but may have been destroyed by the 1982–1983 El Niño; see Glynn, 1989) and probably did so more than 500,000 years ago, although all that is preserved of such a community is chiefly the Villamil molluscan component, *Eucidaris*-like spines, and some coral sedimentary fragments. Coral reefs represent the only well-studied modern habitat on the Galápagos Islands (see Glynn and Wellington, 1983),

Table 4. Taphonomic Patterns on *Terebra frigata* by Size Class[a]

Size Class	n	Clean	Drilled	Shellrep	Pits
1	13	8	3	2	4
2	12	8	1	2	4
Totals	25	16	4	4	8

[a] Size classes represented are: (1) 20.0–29.9 mm; (2) 30.0–39.9 mm. Varibles are the same as for *T.litorea* (Table 3). A Mann-Whitney U test indicates that none of the varibles are significantly different between size classes: Clean, p = .794, 1d.f.; drilled, p = .325, 1d.f.; shell repair, p = .932, 1d.f.; pits, p = .893, 1d.f.

Figure 3. *Cancellaria* cf. *gemmulata* (Sowerby, 1832) with the trace fossil, *Helicotaphrichnus commensalis*, in the columella of the shell (arrow). CAS loc. 61229, shell height 20.2 mm.

yet their fossil record is, as far as is known, almost nonexistent on the islands. Therefore, this fossil molluscan community provides evidence for a coral reef assemblage located offshore of Villamil over 500,000 years ago.

The Villamil fossil community is also unusual because it represents an exceptionally well-preserved subtidal assemblage. Another fossil locality with subtidal fossils welded into volcanic tuff was described from Isla Santa Cruz (Pitt *et al.*, 1986). However, these fossils are not as well-preserved (fragmented and riddled by spionid polychaete trace fossils) and are faunistically less diverse representing a deep water assemblage to depths of 100 meters.

The Villamil fossil assemblage also has a few faunal overlaps with fossils from the northeast side of Santa Cruz island (Dall and Ochsner, 1928, pp. 92–94) (e.g., *Fusinus dupetitthouarsi*, *Polinices uber*, and possibly *Latirus* spp.). Unfortunately,

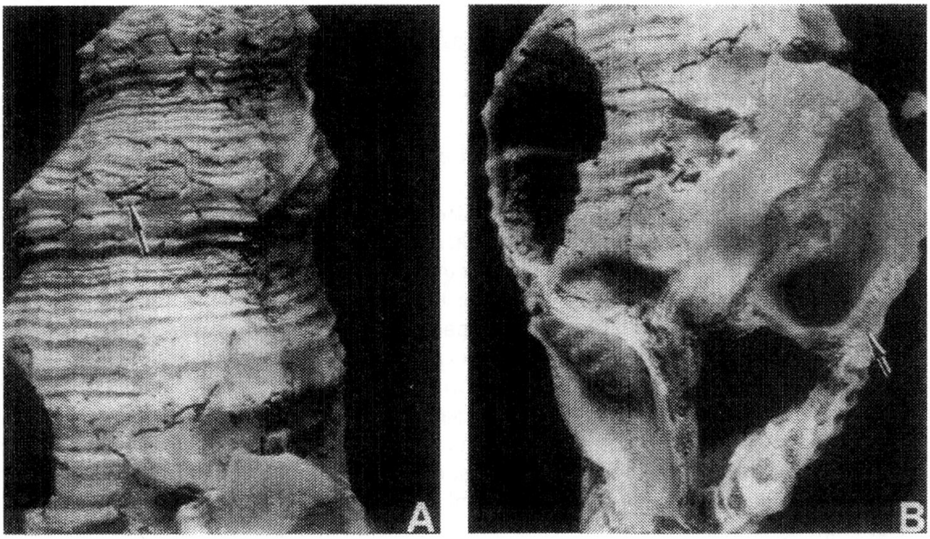

Figure 4. *Fusinus dupetitthouarsi* (Kiener, 1840) with boreholes from spionid polychaetes [arrow in (A)] and encrusting bivalves (*Ostrea* sp.) (B) on aperture side of specimen. CAS loc. 61229, shell height 140 mm.

no recent living subtidal molluscan studies exist in which to compare these paleontological findings.

Many fossil localities from the Galápagos represent mid-to-low intertidal assemblages, rather than subtidal communities, with molluscan species that are associated with rocky basaltic platforms [e.g., *Muricanthus princeps*, *Thais melones*, *Cerithium adustum*: James Island (also called Santiago or San Salvador) CAS loc. 27255; South Seymour Island (Baltra), CAS Loc. 27250; Jervis Island (Rabida), CAS loc. 27547; Sante Fe (Barrington), CAS loc. 61281] (after Dall and Ochsner, 1928; Hertlein, 1972; S. E. Walker, unpubl. data). Because the intertidal fossil fauna is well-represented on these islands, faunal comparisons with living intertidal communities (Hedgpeth, 1969; Houvenaghel and Houvenaghel, 1974; Wellington, 1984) and with communities from the mainland (Panama: Lubchenco *et al.*, 1984) would be important from an evolutionary perspective.

Comparing modern molluscan communities to their fossil counterparts may indicate how community structure has changed through time. Wellington (1984) suggested that extinction and immigration rates for Galápagos mollusks may be rapid based on the percentage of extinct mollusks on the islands (citing the extinct pelecypod, *Pegophysema spherica*, and the terebrid, *T. albemarlensis*; after Hertlein, 1972). For instance, two species of terebrids from Villamil fossil localities are extinct (*T. litorea* and *T. albemarlensis*). Is it possible that terebrids have high species turnover rates? Or, is this an artifact of our taxonomic and taphonomic knowledge? The Galápagos fossil record, then, can provide evidence for the derivation and extinction of various faunas when compared with modern communities and other fossil localities on the mainland.

4.2. Taphonomy and Paleoecology of the Villamil Deposit

The Villamil subtidal mollusks were little affected by predation (drilling and shell repair), bioerosion, or inhabitation by hermit crabs (organisms which can bias the fossil representation of gastropod shells) (see Frey, 1987; Walker, 1986; 1988; 1989).

Few of the shells were bored by naticids and only one shell had an octopus bore hole (Table 2: *Fusinus*). Shell repair, indicated by scars on the shell, were present on the terebrids and were similar to those described by Vermeij *et al.* (1980). Although I do not have direct evidence that these repair marks represent shell-peeling crabs (i.e., verified fossil chelipeds), the marks are quite suggestive that brachyuran crabs were present in this assemblage.

The Villamil fossils are essentially pristine in appearance, chalky to shiny. On such a tectonically active coast such as Isabela, this assemblage was most likely to have then been uplifted with little chance for the shells to be reworked either by hermit crabs, bioeroders, and encrusters, or by storms such as occur today on the east side of Isla Isabela (see description of oceanographic conditions in Glynn and Wellington, 1983).

Darwin also used taphonomic factors, in part, to determine that the occurrence of "perfectly preserved" mollusks embedded in rock above sea level were the product of tectonic uplift (in Valparaiso; Darwin, 1897, pp. 240–241). Although Darwin was on Isla Isabela (Albemarle), he did not discover any fossils. However, if he had discovered the Villamil fossils, he may have reported them in a similar manner to the Valparaiso specimens and most likely would have suggested uplift as a mode of preservation for these "perfectly preserved" specimens.

The presence of lichen or lichen-like organisms on some of the fossils indicates that they were subaerially exposed before collection. These organisms appear to erode the shells, producing pits.

When more is known about other fossil localities on the Galápagos Islands, it may be possible to determine how well the modern fauna is represented in these fossil assemblages, and therefore, the "completeness" of the fossil record in much the same manner as Valentine (1989) did for mollusks from the California Pleistocene. Valentine found that the Pleistocene California record "captured" all of the important extant species and species associations in the families studied. He suggests that it is destruction, and not any lack of accumulation, that accounts for most of the deficiencies in the fossil record of durably skeletonized forms. The processes that affect fossil accumulations are many, from physical processes to biological agents of destruction or accumulation (see Walker, 1990, for brief review).

4.3. Biont Fossils and Trace Fossils Associated with Hermit Crabs

It is unusual that many of the Villamil gastropods would have very few bionts associated with the shell even before death and subsequent uplift. For example, shells from living hermit crabs from the intertidal at Academy Bay and subtidal off the NE sector of Isla Santa Cruz (Canal de Itabaca) contained many encrusting bryozoan shell associates (represented shells: *Cerithium adustum*, *Conus* spp.,

Caducifer cinis, *Thais speciosa*; S. E. Walker and M. Moore, unpubl. data). Encrusting organisms, then, may have lived on the Villamil shells (e.g., like those on the one specimen of *Latiaxis*) but were species with low preservability, and therefore, not represented in the fossil record.

The Galápagos bryozoans may not preserve well because they possess a different (non-etching) shell adhesive or different mineralogy. They do not produce etchings characteristic of the temperate zone bryozoan, *Hippothoa hyalina*, which can be indicative of hermit crab inhabitation (Morris, 1974; Walker, 1988; 1989).

However, the occurrence of the trace fossil, *Helicotaphrichnus commensalis* in the columella of the terebrids and *Cancellaria* cf. *gemmulata*, definitely indicate a hermit crab occupied the shell (see Kern *et al.*, 1974; Walker, 1988; Walker, 1989).

The original maker of the *Helicotaphrichnus* borehole was a spionid polychaete, usually attributed to *Polydora commensalis*, but other species of spionids make the same tube (Kern, 1979). The location of the borehole is ideal for such a "suspension" feeder: the living spionid polychaete is situated directly under the hermit crab's feeding appendages and mouth. In living mollusks, the columellar area is covered by mantle tissue, preventing settlement of this polychaete.

This is the first reported occurrence of the trace fossil, *Helicotaphrichnus* from an archipelago despite its occurrence in mainland assemblages from North America (Kern *et al.*, 1974) and Venezuela (Walker, 1988). It is not known if these spionids occur on the islands today or in other Galápagos fossil deposits.

Hermit crabs may also have influenced the pattern of clionid sponge borings on the Villamil terebrids. For example, a recent *Terebra strigata* shell inhabited by the hermit crab *Paguristes* sp. (from the subtidal of Hood Island) has clionid borings on both sides of the shell (dorsal and aperture side) (S. E. Walker, unpubl. data).

Living terebrids do not have boring clionids on their shells because of their presumed burrowing lifestyle. However, little is known about living terebrids especially those from the Galápagos Islands (systematics treated, in part, by Bratcher and Burch, 1971). Also, it is not known if unoccupied shells remain unburied in the Galápagos habitat, thereby enhancing clionid infestations on the shell.

5. Summary

1. The taxonomic composition of the Villamil assemblage represents a well-preserved subtidal molluscan community associated with coral reefs and reef-sediments. Therefore, a thriving coral reef assemblage composed of molluscan corallivores and snails that preyed on other mollusks was established on the northeast sector of Isla Isabela more than 500,000 years ago. This assemblage will be important in future studies comparing molluscan community structure between marine communities on the Galápagos Islands and the mainland (Panama).

2. The Villamil assemblage lacked intense bioerosion which may indicate that the assemblage was uplifted rapidly with little chance for the shells to be reworked. The lack of encrusting organisms, such as bryozoans, may also be a product of their low preservability.

3. Shell repair marks (from presumably brachyuran crabs) and drill holes from

predatory mollusks (octopus and naticids) indicate a predatory guild existed at Villamil during the pleistocene.

4. The trace fossil *Helicotaphrichnus commensalis*, which is unique to hermit crab-inhabited shells, is reported here for the first time from an archipelago. This finding extends its biogeographic range in the eastern pacific to the Galápagos Islands. The presence of this trace fossil also indicates that hermit crabs were part of this coral reef assemblage.

5. The taxonomic and taphonomic representation paint an impressive paleoecological picture for Villamil: that there was a thriving coral-reef associated community established within a very short time period on Isla Isabela.

Addendum: Surface lava near Villamil fossil locality is at most 150,000 years old (R. Duncan, Personal Communication, May 1991).

ACKNOWLEDGMENTS. The author wishes to thank C. S. Hickman for providing support for this research and M. J. James for the opportunity to present these findings at the Galápagos symposium at Sonoma State University in 1988. I thank M. D. Moore for providing the recent Galápagos input on hermit crabs. I am grateful to A. J. Boucot, M. J. James, and W. D. Pitt, who reviewed the manuscript. Field research was supported by a National Geographic Grant to C. S. Hickman and J. H. Lipps. A University of California Research Grant to C. S. Hickman defrayed costs for my curatorial support. Manuscript preparation was supported by A. Boucot. This is a Contribution from the Museum of Paleontology Number 1531.

References

Bailey, J .H., and Harris, M. P., 1968, Spirorbinae (Polychaeta: Serpulidae) of the Galápagos Islands, *J. Zool. Lond.* **155**:161–184.

Behrensmeyer, A. K., 1984, Taphonomy and the fossil record, *Am. Sci.* **72**:558–566.

Behrensmeyer, A. K., and Kidwell, S. M., 1985, Taphonomy's contributions to paleobiology, *Paleobiology* **11**:105–119.

Bratcher, T., and Burch, R. D., 1971, The Terebridae (Gastropoda) of Clarion, Socorro, Cocos, and Galápagos Islands, *Proc. Calif. Acad. Sci.* **37**:537–566.

Colgan, M. W., and Malmquist, D. L., 1987, The Urvina Bay uplift: A dry trek through a coral community, *Oceanus* **30**:61–66.

Cox, A., and Dalrymple, G. B., 1966, Palaeomagnetism and potassium–argon ages of some volcanic rocks from the Galápagos Islands, *Nature (London)* **209**:776–777.

Dall, W. H., 1924, Note on fossiliferous strata on the Galápagos Islands explored by W. H. Ochsner of the Expedition of the California Academy of Sciences in 1905–1906, *Geol. Mag.* **61**:428–429.

Dall, W. H., and Ochsner, W. H., 1928, Tertiary and Pleistocene mollusca from the Galápagos Islands, *Proc. Calif. Acad. Sci.* **17**:89–139.

Darwin, C., 1897, *Geological Observations on the Volcanic Islands and Parts of South America visited during the Voyage of H. M. S. "Beagle"*, D. Appleton and Company, New York.

Emerson, W. K., 1978, Mollusks with Indo-Pacific faunal affinites in the eastern Pacific ocean, *Nautilus* **92**:91–96.

Frey, R. W., 1987, Hermit crabs: Neglected factors in taphonomy and paleoecology, *Palaios* **2**:313–322.

Glynn, P. W., 1989, Coral mortality and disturbances to coral reefs in the tropical Eastern Pacific, in: *Global ecological consequences of the 1982–1983 El-Niño-Southern Oscillation* (P. W. Glynn, ed.), Elsevier, Amsterdam.

Glynn, P. W., and Wellington, G. M., 1983, *Corals and Coral Reefs of the Galápagos Islands*, University of California Press, Berkeley, California.

Glynn, P. W., Wellington, G. M., and Birkeland, C., 1979, Coral reef growth in the Galápagos: Limitation by sea urchins, *Science* **203**:47–49.
Hedgpeth, J., 1969, An intertidal reconnaissance of rocky shores of the Galápagos, *Wasmann J. Biol.* **27**:1–24.
Hertlein, L. G., 1972, Pliocene fossils from Baltra (South Seymour) Island, Galápagos Islands. *Proc. Calif. Acad. Sci.* **39**:25–46.
Hertlein, L. G., and Strong, A. M., 1939, Marine Plestocene mollusks from the Galápagos Islands, *Proc. Calif. Acad. Sci.* **23**:367–380.
Hickman, C. S., and Lipps, J. H., 1985, Geologic youth of Galápagos islands confirmed by marine stratigraphy and paleontology, *Science* **227**:1578–1580.
Hickman, J., 1985, *The Enchanted Islands, The Galápagos Discovered*, Tanager Books, Dover, New Hampshire.
Houvenaghel, G. T., and Houvenaghel, N., 1974, Aspects ecologiques de la zonation intertidale sur les cotes rocheuses des iles Galápagos, *Mar. Biol.* **26**:135–152.
Jackson, J. B. C., 1983, Biological determinants of present and past sessile animal distributions, in: *Biotic Interactions in Recent and Fossil Benthic Communities* (M. J. S. Tevesz and P. L. McCall, eds.) pp. 39–120, Plenum Press, New York.
James, M. J., 1984, A new look at evolution in the Galápagos: Evidence from the late Cenozoic marine molluscan fauna, *Biol. J. Linn. Soc.* **21**:77–95.
Kaufman, L., 1981, There was biological disturbance on Pleistocene coral reefs, *Paleobiology* **7**:527–532.
Keen, A. M., 1971, *Sea Shells of Tropical West America*, Stanford University Press, Stanford, California.
Kern, J. P., 1979, The ichnofossil *Helicotaphrichnus commensalis* in the Korytnica basin (Middle Miocene; Holy Cross Mountains, Central Poland), *Acta Geol. Pol* **29**:239–242.
Kern, J. P., Grimmer, J. C., and Lister, K. H., 1974, A new fossil spionid tube, Pliocene and Pleistocene of California and Baja California, *J. Paleontol.* **48**:978–982.
Kidwell, S. M., and Behrensmeyer, A. K., 1988, Overview: Ecological and evolutionary implications of taphonomic processes, *Palaeogeogr. Palaeoclimatol. Palaeoecol.* **63**:1–13.
Lubchenco, J., Menge, B. A., Garrity, S. D., Lubchenco, P. J., Ashkenas, L. R., Gaines, S. D., Emlet, R., Lucas, J., and Strauss, S., 1984, Structure, persistence, and role of consumers in a tropical rocky intertidal community (Taboguilla Island, Bay of Panama), *J. Exp. Mar. Biol. Ecol.* **78**:23–73.
Morris, P. A. P., 1974, A comparative study of decalcification of mollusc shells by various bryozoans, *Doc. Lab. Geol. Fac. Sci. Lyon H. S.*, **3**(fasc.1):109–113.
Nesbitt, E. C., and Pitt, W., 1986, *Nassarius* (Gastropoda: Neogastropoda) from the Galápagos Islands, *Veliger* **28**:294–301.
Pitt, W. D., James, M. J., Hickman, C. S., Lipps, J. H., and Pitt, L. J., 1986, Late Cenozoic marine mollusks from tuff cones in the Galápagos Islands, *Proc. Calif. Acad. Sci.* **44**:269–282.
Taviani, M., 1979, I molluschi marini raccolti dalla spedizione "L. mares-G.R.S.T.S." alle isole Galapagos. 1. Gastropoda E bivalvia. Galápagos, studi e ricerche-spedizione "L. Mares-G.R.S.T.S." Museo Zoologico dell' Universita di Firenze [Florence] Gruppo Ricerche e Techiche Subacque, pp. 1–61.
Valentine, J., 1989, How good was the fossil record? Clues from the Californian Pleistocene, *Paleobiology* **15**:83–93.
Vermeij, G., 1978, *Biogeography and Adaptation*, Belknap Press, Amsterdam.
Vermeij, G., Zipser, E., and Dudley, E. C., 1980, Predation in time and space: Peeling and drilling in terebrid gastropods, *Paleobiology* **6**:352–364.
Walker, S. E., 1986, The influence of hermit crabs on gastropod taphonomy. *Fourth North American Paleontological Convention*, p. A48, University of Colorado, Boulder, Colorado.
Walker, S. E., 1988, Taphonomic significance of hermit crabs (Anomura: Paguridea): Epifaunal hermit crab-infaunal gastropod example, *Palaeogeogr. Palaeoclimatol. Palaeoecol.* **63**:45–71.
Walker, S. E., 1989, Hermit crabs as taphonomic agents, *Palaios* **4**:439–452.
Walker, S. E., 1990, Biological taphonomy and gastropod temporal dynamics, Paleon. Soc. Spec. Pub. No. 5, pp. 391–421.
Walker, S. E., Fossil microgastropods of Villamil, Isla Isabela, Galápagos Islands (in prep.).
Weber, W. A., and Gradstein, S., 1984, Lichens and bryophytes, in: *Galápagos Key Environments* (R. Perry, ed.), pp. 71-84, Pergamon Press, Oxford.
Wellington, G. M., Marine environment and protection, in: *Galápagos, Key Environments* (R. Perry, ed.), pp. 247—263, Pergamon Press, Oxford.
Zullo, V. A., 1986, Quaternary barnacles from the Galápagos Islands, *Proc. Calif. Acad. Sci.* **44**:55–66.

Systematic Index

* A page number with "a" refers to an entry in an appendix, "f," to an entry in a figure caption, and "t," to an entry in a table.

Subject Index

*A page number with "a" refers to an entry in an appendix, "f," to an entry in a figure caption, and "t," to an entry in a table.

9 780306 437946

GPSR Compliance

The European Union's (EU) General Product Safety Regulation (GPSR) is a set of rules that requires consumer products to be safe and our obligations to ensure this.

If you have any concerns about our products, you can contact us on ProductSafety@springernature.com

In case Publisher is established outside the EU, the EU authorized representative is:

Springer Nature Customer Service Center GmbH
Europaplatz 3
69115 Heidelberg, Germany

Batch number: 10755284

Printed by Printforce, the Netherlands